AN OVERVIEW OF HEAT TRANSFER PHENOMENA

Edited by **Salim N. Kazi**

INTECHOPEN.COM

An Overview of Heat Transfer Phenomena

http://dx.doi.org/10.5772/2623
Edited by Salim N. Kazi

CBS, Edition 2016

Contributors

Jan Taler, Dawid Taler, Jacob O. Aweda, Michael B. Adeyemi, Mojtaba Dehghan Manshadi, Mohammad Kazemi Esfeh, F.M. Hady, F.S. Ibrahim, S.M. Abdel-Gaied, M.R. Eid, Xiaohui Zhang, Tilak T. Chandratilleke, Nima Nadim, G.A. Rivas, E.C. Garcia, M. Assato, Cheng Lin, Yuzhou Chen, Yoshio Utaka, Zhihao Chen, M.M. Awad, V. Ashoori, M. Shayganmanesh, S. Radmard, Yan-Ping Huang, Jun Huang, Jian Ma, Yan-Lin Wang, Jun-Feng Wang, Qiu-Wang Wang, P. Sivashanmugam, Toshihiko Shakouchi, Mizuki Kito, Armando Gallegos-Muñoz, Nicolás C. Uzárraga-Rodríguez, Francisco Elizalde-Blancas, S. N. Kazi, Hussein Togun, E. Sadeghinezhad

Published by InTech

Janeza Trdine 9, 51000 Rijeka, Croatia

Notice

Publishing Process Manager Mirna Cvijic

Technical Editor InTech Prepress, Novi Sad

Cover InTech Design team

Additional hard copies can be obtained from orders@intechopen.com

An Overview of Heat Transfer Phenomena, Edited by Salim N. Kazi
p. cm.
ISBN 978-953-51-0827-6

INTECH

open science | open minds

Contents

Preface

In the wake of energy crisis due to rapid growth of industries, urbanization, transportation, and human habit, the efficient transfer of heat could play a vital role in energy saving. Industries, household requirements, offices, transportation are all dependent on heat exchanging equipment. Considering these, the present book has incorporated different sections related to general aspects of heat transfer phenomena, convective heat transfer mode, boiling and condensation, heat transfer to two phase flow and heat transfer augmentation by different means. Technique of local heat flux measurement by newly developed devices, heat transfer estimation during metal casting along with numerical analysis, and heat transfer investigation in anemometry are incorporated with analytical and experimental approach.

Natural convection heat transfer in variable engineering aspects is presented in this book. Fluid flow through porous medium and application of nanofluid in different geometrical configurations are discussed with emphasis on numerical simulation. Similarly forced convection heat transfer in different conduit configurations are included in this issue. Numerical approaches are systematically presented and competently compared with the experimental available data.

Droplet impact and evaporation on modified surface, critical heat flux in sub-cooled boiling region and subsequent modeling and details of drop wise condensation with emphasis on droplet movement following temperature gradient are systematically reported.

Two phase flow is a complex phenomenon and it is extensively observed in process industries. Detailed description along with mechanisms and modeling are highlighted here. Spray cooling and boiling phenomena, viscous dissipation effects on heat transfer and heat generation and removal in solid state lasers are presented with great skill.

Intensive investigation is going on to accelerate heat transfer. Some of the approaches are presented in this issue. Enhancement of heat transfer by introducing vortex generator in conduit flow, use of nanofluid as heat transfer liquid, imposing impinging jet on heated surface, use of extended surface and step flow separation are systematically presented.

Though this work was an addition to my prescheduled academic work load, it was very enjoyable. I would like to thank my students for their patience in getting delayed support from me. And lastly, I would like to thank my wife Nilufa and son Mehrab for their sacrifice in daily life during this work.

Salim N. Kazi
Department of Mechanical Engineering, Faculty of Engineering,
University of Malaya, Kuala Lumpur,
Malaysia

Section 1

General Aspects of Heat Transfer

Chapter 1

Measurements of Local Heat Flux and Water-Side Heat Transfer Coefficient in Water Wall Tubes

Jan Taler and Dawid Taler

Additional information is available at the end of the chapter

http://dx.doi.org/10.5772/52959

1. Introduction

Measurements of heat flux and heat transfer coefficient are subject of many current studies. A proper understanding of combustion and heat transfer in furnaces and heat exchange on the water-steam side in water walls requires accurate measurement of heat flux which is absorbed by membrane furnace walls. There are three broad categories of heat flux measurements of the boiler water-walls: (1) portable heat flux meters inserted in inspection ports [1], (2) Gardon type heat flux meters welded to the sections of the boiler tubes [1-4], (3) tubular type instruments placed between two adjacent boiler tubes [5-14]. Tubular type and Gardon meters strategically placed on the furnace tube wall can be a valuable boiler diagnostic device for monitoring of slag deposition. If a heat flux instrument is to measure the absorbed heat flux correctly, it must resemble the boiler tube as closely as possible so far as radiant heat exchange with the flame and surrounding surfaces is concerned. Two main factors in this respect are the emissivity and the temperature of the absorbing surface, but since the instrument will almost always be coated with ash, it is generally the properties of the ash and not the instrument that dominate the situation. Unfortunately, the thermal conductivity can vary widely. Therefore, accurate measurements will only be performed if the deposit on the meter is representative of that on the surrounding tubes. The tubular type instruments known also as flux-tubes meet this requirement. In these devices the measured boiler tube wall temperatures are used for the evaluation of the heat flux q_m. The measuring tube is fitted with two thermocouples in holes of known radial spacing r_1 and r_2. The thermocouples are led away to the junction box where they are connected differentially to give a flux related electromotive force.

The use of the one dimensional heat conduction equation for determining temperature distribution in the tube wall leads to the simple formula

$$q_m = \frac{k(f_1 - f_2)}{r_o \ln(r_1/r_2)}. \tag{1}$$

The accuracy of this equation is very low because of the circumferential heat conduction in the tube wall.

However, the measurement of the heat flux absorbed by water-walls with satisfactory accuracy is a challenging task. Considerable work has been done in recent years in this field. Previous attempts to accurately measure the local heat flux to membrane water walls in steam boilers failed due to calculation of inside heat transfer coefficients. The heat flux can only be determined accurately if the inside heat transfer coefficient is measured experimentally.

New numerical methods for determining the heat flux in boiler furnaces, based on experimentally acquired interior flux-tube temperatures, will be presented. The tubular type instruments have been designed to provide a very accurate measurement of absorbed heat flux q_m, inside heat transfer coefficient h_{in}, and water steam temperature T_f.

Two different tubular type instruments (flux tubes) were developed to identify boundary conditions in water wall tubes of steam boilers.

The first meter is constructed from a short length of eccentric bare tube containing four thermocouples on the fire side below the inner and outer surfaces of the tube. The fifth thermocouple is located at the rear of the tube on the casing side of the water wall tube. First, formulas for the view factor defining the heat flux distribution at the outer surface of the flux tube were derived. The exact analytical expressions for the view factor compare very well with approximate methods for determining view factor which are used by the ANSYS software. The meter is constructed from a short length of eccentric tube containing four thermocouples on the fireside below the inner and outer surfaces of the tube. The fifth thermocouple is located at the rear of the tube (on the casing side of the water-wall tube). The boundary conditions on the outer and inner surfaces of the water flux-tube must then be determined from temperature measurements at the interior locations. Four K-type sheathed thermocouples, 1 mm in diameter, are inserted into holes, which are parallel to the tube axis. The thermal conduction effect at the hot junction is minimized because the thermocouples pass through isothermal holes. The thermocouples are brought to the rear of the tube in the slot machined in the tube wall. An austenitic cover plate with the thickness of 3 mm – welded to the tube – is used to protect the thermocouples from the incident flame radiation. A K-type sheathed thermocouple with a pad is used to measure the temperature at the rear of the flux-tube. This temperature is almost the same as the water-steam temperature.

The non-linear least squares problem was solved numerically using the Levenberg–Marquardt method. The temperature distribution at the cross section of the flux tube was determined at every iteration step using the method of separation of variables.The heat transfer conditions in adjacent boiler tubes have no impact on the temperature distribution in the flux tubes.

The second flux tube has two longitudinal fins which are welded to the eccentric bare tube. In contrast to existing devices, in the developed flux-tube fins are not welded to adjacent water-wall tubes. Temperature distribution in the flux-tube is symmetric and not disturbed by different temperature fields in neighboring tubes. The temperature dependent thermal conductivity of the flux-tube material was assumed. An inverse problem of heat conduction was solved using the least squares method. Three unknown parameters were estimated using the Levenberg-Marquardt method. At every iteration step, the temperature distribution over the cross-section of the heat flux meter was computed using the ANSYS CFX software. Test calculations were carried out to assess accuracy of the presented method. The uncertainty in determined parameters was calculated using the variance propagation rule by Gauss. The presented method is appropriate for membrane water-walls.

The developed meters have one particular advantage over the existing flux tubes to date.The temperature distribution in the flux tube is not affected by the water wall tubes, since the flux tube is not connected to adjacent waterwall tubes with metal bars, referred to as membrane or webs. To determine the unknown parameters only the temperature distribution at the cross section of the flux tube must be analysed.

2. Tubular type heat flux meter made of a bare tube

Heat flux meters are used for monitoring local waterwall slagging in coal and biomass fired steam boilers [5-19].

The tubular type instruments (flux tubes) [10-14,19] and other measuring devices [15-18] were developed to identify boundary conditions in water wall tubes of steam boilers. The meter is constructed from a short length of eccentric tube containing four thermocouples on the fire side below the inner and outer surfaces of the tube. The fifth thermocouple is located at the rear of the tube on the casing side of the water wall tube.

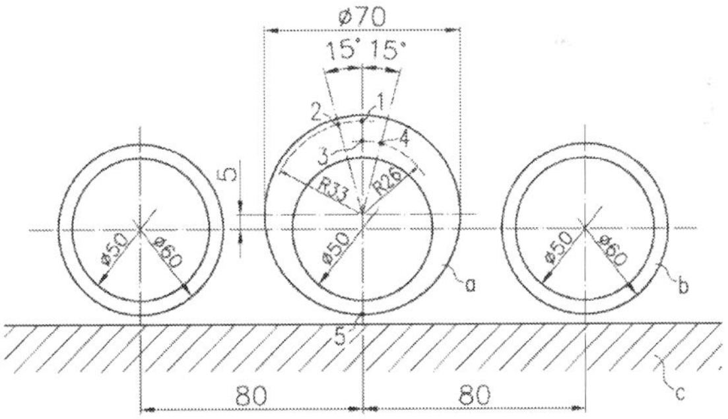

Figure 1. The heat flux tube placed between two water wall tubes, a – flux tube, b – water wall tube, c – thermal insulation

The boundary conditions at the outer and inner surfaces of the water flux-tube must then be determined from temperature measurements at the interior locations. Four K-type sheathed thermocouples, 1 mm in diameter, are inserted into holes, which are parallel to the tube axis. The thermal conduction effect at the hot junction is minimized because the thermocouples pass through isothermal holes. The thermocouples are brought to the rear of the tube in the slot machined in the protecting pad. An austenitic cover plate with the thickness of 3 mm welded to the tube is used to protect the thermocouples from the incident flame radiation. A K-type sheathed thermocouple with a pad is used to measure the temperature at the rear of the flux-tube. This temperature is almost the same as the water-steam temperature. A method for determining fireside heat flux, heat transfer coefficient on the inner surface and temperature of water-steam mixture in water-wall tubes is developed. The unknown parameters are estimated based on the temperature measurements at a few internal locations from the solution of the inverse heat conduction problem. The non-linear least squares problem is solved numerically using the Levenberg–Marquardt method. The diameter of the measuring tube can be larger than the water-wall tube diameter. The view factor defining the distribution of the heat flux on the measuring tube circumference is determined using exact analytical formulas and compared with the results obtained numerically using ANSYS software. The method developed can also be used for an assessment of scale deposition on the inner surfaces of the water wall tubes or slagging on the fire side. The presented method is suitable for water walls made of bare tubes as well as for membrane water walls. The heat transfer conditions in adjacent boiler tubes have no impact on the temperature distribution in the flux tubes.

2.1. View factor for radiation heat transfer between heat flux tube and flame

The heat flux distribution in the flux tube depends heavily on the heat flux distribution on its outer surface. To determine the heat flux distribution q as a function of angular coordinate ϕ, the analytical formulas for the view factor ψ, defining radiation interchange between an infinitesimal surface on the outer flux tube circumference and the infinite flame or boiler surface, will be derived. The heat flux absorbed by the outer surface of the heat flux tube $q(\varphi)$ is given by

$$q(\varphi) = q_m \psi(\varphi). \tag{2}$$

The specific thermal load of the water wall q_m is defined as the ratio of the heat transfer rate absorbed by the water wall to the projected surface area of the water wall. The view factor is the fraction of the radiation leaving the surface element located on the flux tube surface that arrives at the flame surface. The view factor can be computed from

$$\psi = \frac{1}{2}\left(\sin\delta_1 + \sin\delta_2\right). \tag{3}$$

The angles δ_1 and δ_2 are formed by the normal to the flux tube at φ and the tangents to the flux tube and adjacent water-wall tube (Figures 2,4,6). Positive values of δ_1 are measured clockwise with respect to the normal while positive values of δ_2 are measured counterclockwise with respect to the normal. The radial coordinate r_o of the flux tube outer surface measured from the center 0 (Figure 2) is

$$r_o = e\cos\varphi + \sqrt{b^2 - e^2(\sin\varphi)^2}. \quad (4)$$

where: e – eccentric (Figure 2), b – outer radius of flux-tube.

The angle φ_1 can be expressed in terms of the angle φ, flux tube outer radius b, and eccentric e (Figure 2)

$$\varphi_1 = \arcsin\left[\frac{\left(e\cos\varphi + \sqrt{b^2 - (e\sin\varphi)^2}\right)\sin\varphi}{b}\right], \quad \varphi_1 \le \frac{\pi}{2}, \quad (5)$$

$$\varphi_1 = \pi - \arcsin\left[\frac{\left(e\cos\varphi + \sqrt{b^2 - (e\sin\varphi)^2}\right)\sin\varphi}{b}\right], \quad \frac{\pi}{2} \le \varphi_1 \le \pi. \quad (6)$$

First, the view factor for the angle interval $0 \le \varphi_1 \le \varphi_{1,l1}$ was determined

$$\psi = \frac{1+\cos\varphi_1}{2}, 0 \le \varphi_1 \le \varphi_{1,l1}. \quad (7)$$

The limit angle $\varphi_{1,l1}$ (Figures 2 and 3) is given by

$$\varphi_{1,l1} = \arccos\frac{c-e}{b}, \quad (8)$$

where c is the outer radius of the boiler tube.

Next the view factor in the angle interval $\varphi_{1,l1} \le \varphi_1 \le \varphi_{1,l2}$ will be determined. The limit angle $\varphi_{1,l2}$ is: $\varphi_{1,l2} = \varphi_1(\varphi = \pi/2) = (\pi/2) + \arcsin(e/b)$ (Figure3). The view factor ψ is computed from Eq.(2), taking into account that (Figure 4)

$$\delta_1 = \frac{\pi}{2}, \ \delta_2 = \frac{\pi}{2} - (\varepsilon + \varphi_1), \ \varepsilon = \beta + \gamma - \frac{\pi}{2}, \ x_i = b\sin\varphi_1, \ x_i = b\cos\varphi_1,$$
$$\beta = \arcsin\frac{c}{\sqrt{(t-x_i)^2 + (y_i + e)^2}}, \ \gamma = \arcsin\frac{t - x_i}{\sqrt{(t-x_i)^2 + (y_i + e)^2}}, \ \varphi_{l1} \le \varphi_1 \le \varphi_{l2}, \quad (9)$$

where t is the pitch of the water wall tubes.

Next the view factor $\psi(\phi)$ is determined in the angle interval $\varphi_{1,l2} \le \varphi_1 \le \varphi_{1,l3}$ (Figures 3 and 5).

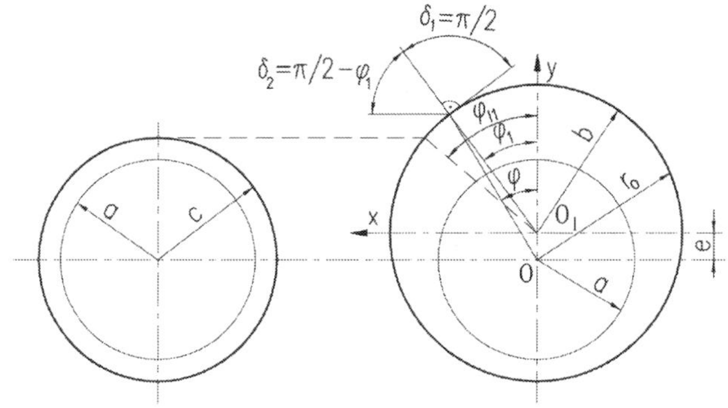

Figure 2. Determination of view factor in the angle interval $0 \leq \varphi_1 \leq \varphi_{1,l1}$

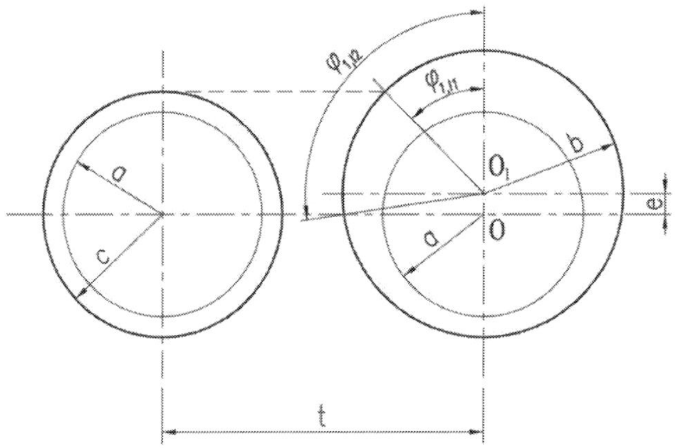

Figure 3. Limit angles $\varphi_{1,l1}$ and $\varphi_{1,l2}$

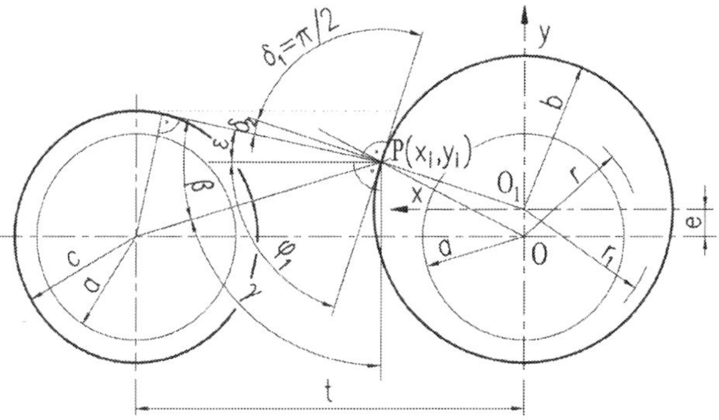

Figure 4. Determination of view factor in the angle interval $\varphi_{1,l1} \leq \varphi_1 \leq \varphi_{1,l2}$

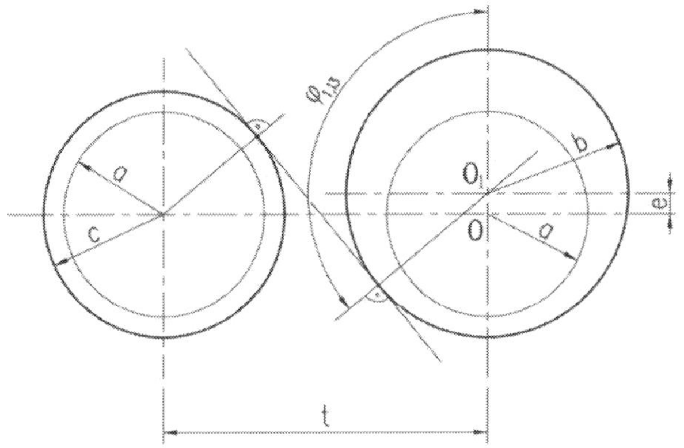

Figure 5. Limit angle $\varphi_{1,13}$

The limit angle $\varphi_{1,13}$ (Figure 5) can be expressed as

$$\varphi_{1,13} = \frac{\pi}{2} + \omega + \kappa, \tag{10}$$

where the angles κ i ω are given by

$$\kappa = \arctan\frac{b-c}{t}, \tag{11}$$

$$\omega = \arccos\frac{b+c}{\sqrt{t^2+e^2}}. \tag{12}$$

The view factor ψ in the interval $\varphi_{1,12} \le \varphi_1 \le \varphi_{1,13}$ is calculated from the following expression (Figure 6)

$$\psi = \frac{1}{2}\left(\sin\delta_2 - \sin\delta_1\right), \varphi_{1,12} \le \varphi_1 \le \varphi_{1,13}, \tag{13}$$

where

$$\delta_1 = \frac{\pi}{2}, \tag{14}$$

$$\delta_2 = \varepsilon + \varphi_1 - \frac{\pi}{2}, \tag{15}$$

$$\varepsilon = \beta + \gamma - \frac{\pi}{2} \tag{16}$$

$$\beta = \arcsin\frac{c}{\sqrt{(t-x_i)^2+(y_i+e)^2}}, \tag{17}$$

$$\gamma = \pi - \arcsin\frac{t-x_i}{\sqrt{(t-x_i)^2+(y_i+e)^2}}, \tag{18}$$

$$x_i = b\sin\varphi_1, \tag{19}$$

$$y_i = b\cos\varphi_1. \tag{20}$$

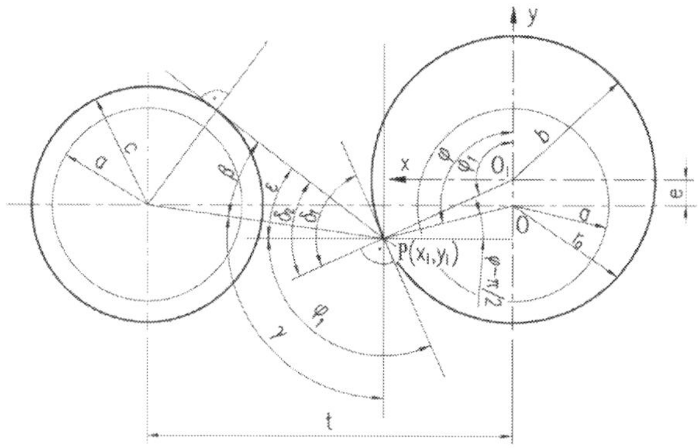

Figure 6. Determination of view factor in the angle interval $\varphi_{1,12} \le \varphi_1 \le \varphi_{1,13}$

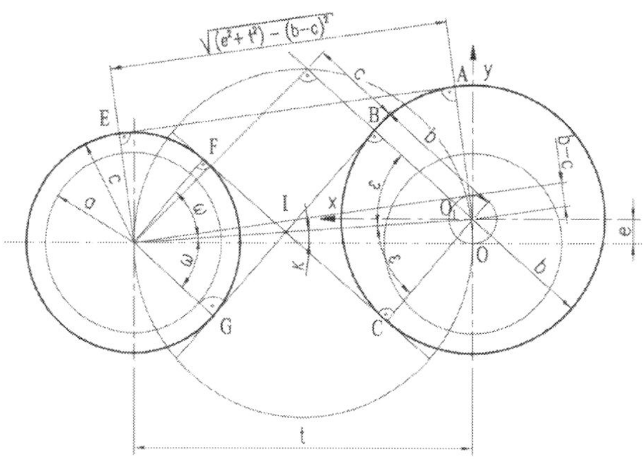

Figure 7. Determination of mean view factor ψ_{bs} for boiler setting over tube pitch t using the crossed string method

Radiation leaving the flame reaches also the boiler setting. The view factor for the radiation heat exchange between boiler setting and rear side of the measuring tube can be calculated in similar way as for the forward part. The mean heat flux q_{bs} resulting from the radiation heat transfer between the flame and the boiler setting can be determined using the crossed-string method [20-21].

The mean value of the view factor ψ_{bs} over the pitch length t is calculated from (Figure7)

$$\psi_{bs} = \frac{1}{2t}\left[\left(FC + BG\right) - \left(FG + BC\right)\right] \tag{21}$$

After substituting the lengths of straight FC and BG and circular segments FG and BC into Eq. (21), the mean value of the view factor over the boiler setting can be expressed as:

$$\psi_{bs} = \frac{b+c}{t}\left(\tan\omega - \omega\right). \tag{22}$$

The mean heat flux over the setting surface is

$$q_{bs} = q_m \psi_{bs}. \tag{23}$$

The angle ω is determined from

$$\tan\omega = \frac{\sqrt{e^2 + t^2 - \left(b+c\right)^2}}{b+c}, \tag{24}$$

If the diameters of the heat flux and water wall tubes are equal, then Eq.(24) simplifies to

$$\tan\omega = \sqrt{\left(\frac{t}{2c}\right)^2 - 1}. \tag{25}$$

The view factor for the radiation heat exchange between boiler setting and rear side of the measuring tube can be calculated in similar way as for the forward part. The view factor in the angle interval $\varphi_{1,14} \le \varphi_1 \le \varphi_{1,15}$ (Figure 8), accounting for the setting radiation, is given by

$$\psi = \psi_{bs} \cdot \frac{1}{2}\left(\sin\delta_2 - \sin\delta_1\right), \ \varphi_{1,14} \le \varphi_1 \le \varphi_{1,15} \tag{26}$$

where the limit angle $\varphi_{1,14}$ is (Figure 8)

$$\psi_{1,14} = \frac{\pi}{2} - \omega + \kappa. \tag{27}$$

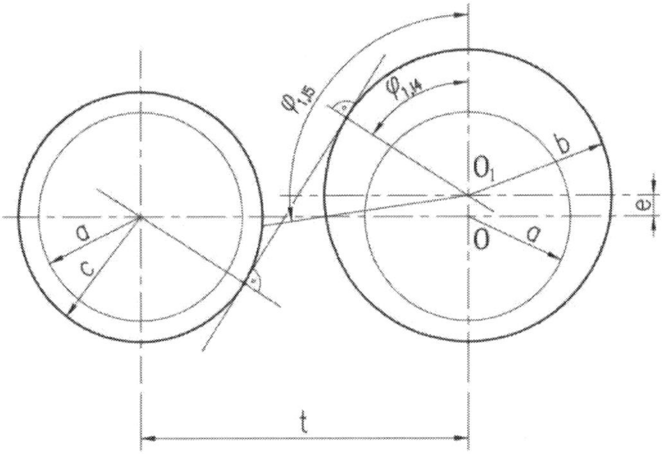

Figure 8. Limit angles $\varphi_{1,I4}$ and $\varphi_{1,I5}=\varphi_{1,I2}=(\pi/2)+\arcsin(e/b)$

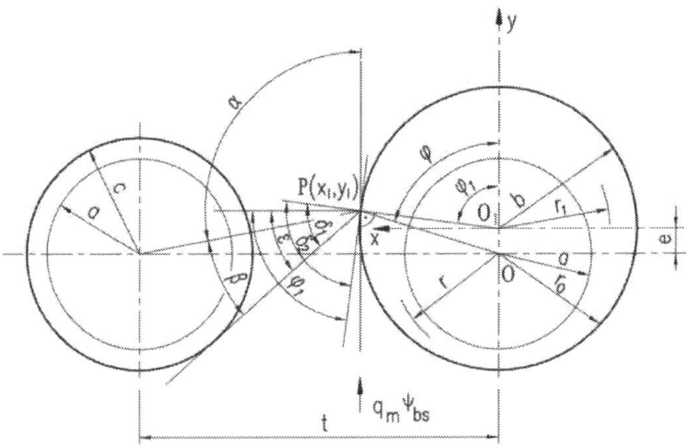

Figure 9. Determination of view factor in the angle interval $\varphi_{1,I4}\leq\varphi_1\leq\varphi_{1,I5}$

The angles δ_1 and δ_2 are (Figure 9)

$$\delta_1=\frac{\pi}{2}+\varepsilon-\varphi_1, \tag{28}$$

$$\delta_2=\frac{\pi}{2}, \tag{29}$$

where

$$\varepsilon = \beta + \gamma - \frac{\pi}{2}, \tag{30}$$

$$\beta = \arcsin\frac{c}{\sqrt{(t - x_i)^2 + (y_i + e)^2}}, \tag{31}$$

$$\gamma = \pi - \arcsin\frac{t - x_i}{\sqrt{(t - x_i)^2 + (y_i + e)^2}}, \tag{32}$$

$$x_i = b\sin\varphi_1, \tag{33}$$

$$y_i = b\cos\varphi_1. \tag{34}$$

The view factor ψ in the interval $\varphi_{1,l5} \leq \varphi \leq \pi$, where $\varphi_{1,l5} = \varphi_{1,l2}$, is given by

$$\psi = \psi_{bs} \cdot \frac{1}{2}\left(\sin\delta_1 + \sin\delta_2\right), \varphi_{1,l5} \leq \varphi \leq \pi, \tag{35}$$

where

$$\delta_1 = \varphi_1 - \varepsilon - \frac{\pi}{2}, \tag{36}$$

$$\delta_2 = \frac{\pi}{2}, \tag{37}$$

$$\varepsilon = \frac{\pi}{2} - (\gamma - \beta), \tag{38}$$

$$\beta = \arcsin\frac{c}{\sqrt{(t - x_i)^2 + (y_i + e)^2}}, \tag{39}$$

$$\gamma = \pi - \arcsin\frac{t - x_i}{\sqrt{(t - x_i)^2 + (y_i + e)^2}}, \tag{40}$$

$$x_i = b\cos\left(\varphi_1 - \frac{\pi}{2}\right), \tag{41}$$

$$y_i = -b\sin\left(\varphi_1 - \frac{\pi}{2}\right). \tag{42}$$

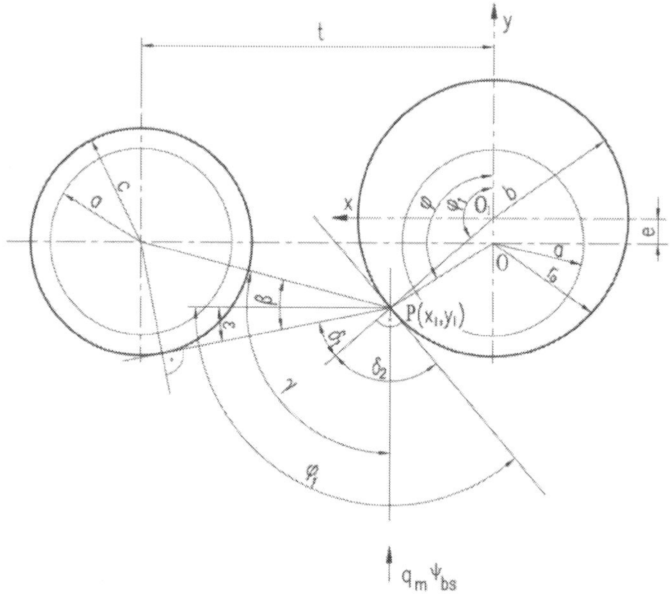

Figure 10. Determination of view factor in the angle interval $\varphi_{1,l5} \leq \varphi_1 \leq \pi$

The total view factor accounts for the radiation heat exchange between the heat flux tube and flame and between the heat flux tube and the boiler setting.

2.2. Theory of the inverse problem

At first, the temperature distribution at the cross section of the measuring tube will be determined, i.e. the direct problem will be solved. Linear direct heat conduction problem can be solved using an analytical method. The temperature distribution will also be calculated numerically using the finite element method (FEM). In order to show accuracy of a numerical approach, the results obtained from numerical and analytical methods will be compared. The following assumptions have been made:

- thermal conductivity of the flux tube material is constant,
- heat transfer coefficient at the inner surface of the measuring tube does not vary on the tube circumference,
- rear side of the water wall, including the measuring tube, is thermally insulated,
- diameter of the eccentric flux tube is larger than the diameter of the water wall tubes,
- the outside surface of the measuring flux tube is irradiated by the flame, so the heat absorption on the tube fire side is non-uniform.

The cylindrical coordinate system is shown in Figure11.

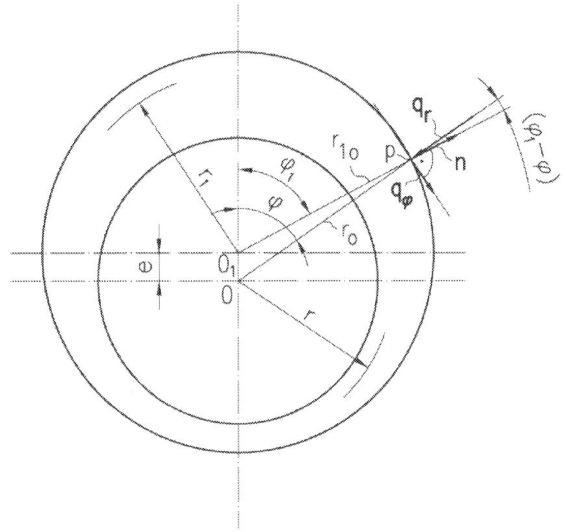

Figure 11. Approximation of the boundary condition on the outer tube surface

The temperature distribution in the eccentric heat flux tube is governed by heat conduction

$$\frac{1}{r}\frac{\partial}{\partial r}\left(kr\frac{\partial\theta}{\partial r}\right)+\frac{1}{r}\frac{\partial}{\partial\varphi}\left(\frac{k}{r}\frac{\partial\theta}{\partial\varphi}\right)=0 \tag{43}$$

subject to the following boundary conditions

$$k\nabla\theta\cdot\mathbf{n}\Big|_{r=r_o}=q_m\psi(\varphi) \tag{44}$$

$$k\frac{\partial\theta}{\partial r}\Big|_{r=a}=h\theta\Big|_{r=a} \tag{45}$$

The left side of Eq. (44) can be transformed as follows (Figure11)

$$\begin{aligned}k\nabla\theta\cdot\mathbf{n}\Big|_{r=r_o}&=\left(\mathbf{q}_{\mathrm{r}}+\mathbf{q}_{\varphi}\right)\cdot\mathbf{n}\Big|_{r=r_o}=\\&=\left[k\frac{\partial T}{\partial r}\cos\left(\varphi_1-\varphi\right)+\frac{k}{r}\frac{\partial T}{\partial\varphi}\sin\left(\varphi_1-\varphi\right)\right]\Bigg|_{r=r_o}\end{aligned} \tag{46}$$

The second term in Eq. (46) can be neglected since it is very small and the boundary condition (44) simplifies to

$$k\frac{\partial\theta}{\partial r}\Big|_{r=r_o}=\frac{q_m\psi(\varphi)}{\cos\left(\varphi_1-\varphi\right)} \tag{47}$$

The heat flux over the tube circumference can be approximated by the Fourier polynomial

$$\frac{q_m \psi(\varphi)}{\cos(\varphi_1 - \varphi)} = q_0 + \sum_{n=1}^{\infty} q_n \cos(n\varphi) \tag{48}$$

where

$$\begin{aligned} q_0 &= \frac{1}{\pi} \int_0^{\pi} \frac{q_m \psi(\varphi)}{\cos(\varphi_1 - \varphi)} d\varphi, \\ q_n &= \frac{2}{\pi} \int_0^{\pi} \frac{q_m \psi(\varphi)}{\cos(\varphi_1 - \varphi)} \cos(n\varphi) d\varphi, \; n = 1, \ldots \end{aligned} \tag{49}$$

The boundary value problem (43, 45, 47) was solved using the separation of variables to give

$$\theta(r,\varphi) = A_0 + B_0 \ln r + \sum_{n=1}^{\infty} \left(C_n r^n + D_n r^{-n} \right) \cos n\varphi. \tag{50}$$

where

$$A_0 = \frac{q_0 r_o(\varphi)}{k} \left(\frac{1}{Bi} - \ln a \right), \tag{51}$$

$$B_0 = \frac{q_0 r_o(\varphi)}{k}, \tag{52}$$

$$C_n = \frac{q_n r_o(\varphi)}{k} \frac{\frac{1}{n} u^n (Bi + n) \frac{1}{a^n}}{Bi\left(u^{2n} + 1\right) + n\left(u^{2n} - 1\right)}, \tag{53}$$

$$D_n = -\frac{q_n r_o(\varphi)}{k} \frac{\frac{1}{n} u^n (Bi - n) a^n}{Bi\left(u^{2n} + 1\right) + n\left(u^{2n} - 1\right)}. \tag{54}$$

The ratio of the outer to inner radius of the eccentric flux tube: $u = u(\phi) = r_o(\phi)/a$ depends on the angle ϕ, since the outer radius of the tube flux

$$r_o = e \cos\varphi + \sqrt{b^2 - (e \sin\varphi)^2} \tag{55}$$

is the function of the angle ϕ.

Eq. (50) can be used for the temperature calculation when all the boundary conditions are known. In the inverse heat conduction problem three parameters are to be determined:

- absorbed heat flux referred to the projected furnace wall surface: $x_1 = q_m$,
- heat transfer coefficient on the inner surface of the boiler tube: $x_2 = h$,
- fluid bulk temperature: $x_3 = T_f$.

These parameters appear in boundary conditions (44) and (45) and will be determined based on the wall temperature measurements at m internal points (r_i,ϕ_i)

$$T(r_i,\varphi_i)=f_i\,,i=1,...,m\,,m\geq 3. \tag{56}$$

In a general case, the unknown parameters: x_1, …, x_n are determined by minimizing sum of squares

$$S=(\mathbf{f}-\mathbf{T_m})^T(\mathbf{f}-\mathbf{T_m}), \tag{57}$$

where $\mathbf{f}=(f_1,\ldots,f_m)^T$ is the vector of measured temperatures, and $\mathbf{T_m}=(T_1,\ldots,T_m)^T$ the vector of computed temperatures $T_i=T(r_i,\varphi_i)$, $i=1,\ldots,m$.

The parameters $x_1 \ldots x_n$, for which the sum (34) is minimum are determined using the Levenberg-Marquardt method [23,25]. The parameters, **x**, are calculated by the following iteration

$$\mathbf{x}^{(k+1)}=\mathbf{x}^{(k)}+\boldsymbol{\delta}^{(k)},\ k=0,1,.... \tag{58}$$

where

$$\boldsymbol{\delta}^{(k)}=\left[\left(\mathbf{J}_m^{(k)}\right)^T\mathbf{J}_m^{(k)}+\mu^{(k)}\mathbf{I}_n\right]^{-1}\times \times\left(\mathbf{J}_m^{(k)}\right)^T\left[\mathbf{f}-\mathbf{T}_m\left(\mathbf{x}^{(k)}\right)\right]. \tag{59}$$

where $\mu^{(k)}$ is the multiplier and $\mathbf{I}_n$ is the identity matrix. The Levenberg–Marquardt method is a combination of the Gauss–Newton method ($\mu^{(k)}\to 0$) and the steepest-descent method ($\mu^{(k)}\to\infty$). The m x n Jacobian matrix of $T(\mathbf{x}^{(k)},\mathbf{r}_i)$ is given by

$$\mathbf{J}^{(k)}=\left.\frac{\partial\mathbf{T}(\mathbf{x})}{\partial\mathbf{x}_T}\right|_{\mathbf{x}=\mathbf{x}^{(k)}}=\left.\begin{bmatrix}\frac{\partial T_1}{\partial x_1} & \cdots & \frac{\partial T_1}{\partial x_n}\\ \cdots & \cdots & \cdots\\ \cdots & \cdots & \cdots\\ \cdots & \cdots & \cdots\\ \cdots & \cdots & \cdots\\ \frac{\partial T_m}{\partial x_1} & \cdots & \frac{\partial T_m}{\partial x_n}\end{bmatrix}\right|_{\mathbf{x}=\mathbf{x}^{(k)}},\quad m=5,\quad n=3, \tag{60}$$

The symbol $\mathbf{I}_n$ denotes the identity matrix of $n \times n$ dimension, and $\mu^{(k)}$ the weight coefficient, which changes in accordance with the algorithm suggested by Levenberg and Marquardt. The upper index T denotes the transposed matrix. Temperature distribution $T(r,\varphi, \mathbf{x}^{(k)})$ is computed at each iteration step using Eq. (50). After a few iterations we obtain a convergent solution.

2.3. The uncertainty of the results

The uncertainties of the determined parameters $\mathbf{x}^*$ will be estimated using the error propagation rule of Gauss [23-26]. The propagation of uncertainty in the independent variables: measured wall temperatures f_j, j=1, ...m, thermal conductivity k, radial and angular positions of temperature sensors r_j, φ_j, j=1, ...m is estimated from the following equation

$$2\sigma_{x_i} = \left[\sum_{j=1}^{m} \left(\frac{\partial x_i}{\partial f_j} \sigma_{f_j} \right)^2 + \left(\frac{\partial x_i}{\partial k} \sigma_k \right)^2 + \sum_{j=1}^{m} \left(\frac{\partial x_i}{\partial r_j} \sigma_{r_j} \right)^2 + \sum_{j=1}^{m} \left(\frac{\partial x_i}{\partial \varphi_j} \sigma_{\varphi_j} \right)^2 \right]^{1/2}, \quad i = 1,2,3 \tag{61}$$

The 95% uncertainty in the estimated parameters can be expressed in the form

$$x_i = x_i^* \pm 2\sigma_{x_i}, \tag{62}$$

where $x_i^*, i = 1,2,3$ represent the value of the parameters obtained using the least squares method. The sensitivity coefficients $\partial x_i / \partial f_j$, $\partial x_i / \partial k$, $\partial x_i / \partial r_j$, and $\partial x_i / \partial \varphi_j$ in the expression (61) were calculated by means of the numerical approximation using central difference quotients:

$$\frac{\partial x_i}{\partial f_j} = \frac{x_i\left(f_1, f_2, \ldots, f_j + \delta, \ldots, f_m\right) - x_i\left(f_1, f_2, \ldots, f_j - \delta, \ldots, f_m\right)}{2\delta}, \tag{63}$$

where δ is a small number.

2.4. Computational and boiler tests

Firstly, a computational example will be presented. "Experimental data" are generated artificially using the analytical solution (50).

Consider a water-wall tube with the following parameters (Figure1.):

- outer radius b = 35 mm,
- inner radius a = 25 mm,
- pitch of the water-wall tubes t = 80 mm,
- thermal conductivity k = 28.5 W/(m·K),

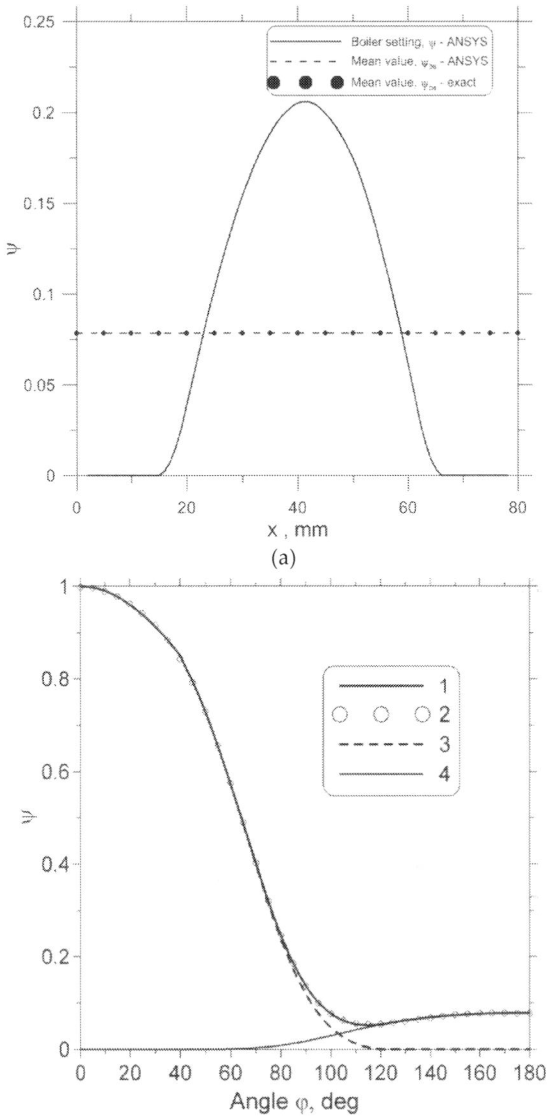

(b)

Figure 12. View factor associated with radiation heat exchange between elemental surface on the boiler setting or flux tube and flame: (a) – view factor for radiation heat transfer between flame and boiler setting, (b) 1 - total view factor accounting radiation from furnace and boiler setting, 2 - approximation by the Fourier polynomial of the seventh degree, 3 - exact view factor for furnace radiation, 4- view factor from boiler setting

- absorbed heat flux $q_m = 200000$ W/m^2,
- heat transfer coefficient $h = 30000$ W/(m^2·K),
- fluid temperature $T_f = 318$ °C.

The view factor distributions on the outer surface of the flux-tube and boiler setting were calculated analytically and numerically by means of the finite element method (FEM) [22]. The changes of the view factor over the pitch length and tube circumference are illustrated in Figures 12 and 13.

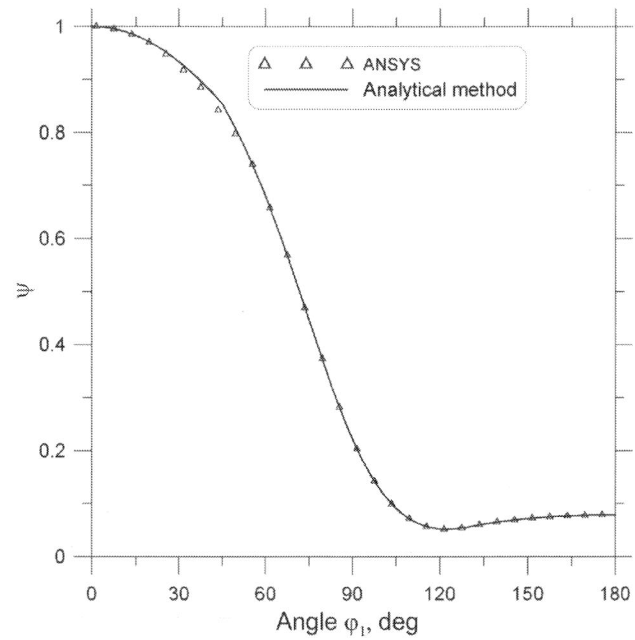

Figure 13. Comparison of total view factor calculated by exact and FEM method

The agreement between the temperatures of the outer and inner tube surfaces which were calculated analytically and numerically is also very good (Figures 14 and 15). The small differences between the analytical and FEM solutions are caused by the approximate boundary condition (47). The temperature distribution in the flux tube cross section is shown in Figure 14.

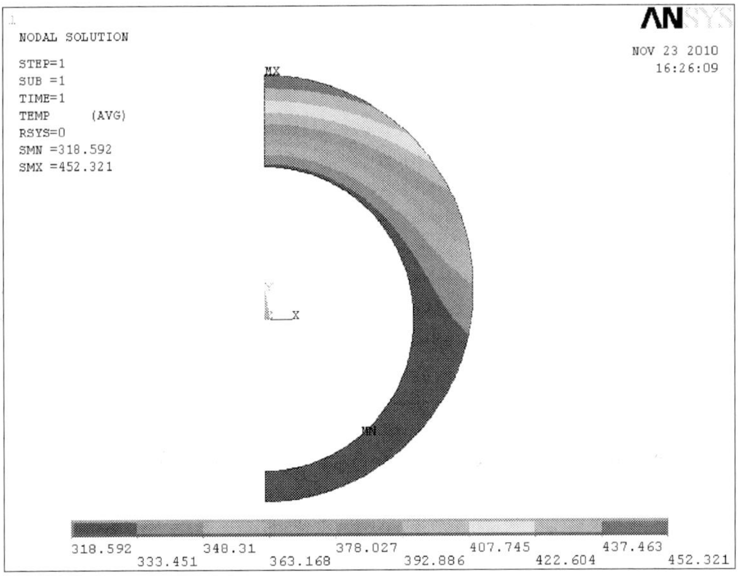

Figure 14. Computed temperature distribution in °C in the cross section of the heat flux tube; q_m = 200000 W/m², h = 30000 W/(m²·K), T_f=318 °C

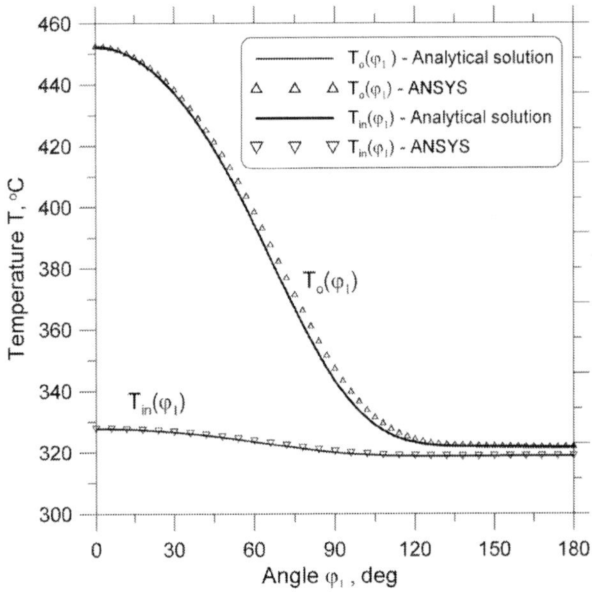

Figure 15. Temperature distribution at the inner and outer surfaces of the flux tube calculated by the analytical and finite element method

The following input data is generated using Eq. (50): $f_1 = 437.98\ ^\circ C$, $f_2 = 434.47\ ^\circ C$, $f_3 = 383.35\ ^\circ C$, $f_4 = 380.70\ ^\circ C$, $f_5 = 321.58^\circ C$.

The following values were obtained using the proposed method: $q_m^* = 200\,000.35\ W/m^2$, $h^* = 30\,001.56\ W/(m^2 \cdot K)$, $T_f^* = 318.00\ ^\circ C$.

In order to show the influence of the measurement errors on the determined thermal boundary parameters, the 95% confidence intervals were calculated. The following uncertainties of the measured values were assumed (at a 95% confidence interval):

$$2\sigma_{f_j} = \pm 0.2K, j = 1, \ldots, 5, 2\sigma_k = \pm 0.5\ W/(m \cdot K), 2\sigma_{r_j} = \pm 0.05mm, 2\sigma_{\varphi_j} = \pm 0.5^\circ, j = 1, \ldots, 5.$$

The uncertainties (95% confidence interval) of the coefficients x_i were determined using the error propagation rule formulated by Gauss.

The calculation using Eq. (61) yielded the following results: $x_1 = 200\,000.35 \pm 3827.72\ W/m^2$, $x_2 = 30\,001.56 \pm 2698.81\ W/(m^2 \cdot K)$, $x_3 = 318.0 \pm 0.11\ ^\circ C$. The accuracy of the obtained results is very satisfactory. There is only a small difference between the estimated parameters and the input values. The highest temperature occurs at the crown of the flux-tube (Figures 14 and 15). The temperature of the inner surface of the flux tube is only a few degrees above the saturation temperature of the water-steam mixture. Since the heat flux at the rear side of the tube is small, the circumferential heat flow rate is significant. However, the rear surface thermocouple indicates temperatures of 2-4 °C above the saturation temperature. Therefore, the fifth thermocouple can be attached to the unheated side of the tube so as to measure the temperature of the water-steam mixture flowing through the flux tube.

In the second example, experimental results will be presented. Measurements were conducted at a 50MW pulverized coal fired boiler. The temperatures indicated by the flux tube at the elevation of 19.2 m are shown in Figure 16. The heat flux tube is of 20G low carbon steel with temperature dependent thermal conductivity

$$k(T) = 53.26 - 0.02376224T, \tag{64}$$

where the temperature T is expressed in °C and thermal conductivity in W/(m·K).

The unknown parameters were determined for eight time points which are marked in Figure 16.

The inverse analysis was performed assuming the constant thermal conductivity $k(\bar{T})$ which was obtained from Eq.(64) for the average temperature: $\bar{T} = (T_1 + T_2 + T_3 + T_4)/4$.

The estimated parameters: heat flux q_m, heat transfer coefficient h, and the water-steam mixture T_f are depicted in Figure 17. The developed flux tube can work for a long time in the destructive high temperature atmosphere of a coal-fired boiler.

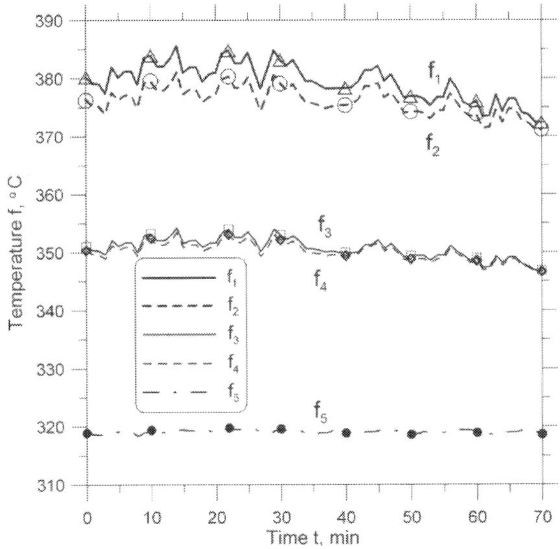

Figure 16. Measured flux tube temperatures; marks denote measured temperatures taken for the inverse analysis

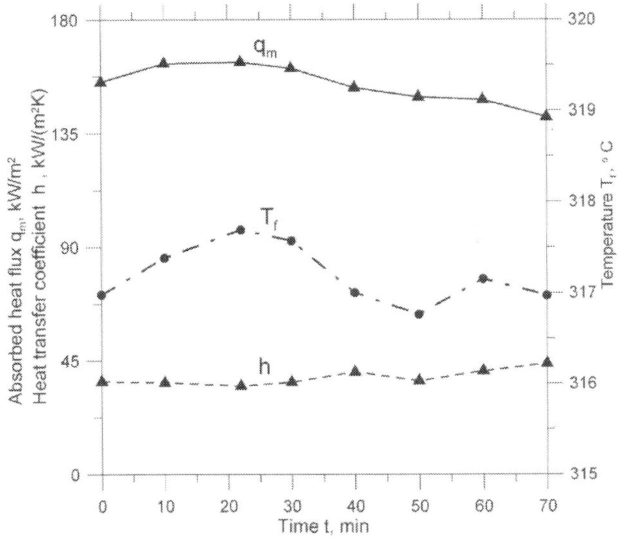

Figure 17. Estimated parameters: absorbed heat flux q_m, heat transfer coefficient h, and temperature of water-steam mixture T_f

Flux tubes can also be used as a local slag monitor to detect a build up of slag. The presence of the scale on the inner surface of the tube wall can also be detected.

3. Tubular type heat flux meter made of a finned tube

In this section, a numerical method for determining the heat flux in boiler furnaces, based on experimentally acquired interior flux-tube temperatures, is presented. The tubular type instrument has been designed (Figure 18) to provide a very accurate measurement of absorbed heat flux q_m, inside heat transfer coefficient h_{in}, and water steam temperature T_f. The number of thermocouples is greater than three because the additional information can help enhance the accuracy of parameter determining. In contrast to the existing devices, in the developed flux-tube fins are not welded to adjacent water-wall tubes. Temperature distribution in the flux-tube is symmetric and not disturbed by different temperature fields in neighboring tubes. The temperature dependent thermal conductivity of the flux-tube material was assumed. The meter is constructed from a short length of eccentric tube containing four thermocouples on the fire side below the inner and outer surfaces of the tube. The fifth thermocouple is located at the rear of the tube (on the casing side of the water-wall tube). The boundary conditions on the outer and inner surfaces of the water flux-tube must then be determined from temperature measurements in the interior locations. Four K-type sheathed thermocouples, 1 mm in diameter, are inserted into holes, which are parallel to the tube axis. The thermal conduction effect at the hot junction is minimized because the thermocouples pass through isothermal holes. The thermocouples are brought to the rear of the tube in the slot machined in the tube wall. An austenitic cover plate with the thickness of 3 mm – welded to the tube – is used to protect the thermocouples from the incident flame radiation. A K-type sheathed thermocouple with a pad is used to measure the temperature at the rear of the flux-tube. This temperature is almost the same as the water-steam temperature. An inverse problem of heat conduction was solved using the least squares method. Three unknown parameters were estimated using the Levenberg-Marquardt method [23, 25]. At every iteration step, the temperature distribution over the cross-section of the heat flux meter was computed using the ANSYS CFX software

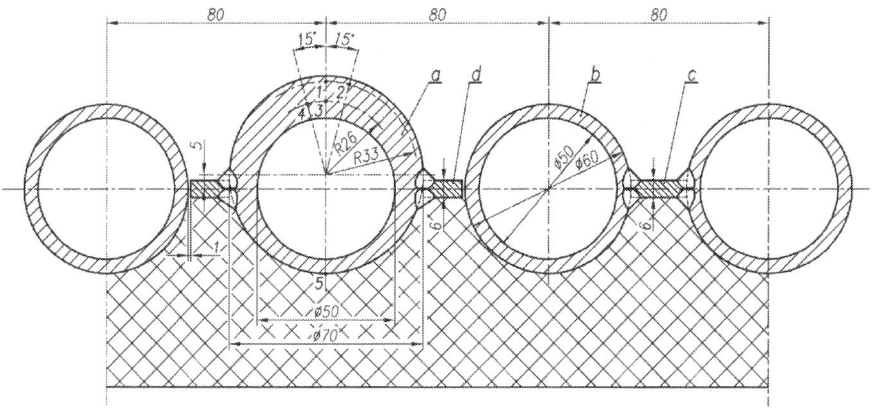

Figure 18. The cross-section of the membrane wall in the combustion chamber of the steam boiler

Test calculations were carried out to assess accuracy of the presented method. The uncertainty in determined parameters was calculated using the Gauss variance propagation rule. The presented method is appropriate for membrane water walls (Figure 18). The new method has advantages in terms of simplicity and flexibility.

3.1. Theory

The furnace wall tubes in most modern units are welded together with steel bars (fins) to provide membrane wall panels which are insulated on one side and exposed to a furnace on the other, as shown schematically in Figure 18.

In a heat conduction model of the flux-tube the following assumptions are made:

- temperature distribution is two-dimensional and steady-state,
- the thermal conductivity of the flux-tube and membrane wall,
- may be dependent of temperature,
- the heat transfer coefficient h_{in} and the scale thickness ds is uniform over the inner tube surface.

The temperature distribution is governed by the non-linear partial differential equation

$$\nabla \cdot \left[k(T) \nabla T \right] = 0, \tag{65}$$

where ∇ is the vector operator, which is called nabla (gradient operator), and in Cartesian coordinates is defined by $\nabla = \mathbf{i}\partial/\partial x + \mathbf{j}\partial/\partial y + \mathbf{k}\partial/\partial z +$. The unknown boundary conditions may be expressed as

$$\left[k(T) \frac{\partial T}{\partial n} \right]\Bigg|_s = q(s), \tag{66}$$

where $q(s)$ is the radiation heat flux absorbed by the exposed flux tube and membrane wall surface. The local heat flux $q(s)$ is a function of the view factor $\psi(s)$ (Figure 19)

$$q(s) = q_m \psi(s), \tag{67}$$

where q_m is measured heat flux (thermal loading of heating surface). The view factor $\psi(s)$ from the infinite flame plane to the differential element on the membrane wall surface can be determined graphically [7], or numerically [22].

In this chapter, $\psi(s)$ was evaluated numerically using the finite element program ANSYS [22], and is displayed in Figure 19 as a function of the extended coordinate s. Because of the symmetry, only the representative water-wall section illustrated in Figure 20 needs to be analyzed. The convective heat transfer from the inside tube surfaces to the water-steam mixture is described by Newton's law of cooling

$$-\left[k(T)\frac{\partial T}{\partial n}\right]\Bigg|_{S_{in}} = h_{in}\left(T\big|_{S_{in}} - T_f\right), \tag{68}$$

where $\partial T/\partial n$ is the derivative in the normal direction, h_{in} is the heat transfer coefficient and T_f denotes the temperature of the water–steam mixture.

The reverse side of the membrane water-wall is thermally insulated. In addition to the unknown boundary conditions, the internal temperature measurements f_i are included in the analysis

$$T_e(\mathbf{r}_i) = f_i, \quad i = 1,\ldots,m, \tag{69}$$

where m = 5 denotes the number of thermocouples (Figure 18). The unknown parameters: $x_1 = q_m$, $x_2 = h_{in}$, and $x_3 = T_f$ were determined using the least-squares method. The symbol r_{in} denotes the inside tube radius, and $k(T)$ is the temperature dependent thermal conductivity. The object is to choose $\mathbf{x} = (x_1, \ldots, x_n)^T$ for n = 3 such that computed temperatures $T(\mathbf{x}, \mathbf{r}_i)$ agree within certain limits with the experimentally measured temperatures f_i.

This may be expressed as

$$T(\mathbf{x},\mathbf{r}_i) - f_i \cong 0, \quad i = 1,\ldots,m, \quad m = 5. \tag{70}$$

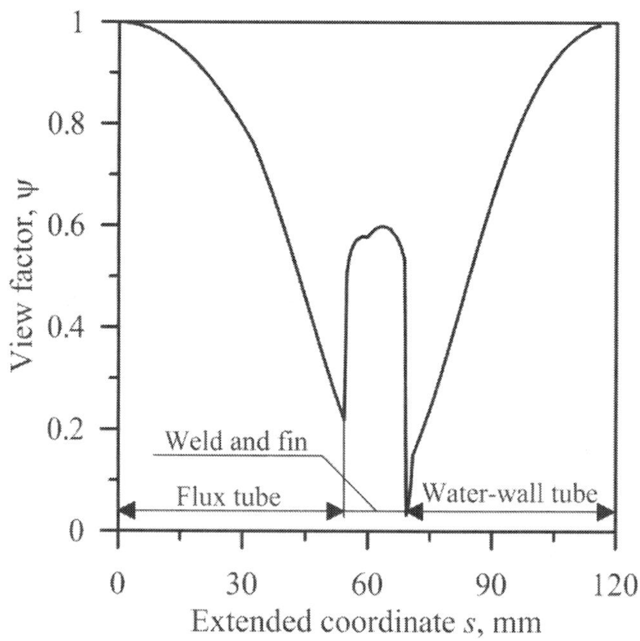

Figure 19. View factor distribution on the outer surface of water-wall tube

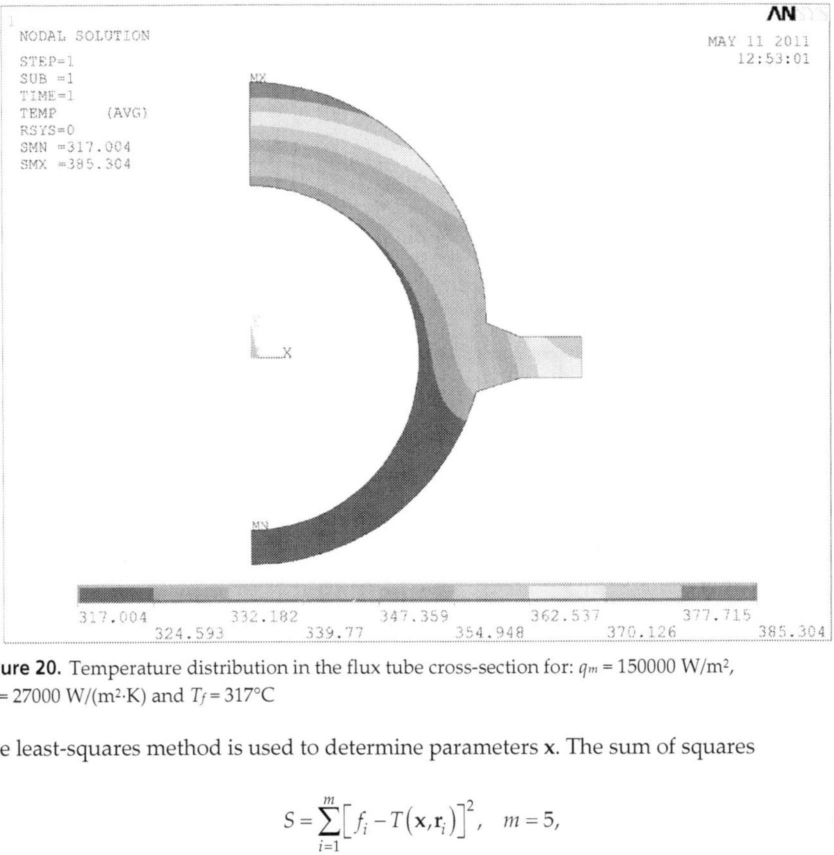

Figure 20. Temperature distribution in the flux tube cross-section for: q_m = 150000 W/m², h_{in} = 27000 W/(m²·K) and T_f = 317°C

The least-squares method is used to determine parameters **x**. The sum of squares

$$S = \sum_{i=1}^{m} \left[f_i - T(\mathbf{x}, \mathbf{r}_i) \right]^2, \quad m = 5, \tag{71}$$

is minimized using the Levenberg–Marquardt method [23, 25].

The uncertainties of the determined parameters $\mathbf{x}^*$ will be estimated using the error propagation rule of Gauss [23-26].

3.2. Test computations

The flux-tubes were manufactured in the laboratory and then securely welded to the water-wall tubes at different elevations in the furnace of the steam boiler. The coal fired boiler produces 58.3 kg/s superheated steam at 11 MPa and 540°C.

The material of the heat flux-tube is 20G steel. The composition of the 20G mild steel is as follows: 0.17–0.24% C, 0.7–1.0% Mn, 0.15–0.40% Si, Max 0.04% P, Max 0.04% S, and the remainder is iron Fe. The heat flux-tube thermal conductivity is assumed to be temperature dependent (Table 1).

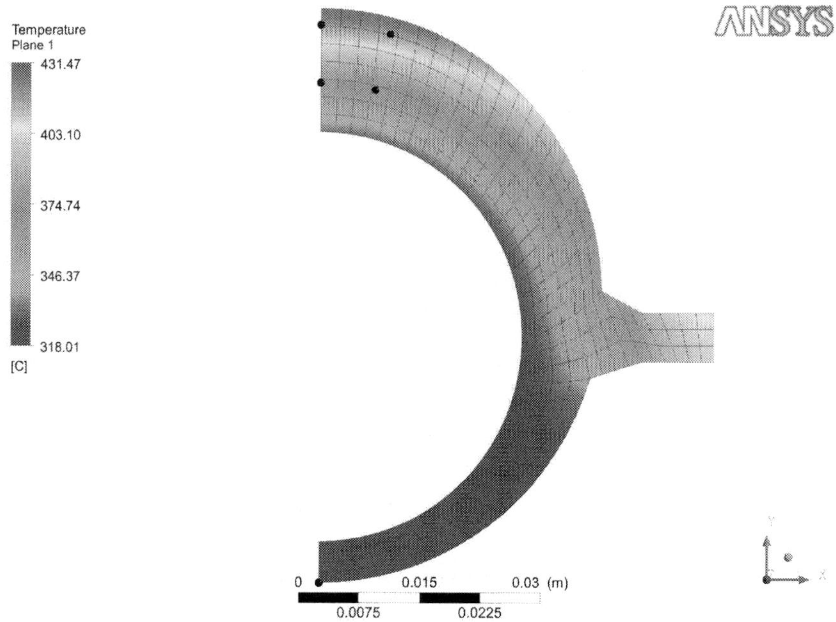

(a)

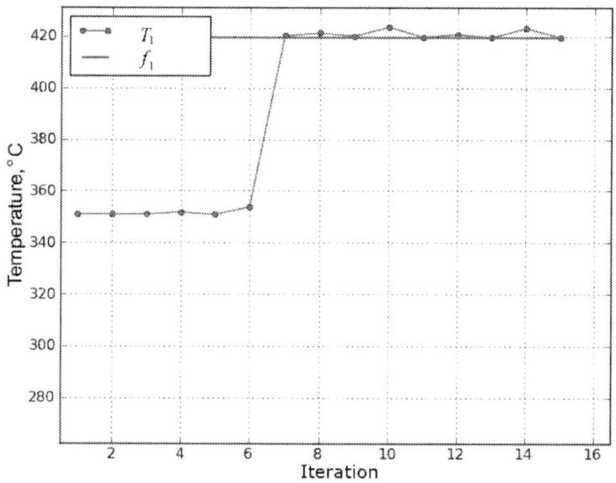

(b)

Figure 21. Solution of the inverse problem for the "exact" data: $f_1 = 419.66$°C, $f_2 = 417.31$°C, $f_3 = 374.90$°C, $f_4 = 373.19$°C, $f_5 = 318.01$°C ; (a) - temperature distribution in the flux-tube, (b) - iteration number for the temperature T_1

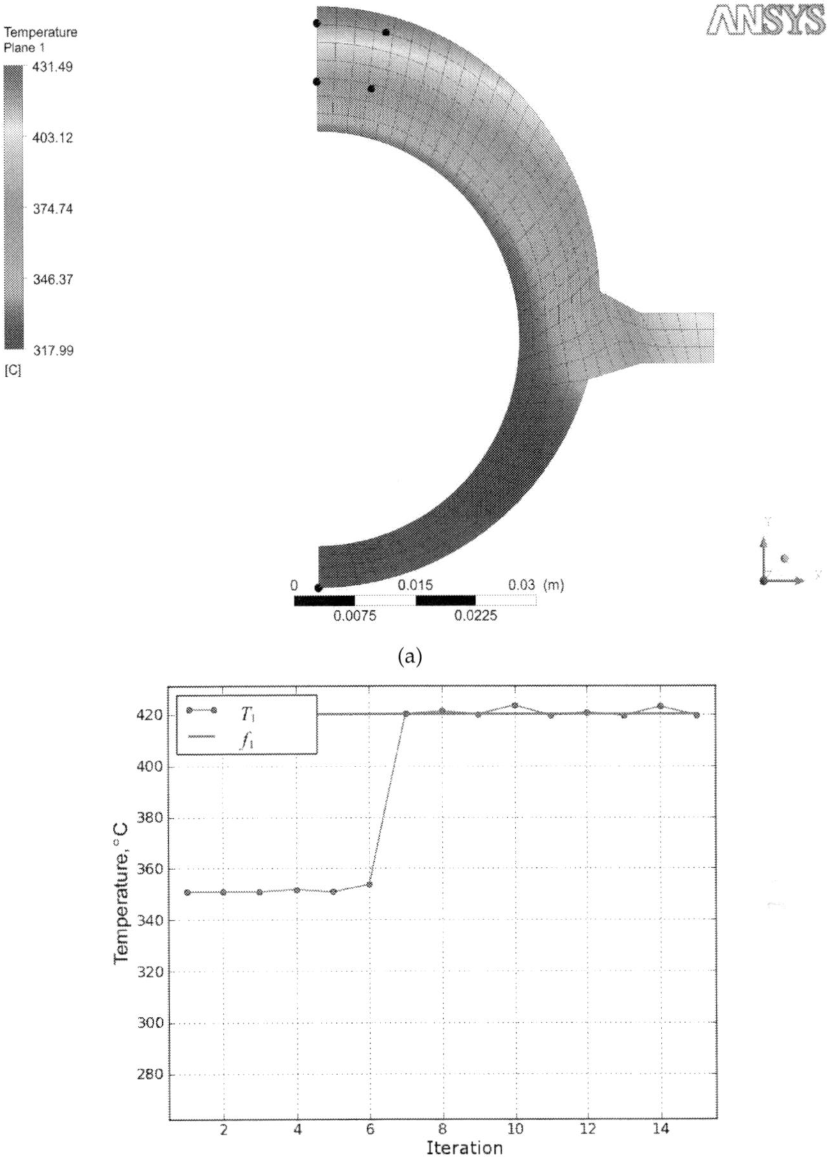

(a)

(b)

Figure 22. Solution of the inverse problem for the "perturbed" data: f_1 = 420.16°C, f_2 = 416.81°C, f_3 = 375.40°C, f_4 = 372.69°C, f_5 = 318.01°C; (a) - temperature distribution in the flux-tube, (b) - iteration number for the temperature T_1

Temperature T, °C	100	200	300	400
Thermal conductivity k, W/(m·K)	50.69	48.60	46.09	42.30

Table 1. Thermal conductivity $k(T)$ of steel 20G as a function of temperature

To demonstrate that the maximum temperature of the fin tip is lower than the allowable temperature for the 20G steel, the flux tube temperature was computed using ANSYS CFX package [22]. Changes of the view factor on the flux tube, weld and fin surface were calculated with ANSYS CFX. The temperature distribution shown in Figure 20 was obtained for the following data: absorbed heat flux, q_m = 150000 W/m^2, temperature of the water-steam mixture, T_f = 317°C, and heat transfer coefficient at the tube inner surface, h_{in} = 27000 W/(m^2·K). An inspection of the results shown in Figure 20 indicates that the maximum temperature of the fin does not exceed 375°C.

Next, to illustrate the effectiveness of the presented method, test calculations were carried out. The "measured" temperatures f_i, i = 1, 2, ..., 5 were generated artificially by means of ANSYS CFX for: q_m = 250000 W/m^2, h_{in} = 30000 W/(m^2·K) and T_f = 318°C. The following values of "measured" temperatures were obtained f_1 = 419.66°C, f_2 = 417.31°C, f_3 = 374.90°C, f_4 = 373.19°C, f_5 = 318.01°C. The temperature distribution in the flux tube cross-section, reconstructed on the basis of five measured temperatures is depicted in Figure 21a.

The proposed inverse method is very accurate since the estimated parameters: q_m = 250000.063 W/m^2, h_{in} = 30000.054 W/(m^2 ·K) and T_f = 318.0°C differ insignificantly from the input values. In order to show the influence of the measurement errors on the determined parameters, the 95% confidence intervals were estimated. The following uncertainties of the measured values were assumed (at 95% confidence interval): $2\sigma_{f_j} = \pm 0.5$ K, j = 1, 2, ..., 5, $2\sigma_k = \pm 1\ \mathrm{W/(m \cdot K)}, 2\sigma_{r_j}$ = ±0.05mm, $2\sigma_{\varphi_j}$ = ±0.5°, j=1,...,5. The uncertainties (95% confidence interval) of the coefficients x_i were determined using the error propagation rule formulated by Gauss [23-26]. The calculated uncertainties are: ±6% for q_m, ±33% for h_{in} and ±0.3% for T_f. The accuracy of the results obtained is acceptable.

Then, the inverse analysis was carried out for perturbed data: f_1 = 420.16°C, f_2 = 416.81°C, f_3 = 375.40°C, f_4 = 372.69°C, f_5 = 318.01°C. The reconstructed temperature distribution illustrates Figure 22a.

The obtained results are: q_m = 250118.613 W/m^2, h_{in} = 30050.041 W/(m^2 ·K) and T_f = 317.99°C. The errors in the measured temperatures have little effect on the estimated parameters. The number of iterations in the Levenberg-Marquardt procedure is small in both cases (Figures 21b and 22b).

4. Conclusions

Two different tubular type instruments (flux tubes) were developed to identify boundary conditions in water wall tubes of steam boilers. The first measuring device is an eccentric tube. The ends of the four thermocouples are located at the fireside part of the tube and the fifth thermocouple is attached to the unheated rear surface of the tube. The meter presented in the paper has one particular advantage over the existing flux tubes to date. The temperature distribution in the flux tube is not affected by the water wall tubes, since the flux tube is not connected to adjacent waterwall tubes with metal bars, referred to as membrane or webs. To determine the unknown parameters only the temperature distribution at the cross section of the flux tube must be analyzed.

The second flux tube has two longitudinal fins. Fins attached to the flux tube are not welded to the adjacent water-wall tubes, so the temperature distribution in the measuring device is not affected by neighboring water-wall tubes. The installation of the flux tube is easier because welding of fins to adjacent water-wall tubes is avoided. Based on the measured flux tube temperatures the non-linear inverse heat conduction problem was solved. A CFD based method for determining heat flux absorbed water wall tubes, heat transfer coefficient at the inner flux tube surface and temperature of the water-steam mixture has been presented. The proposed flux tube and the inverse procedure for determining absorbed heat flux can be used both when the inner surface of the heat flux tube is clean and when scale or corrosion deposits are present on the inner surface what can occur after a long time service of the heat flux tube.

The flux tubes can work for a long time in the destructive high temperature atmosphere of a coal-fired boiler.

Nomenclature

a	inner radius of boiler tube and flux-tube (m)
b	outer radius of flux-tube (m)
Bi	Biot number, $Bi = ha/k$
c	outer radius of boiler tube (m)
e	eccentric (m)
f_i	measured wall temperature at the i-th location (°C or K)
$\mathbf{f}$	vector of measured wall temperatures
h	heat transfer coefficient (W/(m^2·K))
$\mathbf{I}_n$	identity matrix
$\mathbf{J}$	Jacobian matrix of $\mathbf{T}$
k	thermal conductivity (W/(m·K))
l	arbitrary length of boiler tube (m)
m	number of temperature measurement points
n	number of unknown parameters
q_m	heat flux to be determined (absorbed heat flux referred to the projected furnace water wall surface) (W/m^2)
r	coordinate in cylindrical coordinate system or radius (m)
r_i	radial coordinate of the i-th thermocouple (m)
r_{in}	inner radius of the flux-tube (m)
r_o	outer radius of the flux-tube (m)

r	position vector
s	extended coordinate along the fireside water-wall surface (m)
S	sum of the temperature difference squares (K^2)
t	pitch of the water wall tubes (m)
T	temperature (°C or K)
T_f	fluid temperature (°C or K)
T_i	calculated temperature at the location(r_i, ϕ_i) (°C or K)
$\mathbf{T}^m$	m - dimensional column vector of calculated temperatures
$u(\phi)$	ratio of the outer to the inner radius of the tube, $u(\phi) = r_o/a$
x_i	unknown parameter
x	n-dimensional column vector of unknown parameters

Greek symbols

$\alpha, \beta, \gamma, \delta_1, \delta_2, \varepsilon$	angles (rad)
θ	temperature excess over the fluid temperature, $\theta = T - T_f$ (K)
ϕ	angular coordinate (rad)
ϕ_i	angular coordinate of the i-th thermocouple (rad)
μ	multiplier in the Levenberg-Marquardt algorithm
ψ	view factor

Subscripts

in	inner
o	outer
i	*i-th* temperature measurement point
f	fluid

Author details

Jan Taler

Department of Thermal Power Engineering, Cracow University of Technology, Cracow, Poland

Dawid Taler

Institute of Heat Transfer Engineering and Air Protection, Cracow University of Technology, Cracow, Poland

5. References

[1] Segeer M, Taler J (1983) Konstruktion und Einsatz transportabler Wärmeflußsonden zur Bestimmung der Heizflächenbelastung in Feuerräumen. Fortschr.-Ber. VDI Zeitschrift, Reihe 6, Nr 129. Düsseldorf : VDI-Verlag.

[2] Northover EW, Hitchcock JA (1967) A Heat Flux Meter for Use in Boiler Furnaces. J. Sci. Instrum. 44: 371–374.

[3] Neal SBH, Northover EW (1980) The Measurement of Radiant Heat Flux in Large Boiler Furnaces-I. Problems of Ash Deposition Relating to Heat Flux. Int. J. Heat Mass Transfer 23: 1015–1022.

[4] Arai N, Matsunami A, Churchill S (1996) A Review of Measurements of Heat Flux Density Applicable to the Field of Combustion. Exp. Therm. Fluid Sci. 12: 452–460.

[5] Taler J (1990) Measurement of Heat Flux to Steam Boiler Membrane Water Walls. VGB Kraftwerkstechnik 70: 540–546.

[6] Taler J (1992) A Method of Determining Local Heat Flux in Boiler Furnaces. Int. J. Heat Mass Transfer 35:1625–1634.

[7] Taler J (1990) Messung der lokalen Heizflächenbelastung in Feuerräumen von Dampferzeugern. Brennstoff-Wärme-Kraft (BWK) 42: 269-277.

[8] Fang Z, Xie D, Diao N, Grace JR, Lim CJ (1997) A New Method for Solving the Inverse Conduction Problem in Steady Heat Flux Measurement. Int. J. Heat Mass Transfer 40: 3947–3953.

[9] Luan W, Bowen BD, Lim CJ, Brereton CMH, Grace JR (2000) Suspension-to Membrane-Wall Heat Transfer in a Circulating Fluidized Bed Combustor. Int. J. Heat Mass Transfer 43: 1173–1185.

[10] Taler J, Taler D (2007) Tubular Type Heat Flux Meter for Monitoring Internal Scale Deposits in Large Steam Boilers. Heat Transfer Engineering 28: 230-239.

[11] Sobota T, Taler D (2010) A Simple Method for Measuring Heat Flux in Boiler Furnaces. Rynek Energii 86: 108-114.

[12] Taler D, Taler J, Sury A (2011) Identification of Thermal Operation Conditions of Water Wall Tubes Using Eccentric Tubular Type Meters. Rynek Energii 92: 164-171.

[13] Taler J, Taler D, Kowal A (2011) Measurements of Absorbed Heat Flux and Water-side Heat Transfer Coefficient in Water Wall Tubes. Archives of Thermodynamics 32: 77 – 88.

[14] Taler J, Taler D, Sobota T, Dzierwa P (2011) New Technique of the Local Heat Flux Measurement in Combustion Chambers of Steam Boilers. Archives of Thermodynamics 32: 103-116.

[15] LeVert FE, Robinson JC, Frank RL, Moss RD, Nobles WC, Anderson AA (1987) A Slag Deposition Monitor for Use in Coal_Fired Boilers. ISA Transactions 26: 51-64

[16] LeVert FE, Robinson JC, Barrett SA, Frank RL, Moss RD, Nobles WC, Anderson AA (1988) Slag Deposition Monitor for Boiler Performance Enhancement. ISA Transactions 27: 51-57

[17] Vallero A, Cortes C (1996) Ash Fouling in Coal-Fired Utility Boilers. Monitoring and Optimization of On-Load Cleaning. Prog. Energy. Combust. Sci. 22: 189–200.

[18] Teruel E, Cortes C, Diez LI, Arauzo I (2005) Monitoring and Prediction of Fouling in Coal-Fired Utility Boilers Using Neural Networks. Chem. Eng. Sci. 60: 5035–5048.

[19] Taler J, Trojan M, Taler D (2011) Monitoring of Ash Fouling and Internal Scale Deposits in Pulverized Coal Fired Boilers. New York: Nova Science Publishers.

[20] Howell JR, Siegel R, Mengüç MP (2011) Thermal Radiation Heat Transfer. Boca Raton: CRC Press - Taylor & Francis Group.

[21] Sparrow FM, Cess RD (1978) Radiation Heat Transfer. New York: McGraw-Hill.

[22] ANSYS CFX 12. (2010) Urbana, Illinois, USA: ANSYS Inc.

[23] Seber GAF, Wild CJ (1989) Nonlinear regression. New York: Wiley.

[24] Policy on reporting uncertainties in experimental measurements and results (2000). ASME J. Heat Transfer 122: 411–413.

[25] Press WH, Teukolsky SA, Vetterling WT, Flannery BP (2006) Numerical Recipes in Fortran. The Art of Scientific Computing. Cambridge: Cambridge University Press.

[26] Coleman HW, Steele WG (2009) Experimentation, Validation, and Uncertainty Analysis for Engineers. Hoboken: Wiley.

Chapter 2

Experimental Determination of Heat Transfer Coefficients During Squeeze Casting of Aluminium

Jacob O. Aweda and Michael B. Adeyemi

Additional information is available at the end of the chapter

http://dx.doi.org/10.5772/52038

1. Introduction

Casting process is desired because it is very versatile, flexible, and economical and happens to be the shortest and fastest way to transform raw material into finished product. Squeeze casting belongs to permanent mould casting method which offers considerable saving in cost for large production quantities when the size of the casting is not large. Squeeze casting has the advantage of producing good surface finish, close dimensional tolerance and the absence of sand inclusions on the cast surfaces of the products as opined by Das and Chatterjee, (1981).

The solidification process of the molten aluminium metal in the steel mould takes a complex form, (Hosford and Caddell, 1993) and (Potter and Easterling, 1993). During solidification all mechanisms of heat transfer are involved and the solidifying metal undergoes state and phase changes. The final structure and properties of the cast product obtained depend on the casting parameters applied i.e. applied pressures, die pre-heat temperature, delay time and period of applied pressure on the solidifying metal, (Potter and Easterling, 1993), (Bolton, 1989) and (Callister, 1997). The prediction of temperature distribution and solidification rate in metal casting is very important in modern foundry technologies. This helps to control the fundamental parameters such as the occurrence of defects, as well as, the influence on final properties of cast products and the mould wall / cast metal interface contact surface.

Heat transfer coefficients during squeeze cast of commercial aluminium were determined using the solidification temperature versus time curves obtained for varying applied pressures during squeeze casting process. The steel mould / cast aluminium metal interface temperatures versus times curve obtained through polynomial curves fitting and extrapolation was compared with the numerically obtained temperatures versus times curve. Interfacial heat transfer coefficients were determined experimentally from measured values of heating and cooling temperatures of steel mould and cast metal and compared with the numerically obtained values and found to be fairly close in values.

Aluminium is a product with unique properties, making it a natural partner for the building and other manufacturing industries. The commercially pure aluminium metal used for this research work finds extensive use in the building, manufacturing and process industries, both as a material of construction and household goods. Products of squeeze casting are of improved mechanical properties and could be given heat treatment. Heat dissipation from the squeeze cast specimen is fast thus producing products of fine grains as compared to the slow cooling of sand casting, which produces large grains. Products obtained through squeeze casting are with improved mechanical properties.

2. Squeeze casting procedure

A metered quantity of molten metal was poured into the steel mould cavity at a supper-heat temperature of between 40-60 ^{0}C fast but avoiding turbulence. The upper die was then released to close the mould cavity with and without applying any load on the upper die. Thermocouples were inserted into the drilled holes made in the die, which were used to monitor both the die and cast metal temperatures. The terminals of the thermocouples were connected to the chart recorder/plotter (set at the highest speed of 10mm/s and voltage 100mV) through the cold junction apparatus, maintained at 0 ^{0}C throughout the measuring period.

3. Assumptions made

i. Heat transfer in the molten metal cast zone is due to both conduction and convection while conduction heat transfer occurs in the steel mould, it is convection at the outer surface of the steel mould.
ii. The thickness of the cast specimen is much smaller than the diameter (radial dimension), thus giving one dimensional heat transfer process.
iii. Considering the symmetrical nature of the cast specimen, solidification process was assumed symmetrical and only lower half of the specimen's thickness was analysed see figure 1.
iv. The bottom of the squeeze casting rig was and the heat losses to the atmosphere was small and neglected.
v. Heat losses through conduction and convection to the atmosphere at the punch were neglected, as a result of short time of pressure application.
vi. The process of analyses in the cast specimen starts only when the steel mould cavity had been filled with the required quantity of liquid molten metal (i.e. heat transfer processes during pouring of molten aluminium into the steel mould are not considered).
vii. Density of the molten and solidified aluminium metal was assumed to be the same and independent of temperature.

viii. Thermal conductivity and specific heat of aluminium metal were dependent on the cast temperatures.

4. Heat transfer governing equations

4.1. Without pressure application on the cast metal

A measured quantity of molten aluminium metal was poured into the steel mould cavity. The process of solidification begins from the steel mould/cast metal interface and continues inwards into the cast metal. As this process continues, there was an increase in the thickness of the solidified layer and a decrease in the liquid molten metal portion. For the situation when no pressure was applied on the solidified molten metal, the governing heat transfer equations in one dimension are given by equation (1).

$$\rho C \frac{\partial T}{\partial t} = K\left[\frac{\partial^2 T}{\partial r^2} + \frac{1}{r}\frac{\partial T}{\partial r}\right] \tag{1}$$

From figure 1, equation (1) is defined within the region with;

i. Steel mould,

$$L \le r_{st} \le (L+Q)$$

ii. Solidified molten,

$$\left(L - X_r^j\right) \le r_S \le L$$

$$T_S = T_{MM} = 660\ \left({}^0\mathrm{C}\right) \tag{2}$$

iii. Liquid molten metal,

$$0 \le r_L \le \left(L - X_r^j\right)$$

$$K_S \frac{\partial T_S}{\partial r} = 0 \tag{3}$$

$$r = 0 \tag{4}$$

$$T_L = T_P = 720\ \left({}^0\mathrm{C}\right) \tag{5}$$

iv. At the phase change boundary condition;

$$\rho_L L_f \frac{dX_r^j}{dt} = K_L \frac{\partial T_L}{\partial r} - K_S \frac{\partial T_S}{\partial r} \tag{6}$$

where,

$$r = L - X_r^j \tag{7}$$

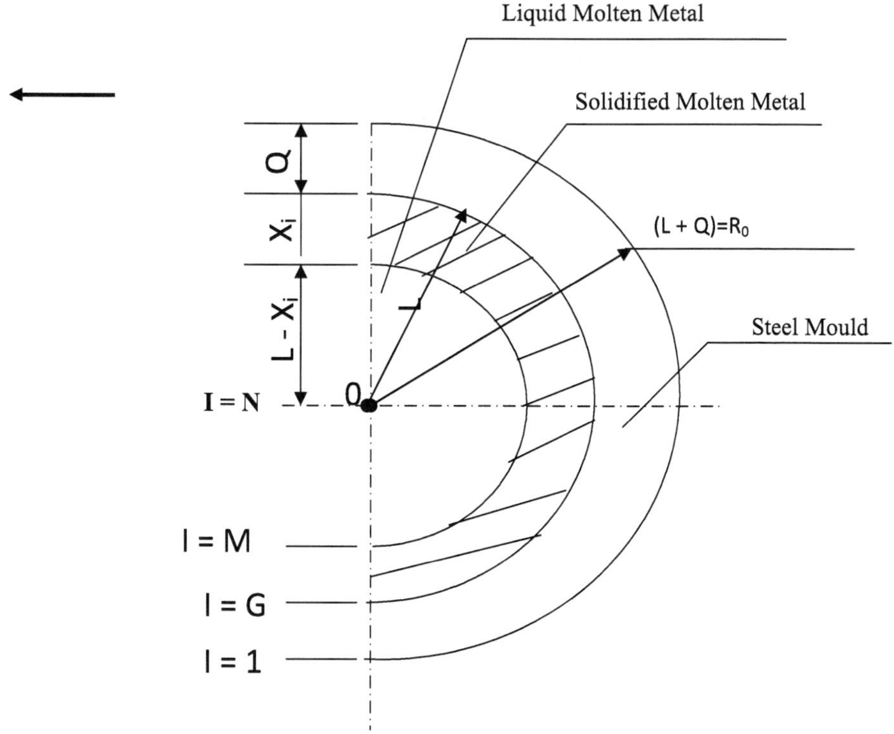

Figure 1. Schematic representation of solidification front in one dimension (radial direction)

4.2. Casting with pressure application on the solidified molten metal

As the cast aluminium metal solidifies, pressure is applied on the specimen, observing lapse or delay time, t_l, while varying the values of pressure applied. The time between the end of pouring of molten metal and pressure application known as lapse time, is recorded. This is necessary such that the cast specimen will not stick to the upper punch or cause the cast metal to tear with pressure application. Due to the applied pressure, an internal energy Δq is generated within the solidified molten metal, (see figure 2).

Inserting the internal energy into the heat transfer equation (1) for solidified molten metal, it becomes equation (8),

$$\rho_S C_S \frac{\partial T_S}{\partial t} = K_S \left[\frac{\partial^2 T_S}{\partial r^2} + \frac{1}{r_S} \frac{\partial T_S}{\partial r} \right] + \Delta q \tag{8}$$

where,

$$\nabla q = \nabla q_P + \nabla q_f \tag{9}$$

Δq -internal energy generated by applied pressure,
Δq_P -energy due to plastic strain within the solidified molten metal material,
Δq_f -frictional energy generated during pressure application,

$$\nabla q_f = \nabla q_{fP} + \nabla q_{fm} \tag{10}$$

Δq_{fP} -frictional energy due to punch / solidified molten metal interface,
Δq_{fm} -frictional energy due to steel mould cylindrical surface / solidified molten metal interface.

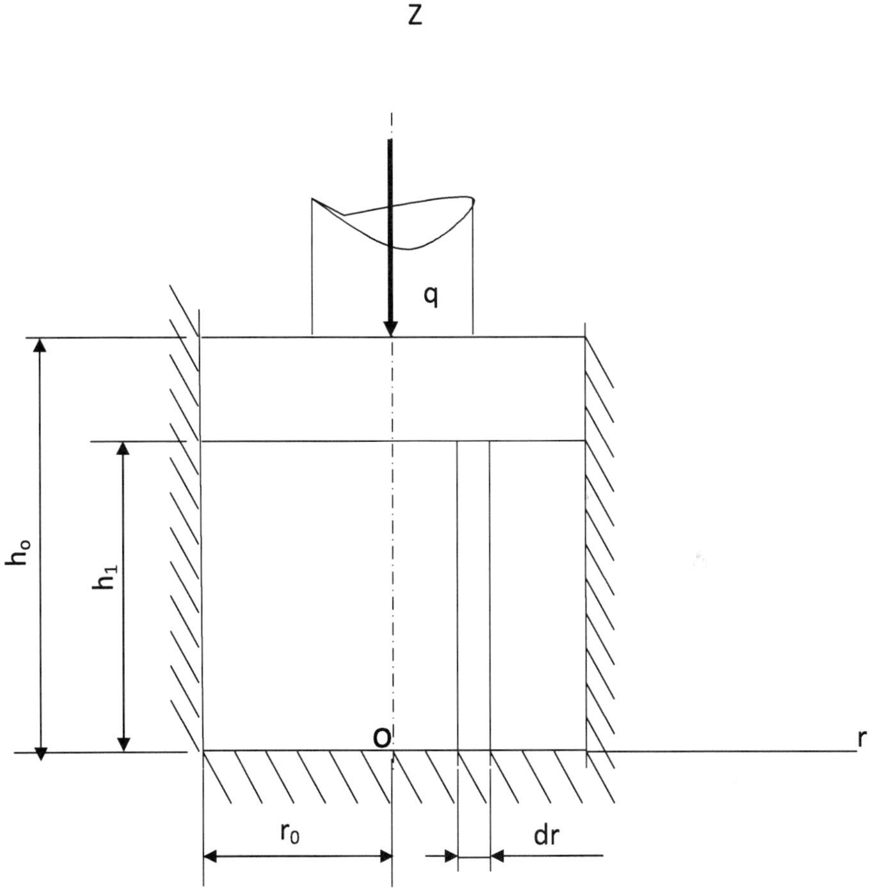

Figure 2. Cast specimen under pressure

5. Formulation of heat transfer equations

Finite difference expressions for the nodal temperatures are obtained by either energy balance within an elemental volume around the node or by substitutions into the governing partial differential equations. Partial time derivative of temperature contained in the equations can be written in terms of a moving gradient as in White (1991) at a velocity of dx/dt. In the heat transfer equation, temperature is defined as a function of distance, r with time, t and represented in equation (11).

$$T = T(r, t) \tag{11}$$

5.1. In the steel mould

$$\frac{dT_{st}}{dt} = \alpha_{st}\frac{\partial^2 T_{st}}{\partial r^2} + \frac{\alpha_{st}}{r_{st}}\frac{\partial T_{st}}{\partial r} \tag{12}$$

where,

$$\frac{dr_{st}}{dt} = 0 \tag{13}$$

$$\alpha_{st} = \frac{K_{st}}{\rho_{st}C_{st}} \tag{14}$$

α_{st} -thermal diffusivity of steel mould material.

5.2. In the solidified molten metal

$$\frac{dT_S}{dt} - \frac{(I-G)}{(M-G)}\frac{dX_i^r}{dt}\frac{\partial T_S}{\partial r} = \alpha_S\frac{\partial^2 T_S}{\partial r^2} + \frac{\alpha_S}{r_S}\frac{\partial T_S}{\partial r} \tag{15}$$

where,

$$\alpha_S = \frac{K_S}{\rho_S C_S} \tag{16}$$

α_S -thermal diffusivity of solidified molten metal material.

5.3. In the liquid molten metal

$$\frac{dT_L}{dt} - \frac{(I-M)}{(N-M)}\frac{dX_i^r}{dt}\frac{\partial T_L}{\partial r} = \alpha_L\frac{\partial^2 T_L}{\partial r^2} + \frac{\alpha_L}{r_L}\frac{\partial T_L}{\partial r} \tag{17}$$

where,

$$\alpha_L = \frac{K_L}{\rho_L C_L} \tag{18}$$

α_L -thermal diffusivity of liquid molten metal material.

6. Nodal divisions

The steel mould, the solidified metal and the molten metal regions were discretized separately. Each of these regions was divided into a fixed number of gridal points as in figure 1.

6.1. In the steel mould

$$d_{st} = \frac{Q}{(G-I)} \tag{19}$$

I = 1, 2, 3,…, G-1

6.2. In the solidified molten metal

$$d_S = \frac{X_r^i}{(M-I)} \tag{20}$$

I = G+1, G+2, G+3, … , M-1

6.3. In the liquid molten metal portion

$$d_L = \frac{(L-X_r^i)}{(N-I)} \tag{21}$$

I = M+1, M+2, M+3 … N-1

6.4. In the phase change boundaries

The phase change is represented by the equations;

$$d_{Ps} = \frac{X_r^i}{(M-G)} \tag{22}$$

$$d_{PL} = \frac{(L-X_r^i)}{(N-M)} \tag{23}$$

6.5. At the completion of solidification

$$d_{SC} = \frac{L}{(N-G)} \tag{24}$$

As solidification time progresses, the boundary locations change, and the thickness of the solidified molten metal in the radial direction increases. The rate of change of boundary location with time is represented, mathematically in equation (25),

$$\frac{dX_r^i}{dt} = \frac{\left(X_r^{j+1} - X_r^{j}\right)}{\delta} \tag{25}$$

Where,

$\left(X_r^{j+1} - X_r^{j}\right)$ -are differences in the thickness of solidified molten metal at a particular time interval as time progresses in the radial direction.

δ -time interval.

7. Boundary conditions

The problem of phase change during solidification is that the location of the solidifying molten metal / liquid molten metal interface is not known and this is determined continuously by appropriate mathematical analysis. This moving interface is normally expressed mathematically by the energy balance equations at the interfaces. In the numerical analysis, as solidification of molten aluminium metal progresses, three boundary interfaces occurred as:

7.1. Steel mould-atmosphere interface (I = 1)

The heat conducted to the steel mould material (from I = 2 to 1 = 1) equals the sum of the change in the internal energy and heat convected from the surface of the steel metal mould material into the atmosphere, mathematically represented in equation (26).

$$\frac{\partial T_{st}}{\partial t} = \frac{2K_{st}}{\rho_{st}C_{st}d_{st}}\frac{\partial T_{st}}{\partial r} - \frac{2H^*}{\rho_{st}C_{st}d_{st}}(T_i^j - T_\infty) \tag{26}$$

I = 1

7.2. In the solidified molten metal-steel mould interface (I = G)

The sum of the heat conducted from the solidifying molten metal/steel mould interface and the change (decrease) in the internal energy at the boundary equal the sum of the heat conducted to the steel mould and the change (increase) in the internal energy at the interface.

$$K_S \frac{\partial T_S}{\partial r} + \frac{1}{2}\rho_S C_S d_S \frac{\partial T_S}{\partial t} = K_{st} \frac{\partial T_{st}}{\partial r} + \frac{1}{2}\rho_{st} C_{st} d_{st} \frac{\partial T_{st}}{\partial t} \tag{27}$$

I = G

7.3. In the liquid molten metal – Solidified molten metal interface, (I = M)

The sum of the heat conducted from the liquid molten metal and the internal energy generated equal the sum of heat conducted to the solidified molten metal and the internal energy generated at the interface as in equation (28).

$$K_L \frac{\partial T_L}{\partial r} - \frac{1}{2}\rho_L C_L d_L \frac{\partial T_L}{\partial t} = K_S \frac{\partial T_S}{\partial r} + \frac{1}{2}\rho_S C_S d_S \frac{\partial T_S}{\partial t} \tag{28}$$

I = M

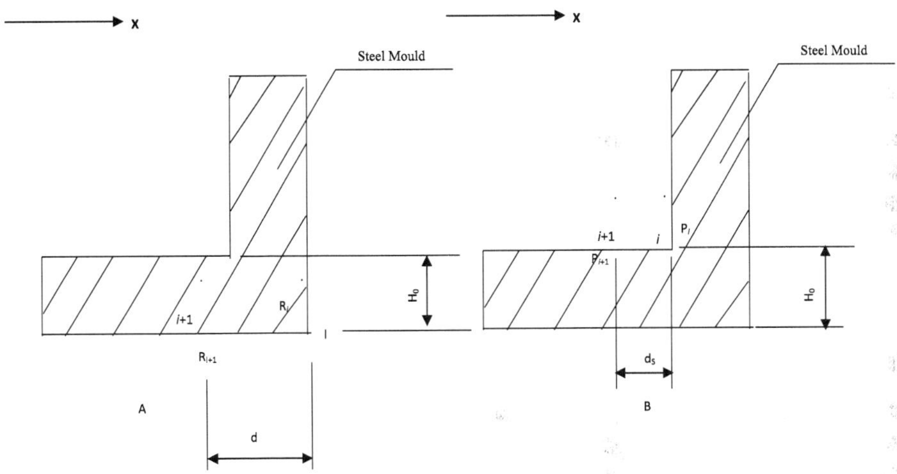

Figure 3. Corner nodes. A) External, B) Internal nodes

7.4. At the corners

The usual one dimensional heat transfer analysis does not take into consideration heat loss at the corners as represented in figure 3. Heat loss at both the external and internal corner nodes of the steel mould have been considered and analysed while the die is lagged at the bottom surface.

7.5. External corner effect

$$q_{i+1} + q_g = q_{conv} \tag{29}$$

The heat conducted from point R_{i+1} to point R_i added to the change in the internal energy equal the amount of heat convected out to the atmosphere at point R_I., Fig. (3a). The finite form of equation (29) is represented by equation (30),

$$\frac{K_{st}}{d_{st}}\frac{\partial T_{st}}{\partial r}-\frac{1}{2}\rho_{st}C_{st}\frac{\partial T_{st}}{\partial t}=\frac{h}{d_{st}}\left(T_i^j-T_\infty^j\right)+\frac{h}{H}\left(T_i^j-T_\infty^j\right) \tag{30}$$

7.6. Internal corner effect

The heat conducted from point P_{I+1} to point P_I added to the change in the internal energy amount to the heat conducted into the steel mould at point I, fig. (3b) thus becoming equation (31);

$$\frac{K_S}{d_{st}}\frac{\partial T_S}{\partial r}+\frac{1}{2}\rho_S C_S\frac{\partial T_{st}}{\partial t}=\frac{K_{st}}{d_S}\frac{\partial T_{st}}{\partial r}+\frac{K_{st}}{H1}\frac{\partial T_{st}}{\partial Z} \tag{31}$$

7.7. First time analysis

Solidification takes place only in the radial direction, a one-dimensional heat solidification problem was assumed numerically to take place in the radial direction only after filling the steel mould cavity with the liquid molten metal.

For the first time analysis, the specimen is considered to be in the molten stage and therefore equation (32) for the liquid molten metal is used for computation. Thus;

$$\rho_L C_L\frac{dT_L}{dt}=K_L\left[\frac{\partial^2 T_L}{\partial r^2}+\frac{1}{r_L}\frac{\partial T_L}{\partial r}\right] \tag{32}$$

This equation is subjected to the boundary conditions with;

$$T_L=T_S \tag{33}$$

The instantaneous radius r_i in the first time analysis is given by equation (34);

$$r_L=\frac{(N-I)}{(N-G)}L \tag{34}$$

I = G, G+1, G+2, G+3,…., N

The boundary velocity of moving coordinate of equation (32) is given as equation (35);

$$\frac{dr_i}{dt}=\frac{d}{dt}\left[\frac{(N-I)}{(N-G)}L\right]=0 \tag{35}$$

7.8. At the completion of solidification

At the completion of solidification, the whole molten region becomes solidified molten metal, and just before pressure is applied, the governing heat transfer equation becomes equation (36) representing the solidified portion;

$$\rho_S C_S \frac{\partial T_L}{\partial t} = K_S \left[\frac{\partial^2 T_S}{\partial r^2} + \frac{1}{r_{SC}} \frac{\partial T_S}{\partial r} \right] \quad (36)$$

The equation is applicable within the region defined by equation (37);

$$r = L \quad (37)$$

The boundary motion at the completion of solidification is as in equation (38);

$$r_{SC} = \frac{(N-I)}{(N-G)} L \quad (38)$$

I = G, G+1, G+2, G+3,…, .N

The boundary velocity at the completion of solidification is expressed as equation (39);

$$\frac{dr_{SC}}{dt} = \frac{d}{dt} \left[\frac{(N-I)}{(N-G)} L \right] = 0 \quad (39)$$

where,

L -constant value

8. Finite difference of governing heat transfer equations

The heat transfer equations generated in the cast metal and interfaces are written in the finite difference forms. These equations are presented in the various regions thus;

8.1. In the steel mould region

$$T_i^{j+1} = \left[1 - \frac{2\delta \alpha_{st}}{d_{st}^2} - \frac{\delta \alpha_{st}}{r_{st} d_{st}} \right] T_i^j + \left[\frac{\delta \alpha_{st}}{d_{st}^2} + \frac{\delta \alpha_{st}}{r_{st} d_{st}} \right] T_{i+1}^j + \frac{\delta \alpha_{st}}{d_{st}^2} T_{i-1}^j \quad (40)$$

where,

$$r_{st} = L + \frac{Q(G-I)}{(G-1)} \quad \text{(a)}$$

$$d_{st} = \frac{Q}{(G-I)} \quad \text{(b)} \quad (41)$$

I = 1,2,3,…………..G

8.2. In the solidified molten metal portion

$$T_i^{j+1} = \left[1 - \frac{(I-G)}{(M-G)}\frac{\left(X_r^{j+1} - X_r^j\right)}{d_S} - \frac{2\delta\alpha_S}{d_S^2} - \frac{\delta\alpha_S}{r_S d_S}\right]T_i^j + \\ + \left[\frac{(I-G)}{(M-G)}\frac{\left(X_r^{j+1} - X_r^j\right)}{d_S} + \frac{\delta\alpha_S}{d_S^2} + \frac{\delta\alpha_S}{r_S d_S}\right]T_{i+1}^j + \frac{\delta\alpha_S}{d_S^2}T_{i-1}^j \tag{42}$$

where,

$$r_S = L - \frac{X_r^j(I-G)}{(M-G)} \quad \text{(a)} \\ d_S = \frac{X_r^j}{(M-I)} \quad \text{(b)} \tag{43}$$

I = G+1, G+2, G+3,………., M

8.3. In the liquid molten metal region

$$T_i^{j+1} = \left[1 - \frac{(I-M)}{(N-M)}\frac{\left(X_r^{j+1} - X_r^j\right)}{d_L} - \frac{2\delta\alpha_L}{d_L^2} - \frac{\delta\alpha_L}{r_L d_L}\right]T_i^j + \\ \left[\frac{(I-M)}{(N-M)}\frac{\left(X_r^{j+1} - X_r^j\right)}{d_L} + \frac{\delta\alpha_L}{d_L^2} + \frac{\delta\alpha_L}{r_L d_L}\right]T_{i+1}^j + \frac{\delta\alpha_L}{d_L^2}T_{i-1}^j \tag{44}$$

where,

$$r_L = \left(L - X_r^j\right)\frac{(I-M)}{(N-M)} \quad \text{(a)} \\ d_L = \frac{\left(L - X_r^j\right)}{(N-I)} \quad \text{(b)} \tag{45}$$

I = M, M+1, M+2, M+3, …, N

8.4. In the phase change boundary condition (I = M)

$$X_r^{j+1} = X_r^j - \left[\frac{\delta K_L}{\rho_L L_f h_{pL}} + \frac{\delta K_S}{\rho_L L_f h_{ps}}\right]T_i^j + \frac{\delta K_L}{\rho_L L_f h_{pL}}T_{i+1}^j + \frac{\delta K_S}{\rho_L L_f h_{ps}}T_{i-1}^j \tag{46}$$

where,

$$h_{PL} = \frac{\left(L - X_r^j\right)}{\left(N - M\right)} \quad \text{(a)}$$
$$h_{Ps} = \frac{X_r^j}{\left(M - G\right)} \quad \text{(b)} \qquad (47)$$

8.5. In the steel mould / atmosphere interface (I = 1)

$$T_i^{j+1} = \left[1 - \frac{2\delta K_{st}}{d_{st}^2 \rho_{st} C_{st}}\right] T_i^j + \frac{2\,\delta K_{st}}{d_{st}^2 \rho_{st} C_{st}} T_{i+1}^j - \frac{2\,\delta H^*}{d_{st} \rho_{st}\, C_{st}} \left(T_i^j - T_\infty^i\right) \qquad (48)$$

I = 1

where;

$$d_{st} = \frac{Q}{\left(G - 1\right)} \qquad (49)$$

I = 1, 2, 3, …, G

8.6. In the solidified molten metal / steel mould interface (I = G)

$$T_i^{j+1} = a\left[\frac{2\delta K_S}{d_S} + d_S \rho_S C_S + \rho_S C_S \frac{\left(I - G\right)}{\left(M - G\right)}\left(X_r^{j+1} - X_r^j\right) + \frac{2\delta K_{st}}{d_{st}} - d_{st}\rho_{st}\, C_{st}\right] T_i^j$$
$$+ a\left[\rho_S C_S \frac{\left(I - G\right)}{\left(M - G\right)}\left(X_r^{j+1} - X_r^j\right) - \frac{2\delta K_S}{d_S}\right] T_{i+1}^j - \frac{2a\delta K_{st}}{d_{st}} T_{i-1}^j \qquad (50)$$

where,

$$a = \frac{1}{\left(d_S \rho_S C_S - d_{st}\rho_{st}\, C_{st}\right)} \quad \text{(a)}$$
$$d_{st} = \frac{Q}{\left(G - I\right)} \quad \text{(b)}$$
$$\mathrm{I} = 1, 2, 3, \ldots, \mathrm{G} \qquad (51)$$
$$d_S = \frac{X_r^i}{\left(M - I\right)} \quad \text{(c)}$$
$$\mathrm{I} = \mathrm{G}+1, \mathrm{G}+2, \mathrm{G}+3, \ldots, \mathrm{M}$$

8.7. In the liquid molten metal / solidified molten metal interface (I = M)

$$T_i^{j+1} = b\begin{bmatrix} d_L\rho_L C_L - \rho_L C_L \dfrac{(I-M)}{(N-M)}\left(X_r^{j+1} - X_r^j\right) - \dfrac{2\delta K_L}{d_L} + d_S\rho_S C_S - \dfrac{2\delta K_S}{d_S} + \\ \rho_S C_S \dfrac{(I-G)}{(M-G)}\left(X_r^{j+1} - X_r^j\right) \end{bmatrix} T_i^j$$

$$+b\left[\frac{2\delta K_S}{d_S} - \rho_S C_S \frac{(I-G)}{(M-G)}\left(X_r^{j+1} - X_r^j\right)\right]T_{i-1}^j \qquad (52)$$

$$+b\left[\frac{2\delta K_L}{d_L} + \rho_L C_L \frac{(I-M)}{(N-M)}\left(X_r^{j+1} - X_r^j\right)\right]T_{i+1}^j$$

where,

$$b = \frac{1}{\left(d_S\rho_S C_S + d_L\rho_L C_L\right)} \quad \text{(a)}$$

$$d_S = \frac{X_r^j}{(M-I)} \quad \text{(b)} \qquad (53)$$

$$d_L = \frac{\left(L - X_r^j\right)}{(N-I)} \quad \text{(c)}$$

I = M, M+1, M+2, M+3,…,N

8.8. External corner effect (I = 1)

$$T_i^{j+1} = \left[1 - \frac{2\delta K_{st}}{\rho_{st}C_{st}d_{st}^2} - \frac{2\delta h}{\rho_{st}C_{st}d_{st}} - \frac{2\delta h}{\rho_{st}C_{st}H}\right]T_i^j +$$

$$+\frac{2\delta K_{st}}{\rho_{st}C_{st}d_{st}^2}T_{i+1}^j + \qquad (54)$$

$$+\left[\frac{2\delta h}{\rho_{st}C_{st}d_{st}} + \frac{2\delta h}{\rho_{st}C_{st}H}\right]T_\infty^j$$

where,

$$d_{st} = \frac{Q}{(G-I)} \qquad (55)$$

8.9. Internal corner effect (I = G)

$$T_i^{j+1} = \left[1 + \frac{2\delta K_S}{\rho_S C_S d_S^2} + \frac{(I-G)}{(M-G)} \frac{\left(X_r^{j+1} - X_r^j\right)}{d_S} + \frac{2\delta K_{st}}{\rho_S C_S d_S d_{st}} + \frac{2\delta K_{st}}{\rho_S C_S H1 d_{st}}\right] T_i^j - \tag{56}$$

$$-\frac{2\delta K_S}{\rho_S C_S d_S^2} T_{i+1}^j - \left[\frac{2\delta K_{st}}{\rho_S C_S d_S d_{st}} + \frac{2\delta K_{st}}{\rho_S C_S H1 d_{st}}\right] T_{i-1}^j$$

where,

$$d_S = \frac{X_r^j}{(M-I)} \quad \text{(a)}$$

$$I = G+1, G+2, G+3, \ldots, M \tag{57}$$

$$d_{st} = \frac{Q}{(G-I)} \quad \text{(b)}$$

$$I = 1, 2, 3, \ldots, G$$

8.10. First time analysis

Finite difference form of equation (32) therefore becomes equation (58);

$$T_i^{j+1} = \left[1 - \frac{2\delta\alpha_L}{d_L^2} - \frac{\delta\alpha_L}{r_L d_L}\right] T_i^j + \left[\frac{\delta\alpha_L}{d_L^2} + \frac{\delta\alpha_L}{r_L d_L}\right] T_{i+1}^j + \frac{\delta\alpha_L}{d_L^2} T_{i-1}^j \tag{58}$$

where,

$$r_L = \frac{(N-I)}{(N-G)} L \quad \text{(a)}$$

$$d_L = \frac{L}{(N-G)} \quad \text{(b)} \tag{59}$$

I = G+1, G+2, G+3, ………., .N

8.11. At the completion of solidification

Finite difference form of equation (36) becomes equation (60);

$$T_i^{j+1} = \left[1 - \frac{2\delta\alpha_S}{d_{SC}^2} - \frac{\delta\alpha_S}{r_{SC} d_{SC}}\right] T_i^j + \left[\frac{\delta\alpha_S}{d_{SC}^2} + \frac{\delta\alpha_S}{r_{SC} d_{SC}}\right] T_{i+1}^j + \frac{\delta\alpha_S}{d_{SC}^2} T_{i-1}^j \tag{60}$$

where;

$$r_{SC} = \frac{(N-I)}{(N-G)}L \quad \text{(a)}$$
$$d_{SC} = \frac{L}{(N-G)} \quad \text{(b)} \qquad (61)$$

I = G+1, G+2, G+3,… , (N-1)

9. Stability criteria

For stability criteria to be achieved, the values of temperature T_i^j in all the heat governing equations should not be negative according to Ozisik (1985) and White (1991) not to negate the law of thermodynamics which could lead to temperature fluctuations. Therefore, for stability to be achieved, the coefficients of T_i^j in each of the equations must be greater than zero.

10. Casting with pressure application and die heating

Pressure was applied only when the cast specimen was solidified, the governing heat transfer equation therefore, takes the form of solidified molten metal (completion of solidification). The finite difference of equation (8) is written as equation (62);

$$T_i^{j+1} = \left[1 - \frac{2\delta\alpha_S}{d_S^2} - \frac{\delta\alpha_S}{r_S d_S}\right]T_i^j + \left[\frac{\delta\alpha_S}{d_S^2} + \frac{\delta\alpha_S}{r_S d_S}\right]T_i^j + \frac{\delta\alpha_S}{d_S^2}T_{i-1}^j + \Delta T \qquad (62)$$

where,

ΔT -temperature change resulting from pressure application

The cast specimen height, h_c, is pressure dependent and the relationship is expressed with the equation (63) below after performing series of experiment with various applied pressure;

$$h_c = -0.00007P + 0.036833 \qquad (63)$$

where coefficient of correlation r = 0.996

The plastic flow stress, $\sigma_{(T)}$, is dependent on both the applied pressure, P, and die temperature, TM, (White, 1991), and expressed with the equation (64);

$$\sigma(T) = 0.244P - 0.0405TM + 18.614 \qquad (64)$$

where coefficient of correlation r = 0.9508

11. Casting with die heating

Aluminium cast specimens were produced with die pre- heating temperatures of between 100- 300^{0}C without applying pressure on the solidifying aluminium metal. The die heating process was carried out, using three electric heater rods (100Watts each) that were connected to a.c supply. The required die temperatures were set and controlled, using a bimetallic thermostat.

12. Heat transfer coefficient evaluations

The method of calculating heat transfer coefficients as reported by Santos et al (2001) and Maleki et al, (2006) is based on the knowledge of known temperature histories at the interior points of the casting or mould together with the numerical models of heat flow during solidification. These temperatures are difficult to measure due to the difficulty in locating accurate position of thermocouple at the interface. Therefore, the inverse heat conduction problem based on non-linear estimation technique of Chattopadhyay, (2007) and Hu and Yu, (2002), has been adopted to determine the values of interface heat transfer coefficients, as a function of time during solidification of squeeze casting. Solidification of squeeze casting of aluminium involves phase change and therefore thermal properties of aluminium are temperature dependent, making the inverse heat conduction problem non-linear.

The governing heat transfer equation in one-dimensional cylindrical coordinates is given by equation (65):

$$\rho c_p \frac{\partial T}{\partial t} = \frac{1}{r}\frac{\partial}{\partial r}\left(K_{al}\, r \frac{\partial T}{\partial r}\right) \tag{65}$$

Equation (65) holds within the boundary condition as expressed in equation (66):

$$q = K_{al}(T)\frac{\partial T}{\partial r} = h_{al}(T)\left[T_{al} - T_M\right] \tag{66}$$

The thermal conductivity K_{al} (Reed-Hill and Abbaschian 1973) and (Elliot, 1988) of aluminium is dependent upon casting temperature, T_{al}, and expressed in equation (67):

$$K_{al}(T) = 241.84 - 0.041T_{al} \tag{67}$$

The heat flow across the casting/mould interface can be characterized by an average interfacial heat transfer coefficient, $h_{al}(T)$ as obtained by Gafur et al (2003) and Santos et al (2004). This is expressed mathematically in equation (68):

$$h_{al}(T) = \frac{q}{\left[T_{al} - T_M\right]} \tag{68}$$

The heat transfer coefficient, h, at the interface is estimated by minimizing the errors between numerically estimated and measured temperatures defined by equation (69):

$$F(h) = \sum_{i=1}^{n}\left(T_{est} - T_{\exp}\right)^2 \tag{69}$$

where,

T_{est} and T_{exp} -are the estimated and experimentally measured temperatures at various thermocouples location and times,

n -iteration stage

13. Numerical simulations of differential equations

Squeeze casting consists of two stages, the first of which is mould filling: - the mould is filled with the required quantity of liquid molten metal; the second is cooling, this continues until the part has solidified completely. Controlling both stages is of major importance for obtaining sound casts with the required geometry and mechanical properties as observed by (Kobryn and Semiatin, (2000), Browne and O'Mahoney, (2001) and Martorano and Capocchi, (2000). When molten metal is poured into the mould cavity, it is initially in the liquid state with a high fluidity. It quickly becomes very viscous, in the early stage of solidification, and later completely solidifies (Gafur et al, 2003). For the numerical analysis of heat transfer problem, the appropriate set of equations were determined that described the heat transfer behaviour in the cast metal (Hearn, 1992). With the boundary conditions, initial conditions, and thermo-physical properties of the materials being known, it is possible to obtain the temperature and variation of the whole casting system (Ozisik, 1985) and (Liu et al (1993). Finite number at discrete points (Adams and Rogers, (1973), Shampire, (1994) and Bayazitoglu and Ozisik, (1988)) within the cast specimen was employed as the numerical method of solution. This method provides the temperature at a discrete number of points in the cast region. In the numerical method, the cast region is defined and divided into discrete number of points. As temperature difference is imposed in the system, heat flows from the high-temperature region to the low-temperature region as shown in figure 1.

To determine the temperature distribution, energy conservation equations were used for each of the nodal points of the unknown temperature at the interfaces and the cast regions ((Incropera and Dewitt, 1985) and (Janna, 1988)). Temperatures were monitored at distance 2mm into the cast metal, represented by grid point M, and at the steel mould/cast metal interface (see figure 4). By using measured temperatures in both the casting and the steel mould, together with the numerical solutions of the solidification problem, heat transfer coefficients were determined based on Beck (1970) solution of the inverse heat conduction problem. The estimation of the surface heat transfer coefficients or heat flux density utilizing a measured temperature history inside a heat-conducting solid is called the inverse heat conduction problem (Cho and Hong 1996). This problem becomes non-linear, as the thermal properties (thermal conductivity, specific heat) are temperature dependent.

14. Experimental procedure

Chromel-Alumel thermocouples TC2, TC3, TC4, TC5 and TC6 were positioned on the sides of the cylindrical steel container, while TC1 and TC7 were positioned in the cast aluminium metal in the cylindrical and bottom flat surfaces respectively as shown in figure 4 below.

Thermocouples of chromel-alumel type, 3mm in diameter were used to determine the solidifying temperatures of the cast molten metal and heating temperatures of the steel mould at the various positions in the cylindrical steel container of figure 4. The solidifying temperatures at both the cylindrical and flat bottom surfaces of the cast molten aluminium metal were monitored at a position 2mm (from the surface of the steel mould –cast aluminium metal interface) into the cast molten aluminium metal.

At the steel mould wall in the cylindrical surface, thermocouples were positioned at X2 = 4mm, X3 = 8mm, X4 = 12mm, X5 = 16mm and X6 = 20mm measured from the cast aluminium metal / steel mould interface to monitor the heating temperatures at these positions of the steel mould wall as shown in fig. 4. From the temperatures versus time curves obtained for each position in the steel mould, the interface heating temperature versus time curve at the cast aluminium metal / steel mould, for position when X = 0 was obtained by using the polynomial curve fitting method.

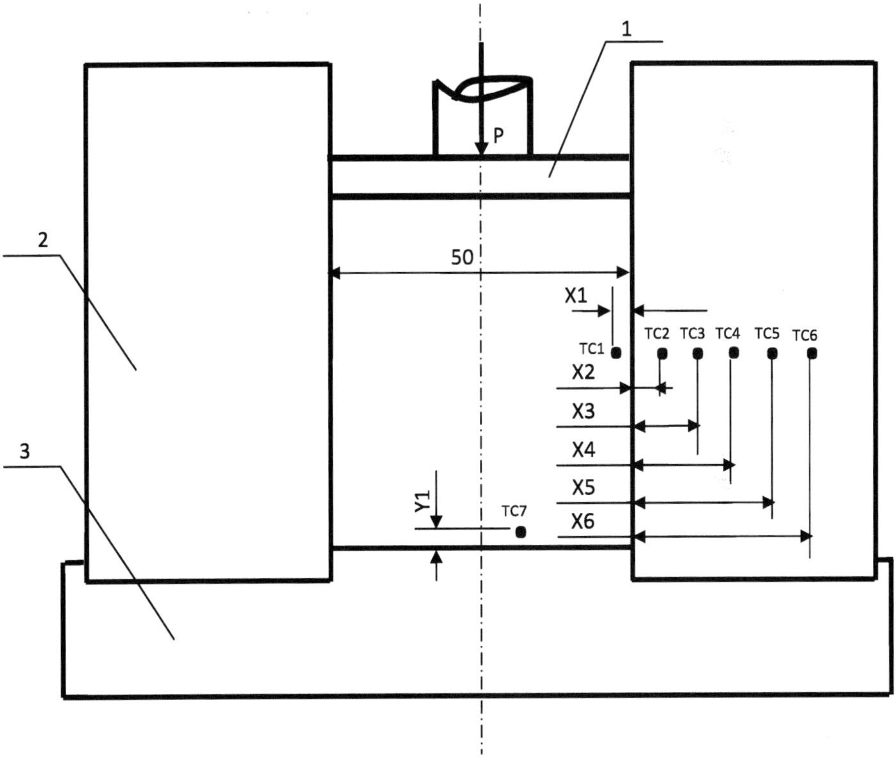

1-upper punch, 2-cylindrical steel mould, 3-lower die
(X1 = 2, X2 = 4, X3 = 8, X4 = 12, X5 = 16, X6 = 20, Y1 = 2 (mm))

Figure 4. Schematic diagram of squeeze casting test rig

This was done by selecting a particular time of heating of steel mould, say t = 10sec. and drawing vertical lines cutting across the heating temperatures versus time curves at various thermocouples' distances within the steel mould. At the point of intersection with each curve, the value of temperature was read against distance X, for the chosen time, t = 10sec. The value of interface steel mould / cast temperature at time say, t = 10sec. was determined at the steel mould / cast metal interface by substituting the value of X = 0 in the polynomial curve fitting equation (70) obtained from the values of temperatures at various distances in the steel mould at a chosen time, t = 10sec.

$$T_{X0} = 0.0031X^4 - 0.168X^3 + 3.263X^2 - 22.812X + 117.8 \tag{70}$$

The temperature obtained by this method corresponds to the interface steel mould / cast metal temperature at a distance X = 0 for the chosen time t = 10sec. If this procedure is repeated for a number of time increments, the temperatures obtained with corresponding times represent the temperature at X=0, for such time increments. The graph of extrapolated temperatures versus time is drawn for position when X = 0 to represent the heating temperatures versus time curve at the steel mould / cast aluminium metal interface is shown in figure 5.

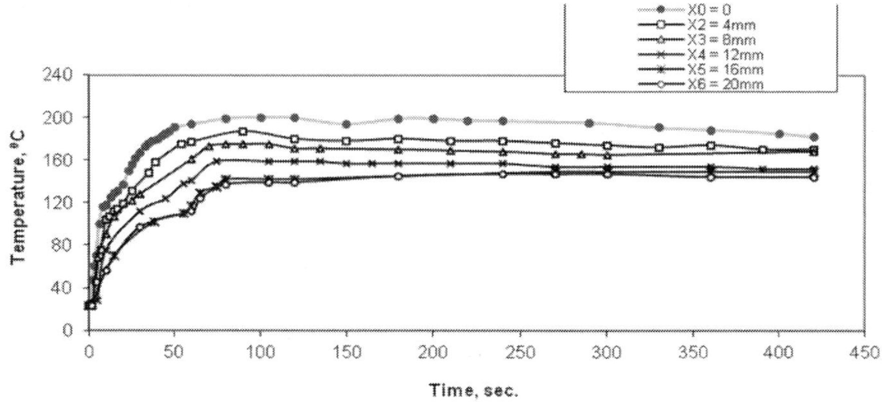

Figure 5. Effect of distance on the heating temperatures of steel mould (extrapolated heating curve at the cast specimen/steel mould interface i.e. X=0)

15. Interface heat transfer coefficients determination

Extrapolated temperature versus time curve of figure 5 for position when X = 0 (i.e. cast aluminium metal / steel mould interface) was used to determine the heat transfer coefficients of solidifying molten aluminium metal. It was used to determine the interface heat transfer coefficients in the cast aluminium metal / steel mould for no pressure and with pressure applications at both the cylindrical and bottom flat surfaces of the steel mould.

The interface heat transfer coefficients between the steel mould and cast aluminium metal at the cylindrical and bottom flat surfaces were determined from the extrapolated experimental heating temperature versus time curve obtained for position X = 0 and aluminium cast solidification temperature versus time curves obtained for the cylindrical and bottom flat surfaces, using equations (66) and (67).

The interface heat transfer coefficients were determined also numerically by the inverse method using the Finite Difference Method (FDM) and the obtained results were compared with the experimentally derived values.

16. Discussions of results

16.1. Temperature-time curves

Figure 6 shows typical temperature versus time curves for solidifying molten aluminium metal and steel mould respectively without the application of pressure on solidifying metal.

This figure shows the comparison of the numerical method usually applied by Cho and Hong (1996) to determine interface steel mould / cast metal temperature versus time curve with the extrapolated experimental method of this present work. The heating curve, as obtained through extrapolations of polynomial curves fitting equations and numerical methods are in close agreement and the deviations from the values obtained numerically varied from between 1.26- 19.31%.

Typical result obtained under pressure is also shown in figure 7, indicating the solidification and heating curves generated for solidifying molten aluminium metal and steel mould which follow the same patterns to the curves in figure 6.

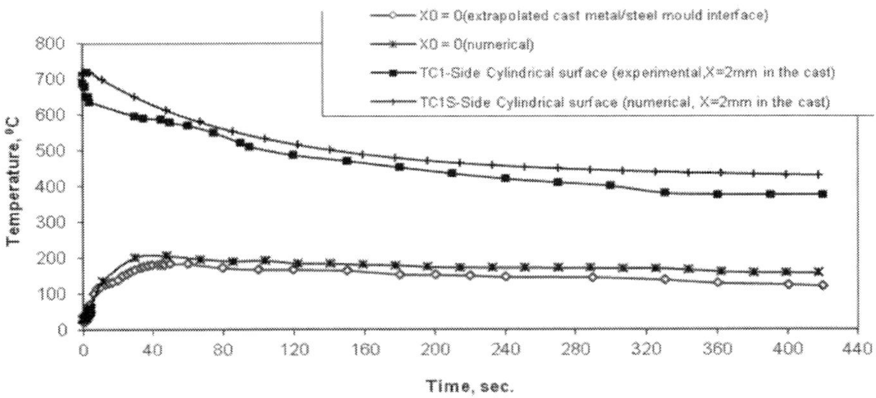

Figure 6. Comparison of experimental measured temperatures with numerical values of aluminium metal without pressure application (P = 0)

With the application of pressure, the peak temperatures recorded are about the same 649°C and 648°C for a pressure of 85.86 MPa at the bottom flat and cylindrical surfaces of the steel mould respectively (see figure 7). The peak temperature (649°C) obtained at the bottom flat surface of the steel mould under applied pressure is found to be higher than that temperature (607°C) without pressure application. This effect may be associated to additional internal heat generated, resulting to higher temperature during pressure application on the solidifying molten aluminium.

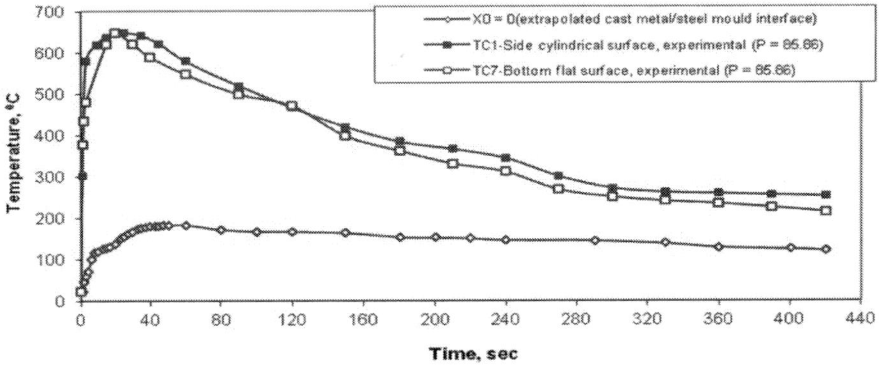

Figure 7. Effect of pressure on the experimental measured temperatures of solidification of aluminium metal (P = 85.86MPa) for side and bottom mould's surface

16.2. Interface heat transfer coefficients with time

From the temperature with time curves of figure 6, the heat transfer coefficients for both cylindrical and bottom flat surfaces were determined for both numerical and calculated values and shown in figure 8. The maximum heat transfer coefficients of 2927.92 W/m²K and 2975.14 W/m²K are obtained at the cylindrical and bottom flat surfaces respectively for no pressure application, which is close to 2900 W/m²K as obtained for pure aluminium by Kim and Lee (1997). The values of heat transfer coefficients decrease rapidly for both the cylindrical and bottom flat surfaces to a level of 866.70 W/m²K and 969.50 W/m²K respectively in 90 seconds. These values further decrease to 361.80 W/m²K and 478.80 W/m²K at these surfaces in another 150 seconds and further decrease then becomes not so noticeable.

From figure 8, the peak values of interface heat transfer coefficients are 2927.92 W/m²K and 2956.73 W/m²K as obtained by experimental and numerical determinations respectively at the cylindrical surface for no pressure application. For times within 40 seconds to 120 seconds the values of the interfacial heat transfer coefficients obtained numerically and experimentally are found to show higher values of about 19.83 % for numerical results to experimental results.

With the application of pressure on the solidifying aluminium metal, the heat transfer coefficients reach maximum values of 3085.34 W/m²K and 3351.08 W/m²K in the cylindrical and bottom flat surfaces respectively (see figure 9). These values also decrease to 847.80

W/m²K and 783.63 W/m²K in 240 seconds in the cylindrical and bottom flat surfaces respectively, while further decrease with time of solidification is no longer pronounced.

16.3. Interface heat transfer coefficients with solidification temperatures

Figures 10 and 12 show the calculated experimental interface heat transfer coefficients for solidifying molten aluminium metal as a function of solidification temperatures of the solidifying molten aluminium metal. Figure 10 shows the variation of heat transfer coefficient with solidification temperatures of aluminium at the cylindrical surface, while figure 12 is the interface heat transfer coefficients with solidifying temperature at the bottom flat surface with and without the application of pressure on the solidifying metal. From the two graphs, the maximum interface heat transfer coefficients obtained without pressure and with pressure application in the bottom flat surface of the steel mould are 2975.14 W/m²K and 3351.08 W/m²K respectively.

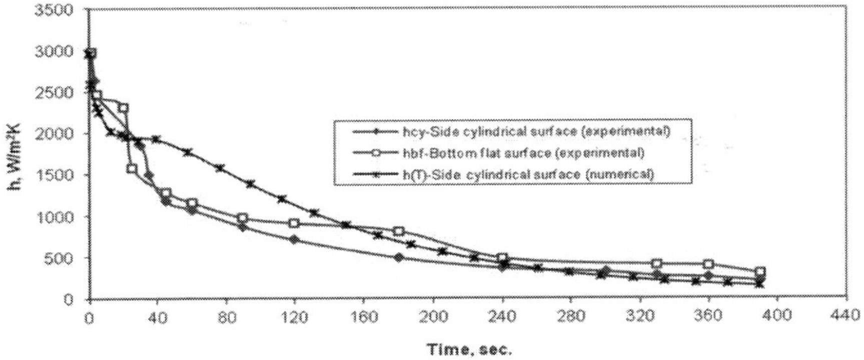

Figure 8. Comparison of numerical values of heat transfer coefficients with calculated experimental values (P = 0)

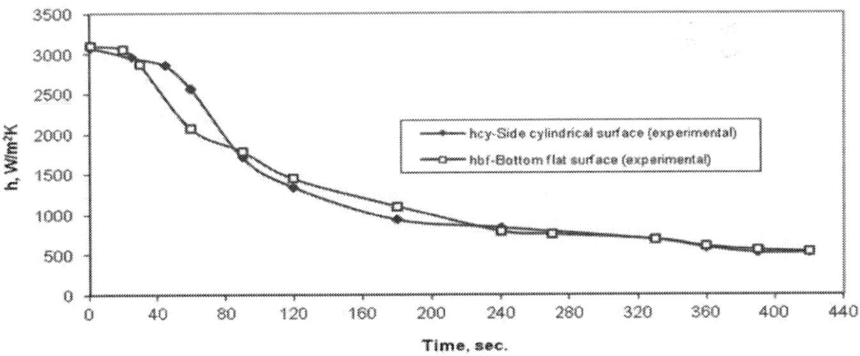

Figure 9. Effect of time of solidification of aluminium metal on heat transfer coefficients with pressure application (P = 85.86MPa) at side and bottom surfaces of steel mould

Figure 11 shows the numerical values of the variation of interface heat transfer coefficients with the solidification temperature of aluminium metal at the cylindrical surface with the application of pressure. The maximum value of heat transfer coefficient of 3397.29 W/m²K at applied pressure of 85.86Mpa as compared to 3351.08 W/m²K obtained through the experimental procedure.

At solidifying temperature above 600°C, a sharp reduction in the interface heat transfer coefficients is noticed at both surfaces as is shown on figures 10, 11 and 12.

For temperatures below 500°C, the interface heat transfer coefficients for both no pressure and pressure applications are close in values. This shows that at temperature below 500°C, the effect of applied pressure is no longer significant on the interface heat transfer coefficient values. The drop in temperature results in solidification of the molten aluminium, which in turn leads to a drop in the heat transfer coefficient values. The effect of applied pressure on the heat transfer coefficients of aluminium becomes more pronounced at solidifying temperatures above 500°C which was also reported by Cho and Hong (1996). Below this temperature, the effect of applied pressure on interface heat transfer coefficient values becomes less pronounced.

Therefore, from figures 10, 11 and 12, it is observed that the effect of applied pressure becomes more significant at temperature close to the liquidus temperature of aluminium as measured along the bottom flat surface of the steel mould (see figure 12). The maximum value of 3397.29 W/m²K is obtained for pressure level of 85.86MPa as compared to 2975.14 W/m²K for no pressure at the bottom flat surface of the steel mould.

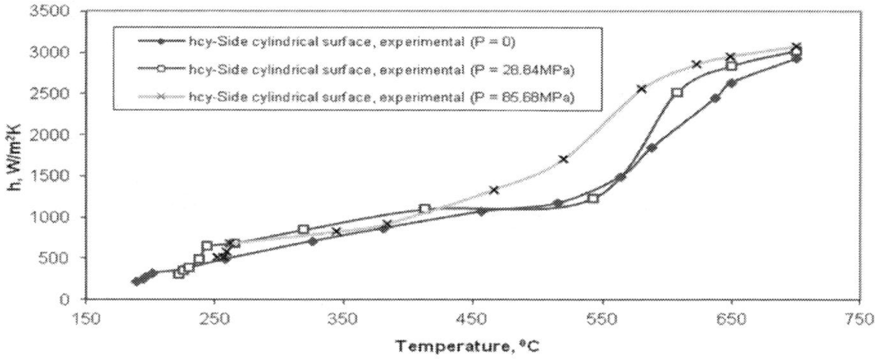

Figure 10. Typical effects of pressure applications on heat transfer coefficients with solidifying temperature at the side cylindrical mould surface by experimental method

16.4. Peak interface heat transfer coefficients with applied pressures

Figure 13 shows the variation of peak values of interface heat transfer coefficients with and without pressure applications. Higher experimental values of heat transfer coefficients are obtained at the bottom flat surface than at the cylindrical surface of the steel mould (see figure 13). This can be associated to greater effect of pressure application experienced at the bottom flat surface than at the side cylindrical surface, thus leading possibly to greater additional internal heat energy generated, and hence obtained higher values of heat transfer coefficients. The results of numerical determination of heat transfer coefficients as in figure 13 shows higher values as compared to the values obtained by experimental method. At applied pressure of 85.86MPa, the obtained heat transfer coefficients are 3397.29 W/m²K and 3351.08 W/m²K by numerical and experimental procedures respectively. From the curves of heat transfer coefficients obtained with temperatures of figures 10, 11 and 12, three distinct portions are noticed. These portions are easily differentiated by the aluminium solidification temperature. These temperatures are below 500°C, solidus phase, 500 to 660°C, liquidus-solidus phase, and above 660°C liquidus phase of solidification of molten aluminium.

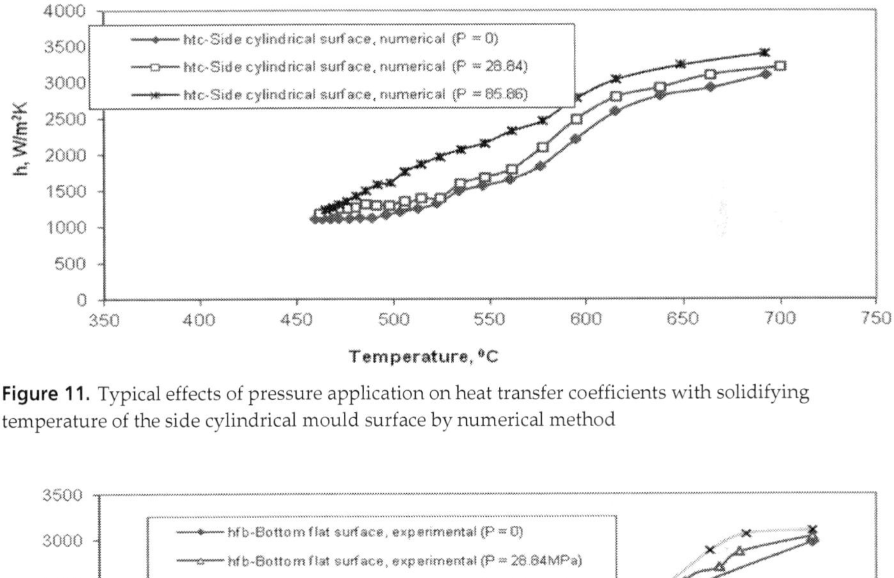

Figure 11. Typical effects of pressure application on heat transfer coefficients with solidifying temperature of the side cylindrical mould surface by numerical method

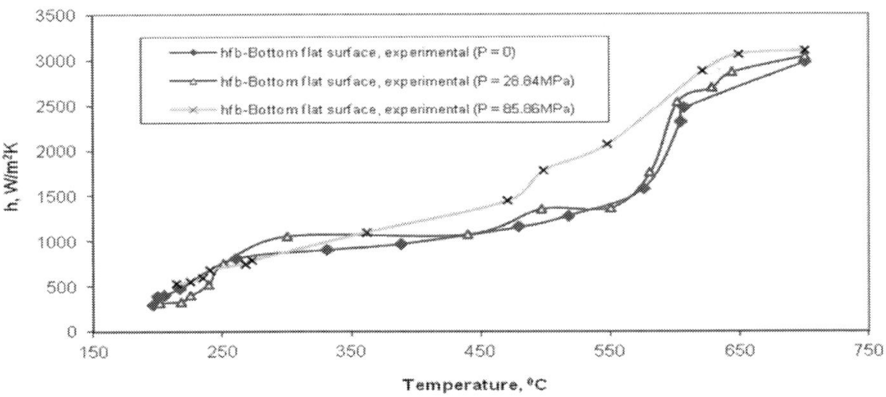

Figure 12. Typical effects of pressure applications on the heat transfer coefficients with solidifying temperature at the bottom flat mould surface

The liquidus-solidus phase, which occurs over a solidification temperature range of 500⁰C to 660⁰C, instead of a constant solidification temperature of 660⁰C can be attributed to the presence of impurities such as silicon, magnesium and manganese in the commercially pure aluminium a fact supported by Higgins (1983).

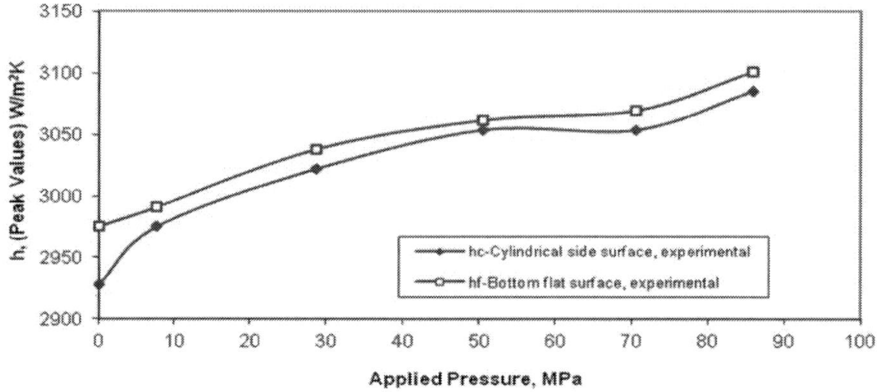

Figure 13. Effect of pressure on the peak values of heat transfer coefficients of aluminium metal at liquidus stage at side and bottom flat moulds' surfaces

The empirical equations for each of the distinct phase changes as a function of applied pressures and solidification temperatures are determined for both the experimental and numerical methods. The empirical equations obtained are for mean average values of heat transfer coefficients based on the experimental method at the cylindrical and flat bottom surfaces. These are equations (71-73):

Temperatures below 500⁰C (solidus phase),

$$h_{S(\exp)} = 3.081T + 1.303P - 232.942 \tag{71}$$

with coefficient of correlation r = 0.9545.

Temperatures between 500⁰C and 660⁰C (liquidus-solidus phase),

$$h_{LS(\exp)} = 10.420T + 5.641P - 4176.022 \tag{72}$$

with coefficient of correlation r = 0.9884.

Temperatures above 660⁰C (super heat, liquidus phase),

$$h_{L(\exp)} = 2.769T + 2.518P + 988.921 \tag{73}$$

with coefficient of correlation r = 0.7825.

The empirical equations (74-76) obtained through the numerical methods are from the results of the computer simulations of heat transfer coefficients at the cylindrical cast metal /

steel mould interface by the application of various applied pressures. These empirical equations are:

Temperatures below 500⁰C (solidus phase),

$$h_{S(num)} = 3.849T + 3.643P - 700.427 \tag{74}$$

with coefficient of correlation r = 0.969

Temperatures between 500⁰C and 660⁰C (liquidus-solidus phase),

$$h_{LS(num)} = 9.027T + 3.414P - 3000.625 \tag{75}$$

with coefficient of correlation r = 0.964

Temperatures above 660⁰C (super heat, liquidus phase),

$$h_{L(num)} = 2.489T + 2.787P + 1342.19 \tag{76}$$

with coefficient of correlation r = 0.772

16.5. Die heating effect

Figure 14 is the effect of die pre-heat temperatures on the values of heat transfer coefficients of aluminium metal without the application of pressure. From the figure, the heat transfer coefficients become lower with increase in die pre-heat temperatures. At die temperature of 95⁰C, the heat transfer coefficient is 3185.34 W/m²K and drop to a value of 2476.73 W/m²K at die temperature of 300⁰C. For all the die temperatures, there is a fall in the heat transfer coefficient's values as solidification temperature decreases.

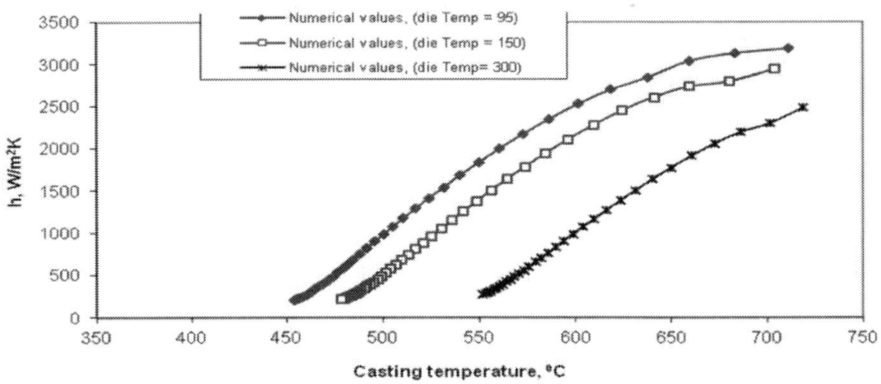

Figure 14. Typical effect of die temperature on heat transfer coefficients without the application of pressure (P=0)

16.6. Comparison of heat transfer coefficient with semi-empirical method

The values of heat transfer coefficients determined using experimental, heat differential equations (numerical), and methods of semi-empirical equations are shown in figure 15 under a pressure application of 85.86MPa. From this graph, the peak values of interface heat transfer coefficient are 3358.19 W/m²K and 3198.79 W/m²K as obtained by heat differential and method of semi-empirical equations respectively for a pressure application of 85.86MPa. The heat transfer coefficients' values for the three methods drop with time and are found to be 1708.03, 1976.81 and 1838.72 W/m²K in 100seconds for experimental, differential and methods of semi-empirical equations respectively.

With die temperature of TM=150⁰C, the peak heat transfer coefficients of 3088.99 W/m²K, 3249.84 W/m²K and 2982.60 W/m²K are obtained for experimental, heat differential and method of semi-empirical equations as shown on figure 16 following the same pattern as in figure 15.

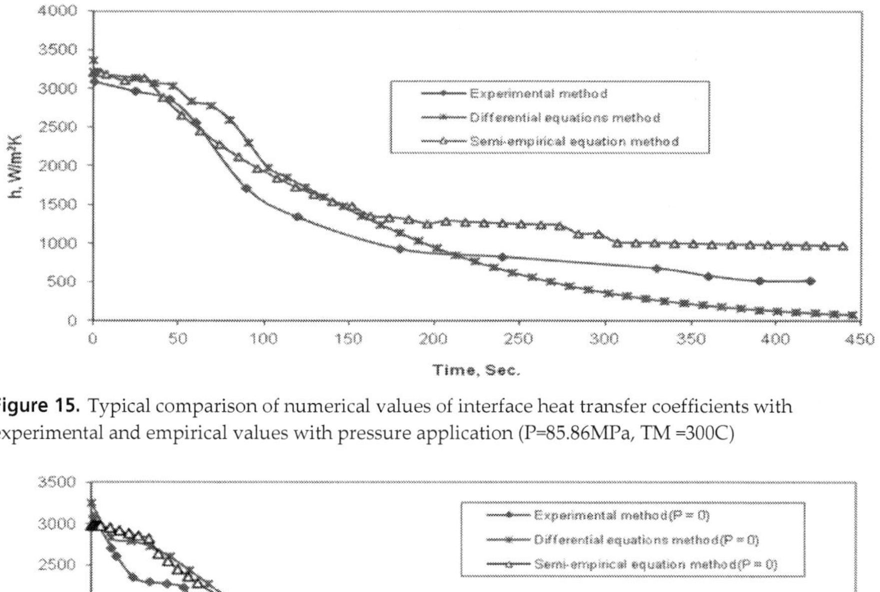

Figure 15. Typical comparison of numerical values of interface heat transfer coefficients with experimental and empirical values with pressure application (P=85.86MPa, TM =300C)

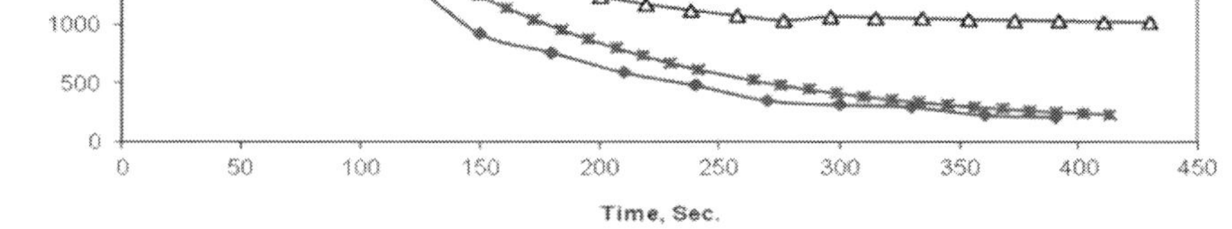

Figure 16. Typical comparison of numerical values of interface heat transfer coefficients with experimental and empirical values with die heating (TM=150°C)

17. Conclusions

The following conclusions can be made from the present investigation:

The graph of temperature against time curves obtained by extrapolating to steel mould / cast metal interface by polynomial curve fitting to heating temperatures graphs with times at various steel mould locations are found to agree in values to the usual numerical methods obtained by previous authors. The interface heat transfer coefficients obtained by the numerical and experimental methods without the application of pressure are found to have values close to that of the numerical methods. The values of the numerical methods were higher by about 19.83%.

Effect of pressure application on the solidifying molten aluminium is more pronounced at casting temperatures above 500^{0}C of the cast aluminium specimen on the values of the interface heat transfer coefficients obtained. Interface heat transfer coefficients are found to decrease with increase in solidification time in both the cylindrical and bottom flat surfaces of the steel mould and thereafter remain fairly constant at temperature below 500^{0}C

Values of experimental peak heat transfer coefficients at the bottom flat surface are found to be higher with pressure application on the solidifying aluminium metal than at the cylindrical surface.

The empirical equations, relating the values of interface heat transfer coefficients with the applied pressures and solidification temperatures at three distinct stages of solidifying molten aluminium are determined and can be applied to determine the heat transfer coefficients.

The values of heat transfer coefficients obtained by heat differential equations incorporating internal heat energy and methods of semi-empirical equations are very close in values. The values as obtained by semi-empirical equations were higher by about 1.7% within the first 5 seconds of solidification.

The semi-empirical equations generated are flexible and could be used to predict the casting temperatures of other metals if the heat transfer coefficient values at the three phase changes are known.

Nomenclature

C_L	- Specific heat of liquid molten aluminium metal,
C_S	-Specific heat of solidified molten aluminium metal,
C_{st}	- Specific heat of steel mould,
d_{sc}	-Spatial step size in the solidified molten metal at completion of solidification,
d_L	-Spatial step sizes at the Liquid portion,
d_{PS}	-Spatial step size at the solidified molten metal phase change,
d_{PL}	-Spatial step size at the solidified molten metal phase change,
d_s	-Spatial step size in the solidified molten metal,
h_0	-Initial height of cast specimen metal (no pressure application),

h_{PL} -Spatial step size at the liquid molten metal phase change,
h_{PS} -Spatial step size at the solidified molten metal phase change,
H^* -Convective heat transfer of the surrounding atmosphere around the steel mould,
I -grid point,
G -grid point at the steel mould-solidified molten metal interface,
M -grid point at the solidified molten metal-liquid molten metal interface,
N -grid point at the centre of the solidifying cast specimen metal,
K_L -Thermal conductivity of liquid molten metal,
K_{st} -Thermal conductivity of steel mould material,
K_S -Thermal conductivity of solidified molten metal,
R_o -external radius of steel mould,
L -internal radius of steel mould,
r -radial axis/coordinate,
T -Temperature,
T_∞ -ambient temperature,
T_i -temperature at a specified grid point(i, j),
T_L -temperature of liquid molten metal,
T_{st}, T_M -temperature of steel mould,
T_{MM} -melting temperature of cast (aluminium) metal,
T_P -pouring temperature of liquid molten metal (super heat temperature),
T_S -temperature of solidified molten metal,
X^i_r -thickness of solidified molten metal at a specified grid point-radial direction,
Q -Thickness of the cylindrical steel mould.

Greek symbols

α_L -thermal diffusivity of the liquid molten metal,
α_{st} -thermal diffusivity of the steel mould material,
α_S -thermal diffusivity of solidified molten metal,
δ -time interval step,
ρ_{st} -density of steel mould material,
ρ_S -density of solidified molten metal,
ρ_L -density of liquid molten metal,
$\sigma_{(T)}$ -Plastic flow stress (temperature dependent).

Author details

Jacob O. Aweda and Michael B. Adeyemi
Department of Mechanical Engineering, University of Ilorin, Ilorin, Nigeria

18. References

Adams, J. Alan & Rogers, David F., (1973), "Computer Aided Heat Transfer Analysis", McGraw-Hill Publishing Company, Tokyo.

Bayazitoglu, Yildiz and Ozisik, M. Necati, (1988), "Elements of Heat Transfer", McGraw-Hill Book Company, New York.

Beck J.V., (1970), "Nonlinear Estimation Applied to the Nonlinear Inverse Heat Conduction Problem", Int. J. Heat Mass Transfer, vol.13, pp703-716.

Bolton, W., (1989), "Engineering Materials Technology", Butterworths-Heinemann Limited, UK.

Browne, David J. and O'Mahoney, D., (2001), "Interface Heat Transfer in Investment casting of Aluminium", Metallurgical and Materials Transactions A, Dec. Vol.32A, pp3055-3063.

Callister, William D. Jr., (1997), "Materials Science and Engineering: An Introduction", 4th Edition, John Wiley and Sons Inc.

Chattopadhyay, Himadri, (2007), "Simulation of transport process in squeeze casting", J. Materials Processing Technology, 186, pp174-178.

Cho I. S. and Hong C. P., (1996), "Evaluation of Heat-Transfer Coefficients at the Casting/Die Interface in Squeeze Casting", Int. J. Cast Metals Res., v.9, pp227-232.

Das A., and Chatterjee S. (1981), "Squeeze Casting of an Aluminium Alloy Containing Small Amount of Silicon Carbide Whiskers", The Metallurgist and Materials Technologist, pp137-142.

Elliott, R., (1988), "Cast Iron Technology", Butterworths, London, U.K.

Gafur, M.A., Nasrul Haque and K. Narayan Prabhu, (2003), "Effects of chill Thickness and Superheat on Casting/Chill Interfacial Heat Transfer During Solidification of Commercially Pure Aluminium", J Materials Processing Technology, 133, pp257-265.

Hearn, E.J. (1992), "Mechanics of Materials", 2nd Edition, Pergamon Press, UK.

Higgins, Raymon A., (1983), "Engineering Metallurgy Part I: Applied Physical Metallurgy", 6th Edition, ELBS with Edward Arnold, UK.

Hosford, William F. and Caddell, Robert M., (1993), "Metal Forming, Mechanics and Metallurgy", 2nd edition, PTR Prentice-Hall Englewood, NJ.

Hu H, and Yu, A., (2002), "Numerical simulation of squeeze cast magnesium alloy AZ91D", Modelling Simul. Mater Sci. Eng., vol.10, pp1-11.

Incropera, Frank P. and Dewitt, David P., (1985), "Fundamental of Heat and Mass Transfer", 3rd Edition, John Wiley and Sons, NY.

Janna, William S., (1988), "Engineering Heat Transfer", SI Edition, Van Nostrand Reinhold (International0, U.K.

Kim, T.G. and Lee, Z.H. (1997), Time-varying heat transfer coefficients between tube-shaped casting and metal mold, Int J, heat mass transfer, 40(15), pp3513-3525.

Kobryn P.A. and Semiatin S. L., (2000), "Determination of Interface Heat-Transfer Coefficients for Permanent Mold casting of Ti-6AL-4v", Metallurgical and materials Transactions, August, Vol.32B, pp685-695.

Liu, A., Voth, T. E. and Bergman, T. L., (1993), "Pure Material Melting and Solidification with Liquid Phase Buoyancy and Surface Tension Forces", Int. J. Heat Mass Transfer, Vol.36 No 2, pp441-442.

Maleki, A., Niroumand, B. and Shafyei, A, (2006), "Effects of squeeze casting parameters on density, macrostructure and hardness of LM13 alloy", Materials Science and Engineering A, 428 pp135-140.

Martorano M. A. and Capocchi J.D.T., (2000), "Heat Transfer Coefficient at the Metal-Mold Interface in the Unidirectional Solidification of Cu-8%Sn alloys", Intl J. Heat Mass Transfer, 43, pp2541-2552.

Ozisik, M. Necati, (1985), "Heat Transfer: – A Basic Approach", McGraw-Hill Publishing, Company U.K.

Potter, D.A. & Easterling, K.E., (1993), "Phase Transformations in Metals", 2nd Edition, Chapman & Hall, London.

Reed-Hill, R E. and Abbaschian, R., (1973), "Physical Metallurgy Principles", 3rd Edition, PWS-KENT Publishing Company, Boston.

Santos, C.A., Quaresma, J.M.V. and Garcia, A., (2001), "Determination of Transient Heat Transfer Coefficients in Chill Mold Castings", Journal of Alloys and Compounds, 139, pp 174-186.

Santos, C.A., Garcia, A., Frick, C.R and Spim J.A., (2004), "Evaluation of heat transfer coefficients along the secondary cooling zones in the continuous casting of steel billets, Inverse problems, Design and Optimization symposium", Rio de Janeiro, pp1-8

Shampire, Lawrence F., (1994), "Numerical Solution of Ordinary Differential Equations", Chapman & Hall, New York.

White, Frank M., (1991), "Heat and Mass Transfer", Addison Wesley Publishing Company, Massachusetts.

Chapter 3

Analytical and Experimental Investigation About Heat Transfer of Hot-Wire Anemometry

Mojtaba Dehghan Manshadi and Mohammad Kazemi Esfeh

Additional information is available at the end of the chapter

http://dx.doi.org/10.5772/51989

1. Introduction

The hot-wire anemometer is a famous thermal instrument for turbulence measurements [1]. The principle of operation of the anemometer is based on the heat transfer from a fine filament where it is exposed to an unknown flow that varies with deviation in the flow rate. The hot-wire filament is made from a special material that processes a temperature coefficient of resistance [2]. Thermal anemometry is the most common method employed to measure instantaneous fluid velocity. It may be operated in one of these two modes, constant current (CC) mode and constant temperature (CT) mode.

Constant-Current (CC) mode: In this mode, the current flow through the hot wire is kept constant and variation in the wire resistance caused by the fluid flow is measured by monitoring the voltage drop variations across the filament.

Constant Temperature (CT) mode: In this mode, the hotwire filament is positioned in a feedback circuit and tends to maintain the hotwire at a constant resistance and hence at a constant temperature and fluctuations in the cooling of the hot wire, filaments are similar to variations in the current flow through the hotwire.

Hot wire anemometers are normally operated in the constant (CTA) mode. The hot-wire anemometry has been used for many years in fluid mechanics as a relatively economical and effective method of measuring the flow velocity and turbulence. It is based on the convective heat transfer from a heated sensing element .Briefly; any fluid velocity change would cause a corresponding change of the convective heat loss to the surrounding fluid from an electrically heated sensing probe. The variation of heat loss from the thermal element can be interpreted as a measure of the fluid velocity changes. In subsonic incompressible flow the heat transfer from a hot wire sensor is dependent on the mass flow, ambient temperature and wire temperature. Since density variations are assumed to be zero, the mass flow variations are only function of velocity changes. The major advantage of maintaining the hot wire at a constant operational temperature and thereby at a constant resistance is that the thermal inertia of the sensing element is automatically adjusted when the flow conditions are varied. The electronic circuit of chosen CTA is shown schematically in Fig. 1. This mode of operation is achieved by incorporating a feedback differential amplifier into the hot-wire anemometer circuit. Such set-up obtains a rapid variation in the heating current and compensates for instantaneous changes in the flow velocity [2]. The sensing element in case studied in this research is a tungsten wire that is heated by an electric current to a temperature of approximately 250 °C. The heat is transferred from the wire mainly through convection. This heat loss is strongly dependent on the excess temperature of the wire, the physical properties of the sensing element and its geometrical configuration. The authors strive to present an analytical solution for heat transfer equation of hotwire for states that can ignore the radiation term for the wire. The fundamental principle of hot-wire anemometer is based on the convective heat transfer, thus in the research, an attempt is made to develop a better perception from the heat transfer of the hot-wire sensor. Also, the effect of air flow temperature variations on the voltage of hot wire, CTA has been studied experimentally. Furthermore, on the basis of air flow velocity and ambient temperature variations, the percentage errors in velocity measurements have been calculated. Finally, based on results, an accurate method has been proposed to compensate for air flow temperature variations.

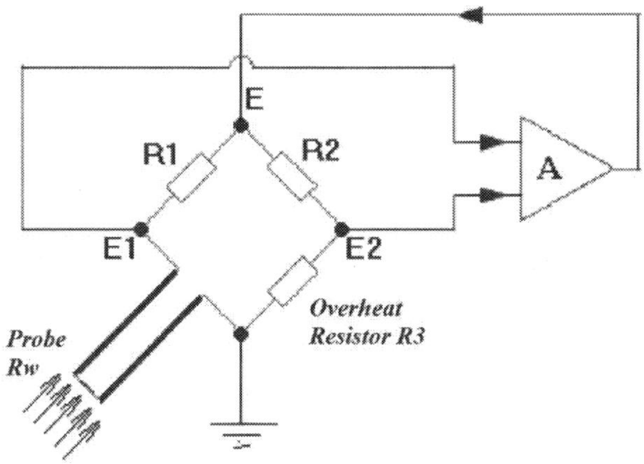

Figure 1. Schematic of a constant temperature anemometer

2. Theoretical background

The hot-wire involves one part of a Wheatstone bridge, where the wire resistance is kept constant over the bandwidth of the feedback loop. The electrical power dissipation $\dot{Q}_{elec}$, when the sensor is heated, is given by:

$$\dot{Q}_{elec} = I^2 R_w \tag{1}$$

I and R_w are the current passing through the sensor and the resistance of the sensor at the temperature Tw, respectively. The convection heat transfer rate to the fluid can be expressed in terms of the heat-transfer coefficient h, as:

$$\dot{Q} = A_w h(T_W - T_a) \tag{2}$$

Where A_w is the surface area of sensor and $(T_w - T_a)$ is the difference between the temperature of the hot-wire sensor and the temperature of the fluid. For steady-state operation, the rate of electric power dissipation equals to the rate of convective heat transfer (assuming the conductive heat transfer to the two prongs is negligible). Thus,

$$I^2 R_W = \pi d L h(T_W - T_a) = \pi L k(T_W - T_a) Nu \tag{3}$$

By introducing the wire voltage $E_W = IR_W$ and using equation (3), one can conclude that (k is the thermal conductivity of the fluid):

$$\frac{E_W^2}{R_W} = \pi L k(T_W - T_a) Nu \tag{4}$$

According to the pioneering experimental and theoretical work by King, the convective heat transfer is often expressed in the following form:

$$Nu = A + BRe^n \tag{5}$$

Where A and B are empirical calibration constants. For long wires in air, King found that A=0.338, B=0.69 and n=0.5. It is interesting to note that King based his derivations on the assumption of potential flow, which is a poor approximation of the real flow around a wire at low Reynolds' numbers, so King's derivation is in a sense approximately erroneous. Nevertheless, King's law has been the considered tool for fitting calibration data in practical hot-wire anemometry for almost a hundred years [3].

By introducing equation (5) into equation (4) can give:

$$\frac{E_W^2}{R_W} = \pi L k(T_W - T_a)(A + B\left(\frac{\rho d}{\mu}\right)^{0.5} U^{0.5}) \tag{6}$$

Equation (6) states that the hot-wire voltage is sensitive both to the velocity and temperature of air. Here, rearranging the equation (6) gives:

$$\frac{E_W^2}{R_W} = (A + BU^{0.5})(T_W - T_a) \tag{7}$$

Where π, l, k, d, ϱ and μ have been included in the constant coefficients A and B.

According to equation (7), Kanevce and Oka [4] introduced the following expression to correct the hot-wire output voltage for the temperature drift:

$$E_{corr} = E_W \left(\frac{T_W - T_a}{T_W - T_{a,r}} \right)^{0.5} \tag{8}$$

$T_{a,r}$ is ambient reference temperature during sensor calibration and T_a is ambient temperature during data acquisition where E_{corr} is corrected voltage. For a hot wire probe with a finite length active wire element, the conductive heat transfer to prongs must be taken into account. In practice this is often achieved by the modifying equation [7] as:

$$\frac{E_w^2}{R_w} = \left(A + BU^n \right)\left(T_w - T_a \right) \tag{9}$$

The values of A, B and n can be determined by a suitable calibration procedure. It should be noted that the term (T_w-T_a) and physical properties of fluid are dependent on the ambient temperature. In the previous related studies, the effect of term (T_w-T_a) is considered only to compensate the ambient temperature variations [5]. In other word, the variations of physical properties of fluid and Nusselt number are ignored. So in this study, the variations of Nusselt number with the fluid temperature have been considered. The following equation for correction of output voltage E. has been proposed by the relations extended in Ref. [6].

$$E_{corr} = E_w \left(\frac{T_W - T_{a,r}}{T_W - T_a} \right)^{0.5(1 \pm m)} \tag{10}$$

In Ref. [7], equation (10) is employed to correct the voltage of CTA output. Results showed that the required error correction factor (m) depends on whether the fluid temperature decreases or increases with respect to the calibration temperature of the CTA.

The CT mode velocity and temperature sensitivities corresponding to equation (9) are:

$$S_u = \frac{\partial E_w}{\partial U} = \frac{nBU^{n-1}}{2} \left[\frac{R_w \left(T_w - T_a \right)}{A + BU^n} \right]^{0.5} \tag{11}$$

$$S_\theta = \frac{\partial E_w}{\partial \theta} = \frac{-1}{2} \left[\frac{R_w \left(A - BU^n \right)}{T_w - T_a} \right]^{0.5} \tag{12}$$

Where θ is a small fluctuation in the fluid temperature. Equations (11) and (12) show that the value of S_u increases and the value of S_θ decreases by increasing the value of (T_w-T_a). A high over-heat ratio (R_w/R_a) is recommended for the measurement of velocity fluctuations [2]. In Ref. [7], it is stated that for an over-heat ratio of 1.4, the error incurred amounts to about 2.5% per degree Celsius temperature change. With the increase in the overheat ratio to 1.6, the error in CTA output is reduced to about 2% per degree Celsius temperature change.

The heat transfer process from a hot-wire sensor is usually expressed in a non-dimensional form where involve a relationship between the Nusselt number, the Reynolds number and the Prandtl number. The Nusselt number is usually assumed to be a function of Reynolds and Prandtl numbers and under most flow conditions, the Prandtl number is constant.

In hot-wire anemometry, the sizes of the sensing element are small, so that the Reynolds number of the flow is very low and the flow pattern over the sensor can be assumed to be symmetrical and quasi-steady. Due to the statement of the flow continuity, the mean free path of the particles is very much less than the diameter of the wire and conventional heat transfer theories are applicable [8]. Furthermore, the length of the sensor is much greater than its diameter. Hence, it may be assumed that the loss conduction through the ends is negligible and the relation for the heat transfer from an infinite cylinder can be applied. Kramers [9] has proposed the following equation based on heat-transfer experimental results for wires (with infinite length-to-diameter ratio), placed in air, water and oil:

$$Nu = 0.42Pr^{0.2} + 0.57Pr^{0.33}Re^{0.5} \tag{13}$$

He selected the film temperature $T_f = (T_w+T_a)/2$ as the reference temperature for the fluid properties.

3. Experimental procedure

An air condition unit was used to carry out the experiments (Fig. 2). A laminar airflow was achieved by means of honeycombs network and screens. The air condition unit is powered by a small variable speed electric fan and four controllable heating elements provide a stable air temperature. The air flow velocity was measured by a pitot tube and a pressure transducer during the calibration and test. The output voltage from the hot-wire, pressure transducer output voltage, and the thermometer (NTC) output voltage are transferred to a computer, via an A/D card, having a 12 bit resolution and up to 100 kHz frequency.

The sensing element in our case is a standard 5μm diameter tungsten wire that is heated by an electric current to a temperature of approximately 250 °C. The active wire length is 1.25 mm. For such probes, the convective heat transfer is about 85 percent of the total heat transfer from the heated-wire element [2].

Before measurements, the hot-wire sensor was calibrated in a wind tunnel and the response of the anemometer bridge voltage was also expanded as a least square fit with a 5th order polynomial ($U=C_0+C_1E+C_2E^2+C_3E^3+C_4E^4+C_5E^5$).

The experiments were carried out on a hot wire sensor operating at an over-heat ratio (R_w/R_a) of 1.8. The sensor, after calibration, was tested at different temperatures. The velocity range was 1-2 m/s, which corresponds to a Reynolds number of 0.18-0.35, and the temperature range, was 17.5-40 °C.

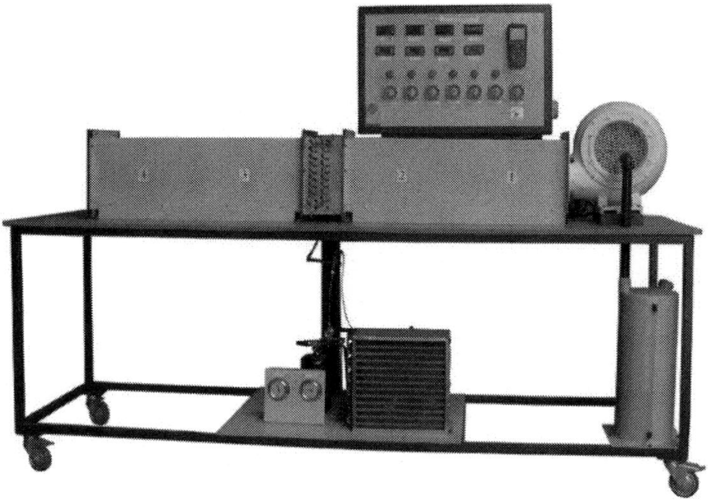

Figure 2. The laboratory air condition unit

4. Results and discussions

To examine the behavior of the hot wire sensor in different conditions and determine the temperature distribution along it, the general hot wire equation must be derived initially. By considering an incremental element of the hot wire, Fig.3, an energy balance can be performed where assume that there is the uniform temperature over its cross-section according to the equation (14).

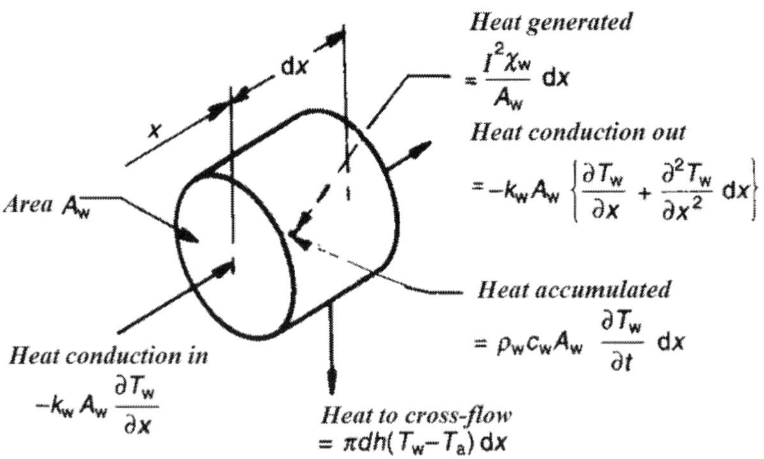

Figure 3. Heat balance for an incremental element [2].

$$\frac{I^2 \chi_w}{A_w} = \pi dh\left(T_w - T_a\right) - K_w A_w \frac{\partial^2 T_w}{\partial x^2} + \pi d\sigma\varepsilon\left(T_w^4 - T_s^4\right) + \rho_w c_w A_w \frac{\partial T_w}{\partial t} \tag{14}$$

Where I is electrical current, χ_w is the electrical resistant of the wire material at the local wire temperature, T_w, and A_w is the cross-sectional area of wire where h is the heat-transfer coefficient, c_w is the specific heat of the wire material per unit mass, k_w is the thermal conductivity of the wire material and d is the diameter of wire. With using the fourth-order Runge-Kutta method, this nonlinear secondary differential equation is solved in two conditions: with radiation term and without radiation term. Fig.4 shows the results for this step. As it is shown, the radiation term does not have any effect on the temperature distribution. The previous results achieved in Ref.[3] indicate that, the error due to radiation is in the range 0.1-0.01% and is quite negligible.

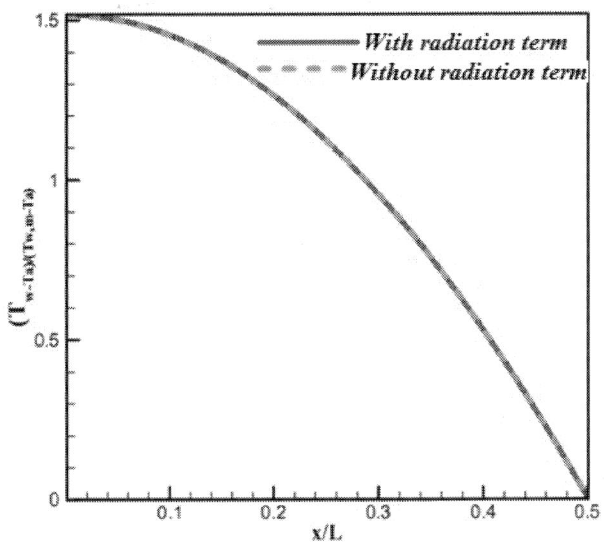

Figure 4. The solution of equation (14) with and without radiation term

Under steady conditions,

$$\frac{\partial T_w}{\partial t} = 0$$

χ_w can be expressed as $\chi_w = \chi_a + \chi_o \alpha_o (T_w - T_a)$. Where χ_a and χ_o are the values of the resistivity at the ambient fluid temperature, T_a, and at 0°C and α_o is temperature coefficient of resistivity at 0°C. Thus, equation (14) can be rewritten as equation 15 [2]:

$$K_w A_w \frac{d^2 T_w}{dx^2} + \left(\frac{I^2 \chi_o \alpha_o}{A_w} - \pi dh \right) \left(T_w - T_a \right) + \left(\frac{I^2 \chi_a}{A_w} \right) = 0 \tag{15}$$

With assuming the ambient temperature is constant along the wire, this equation is of the following form (16):

$$\frac{d^2 T_w}{dx^2} + K_1 T_1 + K_2 = 0 \tag{16}$$

Where

$$T_1 = T_w - T_a$$

And

$$K_1 = \left(\frac{I^2 \chi_o \alpha_o}{A_w} - \pi dh \right)$$

$$K_2 = \left(\frac{I^2 \chi_a}{A_w} \right)$$

The value of K_1 may be negative or positive. Therefore, the solution of equation (16) and temperature distribution along the wire are dependent on the value of K_1. Equation (16) is solved in three states: $K_1<0$, $K_1=0$, $K_1>0$.

4.1. State I: $K_1<0$

For more hot-wire applications, K_1 will be negative [2]. In Ref. [2], it is declared that in this state, the solution for a wire of length L will become:

$$T_w = \frac{K_2}{|K_1|} \left[1 - \frac{\cosh\left(|K_1|^{0.5} x \right)}{\cosh\left(\frac{|K_1|^{0.5} L}{2} \right)} \right] + T_a \tag{17}$$

The mean wire temperature, $T_{w,m}$ is obtained by integrating equation (17):

$$T_{w,m} = \frac{1}{L} \int_{-L/2}^{L/2} T_w(x) dx \tag{18}$$

Inserting equation (17) into equation (18) gives:

$$T_{w,m} = \frac{K_2}{|K_1|}\left[1 - \frac{\tanh\left(|K_1|^{0.5} L/2\right)}{|K_1|^{0.5} L/2}\right] + T_a \tag{19}$$

The non-dimensional steady state wire temperature distribution will be achieved such as equation (20) [2]:

$$\frac{T_w - T_a}{T_{w,m} - T_a} = \frac{\left[\frac{L\times|K_1|^{0.5}}{2}\right]\left[\cosh\left(\frac{L\times|K_1|^{0.5}}{2}\right) - \cosh\left(x\times|K_1|^{0.5}\right)\right]}{\left[\frac{L\times|K_1|^{0.5}}{2}\right]\cosh\left(\frac{L\times|K_1|^{0.5}}{2}\right) - \sinh\left(\frac{L\times|K_1|^{0.5}}{2}\right)} \tag{20}$$

Where T_a is the ambient fluid temperature and $T_{w,m}$ is the mean wire temperature.

The convective and conductive heat transfer rate can be found from the flow conditions and the wire temperature distribution will earn according to the following equations:

$$\dot{Q}_{cond} = 2k_w A_w \left|\frac{dT_w}{dx}\right|_{x=l/2} \tag{21}$$

$$\dot{Q}_{conv} = \pi dhL(T_{w,m} - T_a) \tag{22}$$

To achieve a reasonable accuracy, the ratio of conductive heat transfer to convective heat transfer should be as low as possible.

$$\frac{\dot{Q}_{cond}}{\dot{Q}_{conv}} = 2\frac{k_w A_w |K_1|^{0.5}}{\pi dLh} \times \frac{\tanh\left(0.5L|K_1|^{0.5}\right)}{\left(1 - \frac{2}{L|K_1|^{0.5}}\tanh\left(0.5L|K_1|^{0.5}\right)\right)} \tag{23}$$

According to equation (23), to reduce the effect of the conductive heat transfer rate, the wire should be as long as possible and the thermal conductive of the wire material should have a low value.

The temperature distribution in the form $(T_w-T_a)/(T_{w,m}-T_a)$, is shown in Fig.5. It is shown that the uniformity of the temperature distribution along the wire increases for longer length wires. Also, the value of temperature in different parts of the wire approaches to mean temperature with escalating the length wire.

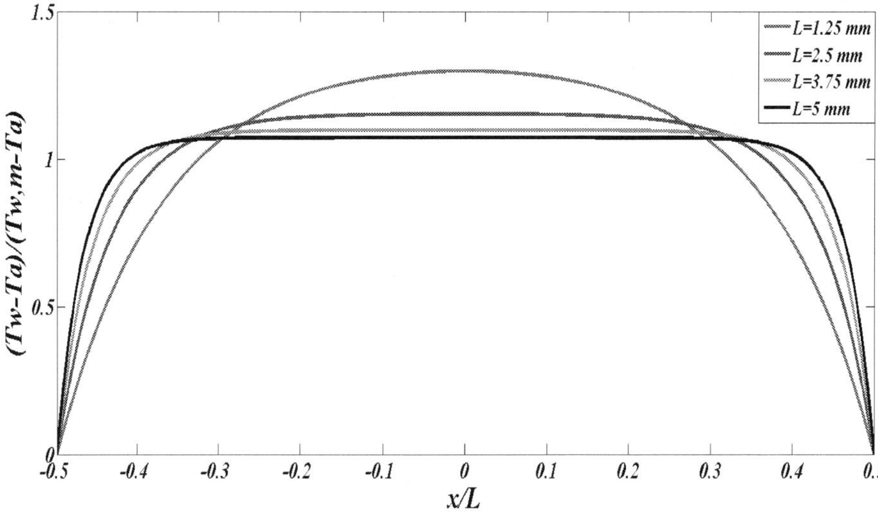

Figure 5. The temperature distribution along a hot wire for various values of L (K_1<0).

The effect of wire length on the percent of conduction and convection heat transfer is shown in Fig. 6. (Diameter of wire is 5μm and the air velocity is equal to 20 m/s). As it is shown the conductive end losses reduces with increasing the wire length but it should be noted, the maximum value of tanh $(0.5L\,|K_1|^{0.5})$ is approximately 1, so exceeding the wire length over $5.3/|K_1|^{0.5}$ will not cause a reduction in the conductive heat transfer rate.

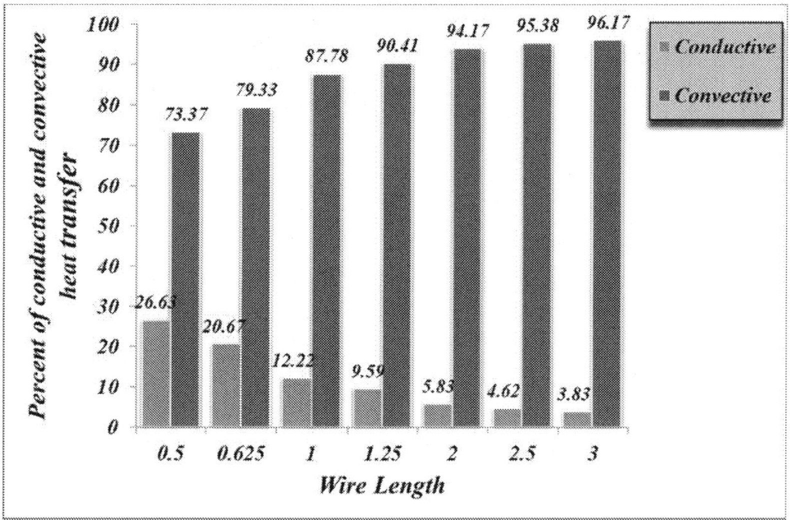

Figure 6. The percent of conduction and convection heat transfer for different wire length.

For hot-wire anemometer applications it is usually advantageous to minimize the rate of conductive heat transfer rate relative to the forced convective heat transfer rate [2]. Fig.7 shows the effect of wire diameter on the non-dimensional temperature distribution. It is shown, the uniformity of temperature distribution decreases with increasing the wire diameter. This variation is due to increasing the wire diameter that will cause the conductive heat transfer rate to the two prongs to be increased as well.

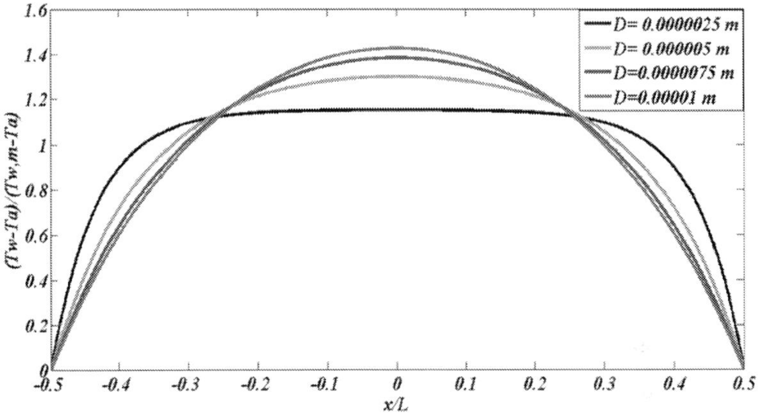

Figure 7. The temperature distribution along the hot wire for various values of d (K1<0).

Comparing between Figs. 6 and 7 shows that the wire diameter has the greater influence on the temperature distribution rather than the wire length.

4.2. State II; K_1=0

In this state, the temperature distribution equation is obtained as:

$$T_w(x) = -\frac{1}{2}K_2L^2\left[\left(\frac{x}{L}\right)^2 - \frac{1}{4}\right] + T_a \tag{24}$$

According to equations (24) and (18), the mean wire temperature is determined as:

$$T_{w,m} = \frac{1}{12}K_2L^2 + T_a \tag{25}$$

By using equations (24) and (25), the non-dimensional wire temperature distribution can be expressed as:

$$\frac{T_w(x) - T_a}{T_m(x) - T_a} = -6\left[\left(\frac{x}{L}\right)^2 - \frac{1}{4}\right] \tag{26}$$

For this state, the non-dimensional temperature distribution is shown in Fig.8. It can be observed that the temperature distribution along the wire is independent of the wire length and for various values of L, all temperature profiles are identical. Also, it can be observed from equation (26) that the non-dimensional temperature distribution does not depend on the wire diameter.

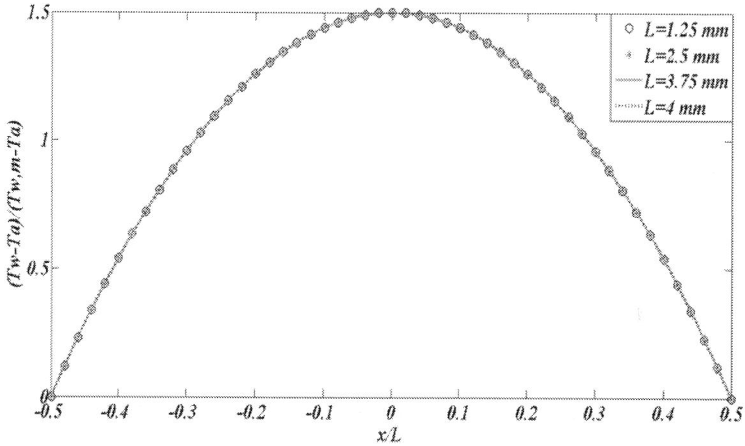

Figure 8. The temperature distribution along a hot wire for various values of L (K1=0).

Furthermore, the ratio of conductive heat transfer rate to the forced convective heat transfer rate can be expressed as:

$$\frac{\dot{Q}_{cond}}{\dot{Q}_{conv}} = \frac{3}{2}\frac{k_w d}{hL} \quad (27)$$

While it is shown, this ratio is directly proportional to the length and it will increase with inverse proportion to the wire diameter.

4.3. State III; K1>0

Using the mathematical analysis, it can be demonstrated that temperature distribution equation is:

$$T_w = \frac{K_2}{K_1}\left[\frac{\cos\left(\left(K_1\right)^{0.5} x\right)}{\cos\left(\frac{\left(K_1\right)^{0.5} L}{2}\right)} - 1\right] + T_a \quad (28)$$

The mean wire temperature and the non-dimensional wire temperature distribution can be expressed as:

$$T_{w,m} = \frac{K_2}{K_1}\left[\frac{\tan\left(\left(K_1\right)^{0.5} L/2\right)}{\left(K_1\right)^{0.5} L/2} - 1\right] + T_a \tag{29}$$

$$\frac{T_w - T_a}{T_{w,m} - T_a} = \frac{\left[\frac{L\times\left(K_1\right)^{0.5}}{2}\right]\left[\cos\left(x\times\left(K_1\right)^{0.5}\right) - \cos\left(\frac{L\times\left(K_1\right)^{0.5}}{2}\right)\right]}{\sin\left(\frac{L\times\left(K_1\right)^{0.5}}{2}\right) - \left[\frac{L\times\left(K_1\right)^{0.5}}{2}\right]\cos\left(\frac{L\times\left|K_1\right|^{0.5}}{2}\right)} \tag{30}$$

Fig.9 shows the non-dimensional temperature distribution ($K_1>0$). As it is shown, in this state some fluctuations appear in the temperature profiles. It can be demonstrated from equation (28) that, with approaching the wire length to $\pi/|k_1|^{0.5}$, these fluctuations increases wherever the amplitude oscillatin decreases with growing the wire length.

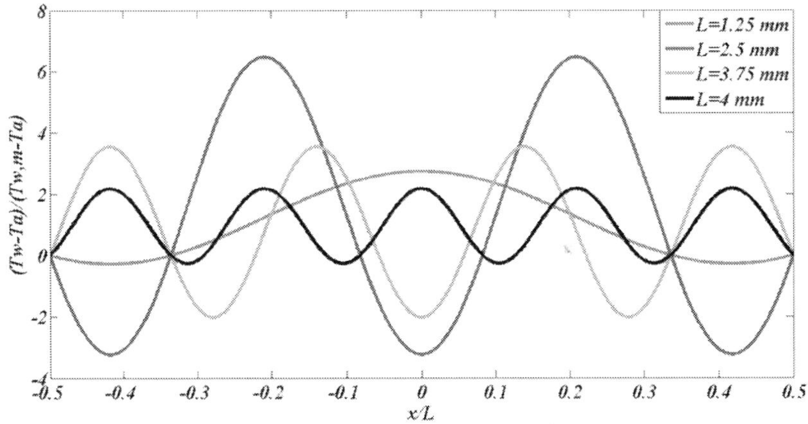

Figure 9. The temperature distribution along a hot wire for various values of L (K1>0).

It should be noted that the temperature profile is strongly dependent on the value of K_1 which is relevant to the heat transfer coefficient (Nusselt number). With setting the value of K_1 to zero, one could determine the critical Nusselt number as the following equation:

$$Nu_{critical} = \frac{I^2 \chi_o \alpha_o}{A_w \pi k} \tag{31}$$

In summary, the authors consider the temperature distribution along the hot-wire in the following cases:

Case I, $Nu > Nu_{critical}$: in this case, increasing the wire length and decreasing the wire diameter will cause the uniformity of temperature distribution to be increased considerably.

Case II, $Nu = Nu_{critical}$:in this case, the temperature distribution is independent of length and diameter of wire.

Case III, $Nu < Nu_{critical}$: here, temperature distribution is non-uniform and there are some fluctuations in temperature distribution.

According to equation (4), by knowing E_w, R_w, T_w and T_a in the anemometer, one can calculate the Nu number. The electrical resistance of the wire's material increases linearly with temperature, so that the resistance can be described as:

$$R_w = R_o\left[1+\alpha_o(T_w - T_o)\right] \tag{32}$$

Where R_0 is the value of the resistance at a reference temperature T_0 and α is the temperature coefficient of resistance. The recommended value for over-heat ratio is equal to 1.8 and the wire temperature of the chosen probe is then 249.22 °C.

In practical application, the hot-wire anemometer output is bridge voltage E (Fig. 1) whereas for determining the Nusselt number, the value of E_w is required. For a balanced anemometer bridge, the relationship between E (bridge voltage) and E_w (hot-wire sensor voltage) is:

$$E_w = \frac{E}{R_1 + R_w} R_w \tag{33}$$

For comparison, the calculated Nusselt number by equation (4) that it is based on the fluid properties evaluated at the film temperature defined as the mean of the upstream flow temperature and temperature on the hot wire versus Reynolds number where based on the fluid properties evaluated at the film temperature is presented in Fig.10 with Kramer's experimental formula.

As it is shown, the data does not collapse to one curve and the deviation increases with increasing Reynolds number. Our results are lower than those given by Kramer's formula and the differences may be caused by the effect of conductive heat transfer to the prongs and three-dimensional effect encountered in experiments. In Ref. [2], it is stated that for a standard probe (d=5μm and l=1.25 mm), the conductive heat transfer to the two prongs is about 15 percent of the total heat transfer from the heated-wire element. Although, the results in this study show that at the high velocity, the percentage of error between the predicted Nusselt number by Kramer's formula and the achieved Nusselt number in this study is 50%. This result confirms that, there is a significant difference between the heat transfer process from finite length hot-wire sensor and infinite length one.

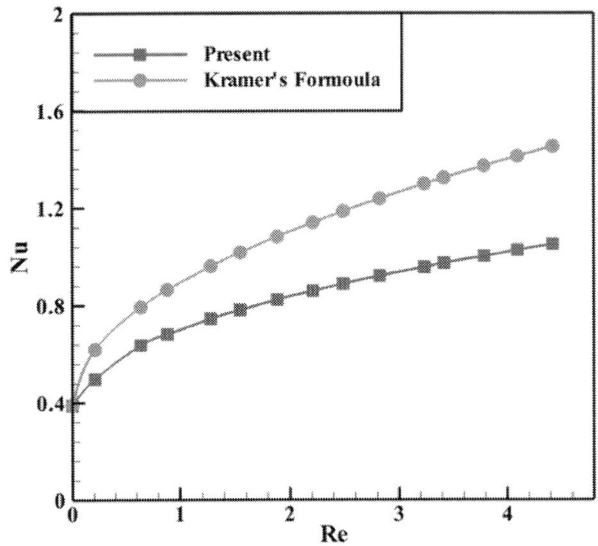

Figure 10. Heat transfer from a hot wire in a free stream flow.

In the next phase, the thermal response of the hot-wire anemometer relative to the velocity and also the air ambient temperature variation is investigated [11]. Figs. 11 and 12 show the variation of convection heat transfer coefficient (h) and Nusselt number versus Reynolds number at different ambient temperatures where fluid properties evaluated at the film temperature $T_f = (T_a + T_w)/2$.

The achieved results indicate convection heat transfer coefficient and Nusselt number vary with variation of air flow temperature and as expected, both of them decrease with increasing the ambient temperature.

Different equations have been proposed to modify the Nusselt number. Lundström et al. [3] claim that it was necessary to evaluate the fluid properties at the air temperature and their results show that evaluating the properties at the film temperature is not enough to achieve a temperature independent calibration law. Collis and Williams [10] realized, using the film reference temperature, that it was necessary to include a temperature loading factor in the Nusselt number King's law according to the following equation;

$$Nu\left(\frac{T_f}{T_a}\right)^{-0.17} = A + BRe^{0.45} \tag{34}$$

Nusselt number (hd/k) includes both the heat-transfer coefficient and the thermal conductivity of the fluid and these parameters are dependent on the ambient temperature. The temperature role on k can be expressed as:

$$\frac{k}{k_r} = \left(\frac{T}{T_r}\right)^a \tag{35}$$

T and T_r are in absolute temperature but the variation of h with ambient temperature is unknown.

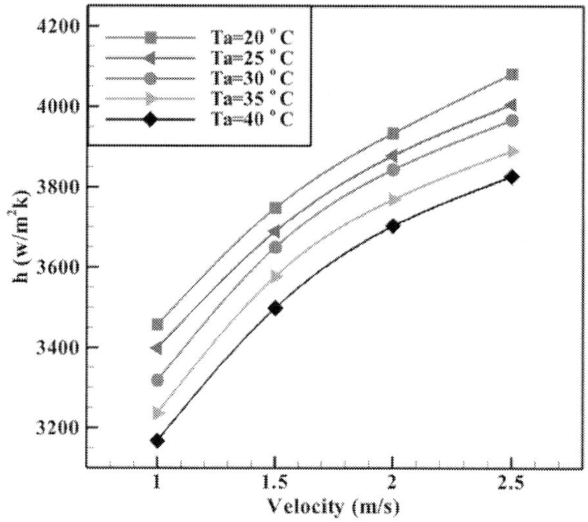

Figure 11. Variations of heat-transfer coefficient Vs. velocity at different ambient temperatures. [11]

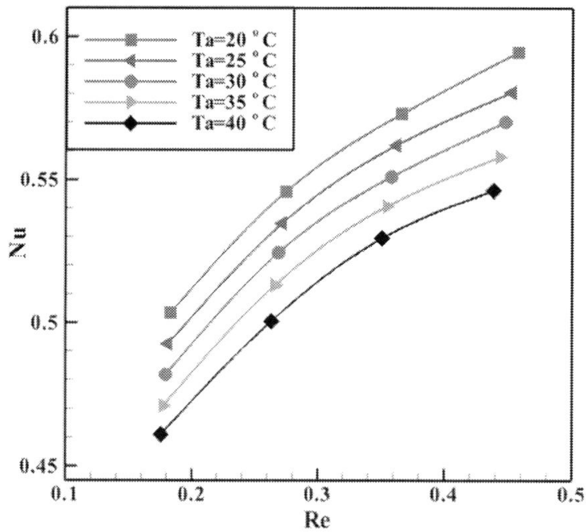

Figure 12. Variations of Nusselt number Vs. Re at different ambient temperatures. [11]

However, the fundamental mechanism for variation of Nusselt number is not known yet but it can be compensated empirically by introducing the modified Nusselt number according to the following equation [11]:

$$Nu_{corr} = Nu\left(\frac{T_a}{T_{a,r}}\right) \tag{36}$$

Where temperatures are in absolute temperature and $T_{a,r}$ is the reference temperature at which the sensor calibration is performed. When the correction is applied to the data in Fig. 12, the data collapse approximately to a single curve as shown in Fig.13.

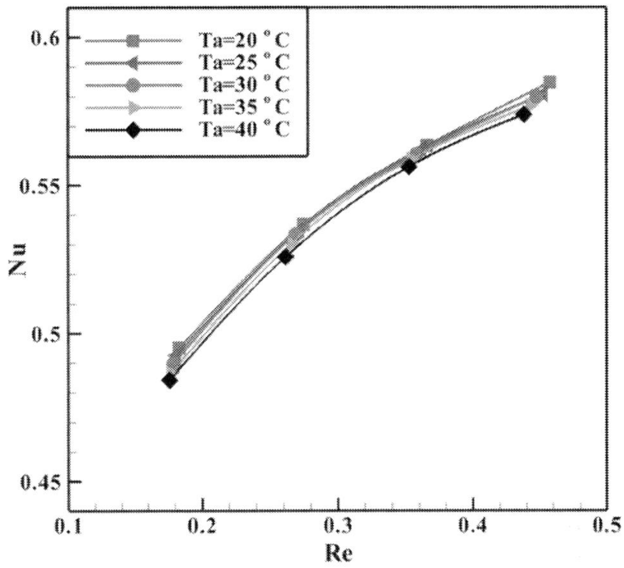

Figure 13. The achieved Nusselt number Vs. Re by employing equation (36) [11].

Fig. 14 presents the response of the CTA on wind speed approximately equal to 1 and 2 m/s at various temperatures. The temperature varies between 22.5 and 37.5°C. As it is shown, the bridge voltage decreases as the higher ambient temperature.

Calibration equations do not include ambient temperature variations, so a correction procedure should be applied. There are three main practical ways [2]:

- Automatic compensation: Use a temperature sensor in the Wheatstone bridge.
- Manual adjustment: Manual adjustment can be made by changing the value of the resistant, R_w, to compensate the changes in T_a.

- Analytical correction: Measure the flow temperature separately and compensate using the heat transfer equation.

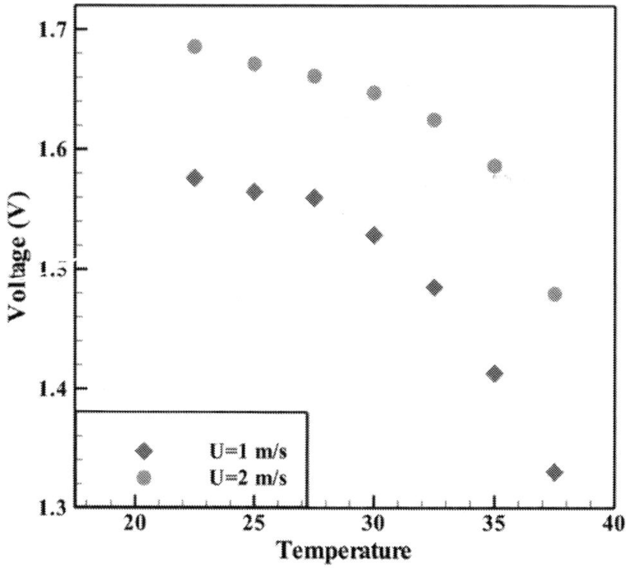

Figure 14. Response of CTA at various ambient temperatures. [11]

In this research, the voltage error due to changes in the ambient temperature is corrected using equation (10) and (37):

$$E_W = E\left(\frac{T_W - T_a}{T_W - T_{a,r}}\right)^{0.5} \times \left(\frac{T_a}{T_{a,r}}\right)^{0.5} \tag{37}$$

Equation (10) only considers the effect of ambient temperature variation but equation (37) regards the effect of Nusselt variation as well the ambient temperature variation.

The percentage error is presented in Fig.15 as a function of flow temperature. The other parameters are reference temperature, Ta,r=25 °C, average sensor temperature, Tw=249.22 °C and flow velocity, U=1.5 m/s.

It can be seen that the achieved results from equation (37) are more reasonable. It should be noted that by increasing the air temperature, the fluid properties will be changed and these changes have to be taken into consideration. This factor is considered in equation (37) so that at high air temperature it can compensate the ambient temperature variations adequately.

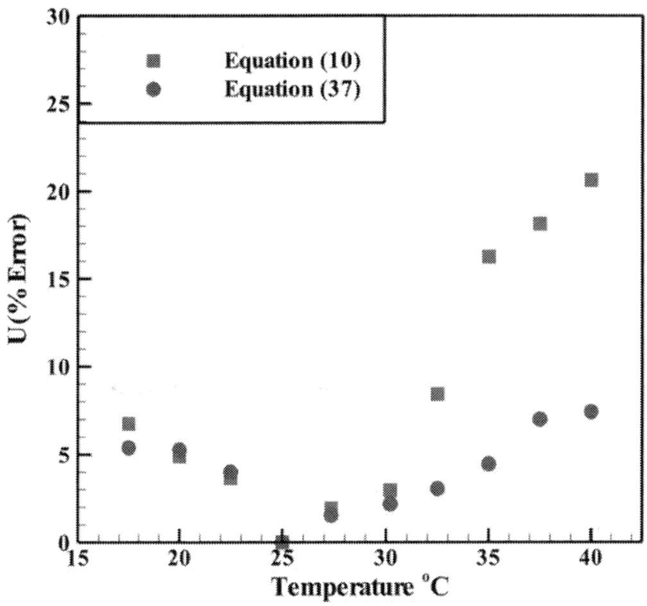

Figure 15. Error in the measurement of flow velocity for the hot wire sensor [11]

5. Conclusion

The analytical solutions for heat transfer equation of hotwire indicate that the temperature distribution along a hot-wire sensor is dependent on the critical Nusselt number. $Nu<Nu_{critical}$ leads to increasing the wire length and decreasing the wire diameter which will cause the uniformity of temperature distribution to be increased considerably. For $Nu=Nu_{critical}$, wire length and diameter don't have any effect on the temperature distribution. If $Nu>Nu_{critical}$, some fluctuations will appear in the temperature profiles.

The results from experimental investigation show that the values of both the air temperature and Nusselt number have influence on the output voltage of the CTA. In this study, two ways have been employed to compensate the ambient temperature changes. In the first case, the effect of ambient temperature variation is only considered but in the second case, the effect of Nusselt variation is also regarded. At low temperature variation, the accuracy of two methods is almost the same whereas by increasing the air temperature, the second method which consider the changes in fluid properties, provide more accurate results in compare with the first method.

6. Appendix

A_w	Surface area of sensor,m^2
C_w	Specific heat of the wire material per unit mass, J/Kg. K
D	Wire diameter, m
E	Bridge voltage ,V
E_{corr}	Corrected voltage, V
E_w	Hot-wire sensor voltage, V
H	heat-transfer coefficient, W/m^2.K
I	Electrical current passing through the sensor, A
K	Thermal conductivity of the fluid, W/m. K
k_w	Thermal conductivity of the wire material, W/m. K
L	Wire length, m
M	Temperature loading factor used in equation (10)
N	Constant used in equation (7)
Nu	Nusselt Number, hd/k
Pr	Prandtl number, $\mu Cp/K$
$\dot{Q}_{elec}$	Electrical power dissipation, W
Re	Reynolds number, ud/ν
R_w	Sensor resistance
T_a	Ambient temperature during data acquisition, K
$T_{a,r}$	Ambient reference temperature during sensor calibration, K

T_w	Temperature of hot-wire sensor, K
$T_{w,m}$	Mean wire temperature, K
U	Instantaneous velocity, m/s
A_o	Temperature coefficient of resistivity at 0°C
E	Emissivity of the sensor
Σ	Stefan-Boltzmann constant
χ_o	Values of resistivity at 0°C
χ_a	Values of resistivity at the ambient fluid temperature
M	Dynamic viscosity, N.s/m^2

Author details

Mojtaba Dehghan Manshadi and Mohammad Kazemi Esfeh
Malekashtar University of Technology, Iran
University of Yazd, Yazd, Iran

7. References

[1] Manshadi M. D. (2011). "The Importance of Turbulence in Assessment of Wind Tunnel Flow Quality", Book chapter No. 12, Wind Tunnels and Experimental Fluid Dynamics Research, Edited by Jorge Colman Lerner and Ulfilas Boldes, Intech publisher.

[2] Bruun H H. Hot-wire Anemometry, Principles and Signal Analysis. Oxford University Press; 1995.

[3] LundstrÖm H, Sandberg M, Mosfegh B. Temperature dependence of convective heat transfer from fine wires in air: A comprehensive experimental investigation with application to temperature compensation in hot-wire anemometry. Experimental Thermal and Fluid Science 2007; 32: 649-657.

[4] Kanevce G, Oka S. Correcting hot-wire readings for influence of fluid temperature variations 1973. DISA Info, No. 15, 21-24.

[5] Benjamin SF, Roberts CA. Measuring flow velocity at elevated temperature with a hot wire anemometer calibrated in cold flow. Heat and Mass Transfer 2002; 45: 703-706.

[6] Improved temperature correction in stream Ware®. DANTEC DYNAMICS. Technical Note of dynamics, Publication No.TN049909, P2, 2002.

[7] Ardakani MA, Farhani F. Experimental study on response of hot wire and cylindrical hot film anemometers operating under varying fluid temperatures. Flow Measurement and Instrumentation 2009; 20: 174_179.

[8] Suminska OA. Application of a constant temperature anemometer for balloon-borne stratospheric turbulence soundings. MSc thesis. University of Rostock; 2008.

[9] Kramers H. Heat transfer from spheres to flowing media. Physica 1946;12:61-80.

[10] Collis DC, Williams MJ. Two-dimensional convection from heated wires at low Reynolds numbers. J. Fluid Mech 1959; 6:357-384.

[11] Dehghan Manshaid M, Kazemi Esfeh M. A new approach about heat transfer of hot-wire anemometer. Accepted to publishing in the Applied Mechanics and Materials Journal, 2012.

Section 2

Convection Heat Transfer

Chapter 4

Boundary-Layer Flow in a Porous Medium of a Nanofluid Past a Vertical Cone

F.M. Hady, F.S. Ibrahim, S.M. Abdel-Gaied and M.R. Eid

Additional information is available at the end of the chapter

http://dx.doi.org/10.5772/53497

1. Introduction

The natural convection flow over a surface embedded in saturated porous media is encountered in many engineering problems such as the design of pebble-bed nuclear reactors, ceramic processing, crude oil drilling, geothermal energy conversion, use of fibrous material in the thermal insulation of buildings, catalytic reactors and compact heat exchangers, heat transfer from storage of agricultural products which generate heat as a result of metabolism, petroleum reservoirs, storage of nuclear wastes, etc.

The derivation of the empirical equations which govern the flow and heat transfer in a porous medium has been discussed in [1-5]. The natural convection on vertical surfaces in porous media has been studied used Darcy's law by a number of authors [6–20]. Boundary layer analysis of natural convection over a cone has been investigated by Yih [21-24]. Murthy and Singh [25] obtained the similarity solution for non-Darcy mixed convection about an isothermal vertical cone with fixed apex half angle, pointing downwards in a fluid saturated porous medium with uniform free stream velocity, but a semi-similar solution of an unsteady mixed convection flow over a rotating cone in a rotating viscous fluid has been obtained Roy and Anilkumar [26]. The laminar steady nonsimilar natural convection flow of gases over an isothermal vertical cone has been investigated by Takhar et al. [27]. The development of unsteady mixed convection flow of an incompressible laminar viscous fluid over a vertical cone has been investigated by Singh and Roy [28] when the fluid in the external stream is set into motion impulsively, and at the same time the surface temperature is suddenly changed from its ambient temperature. An analysis has been carried out by Kumari and Nath [29] to study the non-Darcy natural convention flow of Newtonian fluids on a vertical cone embedded in a saturated porous medium with power-law variation of the wall temperature/concentration or heat/mass flux and suction/injection. Cheng [30-34] focused on the problem of natural convection from a vertical cone in a porous medium with mixed thermal boundary conditions, Soret and Dufour effects and with variable viscosity.

The conventional heat transfer fluids including oil, water and ethylene glycol etc. are poor heat transfer fluids, since the thermal conductivity of these fluids play an important role on the heat transfer coefficient between the heat transfer medium and the heat transfer surface. An innovative technique for improving heat transfer by using ultra fine solid particles in the fluids has been used extensively during the last several years. Choi [35] introduced the term nanofluid refers to these kinds of fluids by suspending nanoparticles in the base fluid. Khanafer et al. [36] investigated the heat transfer enhancement in a two-dimensional enclosure utilizing nanofluids. The convective boundary-layer flow over vertical plate, stretching sheet and moving surface studied by numerous studies and in the review papers Buongiorno [37], Daungthongsuk and Wongwises [38], Oztop [39], Nield and Kuznetsov [40,41], Ahmad and Pop [42], Khan and Pop [43], Kuznetsov and Nield [44,45] and Bachok et al. [46].

From literature survey the base aim of this work is to study the free convection boundary-layer flow past a vertical cone embedded in a porous medium filled with a nanofluid, the basic fluid being a non-Newtonian fluid by using similarity transformations. The reduced coupled ordinary differential equations are solved numerically. The effects of the parameters governing the problem are studied and discussed.

2. Mathematical formulation of the problem

Consider the problem of natural convection about a downward-pointing vertical cone of half angle γ embedded in a porous medium saturated with a non-Newtonian power-law nanofluid. The origin of the coordinate system is placed at the vertex of the full cone, with x being the coordinate along the surface of the cone measured from the origin and y being the coordinate perpendicular to the conical surface Fig (1). The temperature of the porous medium on the surface of the cone is kept at constant temperature T_w, and the ambient porous medium temperature is held at constant temperature T_∞.

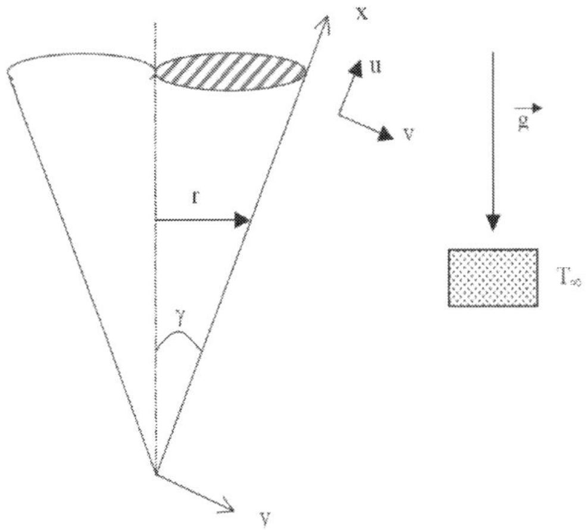

Figure 1. A schematic diagram of the physical model.

The nanofluid properties are assumed to be constant except for density variations in the buoyancy force term. The thermo physical properties of the nanofluid are given in Table 1 (see Oztop and Abu-Nada [39]). Assuming that the thermal boundary layer is sufficiently thin compared with the local radius, the equations governing the problem of Darcy flow through a homogeneous porous medium saturated with power-law nanofluid near the vertical cone can be written in two-dimensional Cartesian coordinates (x,y) as:

$$\frac{\partial(r^m u)}{\partial x}+\frac{\partial(r^m v)}{\partial y}=0, \tag{1}$$

$$\frac{\partial u^n}{\partial y}=\frac{(\rho\beta)_{nf}Kg\cos\gamma}{\mu_{nf}}\frac{\partial T}{\partial y}, \tag{2}$$

$$u\frac{\partial T}{\partial x}+v\frac{\partial T}{\partial y}=\alpha_{nf}\frac{\partial^2 T}{\partial y^2}. \tag{3}$$

Where u and v are the volume-averaged velocity components in the x and y directions, respectively, T is the volume-averaged temperature. n is the power-law viscosity index of the power-law nanofluid and g is the gravitational acceleration. $m=\gamma=0$ corresponds to flow over a vertical flat plate and $m=1$ corresponds to flow over a vertical cone. n is the viscosity index. For the case of $n=1$, the base fluid is Newtonian. We note that $n<1$ and $n>1$ represent pseudo-plastic fluid and dilatant fluid, respectively. Property ρ_{nf} and μ_{nf} are the density and effective viscosity of the nanofluid, and K is the modified permeability of the porous medium. Furthermore, α_{nf} and β_{nf} are the equivalent thermal diffusivity and the thermal expansion coefficient of the saturated porous medium, which are defined as (see Khanafer et al. [36]):

$$\rho_{nf}=(1-\phi)\rho_f+\phi\rho_s,\quad \mu_{nf}=\frac{\mu_f}{(1-\phi)^{2.5}},\quad \alpha_{nf}=\frac{k_{nf}}{(\rho C_p)_{nf}},$$
$$(\rho C_p)_{nf}=(1-\phi)(\rho C_p)_f+\phi(\rho C_p)_s,\quad \frac{k_{nf}}{k_f}=\frac{(k_s+2k_f)-2\phi(k_f-k_s)}{(k_s+2k_f)+2\phi(k_f-k_s)}. \tag{4}$$

Here ϕ is the solid volume fraction.

The associated boundary conditions of Eqs. (1)-(3) can be written as:

$$\begin{aligned} &v=0;\ T=T_w \quad \text{at } y=0;\\ &u=0;\ T\to T_\infty \quad \text{as } y\to\infty, \end{aligned} \tag{5}$$

where μ_f is the viscosity of the basic fluid, ρ_f and ρ_s are the densities of the pure fluid and nanoparticle, respectively, $(\rho C_p)_f$ and $(\rho C_p)_s$ are the specific heat parameters of the base fluid and nanoparticle, respectively, k_f and k_s are the thermal conductivities of the base fluid and nanoparticle, respectively. The local radius to a point in the boundary layer r can be represented by the local radius of the vertical cone $r = x\sin\gamma$.

	$\rho(kg/m^3)$	$C_p(J/kgK)$	$k(W/mK)$	$\beta\times10^5(K^{-1})$
Pure water	997.1	4179	0.613	21
Copper (Cu)	8933	385	401	1.67
Silver (Ag)	10500	235	429	1.89
Alumina (Al_2O_3)	3970	765	40	0.85
Titanium Oxide (TiO_2)	4250	686.2	8.9538	0.9

Table 1. Thermo-physical properties of water and nanoparticles [39].

By introducing the following non-dimensional variables:

$$\eta = \frac{y}{x} Ra_x^{1/2}, \qquad f(\eta) = \frac{\psi(x,y)}{\alpha_f r^m Ra_x^{1/2}}, \qquad \theta(\eta) = \frac{T - T_\infty}{T_w - T_\infty}. \tag{6}$$

The continuity equation is automatically satisfied by defining a stream function $\psi(x,y)$ such that:

$$r^m u = \frac{\partial\psi}{\partial y} \text{ and } r^m v = -\frac{\partial\psi}{\partial x}. \tag{7}$$

where;

$$Ra_x = \left(\frac{x}{\alpha_f}\right)\left[\frac{Kg(\rho\beta)_f \cos\gamma (T_w - T_\infty)}{\mu_f}\right]^{1/n}. \tag{8}$$

Integration the momentum Eq. (2) we have:

$$\frac{\mu_{nf}}{\mu_f} u^n = \frac{(\rho\beta)_{nf} Kg\cos\gamma}{\mu_f}(T - T_\infty). \tag{9}$$

Substituting variables (6) into Eqs. (1)–(5) with Eq. (9), we obtain the following system of ordinary differential equations:

$$\frac{1}{(1-\phi)^{2.5}}(f')^n = \left[1 - \phi + \phi\frac{(\rho\beta)_s}{(\rho\beta)_f}\right]\theta, \tag{10}$$

$$\frac{k_{nf}/k_f}{\left[1-\phi+\phi\frac{\left(\rho C_p\right)_s}{\left(\rho C_p\right)_f}\right]}\theta''+\left(m+\frac{1}{2}\right)f\theta'=0, \tag{11}$$

along with the boundary conditions:

$$\begin{aligned} &f(0)=0,\ \theta(0)=1, \\ &f'(\infty)=0,\ \theta(\infty)=1. \end{aligned} \tag{12}$$

where primes denote differentiation with respect to η , the quantity of practical interest, in this chapter is the Nusselt number Nu_x which is defined in the form:

$$Nu_x=\frac{hx}{k_m}=\frac{-\left.\frac{\partial T}{\partial y}\right|_{y=0}x}{T_w-T_\infty}=-Ra_x^{1/2}\theta'(0). \tag{13}$$

where h denotes the local heat transfer coefficient.

3. Results and discussion

In this study we have presented similarity reductions for the effect of a nanoparticle volume fraction on the free convection flow of nanofluids over a vertical cone via similarity transformations. The numerical solutions of the resulted similarity reductions are obtained for the original variables which are shown in Eqs. (10) and (11) along with the boundary conditions (12) by using the implicit finite-difference method. The physical quantity of interest here is the Nusselt number Nu_x and it is obtained and shown in Eqs. (13) and (14). The distributions of the velocity $f'(\eta)$, the temperature $\theta(\eta)$ from Eqs.(10) and (11) and the Nusselt number in the case of Cu-water and Ag-water are shown in Figs. 2–8. The computations are carried for various values of the nanoparticles volume fraction for different types of nanoparticles, when the base fluid is water. Nanoparticles volume fraction ϕ is varied from 0 to 0.3. The nanoparticles used in the study are from Copper (Cu), Silver (Ag), Alumina (Al_2O_3) and Titanium oxide (TiO_2).

In order to verify the accuracy of the present method, we have compared our results with those of Yih [22] for the rate of heat transfer $\theta'(0)$ in the absence of the nanoparticles $\left(\phi=0\right)$. The comparisons in all the above cases are found to be in excellent agreement, as shown in Table 2. It is clear that as a geometry shape parameter m increases, the local Nusselt number increases. While Table 3 depict the heat transfer rate $\theta'(0)$ for various values of nanoparticles volume fraction ϕ for different types of nanoparticles when the base fluid is water. Figs. 2 and 3 show the effects of the nanoparticle volume fraction φ on the velocity distribution in the case of Cu-water when $\phi=0,0.05,0.1,0.15,0.2,0.3$. It is noted that the velocity along the cone increases with the nanoparticle volume fraction in both of the two cases (i.e. Cu-water and Ag-water), moreover the velocity distribution in the case of Ag-water is larger than that for Cu-water. We can show that the change of the velocity distribution when we use different types of nanoparticles from Fig. 4, which depict the Ag-nanoparticles are the highest when the base fluid is water and when $\phi = 0.1$. Thus the presence of the nanoparticles volume fraction increases the momentum boundary layer thickness.

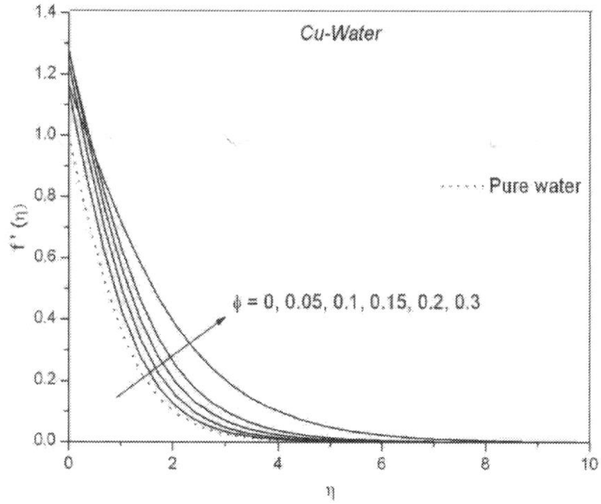

Figure 2. Effects of the nanoparticle volume fraction ϕ on velocity distribution $f'(\eta)$ in the case of Cu-Water.

n	Vertical plate		Vertical cone	
	Yih [22]	Present method	Yih [22]	Present method
0.5	0.3766	0.3768	0.6522	0.6524
0.8	0.4237	0.4238	0.7339	0.7340
1.0	0.4437	0.4437	0.7686	0.7686
1.5	0.4753	0.4752	0.8233	0.8233
2.0	0.4938	0.4938	0.8552	0.8552

Table 2. Comparison of results for the reduced Nusselt number $-\theta'(0)$ for vertical plate $(\lambda = 0)$ and vertical cone $(\lambda = 1)$ when $\phi = 0$.

ϕ	Cu	Ag	Al_2O_3	TiO_2
0.05	0.7423	0.7704	0.6604	0.6725
0.1	0.6931	0.7330	0.5642	0.5852
0.15	0.6301	0.6732	0.4780	0.5057
0.2	0.5591	0.6002	0.4006	0.4331
0.3	0.4052	0.4357	0.2673	0.3062

Table 3. Values of $-\theta'(0)$ for various values of ϕ when $n = 1$.

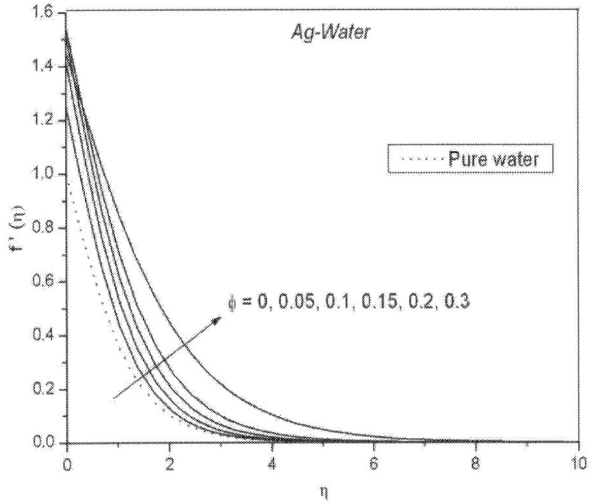

Figure 3. Effects of the nanoparticle volume fraction ϕ on velocity distribution $f'(\eta)$ in the case of Ag-Water.

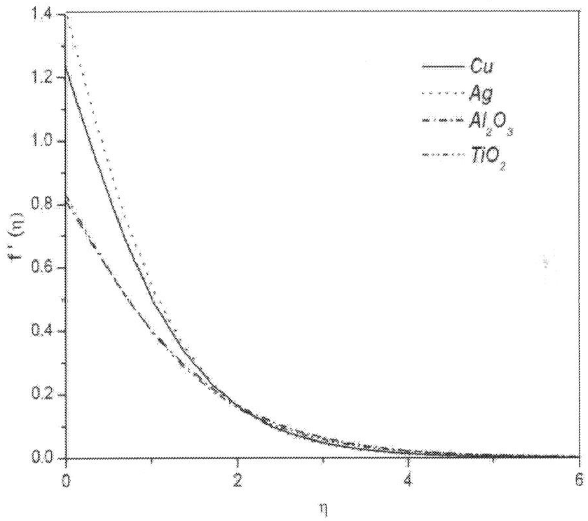

Figure 4. Velocity profiles $f'(\eta)$ for different types of nanofluids when $\phi = 0.1$.

Figs. 5 and 6 are presented to show the effect of the volume fraction of nanoparticles Cu and Ag respectively, on temperature distribution. These figures illustrate the streamline for different values of ϕ, when the volume fraction of the nanoparticles increases from 0 to 0.3, the thermal boundary layer is increased. This agrees with the physical behavior, when the volume of copper and silver nanoparticles increases the thermal conductivity increases, and then the thermal boundary layer thickness increases. Moreover Fig. 7 displays the behavior of the different types of nanoparticles on temperature distribution when $\phi = 0.1$. The figure showed that by using different types of nanofluid as the values of the temperature change and the Ag-nanoparticles are the lower distribution.

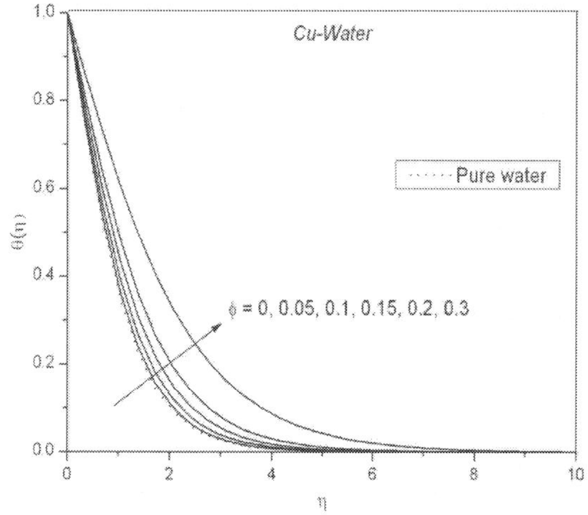

Figure 5. Effects of the nanoparticle volume fraction ϕ on temperature distribution $\theta(\eta)$ in the case of Cu-Water.

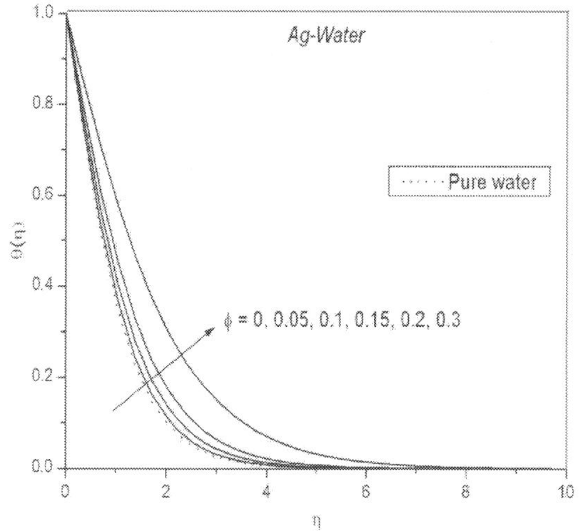

Figure 6. Effects of the nanoparticle volume fraction ϕ on temperature distribution $\theta(\eta)$ in the case of Ag-Water.

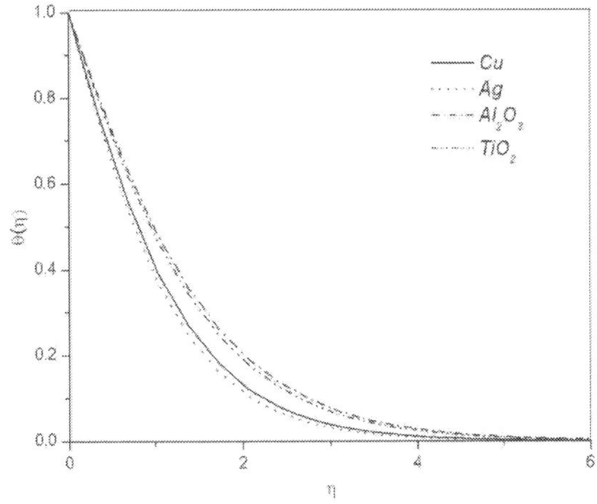

Figure 7. Temperature profiles $\theta(\eta)$ for different types of nanofluids when $\phi = 0.1$.

Fig. 8 shows the variation of the reduced Nusselt number with the nanoparticles volume fraction ϕ for the selected types of the nanoparticles. It is clear that the heat transfer rates decrease with the increase in the nanoparticles volume fraction ϕ. The change in the reduced Nusselt number is found to be lower for higher values of the parameter ϕ. It is observed that the reduced Nusselt number is higher in the case of Ag-nanoparticles and next Cu-nanoparticles, TiO_2-nanoparticles and Al_2O_3-nanoparticles. Also, the Fig. 8 and Table 3 show that the values of $\theta'(0)$ change with nanofluid changes, namely we can say that the shear stress and heat transfer rate change by taking different types of nanofluid. Furthermore this depicts that the nanofluids will be very important materials in the heating and cooling processes.

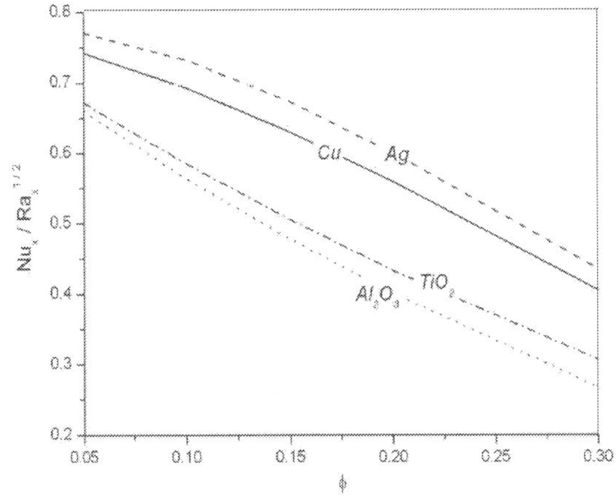

Figure 8. Effects of the nanoparticle volume fraction ϕ on dimensionless heat transfer rates.

4. Conclusions

The problem of the steady free convection boundary layer flow past a vertical cone embedded in porous medium filled with a non-Newtonian nanofluid has been studied and the special case when the base fluid is water has been considered. The effects of the solid volume fraction ϕ on the flow and heat transfer characteristics are determined for four types of nanofluids: Copper (Cu), Silver (Ag), Alumina (Al_2O_3) and Titanium oxide (TiO_2). It has been shown, as expected, that increasing of the values of the nanoparticles volume fraction lead to an increase of the velocity and the temperature profiles and to an decrease of the Nusselt number for the values of the parameter ϕ. It has been found that the Ag-nanoparticles proved to have the highest cooling performance and Alumina-nanoparticles enhanced to have highest heating performance for this problem.

Nomenclature

C_p	specific heat at constant temperature
f	dimensionless stream function
g	acceleration due to gravity
h	local heat transfer coefficient
K	permeability coefficient of the porous medium
k	thermal conductivity
m	geometry shape parameter
Nu_x	reduced Nusselt number
n	viscosity index, $n \geq 0$
Ra_x	modified Rayleigh number
r	local radius of the cone
T	temperature
T_w	temperature at the surface of the cone
T_∞	ambient temperature attained as $y \to \infty$
u,v	Darcian velocity components in x - and y -directions
x,y	Cartesian coordinates

Greek symbols

α	thermal diffusivity
β	volumetric expansion coefficient
γ	half angle of the cone
η	similarity variable
θ	dimensionless temperature
μ	effective viscosity

ρ_f	density of the fluid
ρ_s	nanoparticles mass density
$(\rho C_p)_{nf}$	heat capacitance of the nanofluid
$(\rho C_p)_f$	heat capacity of the fluid
$(\rho C_p)_s$	effective heat capacity of the nanoparticles material
ϕ	nanoparticles volume fraction
ψ	stream function

Subscripts

f	fluid fraction
nf	nanofluid fraction
s	solid fraction
w	condition at the wall
∞	stream function condition at the infinity

Author details

F.M. Hady
Department of Mathematics, Faculty of Science, Assiut University, Assiut, Egypt

S.M. Abdel-Gaied and M.R. Eid*
Department of Science and Mathematics, Faculty of Education, Assiut University, The New Valley, Egypt

F.S. Ibrahim
Department of Mathematics, University College in Jamoum, Umm Al-Qura University, Makkah, Saudi Arabia

5. References

[1] Cheng P (1978) Heat transfer in geothermal systems. *Adv. Heat Transfer* 14: 1–105.
[2] Nield DA, Bejan A (1999) Convection in Porous Media. second ed., *Springer: New York.*
[3] Vafai K (2000) Handbook of Porous Media. *Marcel Dekker: New York.*
[4] Pop I, Ingham DB (2001) Convective Heat Transfer: Mathematical and Computational Modelling of Viscous Fluids and Porous Media. *Pergamon Press: Oxford .*
[5] Ingham DB, Pop I (2002) Transport Phenomena in Porous Media. *Pergamon Press: Oxford*
[6] Cheng P, Minkowycz WJ (1977) Free convection about a vertical flat plate embedded in a porous medium with application to heat transfer from a dike. *J. Geophys. Res.* 82(14): 2040–2044.

* Corresponding Author

[7] Gorla RSR, Tornabene R (1988) Free convection from a vertical plate with nonuniform surface heat flux and embedded in a porous medium. *Transp. Porous Media* 3 : 95–106.

[8] Bakier AY, Mansour MA, Gorla RSR, Ebiana AB (1997) Nonsimilar solutions for free convection from a vertical plate in porous media. *Heat Mass Transfer* 33: 145–148.

[9] Gorla RSR, Mansour MA, Abdel-Gaied SM (1999) Natural convection from a vertical plate in a porous medium using Brinkman's model. *Transp. Porous Media* 36 : 357–371.

[10] Mulolani I, Rahman M (2000) Similarity analysis for natural convection from a vertical plate with distributed wall concentration. *Int. J. Math. math. Sci.* 23(5) : 319 - 334.

[11] Jumah RY, Mujumdar AS (2000) Free convection heat and mass transfer of non-Newtonian power law fluids with yield stress from a vertical plate in saturated porous media. *Int. Comm. Heat Mass Transfer* 27 : 485–494.

[12] Groşan T, Pop I (2001) Free convection over a vertical flat plate with a variable wall temperature and internal heat generation in a porous medium saturated with a non-Newtonian fluid. *Technische Mechanik* 21(4) : 313–318.

[13] Kumaran V, Pop I (2006) Steady free convection boundary layer over a vertical flat plate embedded in a porous medium filled with water at 4°C. *Int. J. Heat Mass Transfer* 49 : 3240–3252.

[14] Magyari E, Pop I, Keller B (2006) Unsteady free convection along an infinite vertical flat plate embedded in a stably stratified fluid-saturated porous medium. *Transp. Porous Media* 62 : 233–249.

[15] Cheng C-Y (2006) Natural convection heat and mass transfer of non-Newtonian power law fluids with yield stress in porous media from a vertical plate with variable wall heat and mass fluxes. *Int. Comm. Heat Mass Transfer* 33 : 1156–1164.

[16] Chamkha AJ, Al-Mudhaf AF, Pop I (2006) Effect of heat generation or absorption on thermophoretic free convection boundary layer from a vertical flat plate embedded in a porous medium. *Int. Comm. Heat Mass Transfer* 33 : 1096–1102.

[17] Magyari E, Pop I, Postelnicu A (2007) Effect of the source term on steady free convection boundary layer flows over an vertical plate in a porous medium. Part I. *Transp. Porous Media* 67 : 49–67.

[18] Nield DA, Kuznetsov AV (2008) Natural convection about a vertical plate embedded in a bidisperse porous medium. *Int. J. Heat Mass Transfer* 51 : 1658–1664.

[19] Mahdy A, Hady FM (2009) Effect of thermophoretic particle deposition in non-Newtonian free convection flow over a vertical plate with magnetic field effect. *J. Non-Newtonian Fluid Mech.* 161: 37–41.

[20] Ibrahim FS, Hady FM, Abdel-Gaied SM, Eid MR (2010) Influence of chemical reaction on heat and mass transfer of non-Newtonian fluid with yield stress by free convection from vertical surface in porous medium considering Soret effect. *Appl. Math. Mech. - Engl. Ed.* 31(6): 675–684.

[21] Yih KA (1997) The effect of uniform lateral mass flux effect on free convection about a vertical cone embedded in a saturated porous medium. *Int. Comm. Heat Mass Transfer* 24(8): 1195–1205.

[22] Yih KA (1998) Uniform lateral mass flux effect on natural convection of non-Newtonian fluids over a cone in porous media. *Int. Comm. Heat Mass Transfer* 25(7): 959–968.

[23] Yih KA (1999) Effect of radiation on natural convection about a truncated cone. *Int. J. Heat Mass Transfer* 42: 4299 - 4305.
[24] Yih KA (1999) Coupled heat and mass transfer by free convection over a truncated cone in porous media: VWT/VWC or VHF/VMF. *Acta Mech.* 137 : 83–97.
[25] Murthy PVSN, Singh P (2000) Thermal dispersion effects on non-Darcy convection over a cone. *Compu. Math. Applications* 40: 1433 - 1444.
[26] Roy S, Anilkumar D (2004) Unsteady mixed convection from a rotating cone in a rotating fluid due to the combined effects of thermal and mass diffusion. *Int. J. Heat Mass Transfer* 47:1673–1684.
[27] Takhar HS, Chamkha AJ, Nath G (2004) Effect of thermophysical quantities on the natural convection flow of gases over a vertical cone. *Int. J. Eng. Sci.* 42: 243–256.
[28] Singh PJ, Roy S (2007) Unsteady mixed convection flow over a vertical cone due to impulsive motion. *Int. J. Heat Mass Transfer* 50: 949–959.
[29] Kumari M, Nath G (2009) Natural convection from a vertical cone in a porous medium due to the combined effects of heat and mass diffusion with non-uniform wall temperature/concentration or heat/mass flux and suction/injection. *Int. J. Heat Mass Transfer* 52: 3064–3069.
[30] Cheng C-Y (2009) Natural convection heat transfer of non-Newtonian fluids in porous media from a vertical cone under mixed thermal boundary conditions. *Int. Comm. Heat Mass Transfer* 36: 693 - 697.
[31] Cheng C-Y (2009) Soret and Dufour effects on natural convection heat and mass transfer from a vertical cone in a porous medium, *Int. Comm. Heat Mass Transfer* 36 : 1020 - 1024.
[32] Cheng C-Y (2010) Soret and Dufour effects on heat and mass transfer by natural convection from a vertical truncated cone in a fluid-saturated porous medium with variable wall temperature and concentration. *Int. Comm. Heat Mass Transfer* 37 : 1031 - 1035.
[33] Cheng C-Y (2010) Soret and Dufour effects on heat and mass transfer by natural convection from a vertical truncated cone in a fluid-saturated porous medium with variable wall temperature and concentration. *Int. Comm. Heat Mass Transfer* 37 : 1031 - 1035.
[34] Cheng C-Y (2010) Nonsimilar boundary layer analysis of double-diffusive convection from a vertical truncated cone in a porous medium with variable viscosity. *Int. Comm. Heat Mass Transfer* 37: 1031 - 1035.
[35] Choi SUS (1995) Enhancing thermal conductivity of fluid with nanoparticles. developments and applications of non-Newtonian flow. *ASME FED 231/MD* 66 : 99–105.
[36] Khanafer K, Vafai K, Lightstone M (2003) Buoyancy-driven heat transfer enhancement in a two-dimensional enclosure utilizing nanofluids. *Int. J. Heat Mass Transfer* 46 : 3639–3653.
[37] Buongiorno J (2006) Convective transport in nanofluids. *ASME J. Heat Transfer* 128 : 240–250.

[38] Daungthongsuk W, Wongwises S (2007) A critical review of convective heat transfer nanofluids. *Ren. Sustainable Energy Rev.* 11: 797–817.

[39] Oztop HF, Abu-Nada E (2008) Numerical study of natural convection in partially heated rectangular enclosures filled with nanofluids. *Int. J. Heat Fluid Flow* 29: 1326–1336.

[40] Nield DA, Kuznetsov AV (2009) The Cheng–Minkowycz problem for natural convective boundary-layer flow in a porous medium saturated by a nanofluids. *Int. J. Heat Mass Transfer* 52: 5792–5795.

[41] Nield DA, Kuznetsov AV (2011) The Cheng–Minkowycz problem for the double-diffusive natural convective boundary-layer flow in a porous medium saturated by a nanofluids. *Int. J. Heat Mass Transfer* 54: 374–378.

[42] Ahmad S, Pop I (2010) Mixed convection boundary layer flow from a vertical flat plate embedded in a porous medium filled with nanofluids. *Int. Comm. Heat Mass Transfer* 37: 987 - 991.

[43] Khan WA, Pop I (2010) Boundary-layer flow of a nanofluid past a stretching sheet. *Int. J. Heat Mass Transfer* 53: 2477–2483.

[44] Kuznetsov AV, Nield DA (2010) Natural convective boundary-layer flow of a nanofluid past a vertical plate. *Int. J. Thermal Sci.* 49: 243 - 247.

[45] Kuznetsov AV, Nield DA (2010) Effect of local thermal non-equilibrium on the onset of convection in a porous medium layer saturated by a nanofluid. *Transp. Porous Media* 83: 425–436.

[46] Bachok N, Ishak A, Pop I (2010) Boundary-layer flow of nanofluids over a moving surface in a flowing fluid. *Int. J. Thermal Sci.* 49: 1663 - 1668.

Chapter 5

Natural Convection Heat Transfer from a Rectangular Block Embedded in a Vertical Enclosure

Xiaohui Zhang

Additional information is available at the end of the chapter

http://dx.doi.org/10.5772/52666

1. Introduction

In many circumstances of practical concern, thermal sources are encapsulated into closed cavities containing a fluid, such as in the case of fuel tanks. In other applications it is the heat source itself which needs to be thermally controlled, such as in electronic packaging, passive cooling, space heating, nuclear design, and geophysics; another example is the natural convection around a horizontally-placed or vertically-positioned radiator, which can be used for a centralized heating and cooling system to regulate the air temperature in a cavity. The location of the radiator affects the temperature distribution and heat transfer in a cavity. Normally, a higher position of the radiator is reasonable when the radiator is used as a cooling device, while a lower position for the radiator is reasonable when the radiator is used for a heating one. It is worth studying the temperature distribution and heat transfer in thermal management and design. Whether the radiator is used as a cooling device or a heating device, the heat transfer of a object in a cavity can be simplified and dominated by natural convection heat transfer mechanism in an enclosure with an isolated plate.

Shyy and Rao [1] conducted an investigation of transient natural convection around an enclosed vertical plate. Numerical simulations and experimental data of natural convection air cooling of an array of two-dimensional discrete flush heaters on a vertical wall of a rectangular enclosure were performed by Ho and Chang [2]. Yang and Tao [3] developed a computational method to deal with the internal isolated islands (set the main diagonal element coefficient big values in velocity discrete equations)for natural convection in an enclosure. Experimental work and numerical simulation were studied by Wang [4] regarding natural convection in an inclined cube enclosure with multiple internal isolated plates. Also, numerical analysis on a 3×3 array of discrete heat sources flush-mounted on one vertical wall of a rectangular enclosure filled with various liquids was done by Tou and

Tso [5]. Natural convection from a discrete bottom flush-mounted rectangular heat source on bottom of a horizontal enclosure was studied by Sezai and Mohamad [6]. Deng et al. [7] investigated numerically the steady state natural convection induced by multiple discrete heat sources (DHSs) in horizontal enclosures. Unsteady convection numerical modelling in a vertical channel with a square cylinder was studied by Saha [8], Static bifurcation was found by Liu and Tao [9,10] who performed numerical computations for the heat transfer and fluid flow characteristics of an internal vertical channel composed by a pair of parallel plates situated in an enclosure. Barozzi and Corticelli [11,12] investigated the two-dimensional buoyant flow in a closed cabinet containing two vertical heating plates with a time-accurate finite method. The predictions showed the long-term behavior of numerical solution is time-dependent. The studies mentioned above have not been concerned with the effect of location for heat source or heat sink on the fluid flow and heat transfer.

Following the pioneering numerical works mentioned above, the present study represents a further effort to extend the studies with numerical simulation. The main objective of this study is to analyze the variation effect of the horizontal and vertical location ratios, a/H and b/H, respectively (defined in the following section) for different cold and hot isolated vertical plates in an enclosure, with respect to the flow configuration, temperature distribution, temperature difference distribution (defined in the following section) and heat transfer characteristics of the natural convection. Effect of Rayleigh number on the fluid flow and heat transfer is also presented.

2. Physical model and numerical method

The physical configuration and boundary conditions of problems investigated in this study are shown in Fig. 1. The two horizontal walls are considered to be insulated, and the two vertical walls which have the same temperatures, as well as the isolated vertical plate are maintained at T_1 and T_2, respectively. In this study, the investigations are carried out through the variation of horizontal location ratio a/H and vertical location ratio b/H in the cases of $T_1 > T_2$ (in this case $T_1 = T_h, T_2 = T_c$) and $T_1 < T_2$ (in this case $T_1 = T_c, T_2 = T_h$).

In the present model, the flow is simulated as a two dimensional phenomenon with the following assumptions or simplifications: a) the fluid (air) is Newtonian, incompressible and the flow is laminar, and; b) the temperature difference $T_1 - T_2$ is small, so that the effect of temperature on fluid density is expressed adequately by the Boussinesq approximation. Next, we consider the following dimensionless variables:

$$X = \frac{x}{H}; Y = \frac{y}{H}; U = \frac{u}{U_R}; V = \frac{v}{U_R}; P = \frac{p}{\rho U_R^2}; \Theta = \frac{T - T_c}{T_h - T_c}; \mathrm{Pr} = \frac{\nu}{a}; Ra = \frac{g\beta H^3 \Delta T}{\nu a}.$$

where the reference velocity is defined as $U_R = (Ra\,\mathrm{Pr})^{1/2} \mathrm{a} / \mathrm{H}$.

The governing equations, that express the conservation of mass, momentum and energy in the fluid domain, become:

$$\frac{\partial U}{\partial X}+\frac{\partial V}{\partial Y}=0 \tag{1}$$

$$U\frac{\partial U}{\partial X}+V\frac{\partial U}{\partial Y}=-\frac{\partial P}{\partial X}+\frac{Pr}{(RaPr)^{1/2}}\cdot\nabla^2 U \tag{2}$$

$$U\frac{\partial V}{\partial X}+V\frac{\partial V}{\partial Y}=-\frac{\partial P}{\partial Y}+\frac{Pr}{(RaPr)^{1/2}}\cdot\nabla^2 V+\Theta \tag{3}$$

$$U\frac{\partial \Theta}{\partial X}+V\frac{\partial \Theta}{\partial Y}=\frac{K}{(RaPr)^{1/2}}\cdot\nabla^2 \Theta \tag{4}$$

where $\nabla^2=\partial^2/\partial X^2+\partial^2/\partial Y^2$ and $K=k/k_f$, with k and k_f being the vertical block location conductivity and thermal conductivity of the fluid, respectively.

No-slip condition is imposed on all the walls for the velocities. Thermal boundary conditions are that $\partial\Theta/\partial Y=0$ for the horizontal insulated walls, and $\Theta=1$ for the inner heat source and $\Theta=0$ for the vertical cold walls or $\Theta=0$ for the inner heat sink and $\Theta=1$ for the vertical hot walls.

The average Nusselt number is given below:

$$Nu=\frac{\mathrm{H}}{2(\mathrm{H}+\mathrm{W})}\left(\int_0^{\mathrm{L/H}}-\frac{\partial T}{\partial X}\Big|_{\mathrm{X=0}}dY+\int_0^{\mathrm{L/H}}\frac{\partial T}{\partial X}\Big|_{\mathrm{X=L/H}}dY\right) \tag{5}$$

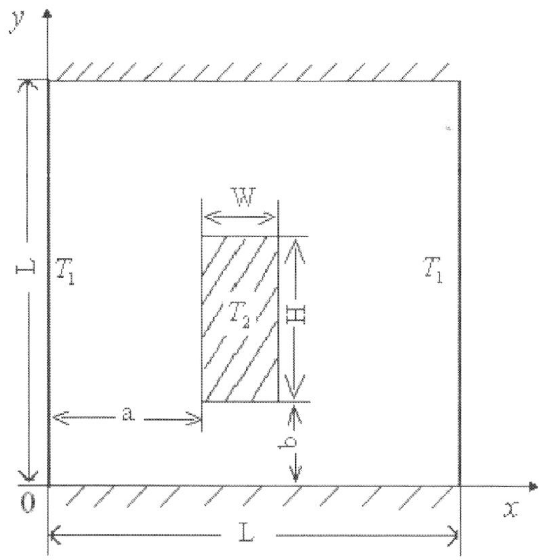

Figure 1. Schematic diagram of configuration

Equations (1) to (5) are solved using a finite volume method (FVM) on a staggered grid system [13]. In the course of discretization, QUICK scheme is adopted to deal with convection and diffusion terms. The equations from the discretization of Eqs. (1) to (4) are solved by the line-by-line procedure, combining the tri-diagonal matrix algorithm (TDMA) and successive over-relaxation iteration (SOR) and the Gauss-Seidel iteration technique with additional block-correction method for fast convergence. The SIMPLE algorithm [13] is used to treat the coupling of the momentum and energy equations. Pressure–correction and velocity-correction schemes are implemented in the model algorithm to arrive at converged solution when both the pressure and velocity satisfy the momentum and continuity equations. The solution is considered to converge when the sum of the normalized residuals for each control equation is of order 10^{-6} .

Special attention is paid to treatment of the isolated solid region. The presence of isolated area is accounted for by a strategy [3, 14] in which a part of solution domain is located in the flow field, therefore, the governing Eqs. (1) to (4) apply to both the fluid and the solid regions. Both the velocity and the dimensionless temperature in it remained zero in iteration process. For details, Ref.[3] may be consulted.

Non-uniform staggered grid system is employed with denser grids clustering near the plate and walls so as to resolve the boundary layer properly. Test runs are performed on a series of non-uniform grids to determine the grids size effects for the Rayleigh numbers 10^4, $10^5$and 10^6 at grid systems 40×40, 60×60 and 80×80, respectively. For each calculation case, a grid independent resolution is obtained. The maximum difference in average Nusselt number between grid (40×40) and grid (60×60) is 5%, the difference in average Nusselt number between grid (60×60) and grid (80×80) is less than 0.2%, so the 60×60 non-uniform grids are used.

The developed computational model is validated against benchmark computational results and is also compared with experimental data. Accuracy of the numerical procedure is validated by the comparison between the predicted results with the benchmark solutions of de Vahl Davis [15] for pure natural convection model in a square cavity with opposite heated and cooled side walls. As shown in Table 1, good agreements are achieved for both the maximum velocity and the average Nusselt number in a broad range of Rayleigh numbers Ra=10^4 to 10^6. Another comparison is also made with respect to the experimental work of Wang [4]. The computationally obtained flow patterns for $Ra = 2\times10^5$ are compared with the flow visualization in Fig. 2. It is seen that the model adequately predicts the flow patterns obtained in the visualizations.

Ra	Present		Benchmark	
	U_{max}	Nu	U_{max}	Nu
10^4	0.192	2.231	0.193	2.245
10^5	0.129	4.50	0.132	4.510
10^6	0.078	8.817	0.077	8.806

Table 1. Comparisons between the predicted results and the benchmark resolutions of de Vahl Davis[15]

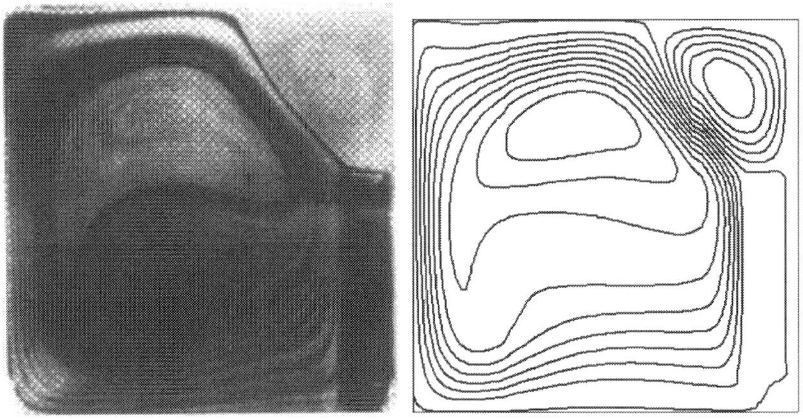

(a) (b)

Figure 2. Flow field (a) Flow visualization [4], (b) Computation

3. Results and discussion

The configuration dimension is in the proportion of $L:H:W = 80:15:3$ .We will examine the fluid flow and heat transfer characteristics under the circumstances of heating or cooling at various locations, followed by the dependence to the Rayleigh number.

3.1. Effect of location on the fluid flow and heat transfer

Numerical simulations have been conducted to elucidate the effect of location ratios variation a/H $(0.3 \leq a/H \leq 2.7,\ b/H = 0.3)$ and b/H $(0.3 \leq b/H \leq 4.0,\ a/H = 0.3)$ on the natural convection of a cold isolated vertical plate and hot isolated vertical plate in an enclosure, Rayleigh number is fixed at $Ra = 2.5 \times 10^4$ in both cases.

3.2. Flow and temperature fields

The buoyancy-driven flow and temperature fields in the enclosure for a cold plate and a hot plate with the variation of location ratios a/H and b/H are illustrated by means of contour maps of streamlines and isotherms, respectively, as exemplified in Figs. 3 to 6.

For Fig.3 (a-f), there are two vortices flow where the directions are different for cold plate and hot plate. With the increase of horizontal location ratio a/H near the middle of the enclosure, the two vertices structure around the vertical plate tend to be more symmetric for cold plate and hot plate. The spaces at the two sides of the plate are large enough for the flow within it to develop, so buoyant convection flow is strong on the upper surface of hot plate, indicating the strong effect of the natural convection, while convection is weak in the case of cold plate relatively.

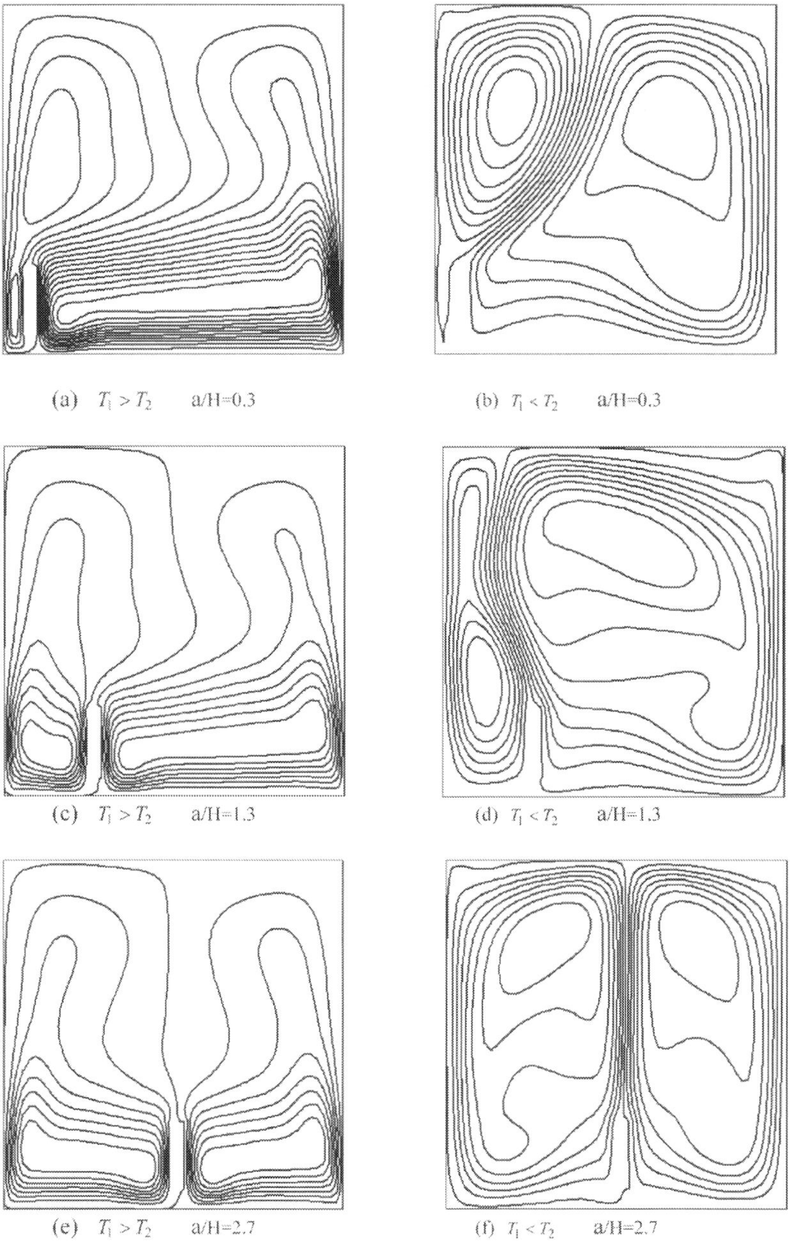

Figure 3. Streamlines for different horizontal locations at $Ra = 2.5 \times 10^4$

Dimensionless temperature distributions, plotted against the variation of the horizontal location ratio a/H, are displayed in Fig.4 (a-f). From this plot, it can be seen that isotherms near walls are almost vertical up to the upper and lower walls. This is due to the conduction effect. The figures reveal also that dimensionless temperature distributions of cold plate are more stratified than those of the hot plate.

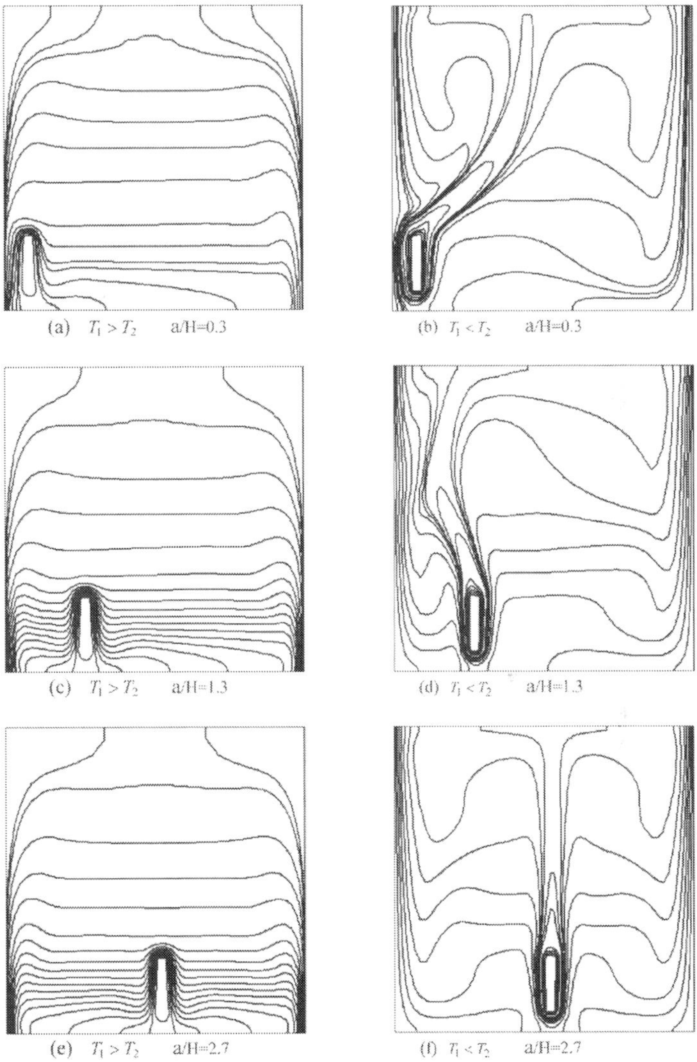

Figure 4. Isotherms for different horizontal locations at $Ra = 2.5 \times 10^4$

For Fig. 3(a and b), Fig. 5 and Fig. 6, we observe significantly different fluid flow and temperature distribution phenomena: the larger of the vertical location ratio b/H, the stronger the flow below the cold plate surface, and the more stratified temperature distribution for the hot plate.

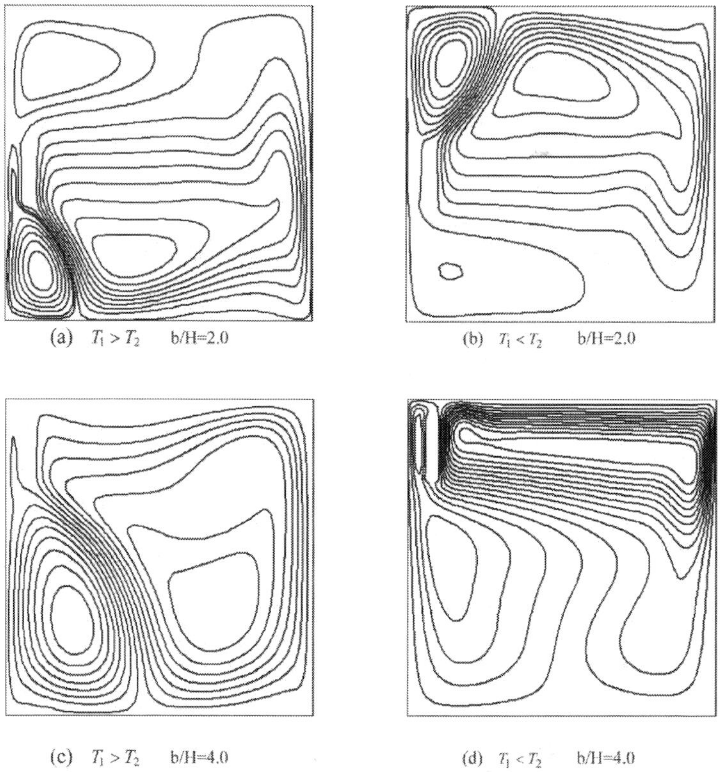

Figure 5. Streamlines for different vertical locations at $Ra = 2.5 \times 10^4$

It is apparent from the comparison of fluid flow configurations and dimensionless temperature distributions of the cold plate and hot plates that the vertical location ratio b/H has a substantial effect on the flow configuration and temperature profile.

3.3. Temperature difference distribution

Uniform temperature difference in an enclosure may have practical importance, particularly in electronics and air conditioning. In order to describe the uniformity of temperature difference in an enclosure we define TD, which means difference of node temperature and average temperature where average temperature is obtained by a weighted area method.

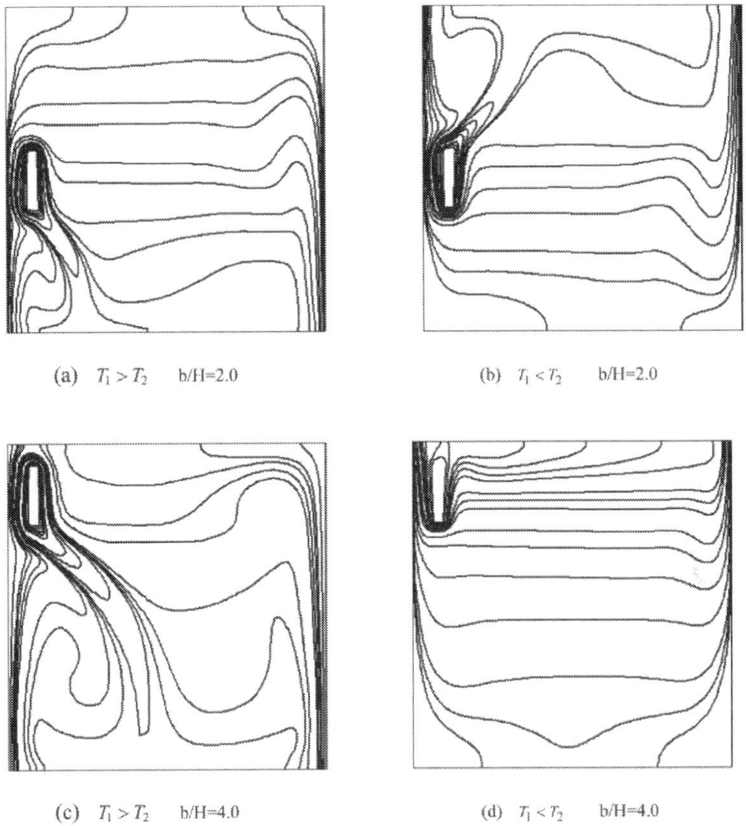

Figure 6. Isotherms for different vertical locations at $Ra = 2.5 \times 10^4$

$$\mathrm{TD} = t_{ij} - \frac{\sum t_{ij} A_{ij}}{\sum A_{ij}} \tag{6}$$

Two cases are studied at $Ra = 2.5 \times 10^4$, case 1: a/H =0.3 and b/H=0.3, and case2: a/H =0.3 and b/H=2.0. Figure 7 shows the temperature difference distribution, marked as TD and plotted against the height of enclosure, while the width is fixed for heating and cooling conditions.

It can be seen that, in case 1, the temperature difference of heating condition along the height of enclosure is more uniform (the maximum dimensionless temperature difference is 0.11) than that for cooling condition (the maximum dimensionless temperature difference is 0.48). While with the increase of height, the dimensionless temperature differences between heating condition (the maximum dimensionless temperature difference is 0.203) and cooling condition (the maximum dimensionless temperature difference is 0.245) are not great along the height of enclosure in case 2. It implies that the requirement of heating and cooling in case 2 for same component can be met.

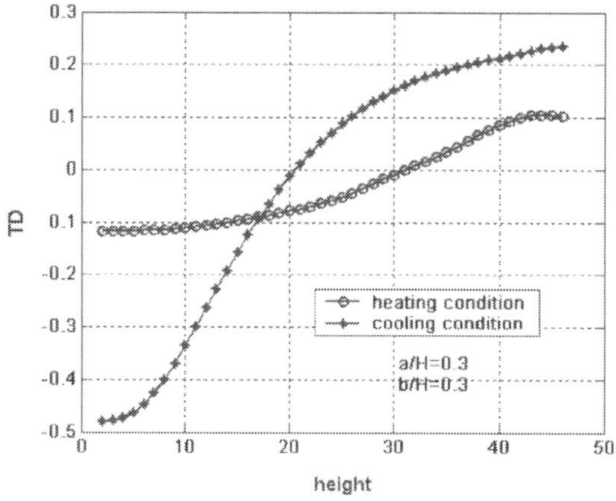

(a) case1

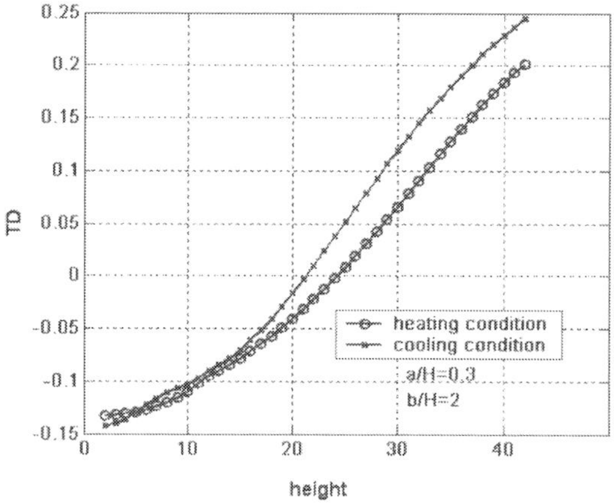

(b) case2

Figure 7. Variation of temperature difference with the height of enclosure for heating and cooling condition

3.4. Heat transfer

Next, attention is focused on the role which the location ratio can play in the heat transfer rate. Figures 8 and 9 present results of the average Nusselt number defined by Equation (5).

For both the cold and the hot plates, b/H is fixed at 0.3. Results show that the same trend occurs with the increase of a/H for both cold plate and hot plate. When the plate is within a wide range of enclosure values, $0.3< a/H <4.8$, the effect of its location on heat transfer is small. The variation character of Nu vs. a/H implies that in a wide range of a/H, the spaces at the two sides of the plate are large enough for the flow within it to develop, therefore, the average heat transfer rate is not sensitive to the distance from the side wall. Indeed, the average hot plate Nusselt number is 20% to 39% higher as compared to that of the cold plate. The fact that the average Nusselt number of the hot plate is higher than the corresponding cold plate, can be explained if we consider the streamlines shown in Fig. 3 and temperature profiles in Fig. 4. From these figures we note that the natural convection occurs easier for the hot plate, due to a higher heat transfer than for the cold plate under the same boundary conditions. The steep increase of average Nusselt number near walls for the cold and hot plates in Fig. 8 is referred to as the "chimney effect". The present results confirm the fact that the stronger chimney effect enhances the heat transfer [16, 17]. For Pr ≈1, the boundary layer thickness for the plate scales as [18],

$$\delta = \delta_T \approx y\left(\frac{g\beta\Delta T y^3}{a\nu}\right)^{-1/4} \tag{7}$$

Which indicates that δ increases as $y^{1/4}$.

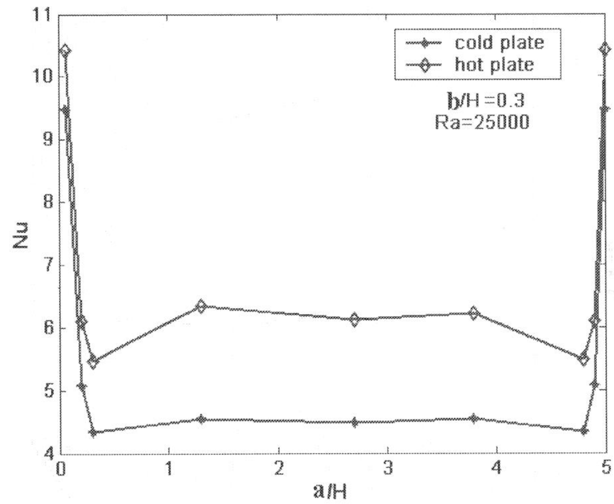

Figure 8. Variation of Nu with horizontal location

The convection is enhanced due to the instability of thermal boundary layers near the left plate and right wall ($0< a/H <0.3$) or right plate and left wall ($4.8< a/H <5$). In this spacing, the thermal boundary layers of wall and plate merge with each other, so the boundary layer thickness is thin leading to the increase of Nusselt number. Therefore, the characteristics of the fluid flow and heat transfer in enclosure are very sensitive to the distance between the wall side and the plate.

Different trends occur when the cold plate and the hot plate are at different vertical location which is shown in Fig. 9. Study of Fig. 9 reveals that for cold plate, as the vertical location ratio increases from 0.3 to 3.3 the average Nusselt number increases from 4.33 to 5.68, and slightly decreases from 5.68 to 5.52 when location ratio increases to 4. That means there exists a maximum average Nusselt number when cold plate location $(b/H)_{opt}$ is 3.3. For the hot plate, average Nusselt number slightly increases from 5.47 to 5.61 when location ratio increase from 0.3 to1.33, and decreases from 5.61 to 4.3 when vertical location ratio increases from 1.33 to 4. There is an appropriate location which corresponds to a maximum heat transfer density for the hot plate also. This can be attributed to two different flow patterns. However, the average Nusselt number variation is not significant for cold and hot plates.

By comparison of Fig. 3(a and b), Fig. 5(c and d), Fig. 6(c and d), and Fig. 9, it can be seen that the symmetry phenomenon appears, *i.e.* the identical problems of natural convection in an enclosure [19].

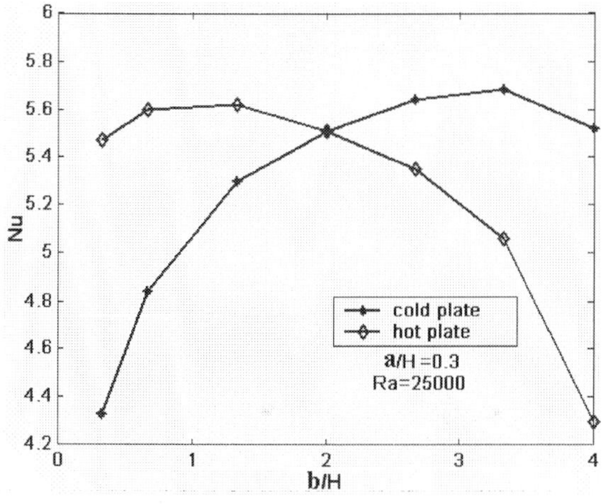

Figure 9. Variation of Nu with vertical location

3.5. Effect of Ra on the fluid flow and heat transfer

The procedure is then repeated over the range $10^2 \le Ra \le 10^8$. The effect of the Rayleigh number on the average Nusselt number by setting a/H =0.3 and b/H =0.3 under the circumstances of heating or cooling the block is shown in Fig. 10.

The flow patterns and isotherms are drawn for two typical Rayleigh numbers: $Ra = 1.0 \times 10^2$ and $Ra = 1.0 \times 10^7$ for one geometrical configuration a/H =0.3 and b/H =0.3. These are presented in Figs. 11 and 12.

For the low Rayleigh numbers (Ra= $10^2 - 10^4$), the flow field consists of a single big vortex in the half domain of cavity. With an increase in Rayleigh number, the heat transfer process is dominated successively by conduction mode, combined mode of conduction-convection and convection mechanism. When the Rayleigh number increases to as high as 1.0×10^7 , it indicates the convection mode is predominated, these can be seen from Figs. 11 and 12.

At the lower Rayleigh numbers (Ra= $10^2 - 10^4$), Nusselt number of heating condition is a little bit higher than that of cooling condition which can be explained that convection of the hot plate is easier to establish than that of cold plate, conduction mechanism is prevailed and hence heat transfer is mainly dominated by conduction. That is why the Nu values shown in Fig. 10 are almost constant in the low Ra number range.

With the increase in Rayleigh number, Eq. (7) shows that at high Rayleigh numbers, the boundary layers on the enclosure right wall and plate left surface become very thin, leading to a significant increase in Nusselt number. It can be seen from Figs. 10 to 12 that in the convection regime ($Ra > 10^4$), the flow fields difference between the cold plate and the hot plate is appreciable, and the deviations of the Nusselt number between the cold plate and the hot plate are greater. The reason is that heat convection of the hot plate is easier to establish than that of the cold plate.

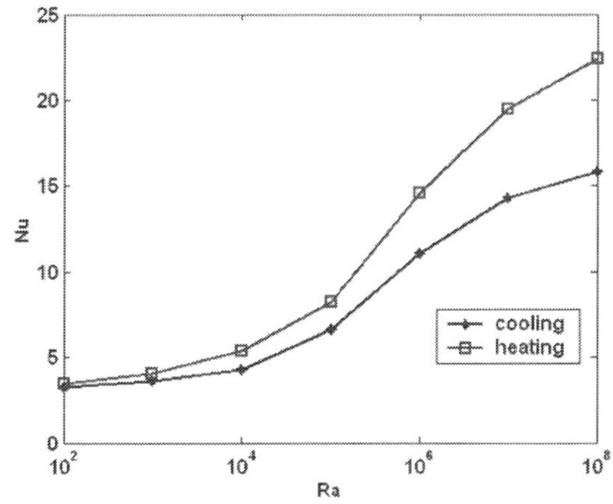

Figure 10. Nu versus Ra for heating and cooling cases

3.6. Special case -non steady state prediction

When the isolate is located in the middle of the enclosure,i.e. *L:H:W:b*=80:15:4:5, After giving definition of dimensionless time $F = \frac{a\tau}{H^2}(Ra\,\mathrm{Pr})^{\frac{1}{2}}$, it is easy get the dimensionless non-steady governing equation based on Equation(1)-(4).

The boundary conditions for this system are as same as steady system above.

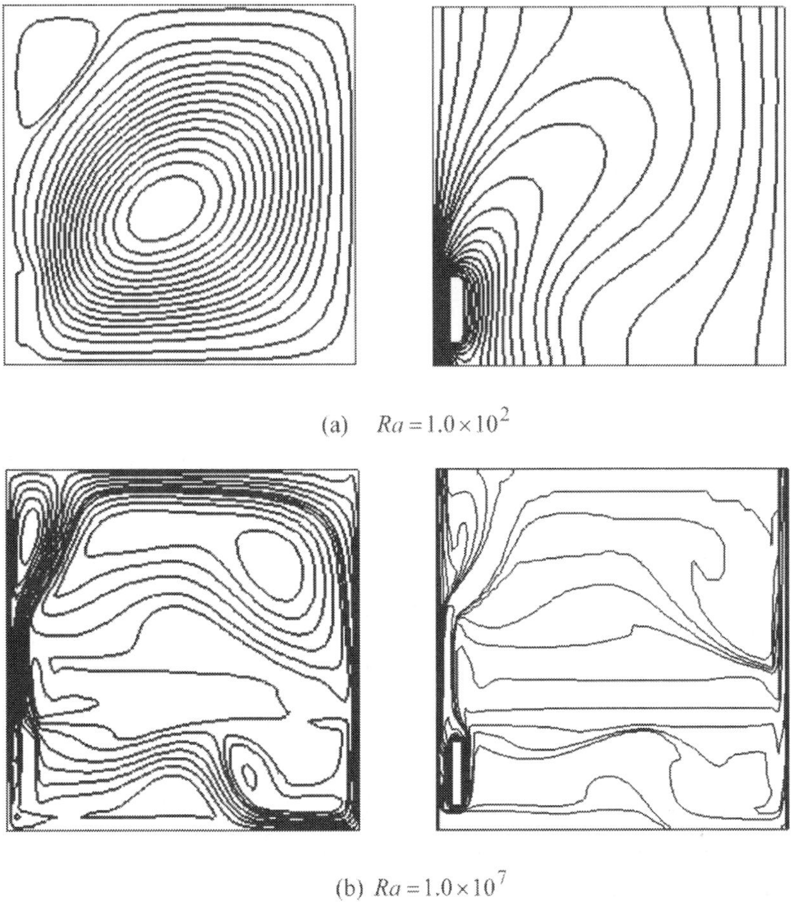

Figure 11. $T_1 < T_2$ flow patterns (left) and isotherms (right) at different Rayleigh numbers

The zero initial conditions set for velocity and temperature fields.

A perfectly time-periodic solution is predicted shown in Fig. 13 for different Raleigh number(Ra=10^4, Ra=5x10^4, Ra=10^5 and Ra=10^6). Fig. 13a reports the time dependent behavior of dimensionless temperature at the monitoring point $(x, y) = (1.3, 1.53)$ of the cavity, and the average Nusselt number result is depicted in Fig. 13c. The first noteworthy feature of Fig. 13 is that, except low value of Ra number(Ra=10000,50000) , after a few time units, a transition to oscillatory flow occurs. the symmetric solution breaks down as instabilities grow, and the time behaviors of quantities relative to geometrically symmetric points begin to differ, The quantities at location $(x, y) = (1.3, 1.53)$ exhibit a clearly periodic behavior for Ra=100000(Fig. 13c,d), the period being about 10 time unit.

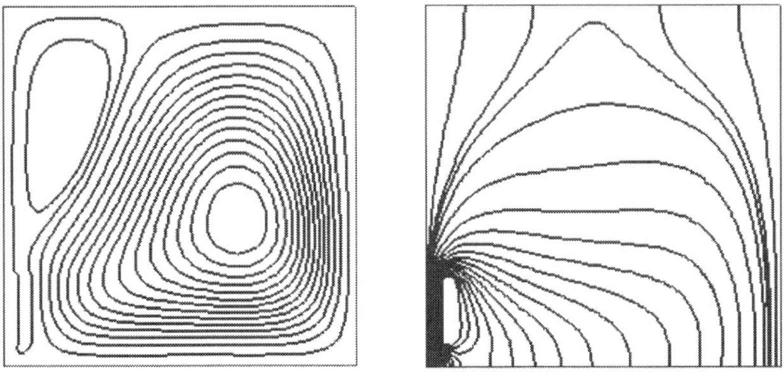

(a) $Ra = 1.0 \times 10^2$

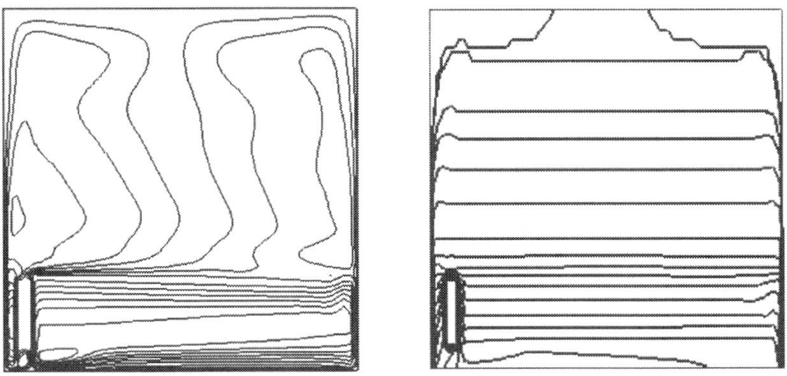

(b) $Ra = 1.0 \times 10^7$

Figure 12. $T_1 > T_2$ flow patterns (left) and isotherms (right) at different Rayleigh numbers

The isotherms are plotted in Fig. 14 for approximately equal time interval in one periodic circle When analyzing the macroscopic nature of the flow configurations, We observe the isotherms are not centrosymmetric at any point in time. The flow configurations manifest themselves in same pattern—the rolls movie to the left or right with the oscillation.

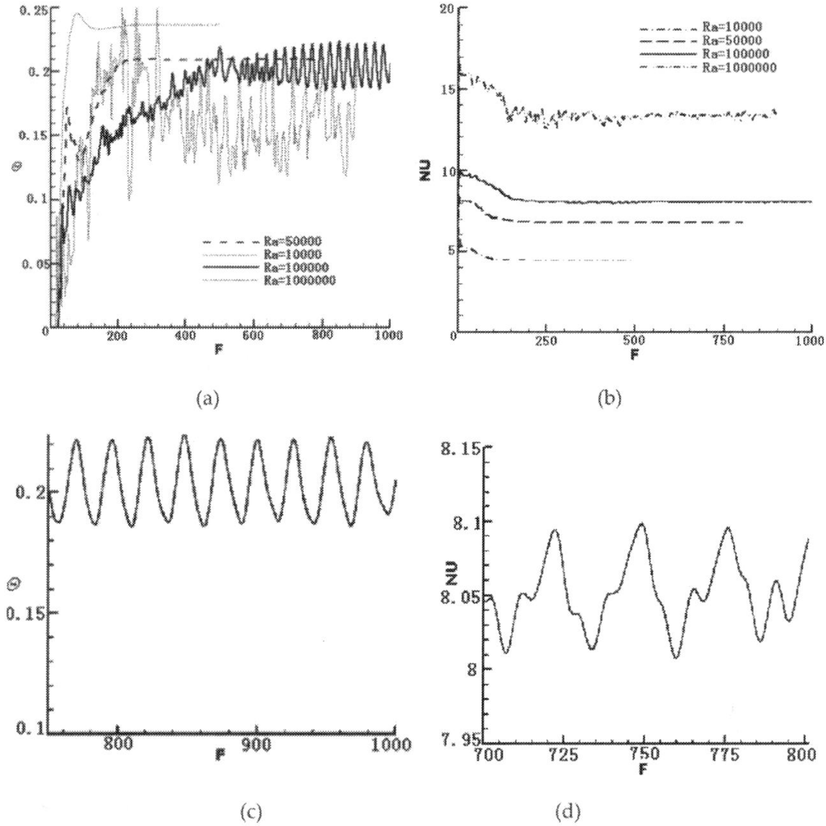

(a)variation of dimensionless temperature with dimensionless time for Ra=10000,50000,100000,1000000 at the monitoring point $(x, y) = (1.3, 1.53)$ of the cavity

(b) variation of average Nusselt number with dimensionless time for Ra=10000,50000,100000,1000000 at the monitoring point $(x, y) = (1.3, 1.53)$ of the cavity

(c) Partial enlargement of $T - F$ at the monitoring point $(x, y) = (1.3, 1.53)$ in the cavity for $Ra=10^5$

(d)Partial enlargement of $Nu - F$ at the monitoring point $(x, y) = (1.3, 1.53)$ in the cavity for $Ra=10^5$

Figure 13. Time history of the dimensionless temperature and Nusselt number at the monitoring point $(x, y) = (1.3, 1.53)$ of the cavity

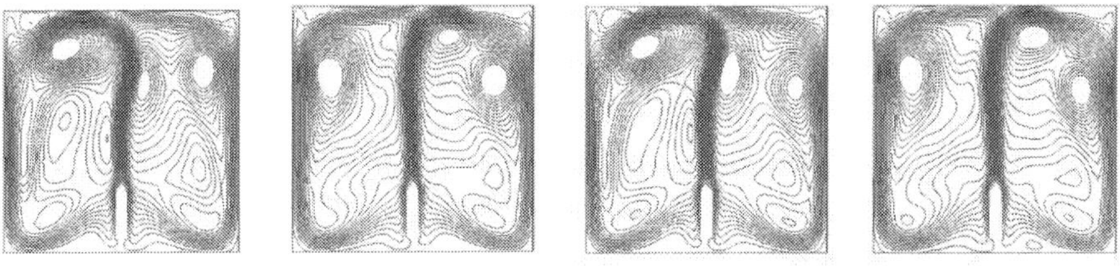

(a) F=749 (b) F=759 (c) F=775 (d) F=786

Figure 14. Series of instantaneous streamlines for Ra $=10^5$

4. Concluding remarks

A numerical study has been presented to unveil primarily the effect of location ratio variation of cold isolated vertical plate and hot isolated vertical plate on the natural convection in an enclosure. We conclude as follows from the numerical results.

a. The flow configurations and dimensionless temperature profiles of cold plate and hot plate are different;
b. The stratification is found to be strong for cold plate, while for the hot plate is relatively weak;
c. The temperature difference along the height of the enclosure in heating conditions is more uniform than that obtained in cooling conditions;
d. The increase of vertical location near the middle of enclosure leads to small temperature differences in heating and cooling conditions;
e. The trend of the average Nu number variation is the same for the cold and hot plates, when the plate is fixed at different horizontal locations. However, the average Nu number of the hot plate is about 20% to 39% larger than that of cold plate at the same Rayleigh number of 25000. For narrow distances between the inner plate and the bounding wall, the inner plate Nusselt number is enhanced, aside from this, the plate average Nusselt number is insensitive to the plate position;
f. The average Nu numbers tend to decrease with the increase of vertical location ratio ($1.33< b/H <4$) for hot plate, but for the cold plate the average Nu number tends to augment with the increase of vertical location ratio ($0.3< b/H <3.3$);
g. An optimum vertical location ratio exists at which the heat transfer is maximum for both cold and hot plates at a specific Rayleigh number;
h. Non-steady modeling simulations reveals that solutions are unique for values of Rayleigh number 10^4 and 5×10^4 where the flow and heat transfer is steady state. While unsteady state flow and heat transfer is appeared as a function of $Ra=10^5$ and $Ra=10^6$ for the rectangular block located in the middle of the cavity.

Nomenclature

a	horizontal location
A	node (i,j) based control volume area
b	vertical location
F	non-dimensional time
g	gravitational acceleration
H	height of vertical plate
k	vertical block location thermal conductivity
k_f	fluid thermal conductivity
K	relative thermal conductivity
L	characteristic length of the enclosure
Nu	Nusselt number
p	pressure
P	non-dimensional pressure
Pr	Prandtl number
Ra	Rayleigh number
t	node (i,j) temperature
T	temperature
TD	temperature difference
u,v	velocity components
U,V	dimensionless velocity
W	width of vertical plate
x,y	Cartesian coordinates
X,Y	dimensionless coordinates

Greek symbols

a	thermal diffusivity
β	coefficient of thermal expansion
δ	velocity boundary layer thickness
δ_T	thermal boundary layer thickness
ρ	density of the fluid
ν	kinematic viscosity of the fluid
Θ	dimensionless temperature

Subscipts

c	cold
h	hot
opt	optimization
R	reference

Author details

Xiaohui Zhang
School of Physical Science and Technology, School of Energy, Soochow University, Suzhou, China

Acknowledgement

This work is supported by National Natural Science Foundation of China（No.51176132）

5. References

[1] Shyy, W., and Rao, M.M., Simulation of Transient Natural Convection around an Enclosed Vertical Plate, *ASME J. Heat Transfer*, vol. 115, pp. 946–953, 1993.

[2] Ho, C.J., and Chang, J.Y., A Study of Natural Convection Heat Transfer in a Vertical Rectangular with Two-dimensional Discrete Heating: Effect of Aspect Ratio, *Int. J. Heat Mass Transfer*, vol. 37, pp. 917–926, 1994.

[3] Yang, M., and Tao, W.Q., Numerical Study of Natural Convection Heat Transfer in a Cylindrical Envelope with Internal Concentric Slotted Hollow Cylinder, *Numerical Heat Transfer*, Part A, vol. 22, pp. 281–305, 1992.

[4] Wang, Q.W., Natural Convection in an Inclined Cube Enclosure with Multiple Internal Isolated Plates, Ph.D. Dissertation., Xi'an Jiaotong University,China,1996.

[5] Tou, S.K.W., and Tso, C.P., 3-D Numerical Analysis of Natural Convective Liquid Cooling of 3x3 Heater Array in Rectangular Enclosure, *Int. J. Heat Mass Transfer*,vol.42, pp.3231-3244,1999.

[6] Sezai, I., and Mohamad, A.A., Natural Convection from a Discrete Heat Source on the Bottom of a Horizontal Enclosure, *Int. J. Heat Mass Transfer*, vol. 42, pp. 2257-2266, 2000.

[7] Deng, Q.H., Tang, G.F., and Li, Y.G., Interaction between Discrete Heat Source in Horizontal Natural Convection Enclosures, *Int. J. Heat Mass Transfer*, vol. 45, pp. 5117-5132, 2002.

[8] Saha, A.K., Unsteady free convection in a vertical channel with a built-in heated square cylinder, *Numerical Heat Transfer Part A*, vol. 38, no. 8, pp. 795-818, 2000.

[9] Liu, J.P., and Tao, W.Q., Numerical analysis of natural convection around a vertical channel in a rectangular enclosure. *Heat and Mass Transfer*; vol. 31, no. 5, pp. 313-321,1996.

[10] Liu, J.P., and Tao, W.Q., Bifurcation to oscillatory flow of the natural convection around a vertical channel in rectangular enclosure, *International Journal of Numerical Methods for Heat and Fluid Flow*, vol. 9, no. 2, pp. 170-85,1999.

[11] Barozzi, G.S., and Corticelli, M.A., Natural convection in cavities containing internal heat sources, *Heat and Mass Transfer*, vol. 36, pp. 473-80, 2000.

[12] Barozzi, G.S., and Corticelli, M.A., Nobile E. Numerical simulation of time-dependent buoyant flows in an enclosed vertical channel, *Heat Mass Transfer*, vol. 35, pp. 89-99, 1999.

[13] Patankar, S.V., Numerical Heat Transfer and Fluid Flow, McGraw-Hill, New York, pp. 146-152, 1980.

[14] Tao, W.Q., Numerical Heat Transfer (2ed edition) Xi'an Jiaotong University Press, pp. 244-245, 2001. (in Chinese)

[15] de Vahl Davis G., Natural Convection of Air in a Square Cavity: a Benchmark Numerical Solution, *Int . J. Meth. Fluids*, vol. 3, pp. 249-264, 1983.

[16] Kazansky, S., Dubovsky, V., Ziskind, G., and Letan, R., Chimney-Enhanced Natural Convection from a Vertical Plate: Experiments and Numerical Simulations, *Int. J. Heat Mass Transfer* ,vol. 46, pp. 497-512, 2003.

[17] Auletta, A., and Manca, O., Heat and Fluid Flow Resulting from the Chimney Effect in Symmetrically Heated Vertical Channel with Adiabatic Extensions, *Int. J. of Thermal Sciences*, vol. 41,pp. 1101-1111, 2002.

[18] Bejan, A., Convection Heat Transfer, Wiley, New York, 1984.

[19] Yang , M., Tao, W.Q. and Wang, Q.W., On the Identical Problems of Natural Convection in Enclosures and Applications of the Identity Character, *Int. J. of Thermal Sciences*, vol. 2, pp. 116–25, 1993 .

Chapter 6

Forced Convective Heat Transfer and Fluid Flow Characteristics in Curved Ducts

Tilak T. Chandratilleke and Nima Nadim

Additional information is available at the end of the chapter

http://dx.doi.org/10.5772/53064

1. Introduction

Curved fluid flow passages are common in most technological systems involving fluid transport, heat exchange and thermal power generation. Some examples are: compact heat exchangers, steam boilers, gas turbines blades, rocket engine nozzle cooling and refrigeration. Such flows are subjected to centrifugal forces arising from continuous change in flow direction and, exhibit unique fluid and thermal characteristics that are vastly different to those within straight passages.

The centrifugal action induced by duct curvature imparts two key effects on the fluid flow. It produces a lateral fluid movement directed from inner duct wall towards the outer wall in the axial flow, thus causing a spiralling fluid motion through the duct. This lateral fluid movement is manifested by large counter-rotating pairs of vortices appearing in the duct cross section and is referred to as secondary flow. The centrifugal action also forms a radial fluid pressure gradient positively biased towards the outer duct wall. The lateral fluid circulation takes place adversely to the radial pressure field and is dampened by the viscous effects. The combined actions of the positive radial pressure gradient and the viscous forces lead to the formation of a stagnant fluid region near the outer wall. Beyond a certain critical axial flow rate, the radial pressure gradient would exceed the equilibrium condition in the stagnant fluid region at outer wall and triggers a localised flow circulation that forms additional pairs of vortices. This flow condition is known as Dean Instability [1] and the additional vortices are called Dean Vortices.

In his pioneering work, Dean [1] proposed the dimensionless Dean Number $K = \left[\frac{D_h}{R}\right]^{\frac{1}{2}} \mathrm{Re}$ for characterising the secondary flow behaviour. Moffat [2] and Eustice [3] have experimentally observed and verified the critical velocity requirements for Dean Instability, while Baylis [4] and Humphery et al. [5] affirmed the use of Dean Number in designating secondary flow behaviour. However, subsequent studies by Cheng et al. [6], Ghia and Shokhey [7], and Sugiyama et al, [8] showed that the duct aspect ratio and curvature ratio also significantly influence Dean Instability in curved rectangular ducts.

For rectangular ducts, Chandratilleke et al. [9,10,11] reported an extensive parametric study, capturing the profound influence from duct aspect ratio, curvature ratio and wall heat flux on the flow behaviour. Using an approach based on the stream function, they developed a two-dimensional model simplified by dynamic similarity assumption in axial direction. This simulation showed very good correlation to the data from their own experimental work [10,11]. The locations of Dean vortex formation within the flow was qualitatively identified through the use of intersecting stream function contours of zero potential. Comprehensive results were presented within the Dean Number range of $25 \leq K \leq 500$, aspect ratio at $1 \leq Ar \leq 8$ and Grashof number at $12.5 \leq Gr \leq 12500$. It was identified that the onset of Dean Instability would strongly be dependent on the duct aspect ratio wherein more Dean vortices are produced in ducts of high aspect ratio. The wall heat flux radically changed the flow patterns and showed a tendency to suppress Dean vortex formation. Subsequently, Yanase [12] and Fellouah et al. [13,14] confirmed and validated the findings by Chandratilleke et al. [11].

The stream function approach is clearly limited to two-dimensional flows and does not accurately applied for real flow situations. In developing improved models, Guo et al. [15] considered laminar incompressible flow and formulated a three-dimensional simulation to explore the interactive behaviour of geometrical and flow parameters on heat transfer and pressure drop. Using flow entropy for hydrothermal assessment, they reported the influence from Reynolds number and curvature ratio on the flow profile and Nusselt number. For curved rectangular ducts, Ko et al. [16] proposed to split the overall flow entropy into individual contributions from heat transfer and fluid viscous friction. The relative magnitudes of these components were appraised to determine the heat or viscous-biased irreversibility in the flow domain. This approach was adopted for achieving thermal optimisation in ducts through minimised overall entropy, and illustrated for enhancing forced convection in curved ducts with longitudinal fins under laminar and turbulent conditions [17,18,19]. However, the entropy approach did not warrant precise identification of Dean Instability.

Fellouah et al. [13,14] also have presented a useful three-dimensional model for rectangular curved ducts of aspect ratio within $0.5 \leq Ar \leq 12$ and curvature ratio at $5.5 \leq \gamma \leq 20$ and. Considering both water and air as working fluids, their results were was experimentally validated using a semi-circular duct. These tests also provided visualisation data on vortex formation at several Dean Numbers. For detecting Dean Instability, they suggested the use of the radial gradient of the axial fluid velocity. In this, a limiting value for the radial gradient of axial velocity is arbitrarily assigned as a triggering threshold for Dean Instability. This approach represents an early attempt to incorporate Dean vortex detection to a simulation process rather than a trail-and-error method as previously practised. In spite of the perceived benefits in computation, this approach cannot be rationalised because the radial gradient of axial velocity is remotely linked with vortex generation. This short fall is reflected when the model of Fellouah et al. [13] failed to detect Dean Instability that was clearly observed in some reported cases [10,11]. With improved capabilities, Chandratilleke et al. [20,21] have suggested other approaches that will be discussed later.

Most experimental and numerical analyses on curved ducts have been performed on rectangular duct geometries. By virtue of shape, such ducts have less wall interference on secondary vortex formation, making it relatively easier for numerical modelling and convenient for experimentation including flow visualisation. Ducts with elliptical and circular cross sections have received less research attention in spite of being a common geometry used in most technological systems. The curved duct flow behaviour in such geometries remains relatively unexplored.

Dong and Ebadian [22], and Silva et al. [23] have numerically simulated the flow through elliptical curved channels and observed that the stagnation regions are vastly different to those in rectangular ducts. Unlike in rectangular ducts, Dean Instability was seen to originate within the flow rather than at the outer duct wall. Papadopoulos and Hatzikonstantinou [24] considered elliptical curved ducts with internal fins and have numerically investigated the effects of fin height on friction factor and heat transfer. They concluded that the appearance of secondary vortices next to the concave wall would make the friction factor dependent on both fin and duct heights. Andrade et al. [25] reported a study with a finite element numerical model and discussed the influence from temperature-dependent viscosity on heat transfer and velocity profile for fully-developed forced convection in elliptical curved tube. They considered both cooling and heating cases wherein the Nusselt Number was found to be lower for cases of variable viscosity than constant properties. This was attributed to the increased viscosity at the cooler inner duct wall dampening the secondary flow and the vortex formation.

The above review of published literature indicates that the available modelling methods have yet to develop for realistic representation of complex secondary flow behaviour and improved predictability for Dean Instability. Parametric influences of duct geometry and flow variables remain unexplored and poorly understood with no decisive approach for defining the onset of Dean vortices and their locations. Significantly improving these shortfalls, this chapter describes an advanced three-dimensional numerical simulation methodology based on helicity, which is congruent with the vortex motion of secondary flow. Facilitating much compliant tracking of vortex flow paths, the model uses a curvilinear mesh that is more effective in capturing the intricate details of vortices and flexibly applied to both rectangular and elliptical ducts. Two intuitive approaches for detecting the onset of Dean Instability are proposed and examined in the study. An extensive parametric investigation is presented with physical interpretation of results for improved understanding of flow behaviour. A thermal optimisation scheme based on flow irreversibilities is developed for curved ducts.

2. Nomenclature

Ar	Aspect Ratio= a/b
a	Height of cross section (mm)
b	Width of cross section (mm)
D_h	Hydraulic diameter = $\dfrac{ab}{\sqrt{\dfrac{(a+b)}{2}}}$ (mm)
F_c	Centrifugal force (N)
g	Gravity (m/s²)
H	Helicity (m/s²)
H^*	Dimensionless Helicity = $\dfrac{HD_h}{U_{in}^2}$
$\hat{i},\hat{j},\hat{k}$	Unit vectors in x,y,z directions
K	Dean number = $\left(\dfrac{D_h}{R}\right)^{\frac{1}{2}}$ Re
p	Static pressure (Pa)
p^*	Dimensionless static pressure = $\dfrac{p}{\frac{1}{2}\rho U_{in}^2}$
R	Radius of curved channel (m)
Re	Reynolds number = $\dfrac{U_{in}D_h}{\upsilon}$
$\hat{S}$	Coordinate along duct cross section for defining secondary flow direction
U_{in}	Velocity at duct inlet (m/s)
u,v,w	Velocities component (m/s)
u^*,v^*,w^*	Dimensionless velocity = $\dfrac{u,v,w}{U_{in}}$
V_r	Axial velocity (m/s)
x,y,z	Coordinates (m)

x^*,y^*,z^* Dimensionless coordinates= $\dfrac{x,y,z}{D_h}$

Greek symbols

γ	Curvature ratio= R/b
θ	Angular position of cross section (deg)
Φ	Dissipation function
ν	Kinematic Viscosity (m²/s)
μ	Dynamic viscosity (Ns/m²)
ρ	Density (kg/m³)
ω	

3. Model description and numerical analysis

Fig. 1 shows the rectangular and elliptical duct geometries used for the three-dimensional model developed in the current study. It also indicates the key geometrical parameters of duct height a, duct width b and the duct radius of curvature R. The geometrical model consists of a semi-circular curved duct test section fitted with a straight inlet passage to ensure fully-developed flow at entry to the curved duct and an outlet passage for smooth flow exit.

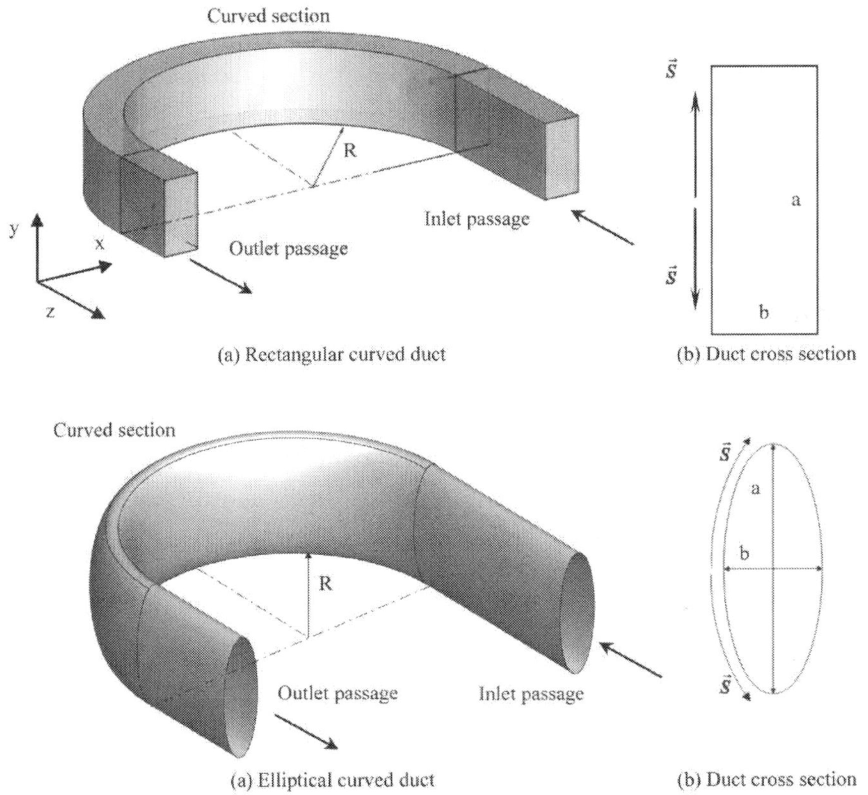

Figure 1. Geometry of rectangular and elliptical curved duct models

The analysis focuses on semi-circular curved ducts having fixed width b of 10 mm and constant radius of curvature R of 125 mm. The duct aspect ratio Ar is changed by varying duct height a. The working fluid air is assumed to be an incompressible Newtonian fluid with temperature-dependent fluid properties. The air enters at an inlet temperature of 300 K and flows steadily through the passage under laminar flow conditions. A uniform heat flux is applied on the outer wall (rectangular duct) or outer half of duct periphery (elliptical duct), while all the other walls being adiabatic. Constant velocity condition is applied to the inlet of the straight duct preceding the curved duct. The flow exit is taken to be a pressure outlet. The duct walls are assumed to have no slip boundary condition.

The numerical model solves the following governing equations:

Time-averaged continuity equation:

$$\nabla.(\vec{V}) = 0 \tag{1}$$

Momentum and Energy conservation equations:

$$\begin{aligned} &\vec{V}.\nabla(\rho\vec{V}) = -\nabla p + \mu\nabla^2\vec{V} + \rho_m\vec{g} + \vec{F}_c \quad &(a) \\ &\vec{V}.\nabla(\rho c_p T) = k\nabla^2 T + S_T \quad &(b) \end{aligned} \tag{2}$$

The magnitude of the centrifugal body force term in radial direction is given by,

$$F_c = \rho\frac{V_r^2}{r} = \rho\frac{(u^2 + w^2)}{\sqrt{x^2 + z^2}} \tag{3}$$

Considering the position and alignment of curved part of geometry, centrifugal source term in Cartesian coordinate system is obtained from,

$$\vec{F}_c = \rho\frac{1 + sign(-z)}{2}\frac{(u^2 + w^2)}{x^2 + z^2}(x\hat{i} + z\hat{k}) \tag{4}$$

In Eq. 4, a Sign Function is used to ensure the centrifugal source term is applied only on the curved side of the geometry (i.e. z ≤ 0). For obtaining the dimensionless parameters, the characteristics length, velocity and pressure are chosen to be D_h, U_{in}, $\frac{1}{2}\rho U_{in}{}^2$, respectively.

The thermally-induced buoyancy is included in the model by relating the density ρ_m in Eq. 2(a) to local fluid temperature. For this, a sixth-order polynomial as given by Eq. 5 is developed, where the coefficients are obtained using the published data. This approach is necessary because the linearity of the Boussinesq approximation caused some discrepancy and is found to be inadequate for evaluating the wall pressure gradient.

$$\rho(T) = 10^{-15}T^6 - 3\times10^{-12}T^5 + 3\times10^{-9}T^4 - 2\times10^{-6}T^3 + 6\times10^{-4}T^4 - 0.1008T + 9.3618 \tag{5}$$

In capturing the helix-like fluid motion of secondary flow, this three-dimensional model incorporates a helicity function, which is defined by Eq. 6. The helicity is non-dimensionalised using the reference helicity based on hydraulic diameter, as given by Eq. 7.

$$H = \vec{V}.\vec{\omega} = u(\frac{\partial w}{\partial y} - \frac{\partial v}{\partial z}) + v(\frac{\partial u}{\partial z} - \frac{\partial w}{\partial x}) + w(\frac{\partial v}{\partial x} - \frac{\partial u}{\partial y}) \tag{6}$$

$$H \approx \frac{U_{in}^2}{D_h} \Rightarrow H^* = H(\frac{D_h}{U_{in}^2}) \tag{7}$$

For identifying the onset of Dean Instability, the model proposes two separate criteria. The first criterion is based on helicity, which is computed from Eq. 7. The second criterion uses the outer duct wall pressure gradient profile for which the non-dimensional wall pressure gradient is obtained from Eq. 8. A sign convention is incorporated in Eq. 8 to designate the opposite rotational directions of the vortices in upper and lower half of duct cross section, following the selection of coordinate $\vec{S}$ along the outer wall of duct cross section, as depicted in Fig. 1(b).

$$\frac{dp^*}{d\vec{S}} = \begin{cases} -\dfrac{D_h}{\frac{1}{2}\rho U^2_{in}} \dfrac{dp}{ds} & \text{at upper half of cross section} \\ \dfrac{D_h}{\frac{1}{2}\rho U^2_{in}} \dfrac{dp}{ds} & \text{at lower half of cross section} \end{cases} \tag{8}$$

For the forced convective heat transfer in the duct, the local Nusselt Number is defined as $Nu = \dfrac{hD}{k}$ where the heat transfer coefficient h, is determined by considering the grid cell temperature difference between the heated wall and the adjacent fluid cell. The average Nusselt number is obtained from the surface integral, $\bar{N}u = \dfrac{\int_A NudA}{A}$.

The model formulates a thermal optimisation scheme using flow irreversibility as a criterion. For this, the overall entropy generation is split in to the components of irreversibility contributed by the wall heat transfer, including thermally-induced buoyancy effects and, that due to fluid flow friction compounded by secondary fluid motion. As such, the components of entropy generation from heat transfer (S_T) and that from fluid friction (S_p) within the solution domain are evaluated using the expressions 9 (a), (b) and (c).

$$S_T = \frac{k}{T^2}\left[(\frac{\partial T}{\partial x})^2 + (\frac{\partial T}{\partial y})^2 + (\frac{\partial T}{\partial z})^2 \right] \tag{a}$$

$$S_P = \frac{\mu}{T}\Phi \tag{b}$$

(9)

where

$$\Phi = 2\,[(\frac{\partial u}{\partial x})^2 + (\frac{\partial v}{\partial y})^2 + (\frac{\partial w}{\partial z})^2] + (\frac{\partial u}{\partial y} + \frac{\partial v}{\partial x})^2 + (\frac{\partial w}{\partial y} + \frac{\partial v}{\partial z})^2 + (\frac{\partial u}{\partial z} + \frac{\partial w}{\partial x})^2 \tag{c}$$

The entropy terms are generalised as volumetric-averaged values using,

$$\begin{aligned} S'''_T &= \frac{\int S_T dv}{dv} && \text{(a)} \\ S'''_p &= \frac{\int S_p dv}{dv} && \text{(b)} \\ S'''_g &= S'''_T + S'''_p && \text{(c)} \end{aligned} \tag{10}$$

Based on the above volumetric entropy generation terms, Bejan number is defined as,

$$Be = \frac{S'''_T}{S'''_g} \tag{11}$$

Bejan Number reflects the relative domination of flow irreversibility by heat transfer with respect to the overall irreversibility [26]. In this, magnitude 1.0 for Bejan Number indicates the entropy generation totally dominated by heat transfer while 0 refers to fluid friction causing all of the flow irreversibility. Thus for curved ducts, the Bejan number performs as a visual map signifying the relative strength of thermal effects within the fluid domain interacted by the secondary flow and Dean vortices. This approach is then utilised for thermally optimising the forced convection process in curved ducts.

Incorporating above governing equations, a finite volume-based CFD model is formulated with the commercial package FLUENT where SIMPLE algorithm is used for pressure-velocity coupling. The momentum and energy equations are discretised by first and second order schemes, respectively. Since the model considers both buoyancy and centrifugal source terms, pressure discretisation is performed with body force-weighted approach. The stability of the solution is monitored through continuity, velocity, energy and dimensionless helicty where the convergence is achieved with values not higher than 10^{-5}. Grid independency is carefully checked paying a special attention to boundary layer modelling at the outer wall.

The simulation is performed to obtain the profiles of velocity, helicity, temperature and Bejan Number at curved duct for a range of flow rates giving Dean Number within 80 to 1600. The wall heat flux is varied up to 1000 W/m^2. The results showing some signs of Dean vortices are further refined by re-running the simulations with closer steps of K to determine the exact point of instability and the critical Dean number. This procedure is repeated for all aspect ratios, flow rates and heat fluxes. Forced convection is evaluated by the local and average Nusselt numbers at the duct wall.

3.1. Mesh generation, grid sensitivity and model validation

In secondary flow simulations, the solution convergence is critically dependant on the grid selection because of the intricate flow patterns and intense flow gradients. For rectangular ducts, a fully structured mesh would be adequate. However, due to the geometrical shape, elliptical ducts tend to have much extreme flow field gradients and would require a more stringent grid arrangement.

In rectangular ducts a progressively reducing mesh is used with a much finer mesh near the outer wall where the onset of Dean Instability is anticipated. This approach has not been attempted in previous studies [11-13, 18-19] wherein it was argued that a mesh size less than 1 mm would not improve the accuracy, but only increase the computational time. The mesh refinement of the present analysis clearly identifies that a finer mesh near the wall is critical for detecting the onset of Dean vortices as accurately as possible. For testing grid dependency, the current study uses five mesh schemes indicated in Table 1. In this, columns A, B and C represent the number of grids over duct width, height and length, respectively, while the column D indicates the progressive reduction of grid size over duct width towards the outer wall.

Scheme	Number of Grids			
	A	B	C	D
Mesh1	26	51	305	1
Mesh2	31	64	305	1
Mesh3	43	84	305	1
Mesh4	50	98	305	1
Mesh5	26	51	305	1.05

Table 1. Grid selection and mesh schemes for rectangular ducts

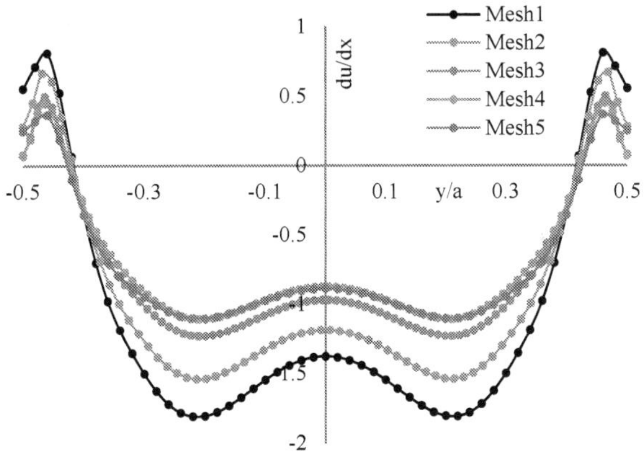

Figure 2. Grid dependancy test for rectangular duct using velocity derivative in y-direction at exit

K=130, Ar = 2

For rectangular ducts, Fig. 2 illustrates the grid dependency test conducted using the velocity derivative in y-direction at the duct outer wall. The Schemes 4 and 5 having mesh size less than 1 mm show better suitability than the other three schemes. However, the Scheme 5 is chosen as the optimum because of its slightly larger cell volume arising from progressively varied mesh size. This approach remarkably improved the vortex capturing ability in the solution domain without excessive computational demand. As such, the present study performed all computations with Scheme 5 of mesh size less than 1 mm, achieving much higher accuracy than any previously reported work.

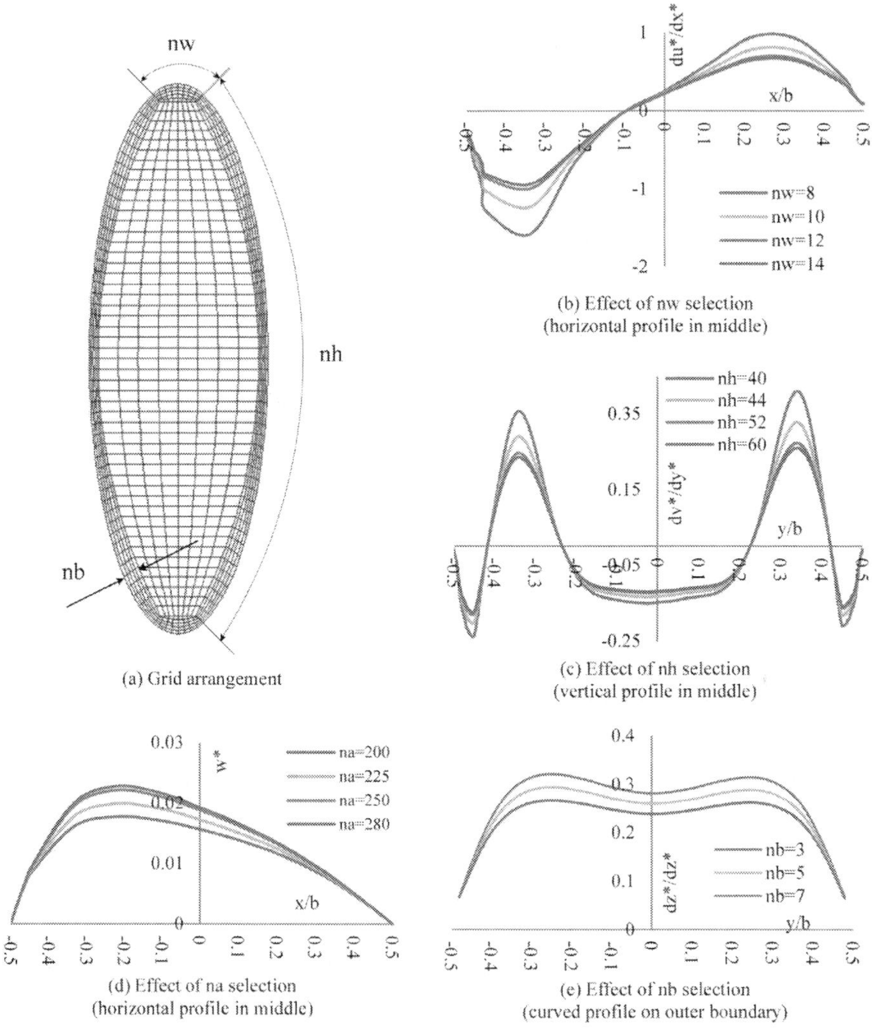

Figure 3. Grid arrangement and grid sensitivity analysis for elliptical curved duct at exit

For handling steeper flow gradients in elliptical ducts, the current analysis uses the different grid arrangement depicted in Fig. 3. This scheme divides the elliptical duct cross section into five regions, as shown in Fig. 3(a) and, the rectangular cells are swept along the axial direction to obtain a fully hexagonal structured mesh. The grid distribution is determined by the parameters nw, nh, nb and na, which are respectively, the number of grids on duct width, duct height and duct periphery while *na* represents the number of layers along the axial direction. This formation ensures uniform grid residual throughout the solution domain. Figs. 3(b), (c) and (d) show the grid sensitivity analysis in terms of nw, nh and nb and na, where 36 permutations are considered. For elliptical duct simulation, the optimal grid selection is taken to be nw = 12, nh = 52, nb = 7 and na = 250.

In validating results for a rectangular duct, Figs. 4(a) and 4(b) show a comparison of the axial flow velocity in x and y directions from the current model with those from Ghia and Shokhey [7] and Fellouah et al. [14]. It is seen that both magnitudes and trends of axial velocity are very favourably matched confirming the integrity of the current numerical process. Similarly for an elliptical duct, Figs. 4(c) and 4(d) provide a comparison of the axial velocity profiles at the mid cross section planes from the current study with those from Dong and Ebadian [22] and Silva et al. [23]. A very good agreement is clearly evident, validating the numerical consistency of the present simulation.

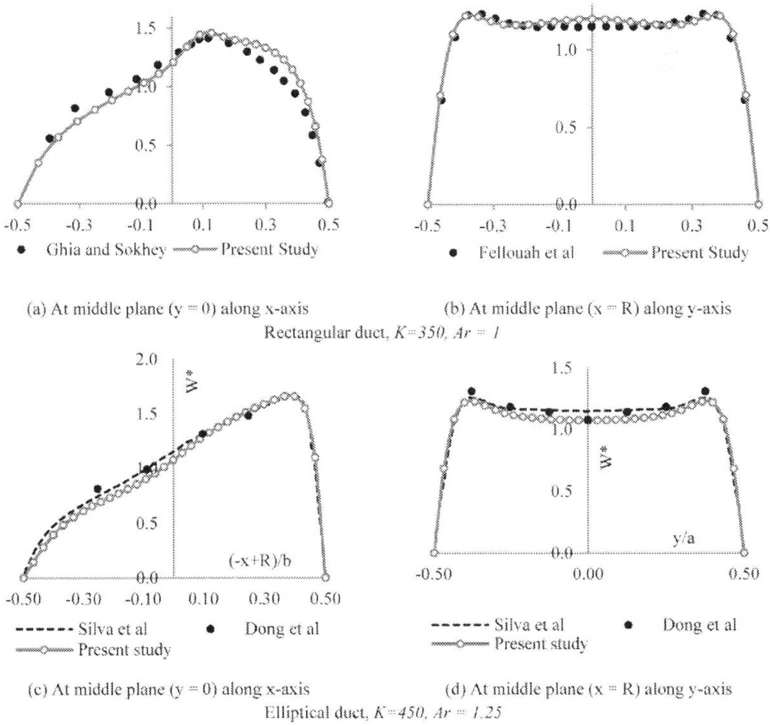

Figure 4. Model validation using dimensionless axial velocity profile at curved duct exit

4. Results and discussion

4.1. Fluid flow characteristics and geometrical influence

Fig. 5 depicts typical dimensionless helicity profiles at the exit of both rectangular and elliptical curved ducts for several Dean numbers with no external wall heating. These patterns show unique flow features that are not present in straight ducts. Initially at low *K*, the flow profiles indicate two large counter-rotating vortices that are attributable to the centrifugal action from the duct curvature.

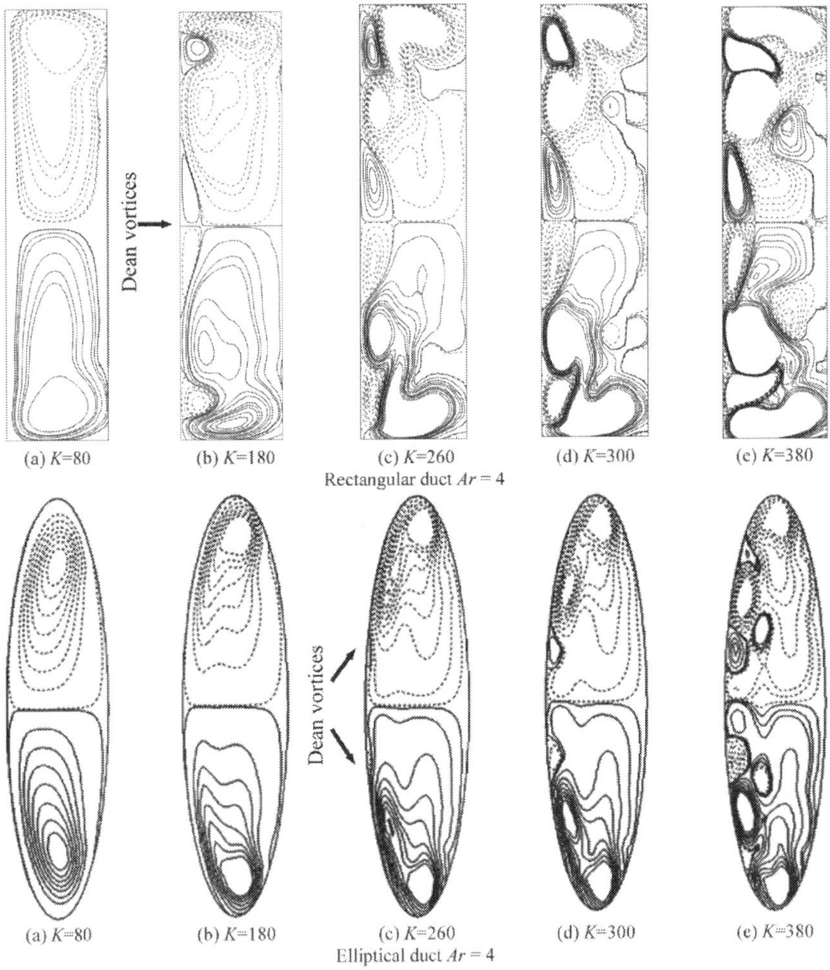

Figure 5. Dimensionless helicity contours in curved ducts for varied flow rates-no wall heating

In curved ducts, the centrifugal action is manifested as two key effects. It generates a positive radial pressure field directed towards the outer duct wall (left wall in Fig. 5). Within this positive (adverse) pressure field, the centrifugal force drives the fluid radially away from the inner wall towards the outer duct wall. This sets up a lateral fluid circulation called secondary flow, which is manifested as the formation of counter-rotating vortices observed in Fig 5 in the duct cross section.

Increased axial flow rate (larger *K*) makes the secondary fluid motion more vigorous to produce stronger vortices. The positive radial pressure field is also Higher flow rate also intensifies that imparts more adverse effect on the secondary fluid motion. This latter effect adversely acts on to slow it down assisted by fluid viscosity. Consequently near the outer duct wall, a stagnant flow region is formed and a state of fluid equilibrium is established. Above a certain critical value of *K*, the radial pressure field becomes far greater for fluid equilibrium within the stagnant region and a weak local fluid recirculation is triggered. This flow situation is referred to as the Dean Instability identified by the critical value of the Dean number. The local fluid circulation triggering flow instability is manifested as pairs of additional vortices called Dean Vortices. In Fig. 5, the onset of Dean Instability occurs within $K = 80$ to 180 for the rectangular duct while for the Dean vortices between $K = 180$ to 260. These Dean vortices are designated as the helicity contour corresponding to $H^* = 0.01$. Upon inception, Dean vortices gradually grow bigger with increasing *K*, as depicted.

It is noticed that in rectangular ducts, the Dean instability tends to occur at a lower *K* than in elliptical ducts of the same aspect ratio. This is because in rectangular ducts, the secondary vortex motion is less steeply deflected at the outer wall by the cross sectional geometry than in elliptical ducts allowing more freedom for fluid movement and Dean vortex formation.

4.2. Dean instability and detection of Dean vortices

In curved ducts, the appearance of Dean vortices is traditionally identified through tedious flow visualisation or by numerical trial-and-error approach. In the latter, simulations are repeatedly performed to gradually narrow down the range of K for determining the critical Dean number and the flow patterns within a chosen tolerance limit. This involves guesswork and significant computational time. Chandratilleke et.al. [11] successfully developed a criterion based on zero-potential stream function contours to identify the locations of Dean vortex generation. Whilst that approach is adequate for two-dimensional cases, it is not applicable for three-dimensional flow. The work of Fellouah et al. [13] used the radial gradient of the axial velocity as a measure of identifying the flow instability. Such technique is not justifiable because the axial velocity change in radial direction is not physically connected with the secondary vortex generation. This inadequacy is clearly reflected in the work of Fellouah et al. [13], where their simulation failed to detect Dean vortices for some basic flow conditions. For example around $K = 180$ for a rectangular duct, Fellouah et al. [13] and Silva et al. [23] did not detect Dean vortices while the current study clearly predicts such vortices.

As of now, a reliable technique for identifying Dean vortices is not available in literature, particularly for three-dimensional simulations. The formulation of a generalised approach is difficult because the inception of Dean vortices are influenced by strongly inter-dependant effects of duct geometry and the flow variables. In forming a technique for Dean vortex detection, the present study proposes two practical criteria that can be integrated into and directly performed within the computational process.

4.2.1. *Criterion 1 - Helicity threshold method*

This criterion utilises dimensionless helicity H^* for tracking the onset of Dean Instability. It assigns the minimum threshold value for H^* by which the helicity contours are demarcated in the flow domain. These contours are designated as Dean vortices. This selection of H^* threshold essentially depends on the contour detection accuracy required, similar to defining the boundary layer thickness with a chosen velocity tolerance in traditional fluid flows.

Through exhaustive simulation runs, the current study has identified and proposes $H^* = \pm 0.01$ to be an excellent choice that warrants precise and consistent detection of Dean vortices for all cases examined. A key advantage of this technique is that the detection precision can be varied to suit the required accuracy. Also, the technique can be readily integrated into the computational process to perform locally in the solution domain rather than as a cumbersome post-processing method. Hence, the determination of Dean vortices is more precise and less time consuming. The achievable precision is demonstrated below in Fig. 6, where the helicity contours obtained by using $H^* = 0.01$ are shown just before and just after the onset of Dean vortices (indicated by arrows).

For the rectangular duct, the Dean vortices are absent in the flow for $K = 95$. However, when K is increased to 102, the helicity contours indicating Dean vortices are first detected. Similarly for the elliptical duct, the appearance of Dean vortices is detectable within $K = 230$ to 235. By adopting H^* threshold less than 0.01, this range can be further narrowed to improve precision of K value that corresponds to the onset of Dean Instability. The critical Dean numbers thus determined are: $K = 100$ for the rectangular duct and $K = 234$ for the elliptical duct. However, evidently such refinements to H^* would only incur a marginal benefit towards precision at the expense of significant increase in computational demand. Therefore, $H^* = 0.01$ is concluded to be a very appropriate threshold for detecting Dean Instability. Previous methods never provided this degree of accuracy, flexibility or ability for vortex detection, signifying that this method is far superior to any reported approach including that by Ghia and Shokhey [7] and Fellouah et al. [13] using the axial velocity gradient.

4.2.2. *Criterion 2 - Adverse wall pressure gradient method*

This criterion is based on the unique features of the fluid pressure distribution along the outer duct wall. The gradient of this pressure profile shows inflection points that initially remain negative at low flow velocities and gradually shifts towards positive magnitudes as the flow rate is increased. Outer wall duct locations indicating negative-to-positive gradient change strongly correlates to the appearance of Dean vortices and forms the basis for a criterion to identify Dean Instability. These characteristics are illustrated in Fig. 7.

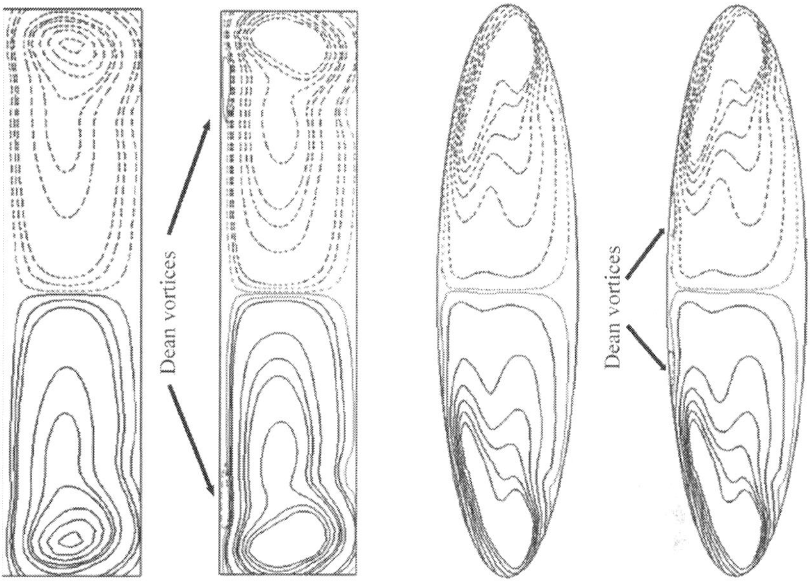

Figure 6. Detection of Dean vortex formation using helicity threshold method $H^* = \pm 0.01$

For several Dean Numbers, Fig. 7 shows typical outer wall pressure gradient profiles evaluated from Eq. 8 for a rectangular duct in Fig. 7(a) and, for elliptical ducts in Figs. 7 (b) and (c). The variable S is the displacement coordinate along the outer duct wall boundary, as indicated in Fig. 1. At low values of K in Fig. 7(a), the pressure gradient remains negative over the entire outer wall. As K increases, the inflection points of the profile gradually shift to towards a positive magnitude, thus creating regions of adverse pressure gradients at the outer wall. The corresponding helicity contours indicate that these localities at the outer wall would develop flow reversal leading to the generation of Dean vortices. At $K = 139$, the inflection points have just become positive, representing the critical Dean Number and the onset of Dean Instability.

Showing similar trends, Fig. 7(b) illustrates the pressure gradient profile for an elliptical duct of aspect ratio 3. At $K = 285$, the entire pressure profile remains negative over the outer wall and the corresponding helicity patterns in Fig. 8(a) show Dean vortices are absent in the flow. At $K = 291$, the pressure gradient just changes from negative to positive at the duct centre and the first appearance (onset) of Dean vortices is noted in the flow profile of Fig. 8(a). For $K = 297$, the pressure gradient profile shows distinctly positive regions that well correlate to Dean vortex locations, as depicted.

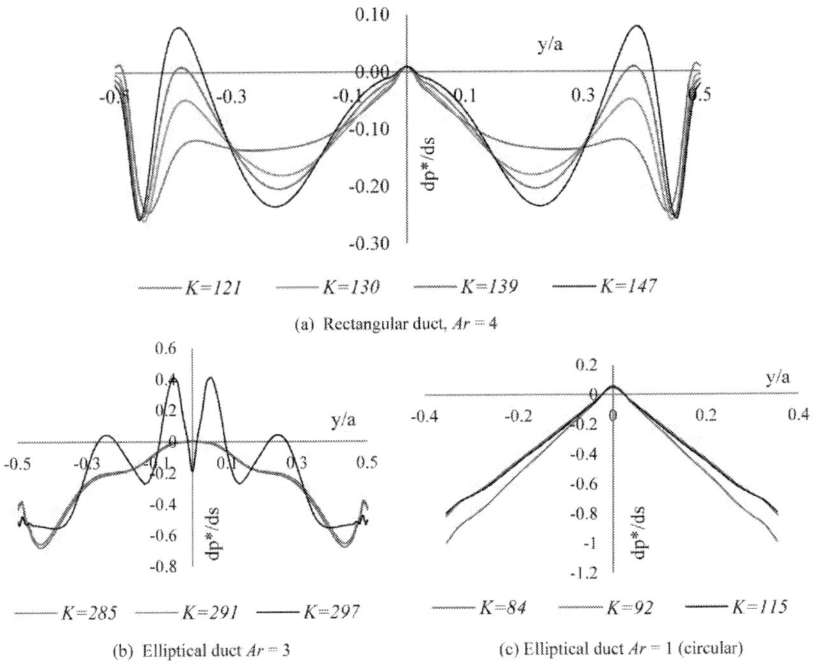

Figure 7. Pressure gradient profiles along the outer wall of curved duct

Above observations conclusively indicate that the adverse (positive) pressure gradients at the outer duct wall could be effectively used for detecting the onset and location of Dean vortices. However, this approach has some drawback as discussed below.

Fig. 7(c) shows the outer wall pressure gradient for an elliptical duct of aspect ratio $Ar = 1$ (circular). This profile does not exhibit the negative-to-positive changeover in the pressure gradient as for elliptical ducts with $Ar > 1$ illustrated in Figs. 7(a) and 7(b), meaning that Dean vortices are not predicted according to the adverse pressure criterion. However to the contrary, Fig. 8(c) indeed shows the onset of Dean vortices at $K = 92$. This may seem a contradiction that is clarified below.

For a duct with $Ar = 1$ (circular duct), Fig. 8(c) clearly shows that the Dean vortices appear in the centre of the flow, but not at the outer wall as with rectangular and elliptical ducts. Therefore, the outer wall pressure gradient has lesser bearing on the flow reversal associated with Dean vortex formation. As such, the pressure gradient criterion tends to over predict the critical Dean Number for instability and has diminished suitability for elliptical ducts with aspect ratio near unity. It would perform satisfactorily for ducts with aspect ratio $Ar > 1$ where Dean vortices are formed at outer duct wall. Upon this overarching scrutiny, the helicity threshold method can be regarded as a precise, reliable and universally applicable technique for detecting Dean vortices in curved ducts of any shape and aspect ratio.

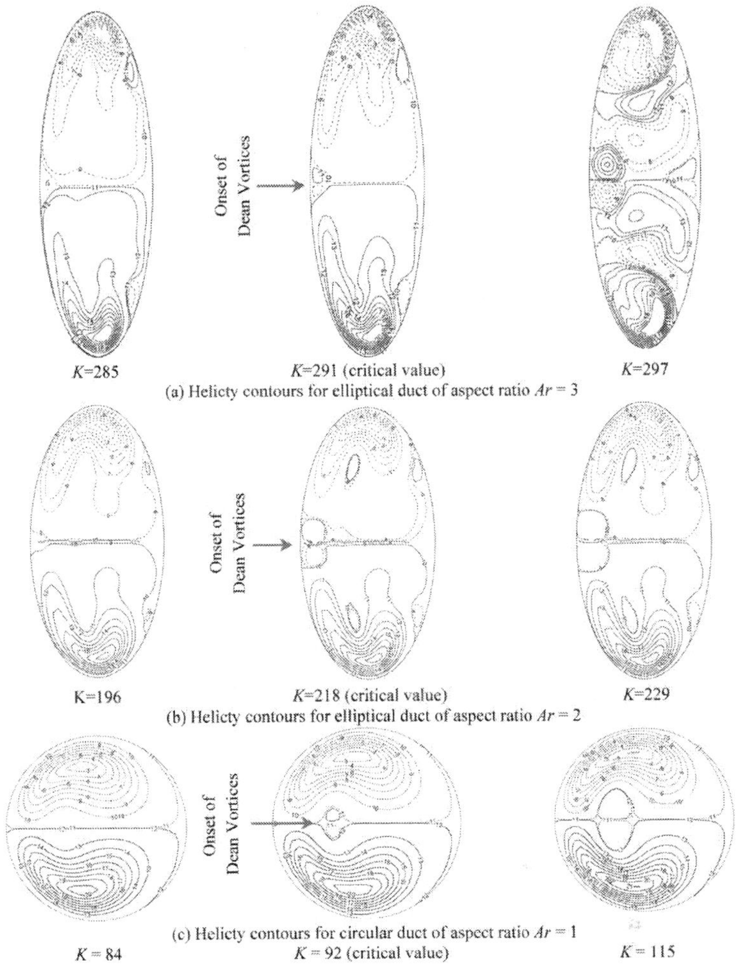

Figure 8. Dean vortex formation in elliptical ducts of different aspect ratios *(detected using $H^* = 0.01$)*

4.2.3. *Effect of duct aspect ratio*

Obtained by using the helicity threshold method in the current study, Fig. 9 illustrates the influence of duct aspect ratio on the critical flow requirement for triggering Dean Instability in both rectangular and elliptical ducts. For comparison, the figure also provides the critical Dean number from the experimental results of Chen et al. [6] for rectangular ducts.

It is noted that the critical Dean number initially increases with the aspect ratio up to a certain value and then falls away for higher *K*. This behaviour conforms to the trend shown in Fig. 8 for elliptical ducts and the observations reported in previous experimental and numerical work. For rectangular ducts, the current model under-predicts the critical requirement for Dean Instability compared to the experimental results of [6]. This is because, the helicity method identifies Dean vortices much early in the growth process, whilst in experiments, the vortices are visually observed after they have grown to a detectable size at a higher Dean number.

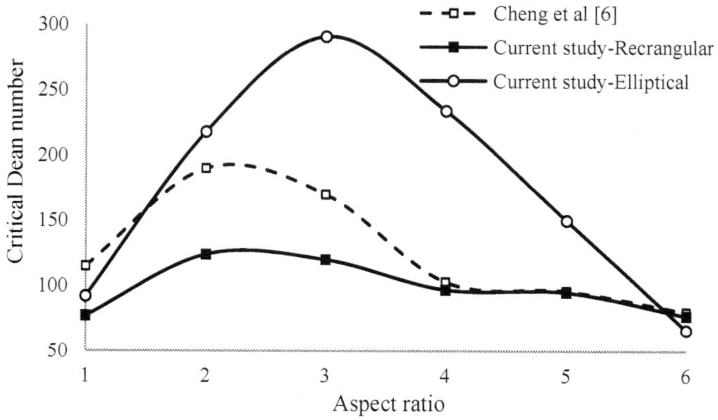

Figure 9. Effect of aspect ratio on critical Dean number

4.3. Thermal characteristics and forced convection

4.3.1. Effects of flow rate

The helicity contours and the fluid temperature fields in Fig. 10 illustrate the effect of outer wall heating in rectangular and elliptical ducts for two selected values of K. The comparison of helicity contours in Figs. 10 and 5 indicates the wall heating essentially sets up a convective fluid circulation that interacts with the secondary flow and significantly alters the flow characteristics. This fluid circulation is driven by the buoyancy forces resulting from thermally-induced density changes.

As evident in Fig. 10 for K =80, at low flow rates, the convective circulation dominates the flow behaviour with less influence from secondary vortices. When the flow rate is increased, the centrifugal action intensifies and more vigorous secondary flow develops in the duct. Consequently, the secondary vortices overcome the thermal buoyancy effects to become more dominant and determine the overall fluid behaviour, as illustrated in Fig. 10 for $K = 380$.

Due to the confined geometry, the fluid flow within elliptical ducts is generally more constrained and tends to have steeper fluid flow gradients than in rectangular ducts. This geometrical effect is more pronounced at low flow rate when the flow patterns are dominated by thermally-induced buoyancy. Fig. 10 with $K = 80$, clearly demonstrates that the elliptical duct has sharper velocity and temperature gradients at the outer wall compared to the rectangular duct. Consequently, the elliptical duct exhibits a higher rate of forced convection than in the rectangular duct within the lower range of K in Fig. 11. For increased flow rate, the secondary vortices begin to dominate the flow behaviour. As shown in Fig. 10 for $K = 380$, the rectangular duct has steeper velocity and temperature gradients at the outer wall compared to elliptical duct. This gives rise to a higher forced convection in rectangular ducts compared to the elliptical duct when K approximately exceeds 350, as depicted in Fig. 11.

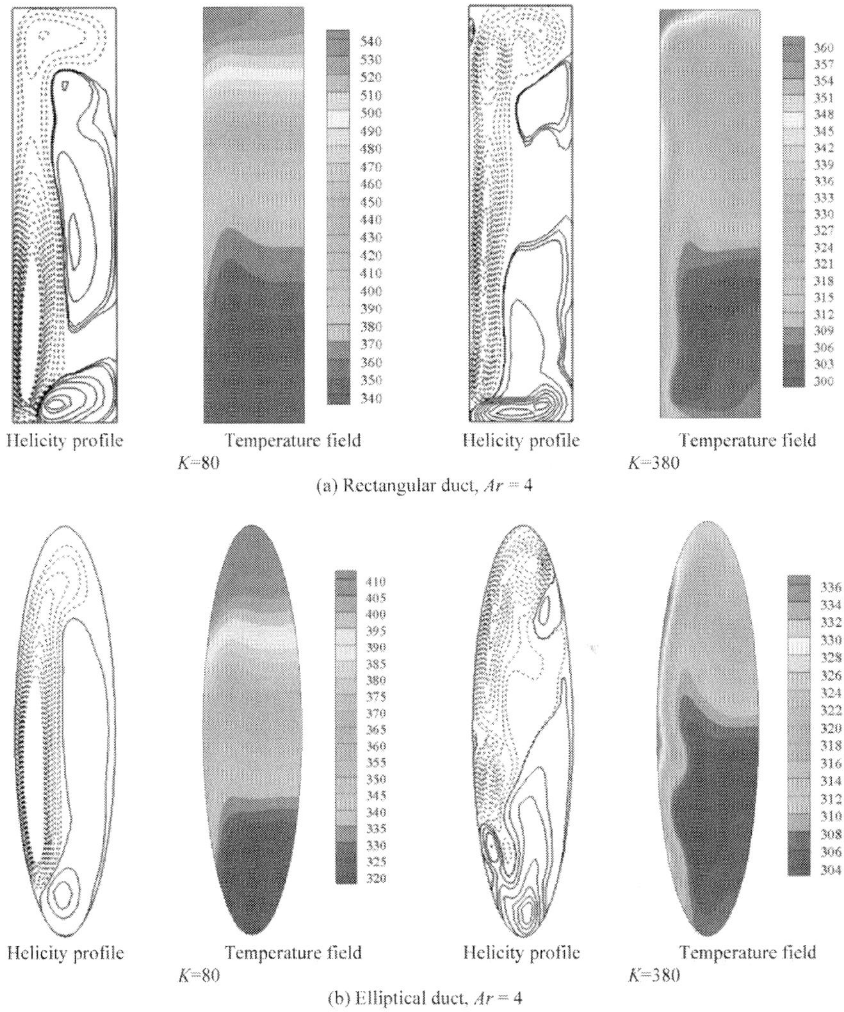

Figure 10. Outer wall heating effect on helicity and temperature profiles, $q = 250$ W/m^2

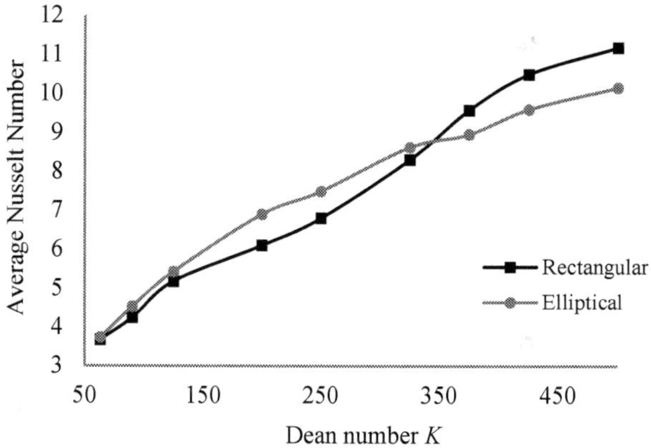

Figure 11. Fig. 11 Variation of average Nusselt number at outer wall with Dean number, $Ar = 4$, $q = 100$ W/m²

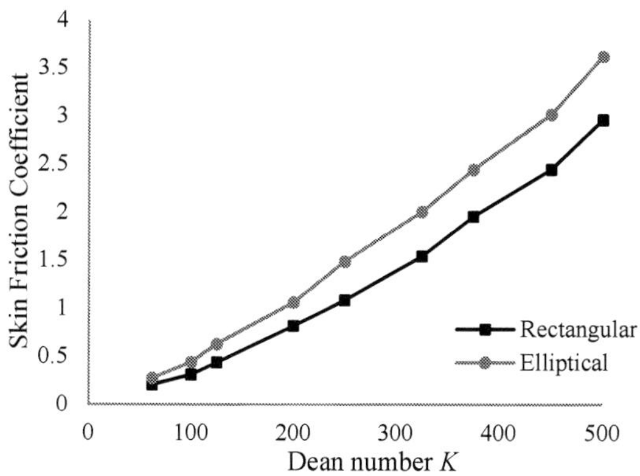

Figure 12. Variation of duct skin friction factor with Dean number, $Ar = 4$, $q = 100$ W/m²

For elliptical and rectangular ducts, Fig. 12 shows the comparison of skin friction factor, which essentially depends on the velocity gradient and is related to the duct pressure drop. With inherently steep fluid profiles, the elliptical duct has slightly larger skin friction compared to rectangular ducts with no crossover in the entire flow range.

4.3.2. *Effects of outer wall heating*

Comparison of helicity contours in Figs. 10 and 5 at $K = 380$ indicates that with outer wall heating, a tendency is developed to impede the formation process of Dean vortices in both types of ducts. This is instigated by the thermal buoyancy-driven convection that continually acts to displace the fluid layer at the outer wall, thus weakening the formation of the stagnant fluid region, where the Dean Instability would occur. Hence, the flow reversal is undermined with diminished potential for Dean vortex generation. High wall heat fluxes would impart more adversity on the triggering flow conditions of Dean Instability. In Fig. 13, these wall heating effects are clearly evident as a gradual decline in Nusselt number with increased outer wall heat flux for both types of ducts.

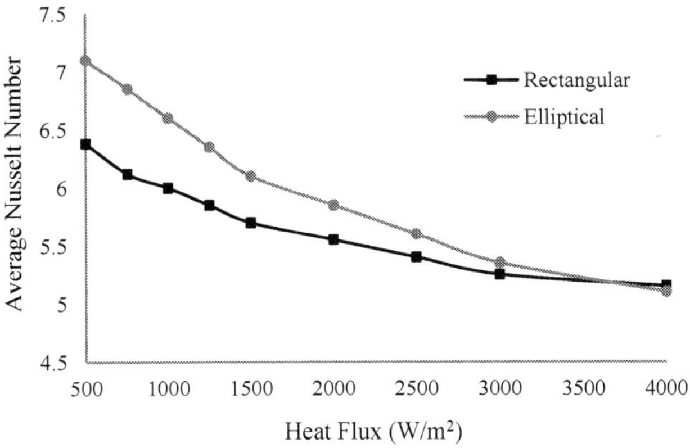

Figure 13. Variation of average Nusselt number at outer wall with outer wall heat flux, $Ar = 4$, $K = 250$

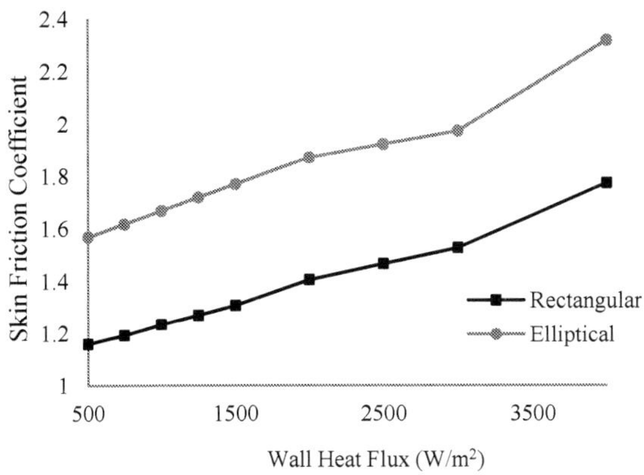

Figure 14. Variation of average duct Skin Friction with outer wall heat flux, $Ar = 4$, $K = 250$

With high wall heat fluxes, thermally induced convective circulation becomes stronger, imparting additional resistance to the axial fluid motion. This effect is more pronounced in elliptical ducts than in rectangular ducts. Consequently, duct skin friction coefficient tends to be relatively higher for elliptical ducts while indicating some increase with the wall heat flux, as evident in Fig. 14.

4.4. Thermal optimisation

Fig. 15 shows the Bejan number (Be) distribution in curved duct cross sections. For both rectangular and elliptical ducts at lower *K*, the flow irreversibility is practically dominated by the entropy generation from wall heat transfer (Be ≈ 1 with a red cast) over the entire flow cross section. As K increases, Bejan Number contours gradually acquire magnitudes less than 1 indicated by the blue cast. This signifies that in curved ducts, the secondary flow provides favourable fluid mixing to transport the hot fluid away from the heated wall, which in turn will improve forced convection. Magnitudes of Be < 1 also signifies an increased contribution to the total irreversibility from viscous effects, which would negate the overall flow benefits. These effects are less pronounced in rectangular ducts than in elliptical ducts. The opposing thermal and hydrodynamic trends identify a potentially useful technique for thermal optimisation of fluid flow through heated curved ducts, as explained below.

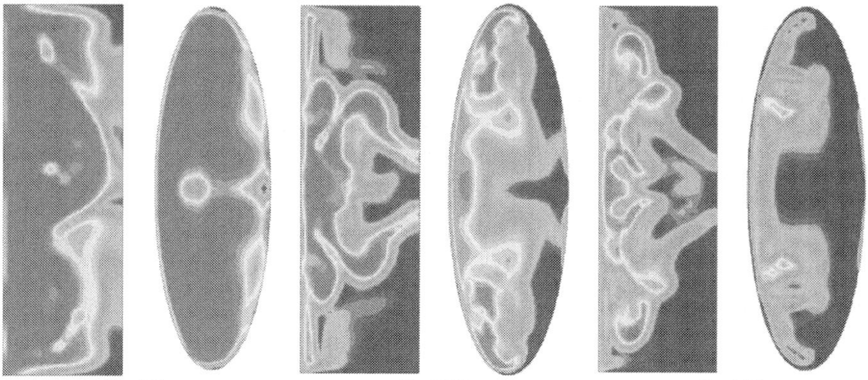

Figure 15. Bejan Number contours for rectangular and elliptical curved ducts, $q = 100$ W/m^2

Fig. 16 illustrates the variation of the flow irreversibility components S_t, S_p and S_g for rectangular and elliptical ducts with *K*, evaluated using Eq. 9 and 10. It is evident from the figure that, when *K* is increased, S_t steadily falls while S_p rapidly grows. Although this falling trend is similar to both types of ducts, the thermal irreversibility component S_t decays more gently in elliptical duct compared to the rectangular duct. This means, the elliptical duct is less efficient in transporting heat from the wall than the rectangular duct. On the other hand, the viscous irreversibility component S_p grows more rapidly in elliptical ducts, signifying that the elliptical shape causes higher frictional pressure loses.

The overall irreversibility S_g, which is the sum of S_t and S_p, is initially dominated by the thermal irreversibility component and steadily falls with K for both duct shapes. However, the falling gradient is steeper for the rectangular duct, which implies more effective heat transfer in the fluid domain with lesser adversity from viscous effects.

At a certain value of *K*, S_g reaches a minimum. This point of inflection is identified as practically the best possible "trade-off" between the highest achievable thermal benefits with the least viscous penalty for curved duct flows. It is noted that the magnitude of the lowest overall irreversibility is higher for elliptical ducts than for rectangular ducts, indicating rectangular ducts would thermally better perform.

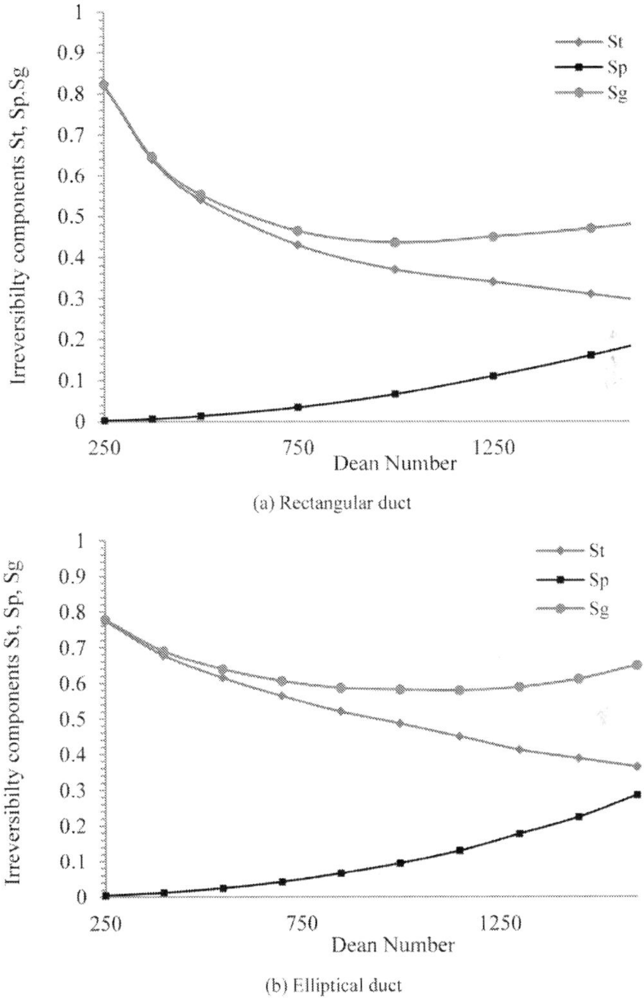

Figure 16. Curved duct thermal optimisation using total entropy generation

5. Conclusions

This book chapter provides a broad overview of the reported findings from the published numerical and experimental studies on fluid flow in curved duct geometries. It then presents and discusses the latest contributions to this field realised through an extensive research programmes of the authors where new analytical techniques and methodologies have been discovered for clearer fundamental understanding on fluid and thermal behaviour in curved ducts.

Improving accuracy over previous methods, the current study formulates a new three-dimensional simulation methodology based on helicity that realistically represents the secondary vortex structures in curved ducts of any shape. The model accurately identifies and predicts unique features of secondary flow and the associated forced convection in both rectangular and elliptical curved ducts. Conforming to the limited data in literature, the predicted results examine the effects of fluid flow rate, duct aspect ratio and heat flux over a wide practical range. The duct geometry and aspect ratio are found to have critical influence on the secondary flow characteristics, indicating a profound effect on Dean vortex formation, Dean Instability and forced convection. Elliptical ducts show marginally inferior thermal performance compared to rectangular ducts under identical flow conditions.

Overcoming a current analytical limitation, the study appraises two novel methods for detecting the onset of Dean Instability in curved ducts, ascertaining their technical feasibility. One such approach uses a defined helicity threshold while the other makes use of features of the outer wall pressure gradient. Both methods offer the process flexibility by integrating into and performed within a computational scheme, for fast, reliable and accurate detection of Dean vortex formation. However, the helicity threshold method is recognised to be more universally applied to all duct shapes and aspect ratios. The pressure gradient approach shows limited functionality for it over-predicts the critical Dean Number in curved ducts of aspect ratio near unity.

The study also presents a useful thermal optimisation technique for heated curved duct based on the Second Law irreversibilities. It identifies the elliptical ducts to have less favourable overall thermal and hydrodynamics characteristics than rectangular ducts.

Author details

Tilak T. Chandratilleke * and Nima Nadim
Department of Mechanical Engineering, Curtin University, Perth, Australia

6. References

[1] Dean, W.R. (1928), Fluid motion in a curved channel, Proc. Roy. Soc. London Ser, A121: 402–420.

---

* Corresponding Author

[2] Moffatt, H.K. (1964), Viscous and resistive eddies near a sharp corner, Journal of Fluid Mechanics, 48 (1): 1–18.

[3] Eustice, J. (1929), Experiments of streamline motion in curved pipes, Proc. Roy. Soc. London Se, 85 (1).

[4] Baylis, J.A. (1971), Experiments on laminar flow in curved channels of square section, Journal of Fluid Mechanics. 48 (3): 417–422.

[5] Humphrey, J.A.C., Taylor, A.M.K., Whitelaw, J.H. (1977), Laminar flow in a square duct of strong curvature, Journal of Fluid Mechanics, 83: 509–527.

[6] Cheng, K.C., Nakayama, J., Akiyama, M. (1977), Effect of finite and infinite aspect ratios on flow patterns in curved rectangular channels, in: Flow Visualization International Symposium, Tokyo, p. 181.

[7] Ghia, K.N., Sokhey, J.S. (1977), Laminar incompressible viscous flow in curved ducts of rectangular cross-section, Trans. ASME I: Journal of Fluids Engineering, 99: 640–648.

[8] Sugiyama, Hayashi, Taro, Yamazaki, Koji.(1983), Flow Characteristics in the Curved Rectangular Channels, Bullletin of JSME, 26 (216): 532-552.

[9] Chandratilleke, T.T.(2001), Secondary flow characteristics and convective heat transfer in a curved rectangular duct with external heating, 5th World Conference On Experimental Heat Transfer, Fluid Mechanics and Thermodynamics, Thessaloniki, Greece.

[10] Chandratilleke, T.T.(1998), Performance enhancement of a heat exchanger using secondary flow effects, in: Proc. of 2nd Pacific—Asia Conf.Mech. Eng., Manila, Philippines.

[11] Chandratilleke, Tilak T., Nursubyakto.(2002), Numerical prediction of secondary flow and convective heat transfer in externally heated curved rectangular ducts, International Journal of Thermal sciences 42: 187–198.

[12] Yanase, S. R. N. M., Kaga, Y.(2005), Numerical study of non-isothermal flow with convective heat transfer in a curved rectangular duct, International Journal of Thermal Sciences 44: 1047–1060.

[13] Fellouah, H., Moctar, A. Ould El, Peerhossaini, H.(2006), A criterion for detection of the onset of Dean instability in Newtonian fluids, European Journal of Mechanics B/Fluids 25: 505–531.

[14] Fellouah, H., Castelain, Ould-El-Moctar, Peerhossaini.(2010), H., The Dean instability in power-law and Bingham fluids in a curved rectangular duct, Journal of Non-Newtonian Fluid Mech. 165: 163–173.

[15] Guo, Jiangfeng, Xu, Mingtian, Cheng, Lin. (2011), Second law analysis of curved rectangular channels, International Journal of Thermal Sciences 50: 760-768.

[16] Ko. T.H, Ting, K.(2006), Entropy generation and optimal analysis for laminar forced convection in curved rectangular ducts: A numerical study, International Journal of Thermal Sciences 45: 138–150.

[17] Ko. T. H.(2005), Numerical investigation on laminar forced convection and entropy generation in a curved rectangular duct with longitudinal ribs mounted on heated wall." International Journal of Thermal Sciences 45: 390–404.

[18] Ko. T. H.(2006), A numerical study on entropy generation and optimization for laminar forced convection in a rectangular curved duct with longitudinal ribs, International Journal of Thermal Sciences 45: 1113–1125.

[19] Ko. T.H, C.P. Wu. (2008), A numerical study on entropy generation induced by turbulent forced convection in curved rectangular ducts with various aspect ratios." International Communications in Heat and Mass Transfer, 36: 25-31.

[20] Chandratilleke, Tilak T, Nadim, Nima and Narayanaswamy, Ramesh. (2010), Secondary flow characteristics and prediction of Dean vortices in fluid flow through a curved duct, 6th Australasian Congress on Applied Mechanics, Perth, Western Australia, Australia.

[21] Chandratilleke, Tilak T, Nadim, Nima and Narayanaswamy, Ramesh. (2011), Entropy-based secondary flow characterisation and thermal optimisation of fluid flow through curved heated rectangular ducts, 9th Astralasian Heat and Mass Transfer Cconference – 9AHMTC, Melbourne, Victoria, Australia.

[22] Dong Z.F., Ebadian M.A.(1991), Numerical analysis of laminar flow in curved elliptic ducts, J. Fluid Eng. 113:555-562.

[23] Silva, Ricardo Junqueira, Valle, Ramon Molina, Ziviani, Marcio.(1999), Numerical hydrodynamic and thermal analysis of laminar flow in curved elliptic and rectangular ducts, International Journal of Thermal Sciences, 38: 585-594.

[24] Papadopoulos, P.K., Hatzikonstantinou, P.M.(2008), Numerical study of laminar fluid flow in a curved elliptic duct with internal fins, International Journal of Heat and Fluid Flow 29: 540–544.

[25] Andrade, Claudia R. and Zaparoli, Edson L.(2001), Effect of temperature-dependent viscosity on fully developed laminar forced convection in curved duct, International communication in heat and mass transfer, 28(2):211-220.

[26] Bejan, Adrian. (1996), Entropy Generation Optimization, CRC Press Inc.

Chapter 7

Forced Turbulent Heat Convection in a Rectangular Duct with Non-Uniform Wall Temperature

G.A. Rivas, E.C. Garcia and M. Assato

Additional information is available at the end of the chapter

http://dx.doi.org/10.5772/52394

1. Introduction

Rectangular ducts are widely used in heat transfer devices, for instance, in compact heat exchangers, gas turbine cooling systems, cooling channels in combustion chambers and nuclear reactors. Forced turbulent heat convection in a square or rectangular duct is one of the fundamental problems in the thermal science and fluid mechanics. Recently, Qin and Fletcher [1] showed that Prandtl's secondary flow of the second kind has a significant effect in the transport of heat and momentum, as revealed by the recent Large Eddy Simulation (LES) technique. Several experimental and numerical studies have been conducted on turbulent flow though of non-circular ducts: Nikuradse [2]; Gessner and Emery [3]; Gessner and Po [4]; Melling and Whitelaw [5]; Nakayama et al. [6]; Myon and Kobayashi [7]; Assato [8]; Assato and De Lemos [9]; Home et al. [10]; Luo et al. [11]; Ergin et al. [12] Launder and Ying [13]; Emery et al. [14]; Hirota et al.[15]; Rokni [16]; Hongxing [17]; Yang and Hwang [18]; Park [19];Zhang et al. [20]; Zheng et al. [21]; Su and Da Silva Neto [22]; Saidi and Sundén [23]; Rokni [24]; Valencia [25]; Sharatchandra and Rhode [26]; Campo et al. [27]; Rokni and Sundén [28]; Yang and Ebadian [29] and others. The Melling and Whitelaw´s [5] experimental work shows characteristics of turbulent flow in a rectangular duct where they have been used a laser-Doppler anemometer in which report the axial development mean velocity, secondary mean velocity, etc. Nakayama et al. [6] show the analysis of the fully developed flow field in rectangular and trapezoidal cross-section ducts; finite difference method was implemented and the model of Launder and Ying [13] has been used. On the other hand, Hirota et al. [15] present an experimental work in turbulent heat transfer in square ducts; they show details of turbulent flows and temperature fields. Likewise, Rokni [16] carried out a comparison of four different turbulence models for predicting the turbulent Reynolds stresses, and three turbulent heat flux models for square ducts. The literature presents various turbulence modeling in which confirm Linear Eddy Viscosity Models (LEVM) can give inaccurate predictions for Reynolds normal stresses: it does not have ability to predict secondary flows in non-circular ducts due to its isotropic treatment. In spite of that, they is one of the most popular model in the engineering due to its simplicity, good numerical stability in which can be applied in a wide variety of flows. Thus, the Nonlinear Eddy Viscosity Model (NLEVM) represents a progress of the classical LEVM in which this last one gives inequality treatment of the Reynolds normal stresses, needs of conditions for calculating turbulence-driven secondary flow in non-circular ducts and it has relatively high cost for solving the necessary two-equation formulation. The Reynolds Stress Model (RSM), also called second order or second moment closure model, is very accurate in the calculation of mean flow properties and Reynolds stresses, for simple to more complex flows including wall jets, asymmetric channel, non-circular duct and curved flows. However, RSM has some disadvantages, such as, very large computing costs. For calculating the turbulent heat fluxes, the Simple Eddy Diffusivity (SED) and Generalized Gradient Diffusion Hypothesis (GGDH) models have been adopted and investigated. Most of the works presented in the literature show results assuming constant temperature on the wall. However, in many engineering applications the heat fluxes and surface temperatures are non-constants around the duct, therefore becoming important the knowledge of the variation of the conductance around the duct, according to Kays and Crawford [30]. According to Garcia's developments [31], it is possible to carry out analysis with non-constant wall temperature boundary conditions. In this case, it is necessary to define a value that represents the mean wall temperatures in a given cross section, in which he has named T_{Wm}. Following this treatment, for the present paper, important results have been computed and they are here being presented for rectangular cross-section ducts. Fluids such as air and water were analyzed under the influence of non-constant wall temperature distributions, with consequent presentations of resulted about turbulent convective heat exchanges and flow temperature profiles.

2. Mathematical formulation

2.1. Governing equations

The Reynolds Averaged Navier Stokes (RANS) equation system is composed of: continuity equation (1), momentum equation (2), and energy equation (3).

$$\frac{\partial}{\partial x_j}(U_j) = 0 \tag{1}$$

$$\frac{\partial}{\partial x_j}(U_i U_j) = -\frac{1}{\rho}\frac{\partial P}{\partial x_i} + \frac{1}{\rho}\frac{\partial}{\partial x_j}\left[\mu\left(\frac{\partial U_i}{\partial x_j} + \frac{\partial U_j}{\partial x_i}\right)\right] + \frac{1}{\rho}\frac{\partial}{\partial x_j}(-\rho\overline{u_i' u_j'}) \tag{2}$$

$$\frac{\partial}{\partial x_j}\left(U_j T_f\right) = \frac{1}{\rho}\frac{\partial}{\partial x_j}\left[\frac{\mu}{\Pr}\frac{\partial T_f}{\partial x_j} + \left(-\rho\overline{u_j' t'}\right)\right] \tag{3}$$

For analyses of fully developed turbulent flow and heat transfer, the following hypothesis has been adopted: steady state, condition of non-slip on the wall and fluid with constant properties. The turbulent Reynolds stress $(-\rho\overline{u_i^{'}u_j^{'}})$ and the turbulent heat flux $(-\rho\overline{u_j^{'}t^{'}})$ were modeled and solved by algebraic and/or differential expressions.

2.2. Turbulence models for reynolds stresses

2.2.1. Nonlinear Eddy Viscosity Model (NLEVM)

The NLEVM Model to reproduce the tensions of Reynolds, it is necessary to include non-linear terms in the basic constitutive equations. This is done by attempting to capture the sensitivity of the curvatures of the stream lines. This model is based on the initial proposal of Speziale [35]. The Reynolds average equations, Equations (1) to (3), are applied for the device presents in the Figure 1(a) and (b).

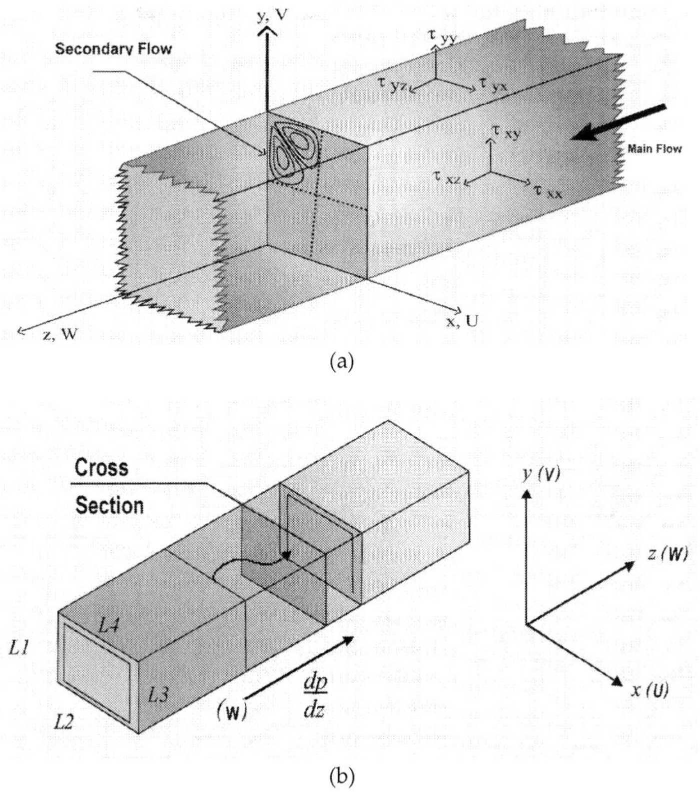

Figure 1. (a) Fully developed turbulent flows in rectangular ducts, (b) Rectangular duct: reference system and transversal section.

The velocity components *U and V* represent the secondary flow, and the axial velocity component *W*, the velocity of the main flow. The transport equations in tensorial form for the turbulent kinetic energy, κ, and the rate of dissipation ε, respectively, they are given by:

$$U_i \frac{\partial k}{\partial x_i} = \frac{\partial}{\partial x_i}\left(\frac{\mu_t}{\rho\sigma_k}\frac{\partial k}{\partial x_i}\right) + P_k - \varepsilon \tag{4}$$

$$U_i \frac{\partial \varepsilon}{\partial x_i} = \frac{\partial}{\partial x_i}\left(\frac{\mu_t}{\rho\sigma_\varepsilon}\frac{\partial \varepsilon}{\partial x_i}\right) + c_1\frac{\varepsilon}{k}P_k - c_2 f_2\frac{\varepsilon^2}{k} \tag{5}$$

The symbols P_k and μ_t, represent the rate of the turbulent kinetic energy production and the turbulent viscosity, respectively, are expressed by:

$$P_k = \tau_{ij}\frac{\partial U_i}{\partial x_j},\ \mu_t = c_\mu f_\mu \rho\frac{k^2}{\varepsilon} \tag{6}$$

In the present work for NLEVM, the formulations of Low Reynolds Number will be assumed for wall treatment. The damping functions f_2 and f_μ observed in Equations (5) and (6) were proposed by Abe et al [36]. These functions and the constant c_1 and c_2 are used in equations $k-\varepsilon$. The subscript *P* refers to the nodal point near to the wall. Thus U_P and k_P are the values of the velocity and kinetic energy in this point, respectively. For the constants c_μ, c_1, c_2, σ_k and σ_ε are assumed the values of 0.09; 1.5; 1.9; 1.4 e 1.3; respectively. New constitutive relation for the tensions of Reynolds in the NLEVM model was assumed in according to the thesis of Assato [8]:

$$\tau_{ij} = \left(\mu_t S_{ij}\right)^L + \left(c_{1NL}\,\mu_t\frac{k}{\varepsilon}\left[S_{ik}S_{kj} - \tfrac{1}{3}S_{kl}S_{kl}\delta_{ij}\right]\right)^{NL} \tag{7}$$

This expression shows that the second term of the right side in Equation (7) represents the nonlinear relation added the original constitutive relation. This quadratic term represents the degree of anisotropy between the normal tensions of Reynolds responsible for predicting the secondary flow in non circular ducts. The values of c_{1NL} proposed by Speziale [35] is equal to 1.68. In this work, c_{1NL} will be analyzed and will adopt values for the formulation of Low Reynolds Numbers. The normal and shear tensions of Reynolds are expressed as:

$$\tau_{xx} = c_{1NL}\mu_t\frac{k}{\varepsilon}\left[\frac{1}{3}\left(\frac{\partial W}{\partial x}\right)^2 - \frac{2}{3}\left(\frac{\partial W}{\partial y}\right)^2\right];\ \tau_{yy} = c_{1NL}\mu_t\frac{k}{\varepsilon}\left[\frac{1}{3}\left(\frac{\partial W}{\partial y}\right)^2 - \frac{2}{3}\left(\frac{\partial W}{\partial x}\right)^2\right] \tag{8}$$

$$\tau_{xy} = c_{1NL}\mu_t\frac{k}{\varepsilon}\left[\frac{\partial W}{\partial x}\frac{\partial W}{\partial y}\right];\ \tau_{xz} = \mu_t\frac{\partial W}{\partial x};\ \tau_{yz} = \mu_t\frac{\partial W}{\partial y} \tag{9}$$

The following differences for the normal tensions of Reynolds are presented and used in order to predict the anisotropy in turbulent flow at non circular ducts,

$$\left(\tau_{yy} - \tau_{xx}\right) = c_{1NL}\mu_t \frac{k}{\varepsilon}\left[\left(\frac{\partial W}{\partial y}\right)^2 - \left(\frac{\partial W}{\partial x}\right)^2\right] \tag{10}$$

Therefore, the Equation (6), including the tensions of Reynolds given in Equation (9), the turbulence production term is expressed as:

$$P_k = \tau_{xz}\frac{\partial W}{\partial x} + \tau_{yz}\frac{\partial W}{\partial y} \tag{11}$$

2.2.2. *Reynolds Stress Model (RSM)*

The most complex turbulence model is the Reynolds Stress Model (RSM), also known as second order model. It involves calculations of Reynolds stresses to an individual form, $\rho\overline{u_i^{'}u_j^{'}}$. These Reynolds stresses are used for formulating the differential equations of turbulent flow transport. The individual Reynolds stresses are utilized to close the average Reynolds equations of the momentum conservation. This model has shown superiority in relation to the two equation models (for example, $k-\varepsilon$) in simulating of complex flows that involve swirl, rotation, etc. The exact transport equations for the Reynolds stresses, $\rho\overline{u_i^{'}u_j^{'}}$, can be written as:

$$\begin{aligned}&\frac{\partial}{\partial t}(\rho\overline{u_i'u_j'})(a) + \frac{\partial}{\partial x_k}(\rho u_k\overline{u_i'u_j'})(b) = -\frac{\partial}{\partial x_k}\left[\rho\overline{u_i'u_j'u_k'} + \overline{p(\delta_{kj}u_i' + \delta_{ik}u_j')}\right](c) + \\ &\frac{\partial}{\partial x_k}\left[\mu\frac{\partial}{\partial x_k}(\overline{u_i'u_j'})\right](d) - \rho\left(\overline{u_i'u_k'}\frac{\partial u_j}{\partial x_k} + \overline{u_j'u_k'}\frac{\partial u_i}{\partial x_k}\right)(e) - \rho\beta\left(g_i\overline{u_j'\theta} + g_j\overline{u_i'\theta}\right)(f) + \\ &\overline{p\left(\frac{\partial u_i'}{\partial x_j} + \frac{\partial u_j'}{\partial x_i}\right)}(g) - 2\mu\overline{\frac{\partial u_i'}{\partial x_k}\frac{\partial u_j'}{\partial x_k}}(h) - 2\rho\Omega_k\left(\overline{u_j'u_m'}\varepsilon_{ikm} + \overline{u_i'u_m'}\varepsilon_{jkm}\right)(i) + S(j)\end{aligned} \tag{12}$$

Where the letters represent: (a) Local derivative of the time; (b) $C_{ij} \equiv$ Convection; (c) $D_{T,ij} \equiv$ Turbulent diffusion; (d) $D_{L,ij} \equiv$ Molecular diffusion; (e) $P_{ij} \equiv$ Production term of stresses; (f) $G_{ij} \equiv$ Buoyancy production term; (g) $\phi_{ij} \equiv$ Pressure-stress (redistribution); (h) $\varepsilon_{ij} \equiv$ Dissipation term; (i) $F_{ij} \equiv$ Production term for the rotation system; (j) $S_j \equiv$ Source term. The terms of the exact equations presented previously, C_{ij}, $D_{L,ij}$, P_{ij} and F_{ij} do not require modeling. However, the terms $D_{T,ij}$, G_{ij}, ϕ_{ij} and ε_{ij} need to be modeled to close the equations. For the present analysis, the model LRR (Launder, et al [37]) is chosen, which assumes that the correlation of velocity- pressure is a linear function of anisotropy tensor in the phenomenology of the redistribution, ϕ_{ij}. For the wall treatment, it is also assumed the Low Reynolds numbers and periodic conditions in according to Rokni [16]. This model had been simulated in the commercial code Fluent 6.3.

2.3 Turbulence models for turbulent heat flux

2.3.1. *Simple Eddy Diffusivity (SED)*

This method is based on the Boussinesq viscosity model. The turbulent diffusivity for the energy equation can be expressed as: $\alpha_t = {}^{\mu_t}\!/_{\rho\sigma_t}$, where the turbulent Prandtl number, σ_t for the SED model assumes a value constant in the entire region. For the air, σ_t it assumes a value equal to 0.89. The turbulent heat flux is given by,

$$\rho\overline{u_j t} = -\frac{\mu_t}{\sigma_T}\frac{\partial T_f}{\partial x_j} \tag{13}$$

2.3.2. *Generalized Gradient Diffusion Hypothesis (GGDH)*

Daly and Harlow [38] present the following formulation to the turbulent heat flux:

$$\rho\overline{u_j t} = -\rho C_t \frac{k}{\varepsilon}\left(\overline{u_j u_k}\frac{\partial T_f}{\partial x_k}\right) \tag{14}$$

The constant C_t, assumes the value of 0.3. The main advantage of this model is in considering the anisotropic behavior of the fluid heat transport in ducts.

2.3.3. *Dimensionless energy equation for SED and GGDH models*

For a given cross section of area "*A*", it is possible to define a mean velocity "U_b" and a bulk temperature "T_b", expressed as:

$$U_b = \frac{1}{A}\iint W.dx.dy \tag{15}$$

and

$$T_b = \frac{\int_A W.T_f.dA}{U_b.A} = \frac{1}{A.U_b}\iint W.T_f.dx.dy \tag{16}$$

Kays and Crawford [30] developed a formulation to rectangular cross section ducts. They considered the boundary conditions with prescribed uniform wall temperatures at the cross section, and at the duct length. According to Garcia [31], it is possible to carry out an analysis with non-uniform wall temperature boundary conditions. In this case, it is necessary to define a value that represents the mean wall temperatures in a given cross section, "T_{Wm}", given as:

$$T_{Wm}=\frac{\left[\frac{1}{L}.\int_0^L T_1(0,y).dy+\frac{1}{D}.\int_0^D T_2(x,0).dx+\frac{1}{L}.\int_0^L T_3(D,y).dy+\frac{1}{D}.\int_0^D T_4(x,L).dx\right]}{2(L+D)} \tag{17}$$

It is possible to develop a formula similar to Kays and Crawford [30], and a new expression for the turbulent energy equation can be presented as:

$$U\frac{\partial T_f}{\partial x}+V\frac{\partial T_f}{\partial y}+W\frac{\partial T_f}{\partial z}-\left[\frac{\partial}{\partial x}\left(\alpha\frac{\partial T_f}{\partial x}-\overline{ut}\right)+\frac{\partial}{\partial y}\left(\alpha\frac{\partial T_f}{\partial y}-\overline{vt}\right)+\frac{\partial}{\partial z}\left(\alpha\frac{\partial T_f}{\partial z}-\overline{wt}\right)\right]=0 \tag{18}$$

The following considerations are applied to obtain the variables in dimensionless form:

$$X=\frac{x}{D_h}, \tag{19}$$

$$Y=\frac{y}{D_h} \tag{20}$$

and

$$\phi=\frac{\alpha.\left(T_{Wm}-T_f\right)}{U_b.D_h^2.\left(\frac{dT_b}{dz}\right)} \tag{21}$$

Replacing the Equations (13) or (14), (19)-(21) in Equation (18), dimensionless energy equations for SED and GGDH are obtained, respectively, expressed as:

$$\frac{\partial}{\partial X}\left\{(\alpha+\alpha_t)\frac{\partial\phi}{\partial X}\right\}+\frac{\partial}{\partial Y}\left\{(\alpha+\alpha_t)\frac{\partial\phi}{\partial Y}\right\}-(D_h)\left(U\frac{\partial\phi}{\partial X}+V\frac{\partial\phi}{\partial Y}\right)=-\frac{W}{U_B}\alpha\left(\frac{\phi}{\phi_B}\right) \tag{22}$$

$$\begin{aligned}&\frac{\partial}{\partial X}\left[(\alpha_{ex})\frac{\partial\phi}{\partial X}\right]+\frac{\partial}{\partial Y}\left[(\alpha_{ey})\frac{\partial\phi}{\partial Y}\right]-D_h\left(U\frac{\partial\phi}{\partial X}+V\frac{\partial\phi}{\partial Y}\right)=-\frac{W}{U_B}\alpha\left(\frac{\phi}{\phi_B}\right)-C_t\frac{\partial}{\partial X}\left[\Gamma_X\frac{\partial\phi}{\partial Y}\right]\\&-C_t\frac{\partial}{\partial Y}\left[\Gamma_Y\frac{\partial\phi}{\partial X}\right]\end{aligned} \tag{23}$$

The fluid temperature field "T_f" can be replaced by "T_b" and the Equation (21) can be expressed as:

$$\phi_b=\frac{\alpha.\left(T_{Wm}-T_b\right)}{U_b.D_h^2.\left(\frac{dT_b}{dz}\right)}, \tag{24}$$

and

$$T_b = T_{Wm} - \frac{D_h^2.U_b}{\alpha}.\frac{dT_b}{dz}.\phi_b \tag{25}$$

From Equation (21), the Equation (26) is obtained, and applying this in Equation (16), the bulk temperature is obtained and expressed in Equation (27):

$$T_f = T_{Wm} - \frac{\phi.U_b.D_h^2.\left(\frac{dT_b}{dz}\right)}{\alpha}, \tag{26}$$

and

$$T_b = T_{Wm} - \frac{D_h^2}{\alpha.A}.\frac{dT_b}{dz}.\iint W.\phi.dx.dy \tag{27}$$

Replacing Equation (27) in Equation (24), and using Equations (19) and (20), the dimensionless bulk temperature is given as:

$$\phi_b = \frac{D_h^2}{A.U_b}.\iint W.\phi.dX.dY \tag{28}$$

It is possible to compute the heat transfer rate per unit length on the wall surface, " q' ", as shown in Equation (29) in function of "T_{Wm}", "T_b", and the average heat convection coefficient, " $\bar{h}$ ". From fluid enthalpy derivative gradient [$dh_b = c_p.dT_b$], the heat transfer rate per unit length in the fluid, " q'_f ", can be expressed by Equation (30).

$$q' = P_e.\bar{h}.(T_{Wm} - T_b), \tag{29}$$

and

$$q'_f = \rho.U_b.A.c_p.\frac{dT_b}{dz} \tag{30}$$

Equation (30) can be integrated to two cross sections (inlet, z_1, and outlet, z_2), thus, the following expression is obtained,

$$T_{b_{z2}} = T_{Wm} - \left(T_{Wm} - T_{b_{z1}}\right).e^{-\left(\frac{P_e}{2A}\right)^2.\frac{\alpha}{U_b}.Nu.(z_1 - z_2)} \tag{31}$$

From Equation (31), the bulk temperature longitudinal (*z-axis*) variation "ΔT_b" is obtained. It is done by "cutting" the duct into a lot of segments and applying the numerical method to find "ΔT_b" at each finite cross section. For a given bulk temperature at the duct inlet section (T_{b1}), after solving the equation system, duct outlet bulk temperature (T_{b2}) is calculated from Equation (31). The Dimensionless boundary conditions are given by the following equations:

$$\phi(0,Y)=\frac{\alpha.(T_{Wm}-T_1)}{U_b.D_h^2.\left(\frac{dT_b}{dz}\right)} \tag{32a}$$

and

$$\phi(X,0)=\frac{\alpha.(T_{Wm}-T_2)}{U_b.D_h^2.\left(\frac{dT_b}{dz}\right)} \tag{32b}$$

$$\phi\left(\tfrac{D}{D_h},Y\right)=\frac{\alpha.(T_{Wm}-T_3)}{U_b.D_h^2.\left(\frac{dT_b}{dz}\right)},\text{ and }\phi\left(X,\tfrac{L}{D_h}\right)=\frac{\alpha.(T_{Wm}-T_4)}{U_b.D_h^2.\left(\frac{dT_b}{dz}\right)} \tag{33}$$

When considering uniform wall temperature, the Equations (32) and (33) are equal to zero, and for these particular conditions, it is possible to notice that these boundary conditions are not functions of " dT_b/dz ". That simplification becomes equal to the one studied by Patankar [32]. Equations (24)-(26) , as well as the boundary conditions from Equations (32) and (33), form a set of differential equations, in which "ϕ" and "dT_b/dz" parameters are unknown. When that equation system is solved, it is possible to obtain "T_f".

2.3.4. *Additional equations*

Additional equations were utilized for the calculation of the factor of friction Moody, f ; coefficient of friction Fanning, C_f ; Prandtl law; local Nusselt number for the Low Reynolds formulation (Rokni [16]), Nu_{xp} and Correlation of Gnielinsky, respectively. Thus, the additional equations are given by the following equations:

$$f\equiv\frac{-(dP/dz).D_h}{\rho U_B^2/2}, \tag{34}$$

$$C_f=\frac{f}{4} \tag{35}$$

$$\frac{1}{\sqrt{f}}=2\log\left(\mathrm{Re}\sqrt{f}\right)-0.8, \tag{36}$$

$$Nu_{xp}=D_h\frac{(T_w-T_P)}{\eta(T_w-T_b)} \tag{37}$$

$$Nu = \left[\frac{(f/8)(\mathrm{Re} - 1000)\mathrm{Pr}}{\left(1 + 12.7(f/8)^{1/2}\left(\mathrm{Pr}^{2/3} - 1\right)\right)} \right] \tag{38}$$

3. Numerical implementation

After applying the method of finite differences to the algebraic equations, to obtain the temperature fields, the following five steps indicate the developed methodology in the numerical solution. (Garcia [31]):

Step 1. To define the function value of the non uniform temperatures in the walls of the duct $T_{Wm} = f\left(T_1(0,y), T_2(x,0), T_3(D,y), T_4(x,L)\right)$, This function can be expressed by a Fourier expansion;

Step 2. To obtain velocity field and estimated values for " U_b ""T_{Wm}" *and* "dT_b/dz";

Step 3. Equations for the boundary conditions are evaluated (Equations 32 and 33);

Step 4. Dimensionless energy equation (Temperature field, "Ø") the Equation (23) is solved and "$Ø_b$" is computed according to Equation (28), until convergence is obtained ($Ø_b$ < tolerance). This is the end of the first iterative loop;

Step 5. A value for "dT_b/dz" is computed in accordance with Equation (25). Boundary conditions are updated (step 3) to obtain a solution for the new temperature field (step 4), until convergence is obtained (dT_b/dz < tolerance). This is the end of the second iterative loop;

For all steps, "tolerance of 10^{-7}" is the value to be accomplished by the convergence criteria, which is applicable to "$Ø_b$" (dimensionless bulk temperature), "dT_b/dz" and "Ø" (dimensionless temperature field).

4. Results and discussion

4.1. Fluid flow and heat transfer field

The Figure 2(a) shows the utilized grid (120X120) in the numerical simulation for the formulations of Low Reynolds, the Figure 2(b) it represents the secondary flow contours and comparisons of the velocity profile (NLEVM, Assato [8]) with the experimental work of the Melling and Whitelaw [5] for fluid water and Re=42000.

The predicted distributions of the friction coefficient (NLEVM and RSM) and Nusselt number (SED and GGDH) dependence on Reynolds number for fully developed flow and heat transfer in a square duct is shown in Figure 3(a) and 3(b), respectively.

Figure 4(a): comparisons of the Results (RSM-SED) numerical with the experimental for temperature profile (wall constant temperature) $(T_{Wm} - T_f)/(T_{Wm} - T_C)$ with fluid air and *Re*=65000 (Hirota [15]) are shown, the figure 4(b) shows the variation of the temperature profile with non-uniform wall Temperature: south=400K, north=373K, east=393K, west=353K; presented as Case I.

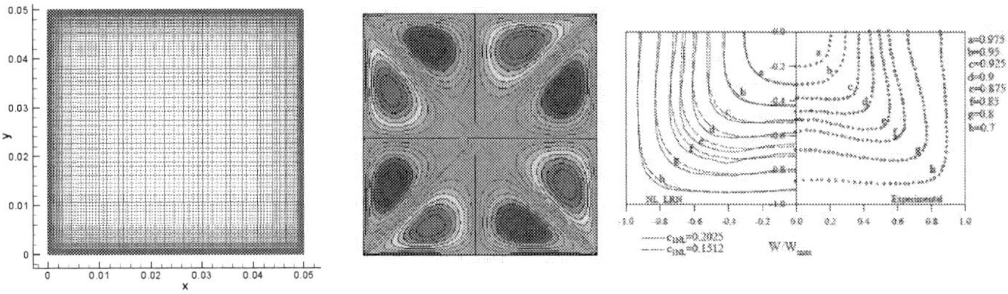

(a) (b)

Figure 2. (a) Grid 120X120 for numerical simulation (b) Secondary flow contours and comparisons of the axial mean velocity with Melling and Whitelaw [5] for water utilizing NLEVM Model.

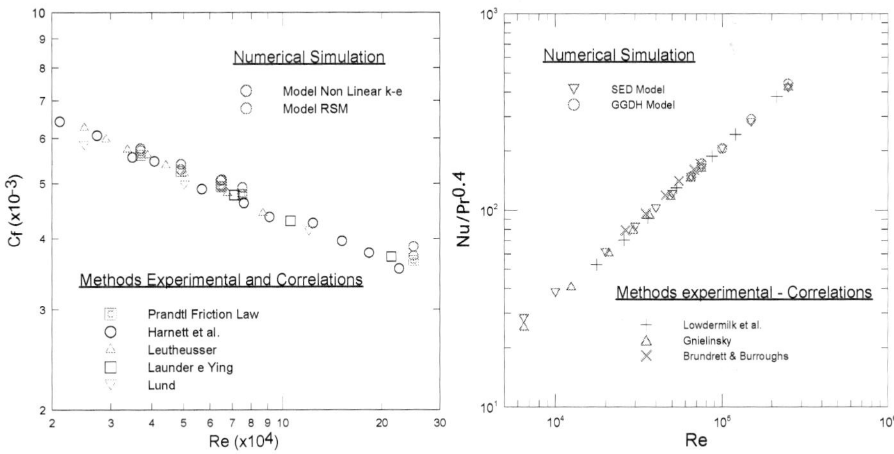

(a) (b)

Figure 3. (a) Friction coefficient for fully developed flow, (b) Nusselt number dependence on Reynolds number for fully developed flow

Already the Figure 5(a) shows: The variation of the temperature profile with non-uniform wall temperature, represented by means of functions sine (Case II), south=(350-20Sin(ζ))K, north=(400-50Sin(ζ))K, east=(330+20Sin(ζ))K, west=(350+50Sin(ζ))K. where ζ is function of the radians (0-π/2) and *i,j* (points number of the grid in the direction x and y, respectively). The Figure 5(b) shows the behavior of the "T_b" and "DTb/dz" for different square cross sections in the direction of the main flow, according to Equation (31).

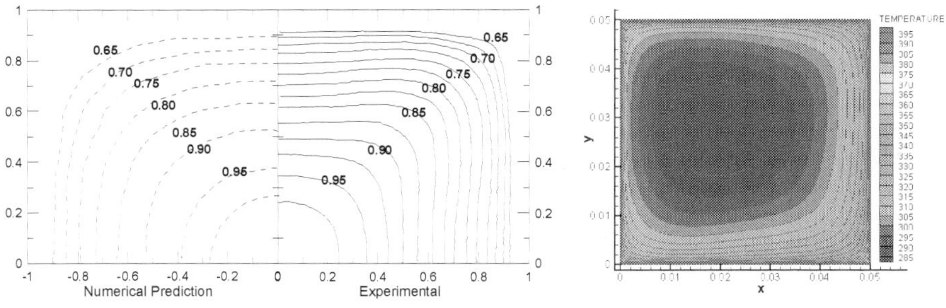

(a) (b)

Figure 4. (a) Results (RSM-SED) numerical and (Hirota et al [15]) experimental for mean temperature (uniform wall temperature) (b) Fluid temperature with non-uniform wall Temperature (Case I).

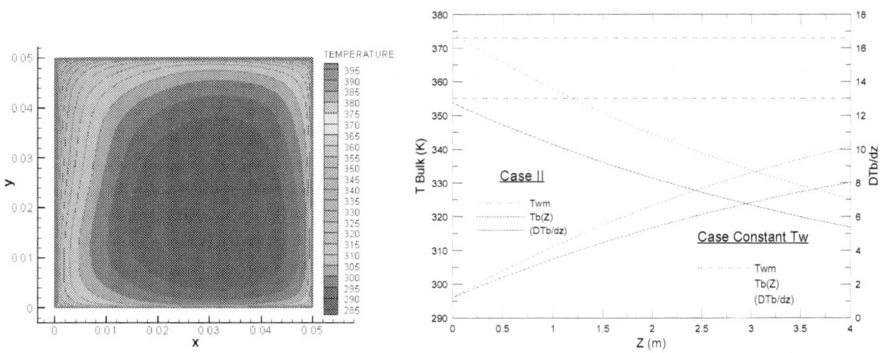

(a) (b)

Figure 5. (a) Fluid temperature with non-uniform wall temperature (Case II) (b) Behavior "T_b" for different square cross sections and cases in the direction of the main flow.

The Figures 6 (a) and (b), shown the temperature distribution for a rectangular duct aspect ratio (1:2) represented by means of a function sine (Case II). A third case denominated Case III is represented by: south= $405+10\left[\frac{\text{nx-1}}{\text{nx}_{\text{máx}}-1}\right]$ *K;* North= $395+10\left[\frac{\text{nx-1}}{\text{nx}_{\text{máx}}-1}\right]$ *K;* east= 405-10 $\left[\frac{\text{ny-1}}{\text{ny}_{\text{máx}}-1}\right]$ *K;* west= $415-10\left[\frac{\text{ny-1}}{\text{ny}_{\text{máx}}-1}\right]$ *K*. Some results for rectangular ducts are shown in the Table 1.

In the doctoral thesis Garcia (1996) was analyzed the laminar flow coupled to the conduction and radiation in rectangular ducts and concluded that as increases the aspect ratio, the Nusselt number found in the coupling, differs from that found for ducts with constant temperature imposed around the perimeter of the section. which shows that could be making a mistake to consider the literature results without calculating the energy equation.

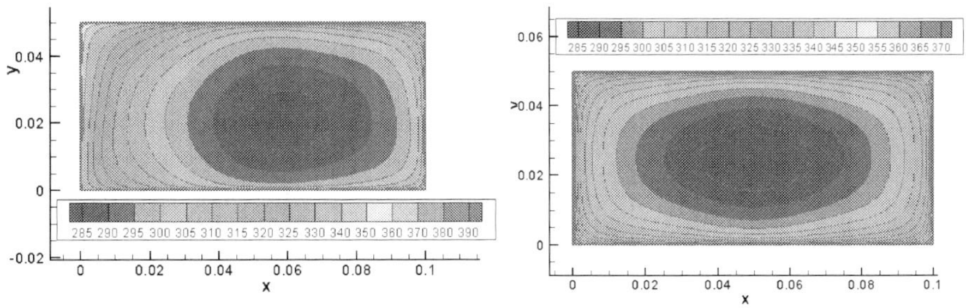

(a) (b)

Figure 6. (a) Retangular duct aspect ratio (1:2) case II, Re=65000 with *Tb*=300 *K*, utilizing the SED model (b) Retangular duct aspect ratio (1:2) with constant temperature in the perimeter *Twm*=373 *K*, *Re*=65000 with *Tb*=300 *K*, utilizing the SED model.

In the present study, the variations of the average Nusselt number for a square duct and different cases analyzed (uniform and non uniform temperature in the perimeter) are minimal. Already in the case of rectangular duct with an aspect ratio (1:2), the variations should be taken into account as shown in Table 1.

Cases Analyzed	Reynolds number (Re)	Nusselt number calculated *(Nu)*	Correlation Dittus Boelter *(Nu)*.
Temperature Constant	65000	145, 910	142,89
Case II	65000	139, 682	-
Case III	65000	145, 059	-
Temperature Constant	28853	79, 101	77,1
Case III	28853	76, 769	-

Table 1. Numerical results obtained through RSM-SED model [34], for the averaged Nusselt number in a rectangular duct with aspect ratio (1:2).

5. Conclusions

The results shown what for the friction factor and Nusselt number in a wide range of the Reynolds number with uniform wall temperature have a reasonable approach with the experimental works and correlation of the literature, (Figure 3(a), (b)). The Figures 4(b) and 5(a) shown new results investigated in present study, which is observed a distortion of the temperatures field and as consequence a variation of the Nusselt number caused mainly by the distribution of the non-uniform wall temperature (Case I and II, with fluid air and *Re*=65000, respectively). Most applications can be approximated by the functions sine and cosine in the wall, but we are able to resolve by means of the methodology presented, any peripheral heat flux variation that can be expressed by a Fourier expansion (Kays and Crawford [30]). The Figure 5(b) shows the comparisons of the behavior of the curves "T_b" and "DTb/dz" in the direction of the main flow for Case II and Case uniform wall temperature. The variations of the average Nusselt number for a square duct and different cases analyzed (uniform and non uniform temperature in the perimeter) are minimal. Already in the case of duct with an aspect ratio (1:2) the variations should be taken into account. These results can be helpful in the project of thermal devices as in heat transfer and secondary flows in cavities, seals, channel of gas turbines and others.

Nomenclature

C_{ij} convection
c_p specific heat at constant pressure
D duct height
D_h hydraulic diameter, $D_h = 4.A/P_e = 2.L.D/(L+D)$
$D_{L,ij}$ molecular diffusion
$D_{T,ij}$ Turbulent diffusion
dP/dz pressure gradient at z direction (longitudinal axis)
f, C_f factor of Moody´s friction, and Fanning´s friction coefficient, respectively.
F_{ij} Term production for the rotation system
G_{ij} buoyancy production Term
k_f fluid thermal conductivity
L duct width
Nu Nusselt number
P_e perimeter
P_{ij} production Term of tensions
P_k turbulence production term
Re Reynolds number
S_j Source term
T temperature
T_b internal flow bulk temperature
T_{Wm} wall mean temperature
T_1, T_2, T_3 and T_4 temperature distributions at duct wall (bottom, right side, top and left side)
U_b internal flow bulk velocity
U, V and W average velocity in the direction x, y and z; respectively
y^+ dimensionless wall distance

Greek Symbols

α	thermal diffusivity, $\alpha = k_f / (\rho . c_p)$
ε_{ij}	Term of dissipation
η	distance normal to the wall
ϕ	dimensionless temperature distribution
ϕ_{ij}	Term of pressure-tension (redistribution)
μ	dynamic viscosity
μ_t	turbulent viscosity
ρ	density
σ_t	turbulent Prandtl number

Author details

G.A. Rivas*, E.C. Garcia and M. Assato
Instituto Tecnologico de Aerŏnautica (ITA),Brazil

6. References

[1] Z.H. Qin; R.H. Plecther, Large eddy simulation of turbulent heat transfer in a rotating square duct, International Journal Heat Fluid Flow. 27 (2006) 371-390.

[2] J. Nikuradse, 1926, Untersuchung uber die Geschwindigkeitsverteilung in turbulenten Stromungen, Diss. Göttingen, VDI - forschungsheft 281.

[3] F.B. Gessner and A.F. Emery, A Reynolds stress model for turbulent corner flows – Part I: Development of the model, Journal Fluids Eng. 98 (1976) 261-268.

[4] F.B. Gessner, and J.K. Po, A Reynolds stress model for turbulent corner flows – Part II: Comparison between theory and experiment, Journal Fluids Eng. 98 (1976) 269-277.

[5] A. Melling and J.H. Whitelaw, Turbulent flow in a rectangular duct, Journal Fluid Mechanical. 78 (1976) 289-315.

[6] A. Nakayama, A.; W.L. Chow and D. Sharma, Calculation of fully development turbulent flows in ducts of arbitrary cross-section, Journal Fluid Mechanical, 128 (1983) 199-217.

[7] H.K. Myon and T. Kobayashi, Numerical Simulation Of Three Dimensional Developing Turbulent Flow in a Square Duct with the Anisotropic κ-ε Model, Advances in Numerical Simulation of Turbulent Flows ASME, Fluids Engineering Conference, 1991. Vol.117, Portland, United States of America, pp. 17-23.

[8] M. Assato, Análise numérica do escoamento turbulento em geometrias complexas usando uma formulação implícita, Doctoral Thesis, Departamento de Engenharia

* Corresponding Author

Mecânica, Instituto Tecnológico de Aeronáutica - ITA, São José dos campos - SP, Brazil, 2001.

[9] M. Assato; M.J.S. De Lemos, Turbulent flow in wavy channels simulated with nonlinear models and a new implicit formulation, Numerical Heat Transfer – Part A: Applications. 56 (4) (2009) 301-324.

[10] D. Home; M.F. Lightstone; M.S. Hamed, Validation of DES-SST based turbulence model for a fully developed turbulent channel flow problem, Numerical Heat Transfer – Part A: Applications. 55 (4) (2009) 337-361.

[11] D. D. Luo; C.W. Leung; T.L. Chan; W.O. Wong, Simulation of turbulent flow and forced convection in a triangular duct with internal ribbed surfaces, Numerical Heat Transfer – Part A: Applications. 48 (5) (2005) 447-459.

[12] S. Ergin; M. Ota; H. Yamaguchi, Numerical study of periodic turbulent flow through a corrugated duct, Numerical Heat Transfer – Part A: Applications. 40 (2) (2001) 139-156.

[13] B.E. Launder and W.M.Ying., Prediction of flow and heat transfer in ducts of square cross section", Proc. Inst. Mech. Eng., 187 (1973) 455-461.

[14] A.F. Emery, P.K. Neighbors and F.B. Gessner, The numerical prediction of developing turbulent flow and heat transfer in a square duct, Journal Heat Transfer, 102 (1980) 51–57.

[15] M. Hirota, H. Fujita, H. Yokosawa, H. Nakai, H. Itoh, Turbulent heat transfer in a square duct, International Journal Heat and fluid flow, 18 (1997) 170-180.

[16] M. Rokni, Numerical investigation of turbulent fluid flow and heat transfer in complex duct, Doctoral Thesis, Department of Heat and Power Engineering. Lund Institute of Technology, Sweden, 1998.

[17] Y. Hongxing, Numerical study of forced turbulent heat convection in a straight square duct, International Journal of Heat and Mass Transfer, 52 (2009) 3128-3136.

[18] Y.T. Yang; M.L. Hwang, Numerical simulation of turbulent fluid flow and heat transfer characteristics in a rectangular porous channel with periodically spaced heated blocks, Numerical Heat Transfer – Part A: Applications. 54 (8) (2008) 819-836.

[19] T.S. Park, Numerical study of turbulent flow and heat transfer in a convex channel of a calorimetric rocket chamber, Numerical Heat Transfer – Part A: Applications. 45 (10) (2004) 1029-1047.

[20] J. Zhang; L. Dong; L. Zhou; S. Nieh, Simulation of swirling turbulent flows and heat transfer in a annular duct, Numerical Heat Transfer – Part A: Applications. 44 (6) (2003) 591-609.

[21] B. Zheng; C.X. Lin; M.A. Ebadian, Combined turbulent forced convection and thermal radiation in a curved pipe with uniform wall temperature, Numerical Heat Transfer – Part A: Applications. 44 (2) (2003) 149-167.

[22] J. Su; A.J. Da Silva Neto, Simultaneous estimation of inlet temperature and wall heat flux in turbulent circular pipe flow, Numerical Heat Transfer – Part A: Applications. 40 (7) (2001) 751-766.

[23] A. Saidi; B. Sundén, Numerical simulation of turbulent convective heat transfer in square ribbed ducts, Numerical Heat Transfer – Part A: Applications. 38 (1) (2001) 67-88.

[24] M. Rokni, A new low-Reynolds version of an explicit algebraic stress model for turbulent convective heat transfer in ducts, Numerical Heat Transfer – Part B: Fundamentals. 37 (3) (2000) 331-363.

[25] A. Valencia, Turbulent flow and heat transfer in a channel with a square bar detached from the wall, Numerical Heat Transfer – Part A: Applications. 37 (3) (2000) 289-306.

[26] M.C. Sharatchandra; D.L. Rhode, Turbulent flow and heat transfer in staggered tube banks with displaced tube rows, Numerical Heat Transfer – Part A: Applications. 31 (6) (1997) 611-627.

[27] A. Campo; K. Tebeest; U. Lacoa; J.C. Morales, Application of a finite volume based method of lines to turbulent forced convection in circular tubes, Numerical Heat Transfer – Part A: Applications. 30 (5) (1996) 503-517.

[28] M. Rokni; B. Sundén, Numerical investigation of turbulent forced convection in ducts with rectangular and trapezoidal cross section area by using different turbulence models, Numerical Heat Transfer – Part A: Applications. 30 (4) (1996) 321-346.

[29] G. Yang; M.A. Ebadian, Effect of Reynolds and Prandtl numbers on turbulent convective heat transfer in a three-dimensional square duct, Numerical Heat Transfer – Part A: Applications. 20 (1) (1991) 111-122.

[30] W.M. Kays; M.Crawford, Convective Heat and Mass Transfer, McGraw-Hill, New York, USA, 1980, pp. 250-252.

[31] E.C. Garcia, Condução, convecção e radiação acopladas em coletores e radiadores solares, Doctor degree thesis, ITA - Instituto Tecnológico de Aeronáutica, São José dos Campos, SP, Brasil, 1996.

[32] S.V. Patankar, Computation of Conduction and Duct Flow Heat Transfer, Innovative Research, Maple Grove, USA, 1991.

[33] M.J. Moran, et al, Introdução à Engenharia de Sistemas Térmicos: Termodinâmica, Mecânica dos Fluidos e Transferência de Calor, LTC Ed., Rio de Janeiro-RJ, Brasil, 2005.

[34] G. A. Rivas Ronceros, Simulação numérica da convecção forçada turbulenta acoplada à condução de calor em dutos retangulares, Doctor degree thesis, ITA - Instituto Tecnológico de Aeronáutica, São José dos Campos, SP, Brasil, 2010.

[35] C. G. Speziale, On Nonlinear k-ε and k-l Models of Turbulence, J. Fluid Mech., vol. 176, pp. 459-475, 1987.

[36] K. Abe, et al, An Improved k-ε Model for Prediction of Turbulent Flows with Separation and Reattachment, Trans. JSME, Ser. B, vol. 58, pp. 3003-3010, 1992.

[37] B.E. Launder, G.J. Reece, and W. Rodi: Progress in the development of a Reynolds-stress turbulence closure, J. Fluid Mech., vol.68, 537-566, 1975.

[38] Daly and Harlow: transport equations in turbulence, Phys. Fluids., vol.13, 2634-2649, 1970.

Section 3

Boiling and Condensation

Chapter 8

Droplet Impact and Evaporation on Nanotextured Surface for High Efficient Spray Cooling

Cheng Lin

Additional information is available at the end of the chapter

http://dx.doi.org/10.5772/51826

1. Introduction

Resulting from Moore's law in semiconductor technology, the progresses such as shrinking feature size, increasing transistor density, and improving circuit speeds, lead to higher chip power dissipations and heat fluxes. Consequently, new and novel cooling techniques are of interest. Bar-Cohen et al. [1] have reviewed several techniques for direct liquid cooling, such as pool boiling, gas-assisted evaporative cooling, jet impingement, spray cooling and synthetic jets, emphasizing the important implications of a direct liquid approach in the thermal management of hot spots, where heat fluxes can be as high as 1~2 kW/cm². Nanotubes have unique properties as discussed by Berber et al. [2] are reported to have measured high thermal conductivities around 6600 W/mK at room temperature for carbon nanotubes. These can be placed in the thermal interface material to provide a low heat resistance path through the thermal interface material, significantly improving the thermal conductivity of the TIM. Two-phase heat transfer involving the evaporation of a liquid in a hot region and the condensation of the resulting vapor in a cooler region can provide large heat fluxes needed for microelectronic packages to operate at acceptable temperature levels. Spray cooling, which involves the boiling of a working fluid on a heated surface, is an example of efficient heat transfer scheme that exploit the benefits of two-phase heat transfer. Nanotextured surfaces provide new opportunities to improve the controllable fluid and heat transport for thin film evaporation. Significant efforts in previous work focused on spray cooling of microstructured surfaces [3,4] and the employment of nano-textured surfaces to achieve an enhanced boiling heat transfer [5]. Carbon nanotubes (CNTs) forests have more recently been investigated to enhance nucleate boiling and film boiling [6]. Four regimes have been identified: flooded, thin film, partial dry-out and dry-out. The heated surfaces of micro-structure are most suited in the thin film and partial dry-out regimes because of the wetting enhancement. Sodtke and Stephan [7] demonstrated that micro-structured surfaces lead to an increased contact line length and thereby increase the overall heat flux. The proposed enhanced boiling mechanism is the integration of spray cooling on nanotextured surfaces, which is expected to improve the heat transfer coefficient over 10 times (up to 1000 W/cm^2). Numerous studies have been proposed cooling methods using spray cooling and pooling boiling with nanostructure have achieved very high heat fluxes; they are listed in Table 1 [8-13]. Amon *et al.*[8] and Hsieh and Yao[9] studied the heat transfer on square microstuds (160~480 mm size, groove depths 333~455 μm, groove widths 120~360 μm) manufactured on silicon. Two full-cone pressurized spray nozzles (60° and 80° cone angles, flow rates up to 4.41 g/cm^2 min, d_{32} between 75 and 100 μm) were used to spray water at very low flow rates onto the surfaces. Surface texture was found to have little effect in the single-phase and dryout regimes. The authors attributed the higher heat transfer observed for the microtextured silicon surfaces in the intermediate regimes to more effective spreading of the liquid by capillary forces. A plain aluminum surface was found to have higher heat transfer than a silicon surface, but this disadvantage could be overcome by surface texturing. The maximum heat flux achieved was just over 50 W/cm^2. Visaria and Mudawar [10] developed a relatively new universal CHF correlation that combines the spray inclination functions and corrected subcooling constant. The CNT coating, the parallel vertical CNTs as well as the mesh of CNTs create deep, near-zero-angle cavities that are ideal for embryo formation, especially for the low contact angle coolants. Ujereh *et al.*[11] also demonstrated significant CHF enhancement with CNTs for both silicon and copper walls because of the aforementioned increase in surface area. Li *et al.* [12] fabricated Cu nanorods having 50 nm diameter and 450 nm height using oblique-angle deposition. In a pool boiling experiment of water, the wall superheat was decreased, and the CHF was improved by about 10% compared to a flat Cu surface. The authors showed that multiple scales from nano to micro play a key role in enhancing the nucleate boiling performance. Chen *et al.* [13] synthesized Cu and Si nanowires by electroplating Cu into nanoscale pores and aqueous electroless etching (EE) techniques respectively. The pool boiling experiments with water showed more than 100% increase in the CHF value. The present work focuses on pool boiling of saturated water on nanowires, made of Si or Cu, and observe significant enhancement of both critical heat flux and thermal conductance on nanowires compared to plain surface. The reported CHF (~200 W/cm^2) is among the highest values for pool boiling heat transfer.

The remarkable boiling heat transfer performance of the proposed scheme is mainly attributed to the following four characteristics: (1) the large latent heat of vaporization which makes boiling a very efficient mode of heat transfer, (2) the nanostructures which achieve higher cooling performance in the thin film and partial dry-out regime, because more water will be retained on the heat transfer surface due to the capillary force, (3) the enhancement in critical heat flux (CHF) which is realized by the increase of boiling area on nano- textured surfaces, and (4) the super-hydrophilic nano-textured surface which is expected to result in much higher evaporative heat transfer rate. Recently, we have demonstrated manipulation schemes that can passively drive water droplets undergoing spontaneous self-directed motion upon contact with a chemically patterned nano-textured surface (nano-wetting effect) and with a surface tension gradient [14,15]. As such, small droplets can be transported at high rates to quickly remove dissipated heat from the surface. Following the presentation of a recent conference paper [16], this work proposes enhanced boiling mechanism is the integration of spray cooling on nanotextured surfaces, which is expected to improve the heat transfer coefficient over multiple times.

Reference	Working fluid	Droplet size	Heater size	Dominant heat transfer mechanism	Critical Heat Flux (CHF)
2005 ASME JHT (Amon et al.) [8]	HFE-720	50~100 μm	25.2×25.2 mm^2	Spray cooling (Evaporation and boiling)	45 W/cm^2
2006 IJHMT (Hsieh et al.) [9]	Water	75~100 μm	25.2×25.2 mm^2	Spray cooling (Evaporation and boiling)	50 W/cm^2
2009 IEEE CPAT (Visaria and Mudawar) [10]	Water, FC-72, FC-77, FC-87 and PF-5052	111~249 μm	10×10 mm^2	Spray cooling (Evaporation and boiling)	100 W/cm^2
		Nanowire size/ Contact angle(CA)			
2007 IJHMT (Ujereh et al.) [11]	FC-72	50 nm(dia.) 20~30 μm(height) Very small CA	12.7×12.7 mm^2	Pool boiling (Nucleate Boiling)	~30 W/cm^2
2008 Small (Li et al.) [12]	Water	40~50 nm (dia.) 450 nm(height) CA=38.5 °	10×10 mm^2	Pool boiling (Nucleate Boiling)	~160 W/cm^2
2009 Nano Letter (Chen et al.) [13]	Water	200 nm(dia.) 40~50 μm(height) CA∼0 °	10×10 mm^2	Pool boiling (Nucleate Boiling)	~200 W/cm^2

Table 1. Summary of previous works on evaporation/boiling on micro/nano-structure coated surface

2. Experiment set-up and procedure

Deionized water impinging on solid surface was studied by recording the impingement process with a high speed camera at 2000 frames/s. A schematic of the experiment set-up is shown in figure 1. The component include the following: droplet generation system, nano-textured surface and heating system, high-speed charged coupled device (CCD) camera (Ultima APX Fastcam, Photron Ltd., Japan) and lighting system. All components were synchronized so as to achieve simultaneous droplet imaging and thermal measurements.

The water droplets used in the experiment had diameter of ~2 mm and velocity of ~0.3 m/s. Three different impingement surfaces were used: an oxidized silicon, a vertical CNTs and curved CNTs surface. The range of diameters and lengths of the CNTs is about 57~105 nm and 4.28~8.48 μm, respectively. In table 2, the detail specification of the growth of CNTs on the silicon substrate, and test section are presented.

MWCNTs have been synthesized on Ni/Ti multilayered metal catalysts by thermal CVD at 400 °C, in which acetylene was used as carbon source in a 3-inch diameter quartz tube of the furnace. The Ni/Ti multilayered metal catalysts were found to decrease CNT growth temperature effectively. We used a temperature ramp to 800 ◎ at a rate of 20 ◎/min with a constant argon flow rate of 400 sccm (standard cubic centimeter per minute). At the stable temperature, followed by Ar (200 sccm) and NH_3 (200 sccm) were added to the furnace for 10 minutes. Finally, the chip cooled down at a rate 5 °C/ min. The coating thickness and nanowire size were measured with a micrometer and shown in an SEM images. Figure 2 shows the cross sectional view of the carbon nanotubes coated surface. The as-grown CNTs are superhydrophobic with a measured contact angle as high as 146°. The carbon nanotubes were functionalized using two methods: (1) H_2O plasma was utilized during the treatment, the flow rate of H_2O plasma was set at 1 sccm with duration of 30 sec.(2)aqueous solution of 6M H_2SO_4 at 80°C for 1 hour as shown in Figure 3. Water contact angle measurement using a FTA200 system (First Ten Angstroms Inc.) indicated the contact angle of 4.7°±1.0° (superhydrophilic) for the vertical CNTs surfaces and the contact angle of 1.1°±0.3° (superhydrophobic) for the curved CNTs surfaces. This functionalization of CNTs is superhydrophilic with a measured contact angle as low as 5° are shown in Figure 4.

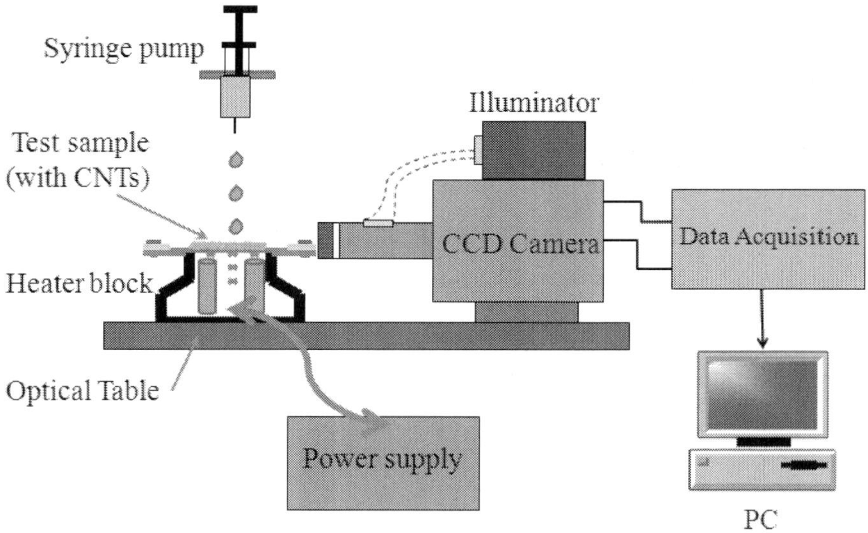

Figure 1. Schematic of an experimental apparatus.

Parameter \ Type	*Vertical CNTs*	*Curved CNTs*
Height h(μm)	7.28~8.48	4.28~7.14
Diameter D(nm)	70.9~92.1	57.3~105.4
Silicon substrate W×L×H (mm)	10×10 ×0.4	10×10×0.4

Table 2. The structural parameter of test sections

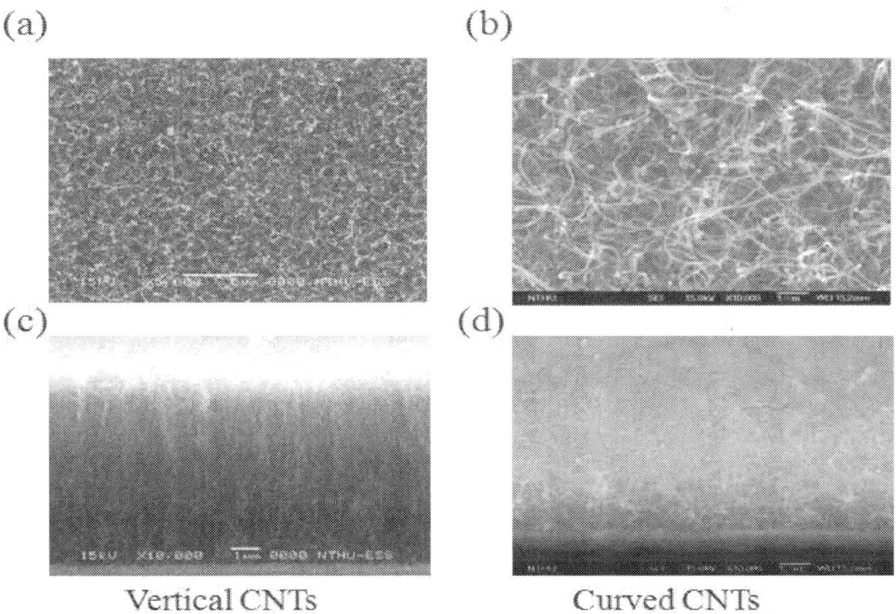

Figure 2. SEM images of the different testing surfaces:(a) top view of the vertical CNTs,(b) top view of the curved CNTs, (c)side view of the vertical CNTs, (d) side view of the curved CNTs.

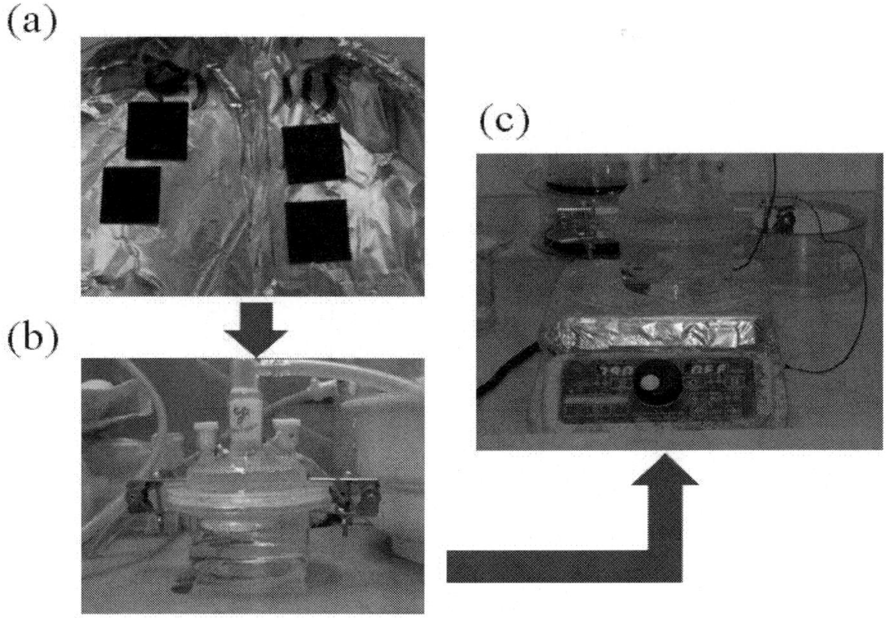

Figure 3. Schematic of the samples with CNTs surface treatment sequence: (a) top view of the CNTs samples,(b) vacuum for 5minutes using pump, (c) 80 °C, H_2SO_4 solution for 1hr.

The heat flux and the heat transfer coefficient of the heated surface were evaluated with the temperature difference (ΔT), by measuring the mean temperature between the surrounding temperature around the heated surface and the wall temperatures read by the thermocouples in the copper block. The heat flux is obtained as follows,

$$q^{''} = \frac{IV}{A} = h(T_s - T_a) = h\Delta T \tag{1}$$

Note that the heat transfer coefficient in Eq. 1 can also be written in the form

$$h = \frac{q^{''}}{(T_s - T_a)} \tag{2}$$

where q" is the heat flux (W/cm²), I is the current (Ampere), V is the voltages, and A is the heating surface area of the heated surface. The total amount of heat supplied to the heater was measured by using a power meter system, and the data for all temperatures and powers were recorded by using a data acquisition system. Heat flux was regulated successively by changing the voltage input through the DC power supply. When steady state was reached (the variation of ΔT in the range of approximately 0.2 °C), all required data were stored, and the next step was executed at a higher voltage.

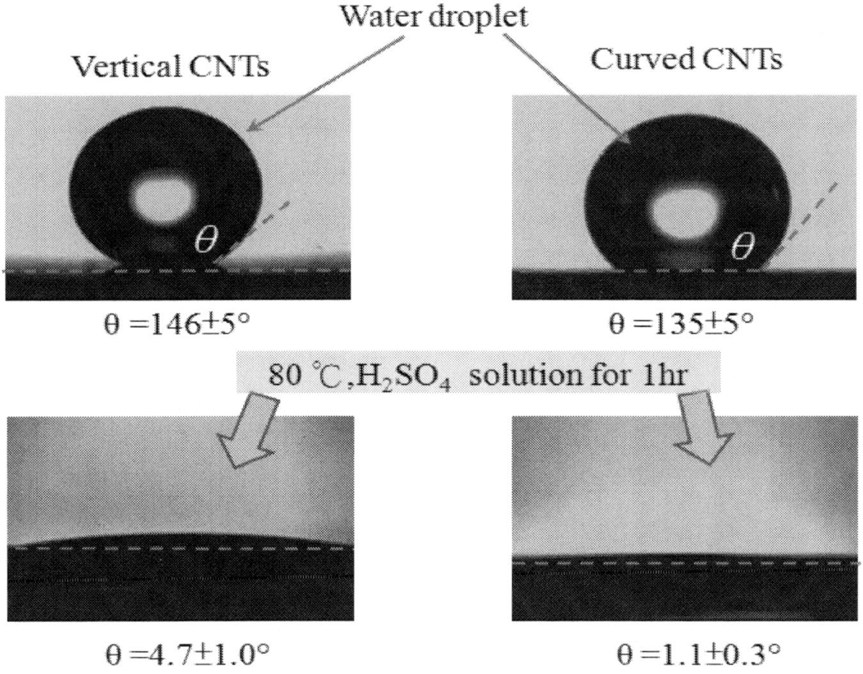

Figure 4. The contact situations of water droplets on carbon nanotubes surfaces without (upper row) and with (bottom row) H_2SO_4 treatment

3. Theoretical model

One aspect of spray cooling is basic boiling phenomena, which should be considered as a prerequisite to the more complex process of spray cooling. Therefore, consider boiling in the absence of an incoming spray. Boiling is a highly efficient means of heat transport in which liquid is vaporized due to the temperature of the liquid exceeding the saturation vapor pressure. In boiling, heat can be removed by increasing the temperature of the liquid (sensible heating) as well as the vaporization process (latent heating). As shown in Figure 5 [17], the boiling curve is a plot of surface heat flux versus excess surface temperature above saturation. When a preheated alloy exits the die in an extrusion, forging, or continuous casting process, it is typically at a temperature above the Leidenfrost point (D point) and the surface experiences film boiling. This boiling regime is characterized by a thermally-insulating layer of steam forming between the surface and individual impinging drops, resulting in poor heat transfer. When the Leidenfrost temperature is reached, the vapor film is interrupted by partial contact of liquid with the surface, causing the surface heat flux to increase with decreasing temperature until the point of critical heat flux (C point: CHF). After CHF, the surface is cooled by nucleate boiling until the temperature falls into the single-phase regime. Spray cooling is preferred to quenching in stagnant liquid because it raises the Leidenfrost temperature and enhances significantly the heat transfer rate even in the film boiling regime.

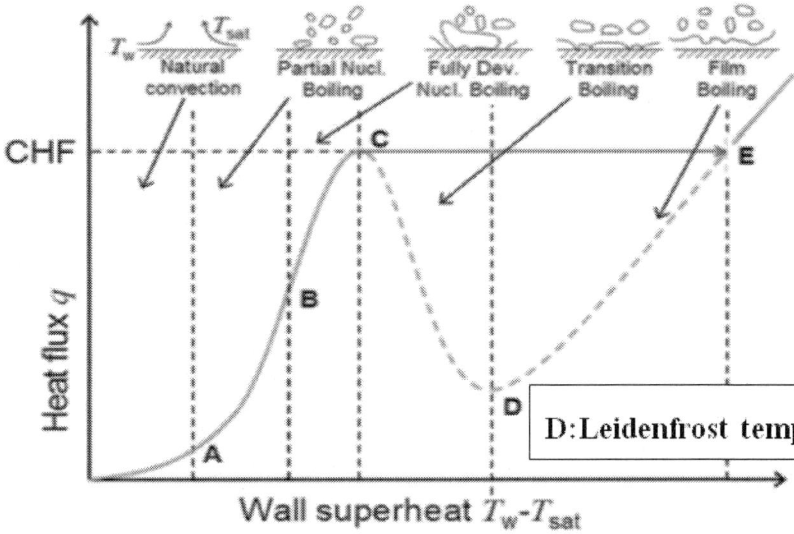

Figure 5. Typical boiling curve and associated boiling regimes showing heat flux vs. wall superheat ($T_w - T_{sat}$) (adapted from ref. 17)

3.1. Impinging droplets on wetting surface modeling

The dynamics of droplet impact on the two types CNTs surfaces modified by aqueous solution of 6M H_2SO_4 at 80°C for 1 hour were characterized to demonstrate how the surface energy influences drop impact transitions from bouncing to wetting. The antiwetting pressure (P_A) and wetting pressure (P_w) influenced the wetting states of impinging droplets. When Pw was larger than Pa, the droplet struck the surface in a wetting state. The wetting pressure (Pw) is given by

$$P_W = \frac{1}{2}\rho\, V^2 \tag{3}$$

where ϱ and V are the density and the velocity of the impinging liquid, respectively. For nanotextured surface, the antiwetting pressure is capillary pressure. The maximum value of P_A is calculated as the Laplace pressure of the maximum deformation of the water-air interface between posts of nanostructures, the capillary pressure is defined as

$$P_A = -2\sqrt{2}\gamma_{lv}\cos\theta_A / D \tag{4}$$

where γ_{lv} is the surface energy of the water at the liquid-vapor interface (0.073 N/m), θ_A is the advancing CA of the water droplet on the flat surface, and D is the spacing between the nanoscale posts of the nanostructure surface. However, this model has limitations in that only two wetting states are available: bouncing or wetting. Thus, for example, the wetting state of a droplet partially pinned at the contact area cannot be sufficiently explained by these two pressures. A new wetting pressure, the effective water hammer pressure(P_e) at the contact stage, is therefore introduced. At the moment of impact on the nanostructure, the compressed region of the droplet with pressure (P_e) pushes on the liquid-air interface between the nanoposts and induces the penetration of liquid in the contact area. The maximum possible P_e for a spherical droplet impinging at low impact speed is defined by

$$P_e \approx 0.2\rho CV \tag{5}$$

where C is the speed of sound in water (1497 m/s). When P_A is smaller than Pe but larger than Pw, the droplet is in a partial wetting state. The water droplets used in the experiment had diameters of ~2 mm and velocity of ~0.3 m/s. A high-speed camera (up to 4 K frames/s) was used in the study. Relative magnitude of the wetting and antiwetting pressures decides the wetting states of impinging droplets: (1) total wetting state (P_e>P_W>P_A) as water penetrates in both contact and spreading stage.(2) partial wetting state(P_e> P_A >P_W) as water penetrates only during contact stage.(3) Total nonwetting state (P_A > P_e> P_W) as the structure resist wetting in both stages [18]. In this experiment the droplet completely wetting state within the modified CNTs surfaces, with penetration of water in both contact and spreading stages.

3.2. Key parameter of evaporative spray cooling

The surface enhancement is realized by an array of square pins. The microstructures on the heating surface not only increase the heat transfer area but also provide a driving force for liquid spreading. To characterize the capillary force between micro-pins, a dimensionless parameter, Bond number (Bo), be expressed as

$$Bo = \frac{G}{\sqrt{\gamma / (\rho_l - \rho_v) g}} \tag{6}$$

The Bond number was defined as the ratio of the gravitational force to the surface tension force. Generally, narrower grooves (smaller G; smaller Bo) are more desirable in evaporative heat transfer because of the better liquid spreading ability. Hence, there may exist on an optimal groove width for the evaporative water spray cooling on enhanced silicon surfaces. Another influential surface factor in current study is the bottom surface area of the grooves, because it is directly related to the surface area available for the liquid evaporation in the grooves.

Several control parameters that influence on the impact dynamics include: droplet size, liquid viscosity μ, and impact velocity V_0 .We describe these effects in terms of dimensionless numbers: the Weber number We_d, the ratio of kinetic energy to surface energy, characterizing the deformability of the droplet; the Reynolds number Re, the ratio of inertia to viscosity effect:

$$We_d = \frac{\rho_f V_0{}^2 d}{\sigma}; \ \mathrm{Re}_d = \frac{\rho_f V_0 d}{\mu} \tag{7}$$

Here d the diameter of the liquid drop, V_0 is the impact velocity, ρ is the liquid density, σ is the surface tension, and μ is the liquid viscosity.

At wetting heat transfer, the droplets can be in continuous or semi-continuous direct contact with heat surface. After an initial period of transient conduction heat transfer, the droplets enter into either nucleate or transition boiling regimes. In this case, the droplet incoming Weber number may have a week effect on enhancing the droplet breakup. Wet (hydrophilic surface) cooling results in a significant drop in the surface temperature due to its highly cooling efficiency. In non-wet cooling, also referred to as film boiling, a significant amount of water vapor is generated between the heat surface and the droplet, thus preventing direct contact. Since vapor has a very low thermal conductivity, it acted as insulation between the surface and the incoming spray; therefore, lower cooling efficiency. In this boiling regime, the impact droplet velocity (or Weber number) has a significant influence on the cooling efficiency. For low Weber number, droplets cannot penetrate through the film layer. For high Weber number, droplets can penetrate through the film layer, and more surface contact can be established. The impact velocity for this case is ~30 cm/s and the corresponding Weber number (We_d) is 2.47.

3.3. Empirical model of evaporative spray cooling

In spray cooling, empirical models have been developed with the continuous expansion of experimental data based on system of interests. Mudawar and Estes (19) first attempted an empirical model to predict CHF in spray cooling by correlating CHF with the volumetric flux of liquid and the Sauter Mean Diameter of droplets, as following:

$$\frac{q_{CHF}}{\rho_v h_{lv} V} = 1.467 \left[(1+\cos(\frac{\theta}{2}))\cos(\frac{\theta}{2}) \right]^{0.3} \left[\frac{\rho_l}{\rho_v} \right]^{0.3} \left[\frac{\rho_l V^2 d_{32}}{\sigma} \right]^{-0.35} \left[1 + 0.0019 \frac{\rho_l C_{pl} \Delta T_{sub}}{\rho_v h_{lv}} \right] \tag{8}$$

where θ is the spray cone angle, d_{32} is the Sauter Mean Diameter, σ is the surface tension, $\Delta T_{sub}=T_{sat}-T_l$ is liquid subcooling at nozzle inlet, V is average volumetric flux over spray impact area, and h_{lv} is the evaporative latent heat. To predict CHF using Eq.(8), the nozzle parameters and droplet parameter (pressure drop across the nozzle, volumetric flow rate, inclined angle, and Sauter Mean Diameter of droplets) have to be tested. In addition, the distance between the nozzle orifice and the heated surface needs to be chosen carefully, so that the spray cone exactly covers the heated surface. This model was validated by a set of experiments of the spray cooling on a rectangular 12.7x12.7 mm^2 flat surface using refrigerants (FC-72 and FC-87). The volumetric flow rate was regulated inside the range of (16.6~216)x10^3 $m^3 \cdot s^{-1}/m^2$. The Sauter Mean Diameter of droplets was inside the range of 110~195 μm. The superheat temperature was below 33 °C. The accuracy of this model was claimed to be within ±30%.

Fluid flow for droplet impact was modeled using a finite-difference solution of the Navier–Stokes equations in a three-dimensional Cartesian coordinate system. The liquid is assumed to be incompressible and any effect of the ambient air on droplet impact dynamics is neglected. The fluid flow is assumed to be laminar. The free surface of the deforming droplet is tracked using the volume of fluid (VOF) method which solves the time-dependent Navier-Stokes equations, and interface evolution equation to simulate a mixture of two immiscible fluids with surface tension described in [20].

4. Results and discussion

During the entire impingement process, the droplet goes through impact, advancing, maximum spreading, receding and a period of oscillation with pinned contact line. For the non-heating (T_s=24°C) case once the oscillation decrease the drop gradual equilibrium at a 55°contact angle on oxidized silicon surface in Figure 6(a). Figure 6(b) shows time sequence images of a water droplet falling on superhydrophilic surface. Figure 6 indicates the different droplet dynamics variables. From the figure 7 prior to impact, the droplet diameter (d_0) and impact velocity (V_0) were measured. The time is determined to be zero at the moment of impact and the spreading time can be calculated relative to the impact time.

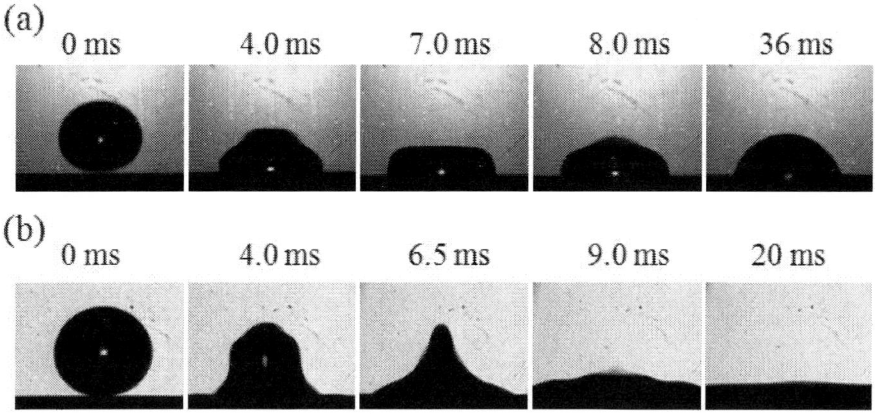

Figure 6. Illustration of droplet dynamic impinging process : (a)on the oxidized flat surface, (b)on hydrophilic curved-CNTs surface.

Images of a typical impingement with boiling process are shown in figure 8, where (a)-(e) are the images corresponding to the five stages listed. During initial impact, the droplet rapidly spreads, rebounds and then oscillates at a near-constant wetting diameter. Boiling involves initial nucleation followed by severe boiling including droplet ejection and expansion, and evaporative cooling with the absence of boiling. At non-boiling heating levels the droplet experiences evaporation once the oscillation period ends. Droplet height decreases while the wetting diameter remains nearly constant. When a critical angle is reached, the droplet diameter diminishes rapidly and then dry-out occurs.

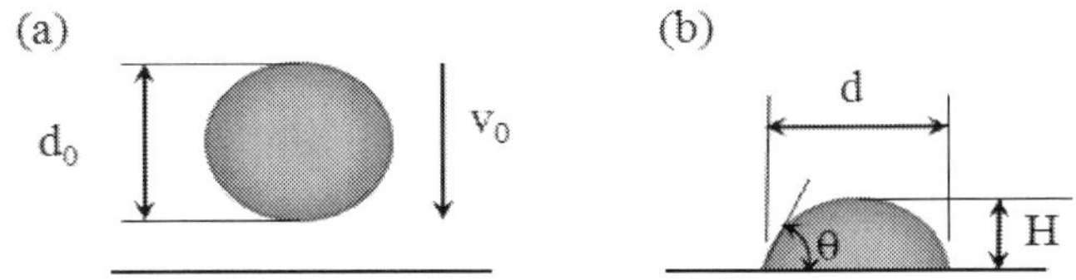

Parameter	Non-dimensional
d^*	d/d_0
H^*	H/d_0
t^*	$t/(d_0/V_0)$

Figure 7. Droplet hydrodynamics parameters: (a) pre-impact, (b) post-impact

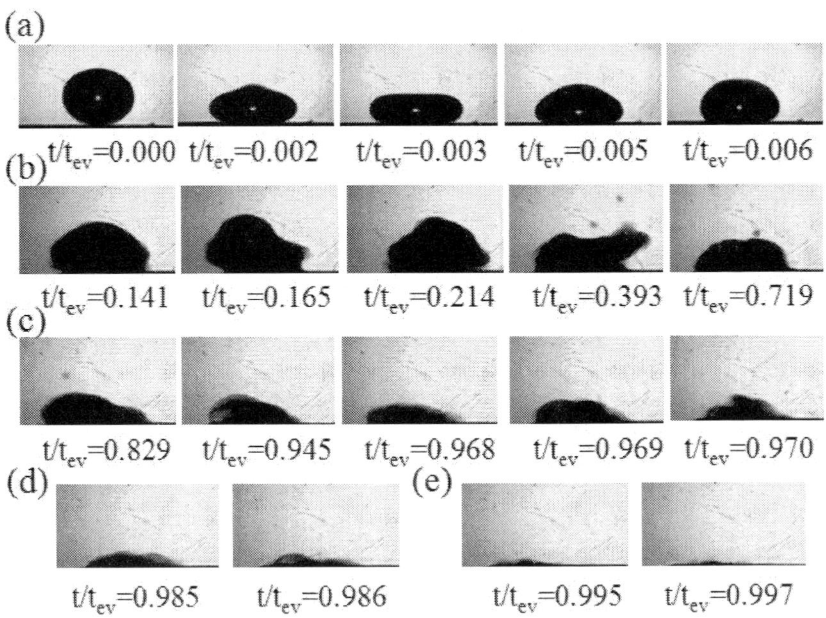

Figure 8. Droplet images of the five stages of impingement process for water impinging on the curved CNTs surface at Ts = 187°C: (a) initial impact, (b) boiling, (c) near-constant wetting diameter evaporation, (d) fast receding contact line evaporation, (e) final dry-out

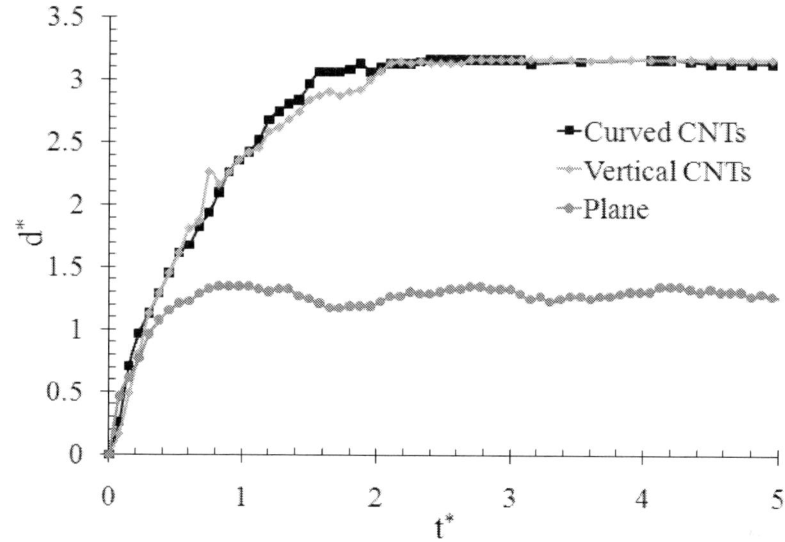

(a)

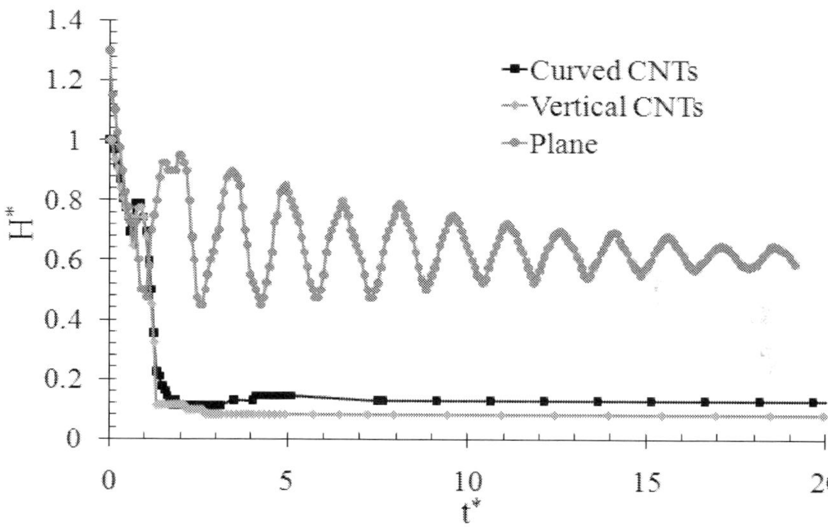

(b)

Figure 9. Initial impact non-dimensional parameter time evolution illustration at T_s=24 °C:(a)d^* [d/d_o] versus t^*[$t/(d_o/V_o)$],(b) H^*[H/d_o] versus t^*[$t/(d_o/V_o)$]

Figure 9(a) shows a rapid initial advancing to a maximum diameter followed by a rebound. The maximum spreading for all cases occurs within $t^*<2$ at T_s = 24°C. Figure 9(b) show that the droplet height at essentially the same phase oscillating at a 55°contact angle on oxidized silicon surface, where a higher droplet height results in larger contact angles. These results indicate that the surface energy differences are enhanced somewhat for the nanotextured surface with a reduction of the relative advancing and receding contact angles during the initial impact. Also, the contact angle oscillations that occur once spreading has come to equilibrium seem to be damped. Image of water impingement on three surfaces with boiling are shown in figure 10, where (a)-(c) are the images corresponding to the full-drop bouncing back. The rebounding surface temperature of oxidized silicon, vertical CNTs and curved CNTs comparison are 165°C, 247°C and 257°C, respectively, for surface temperatures at a heating (T_s) and rebounding (T_r) condition.

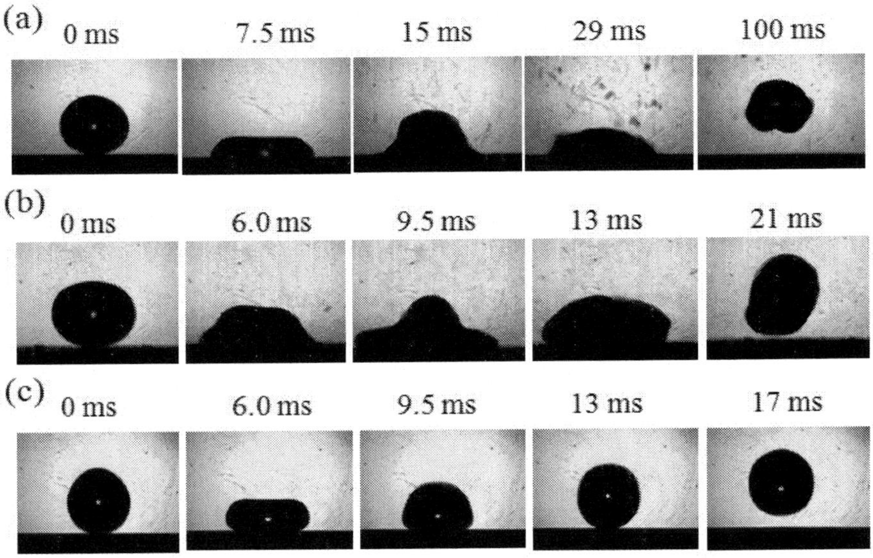

Figure 10. Snapshots of free fall droplets released from various surfaces at rebounding temperature (T_r):(a) on the oxidized silicon at T_r=165°C,(b) on the vertical CNTs surface at T_r=247°C, (c)on the curved CNTs surface at T_r=257°C .

Results show that the curved wire nanostructured surface enhanced capillary pumping effect and retained water at higher surface temperatures (T_s=226°C), as seen in figure 11. During the advancing phase the contact line length is nearly identical for the two nanotextured surfaces, however, during the retraction phase the contact line for the droplet on the curved CNTs at a surface temperature of 248°C becomes pinned, while the droplet on the vertical CNTs at a surface temperature of 248°C retracts and then rebounds off the surface is shown in figure 12. The increased spreading area over the curved CNTs surface suggests a greater potential for increased liquid lateral transportation and retention, thus bring out the consequence in improving high efficiency cooling.

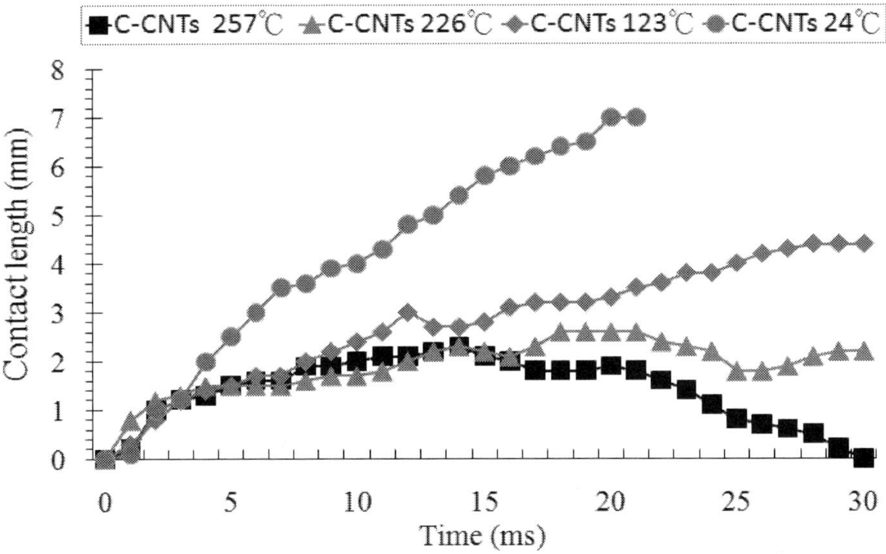

Figure 11. Contact line length versus time for droplets impinging on the curved CNTs surface at different temperatures

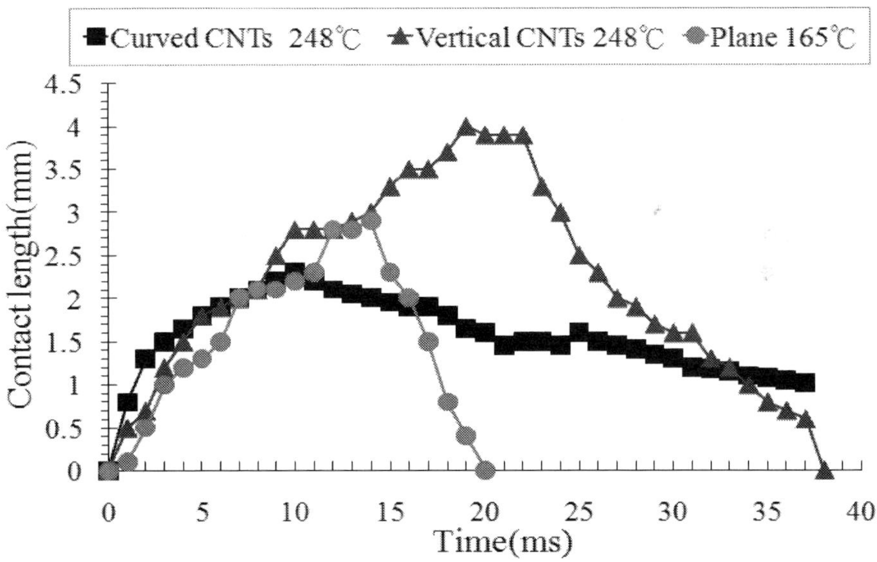

Figure 12. Contact line length versus time for droplets impinging on three surfaces at different temperatures

A digital pipette dispensed 0.1~2 μl of water onto CNTs nanotextured surface (hydrophilic surface) because the CNTs surfaces were hydrophilic modification. A 2.0 μl droplet of distilled water is placed on the heat surface at T_s=125 °C with three samples of oxidized silicon, vertical CNTs , or curved CNTs, respectively, and the water droplets were extended to a film as a function of very smaller CA was larger. The impact of a distilled water droplet upon a heated surface was investigated experimentally using a high-speed digital camera. Figure 13 illustrates images corresponding to the different characteristic stages of water droplet of 2.0μl evaporation on the oxidized silicon surface for a typical boiling condition. The time of initial solid-liquid contact, i.e. the first significant frame of a series, was taken to be the origin of the time axis (t_0 = 0). Thus, the last frame with the droplet still being visible corresponds to the droplet evaporation time. The droplet immediately boils when it contacts the surface, large bubble can be divided into film evaporation and nucleate boiling consequently dry-out are shown in figure 14. The curved CNTs surface enhanced capillary pumping effect and retained water during higher surface temperatures, evaporation rates are high and consequently the droplet lifetimes are short as shown in figure 15.

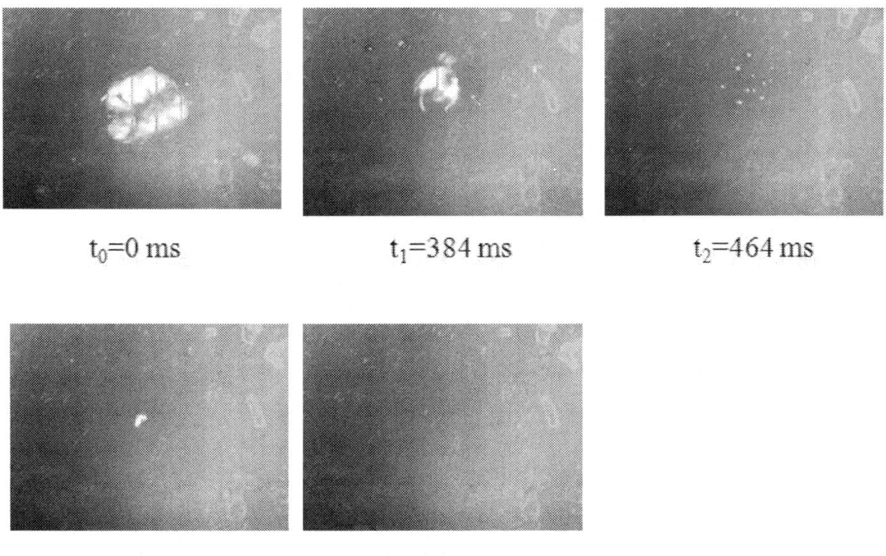

Figure 13. Images of top view corresponding to the characteristic stages of the droplet evaporation process for 2 μL water droplet impinging on plane oxidized silicon, typical boiling condition, Ts = 125 °C.

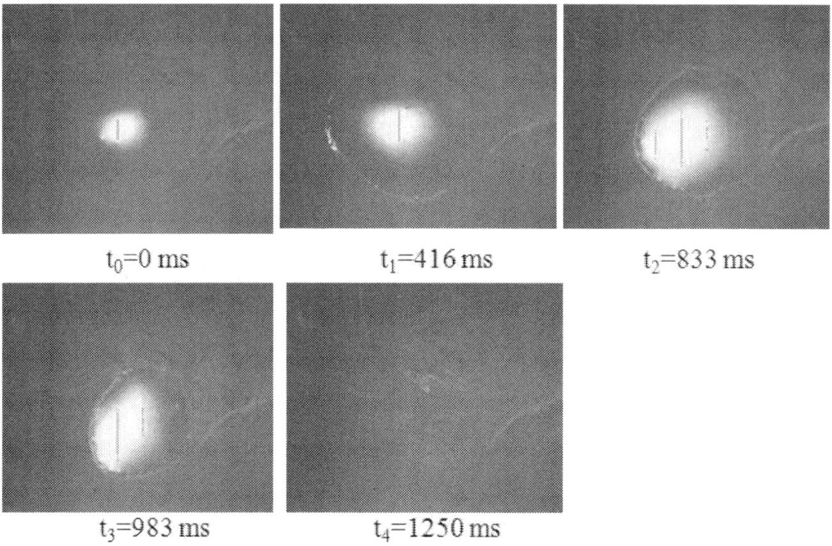

Figure 14. Images of top view corresponding to the characteristic stages of the droplet evaporation process for 2 μL water droplet impinging on the vertical CNTs surface, typical boiling condition, Ts = 125 °C.

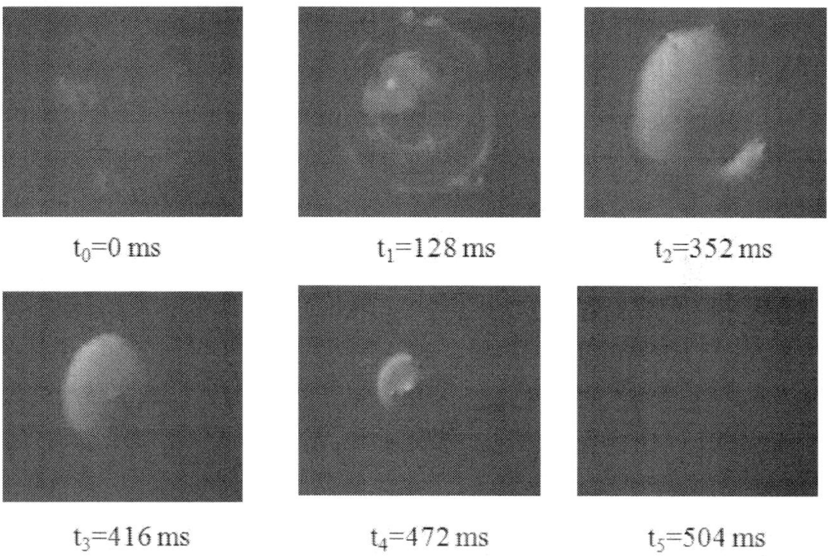

Figure 15. Images of top view corresponding to the characteristic stages of the droplet evaporation process for 2 μL water droplet impinging on the curved CNTs surface, typical boiling condition, Ts = 125 °C.

Figure 16 show the heat transfer curves of water spray cooling on the oxidized silicon and nanotextured surfaces when the cooling liquid is sprayed at a constant flow rate (11.7 ml/min) and standard ambient pressure (1 atm). In the evaporative zone, the heat flux range of 30~40 W/cm², the mean heat transfer coefficient of the curved CNTs surface was approximately 140% higher than that of the plane surface. Figure 17 shows the heat transfer coefficient with heat flux. The heat transfer coefficient derived from Eq. 1. The heat transfer coefficient of the spray cooling curve has a peak in the evaporative zone, and then, near 15 W/cm², it rapidly decreases. The maximum heat transfer coefficient occurs at curved CNTs surface because the effect of the evaporative latent heat increases to enhance heat transfer. In addition, the heat transfer coefficients of the CNTs nanotextured surface are in the range of 10~25 W/cm², which is higher than that of the plain surface due to the greater level of liquid wetting in the evaporative zone by spray cooling.

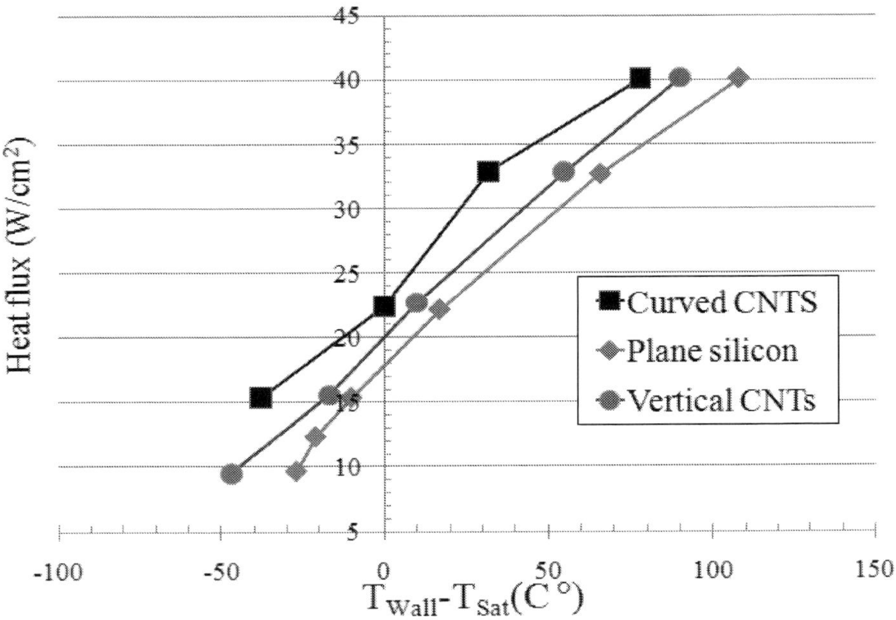

Figure 16. Comparison of heat flux between the oxidized silicon and nano-textured surfaces for spray cooling.

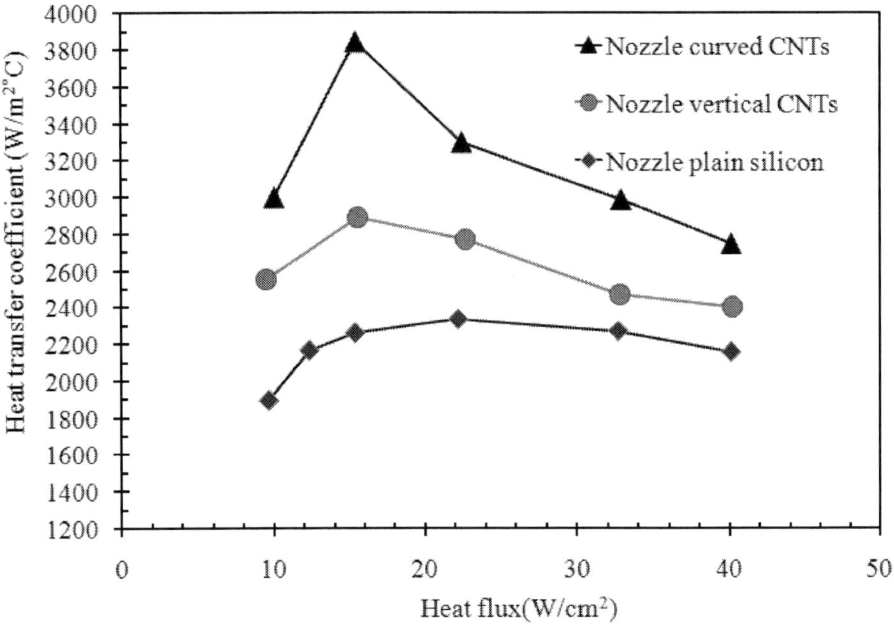

Figure 17. Heat transfer coefficients with variations in heat flux between the oxidized silicon and nano-textured surfaces for spray cooling.

5. Conclusions

The present study demonstrates the hydrodynamic characteristics of droplet impinging on heated surfaces using high-speed imaging and evaluates the heat transfer performance of surface temperature for water on both plane and nanotextured surfaces. Use of the nanotextured surface have resulted in a more uniform temperature profile at and near the impact area exhibiting lower minimum wall temperature especially at higher heat flux values. Nanotextured surfaces also yield lower static contact angle and enhanced film dynamics resulting in a noticeable enhanced heat transfer behavior. Results show that the curved CNTs surface enhanced capillary pumping effect and retained water during higher surface temperatures. This increased spreading over the nanotextured surface suggests a greater potential for increased liquid transport. In the evaporative zone, the heat flux range of 30~40 W/cm², the mean heat transfer coefficient of the curved CNTs surface was approximately 140% higher than of the oxidized silicon surface. This study offers insight into the design of nanotextured surface and new opportunities to achieve capillary pumping capability for microfluidic- based thermal management systems.

Nomenclature

C_p	heat capacity (J/kg•k)
d	droplet diameter (mm)
d_o	droplet diameter prior to impact (mm)
d^*	non-dimensional droplet diameter, d/d_o
G	groove width (mm)
g	gravitational constant (N/s^2)
I	Ampere (A)
H	droplet height (mm)
H^*	non-dimensional droplet height, H/d_o
t	time (s)
t^*	non-dimensional time, $t/(d_o/V_o)$
t_{ev}	droplet evaporation time (s)
T	temperature (°C)
T_r	rebounding temperature(°C)
T_s	surface temperature(°C)
V	voltage (V)
V_o	droplet impact velocity (m/s)

Greek symbols

ρ_f	fluid density (kg/m^3)
μ	dynamic viscosity (Pa•s)
γ	surface tension (N/m)
θ	contact angle

Subscript

a	ambient
l	liquid phase
s	surface
v	vapor phase
sat	saturation
sub	subcooling
CHF	critical heat flux

Author details

Cheng Lin
The Second Research Division, Chung-Shan Institute of Science & Technology, Lung-Tan, Taoyan County, Taiwan, R.O.C

Acknowledgement

I would like to thank the National Science Council of the Republic of China, Taiwan, for financially supporting this research under Contract No. NSC 99-2221-E-007-028-MY2. The author also wishes to thank the NTHU ESS MEMS Laboratory for use of their facilities.

6. References

[1] Bar-Cohen A., Arik, M., and Ohad, M.," Direct liquid cooling of high flux micro and nano electronic components," Proceedings of the IEEE, Vol. 94, No. 8, pp. 1549-1570, 2006.

[2] Berber, S., Kwon, Y.-K., and Tománek, D., "Unusually high thermal conductivity of nanotubes," Phys. Rev. Lett., 84, pp. 4613-4616, 2000.

[3] Silk, E. A., Kim, J. G., Kiger, K. T., "Spray cooling of enhanced surfaces: Impact of structured surface geometry and spray axis inclination", Int. J. Heat Mass Transfer, Vol. 49, pp. 4910-4920, 2006.

[4] Coursey, J. S., Kim, J. G., Kiger, K. T. "Spray cooling of high aspect ratio open microchannels", J. Heat Transfer, Vol. 129, no. 8, pp.1052-1059, 2007.

[5] Pal, A., Joshi, Y.," Boiling of water at sub-atmospheric conditions with enhanced structures: Effect of liquid fill volume", J. Electron. Packag., Vol. 130, no. 1, 011010, 2008.

[6] Ahn, H. S., Sinha, N., Zhang, M., Banerjee, D., Fang, S. K., Baughman, R. H., " Pool boiling experiments on multiwalled carbon nanotube (MWCNT) forests", J. Heat Transfer, Vol. 128, pp. 1335-1342, 2006.

[7] Sodtke C. , Stephan P. ,"Spray cooling on micro-structured surfaces," Int. J. Heat Mass Transfer, Vol. 50, pp. 4089–4097, 2007.

[8] Amon, C.H., Yao, S. C., Wu, C. F. Wu, and Hsieh, C. C., " Microelectro- mechanical system-based evaporative thermal management of high heat flux electronics," ASME Journal of Heat Transfer, Vol. 127, pp. 66-75, 2005.

[9] Hsieh, C. C., Yao, S. C.," Evaporative heat transfer characteristics of a water spray on micro-structured silicon surfaces," Int. J. Heat Mass Transfer, Vol. 49, pp. 962–974, 2006.

[10] Visaria, M. and Mudawar, I.," Application of two-phase spray cooling for thermal management of electronic devices," IEEE Trans. Compon. Packag Technol., vol. 32, no. 4, pp.784–793, 2009.

[11] Ujereh, S., Fisher, T., Mudawar, I.," Effects of carbon nanotube arrays on nucleate pool boiling," Int. J. Heat Mass Transfer, Vol. 50, pp. 4023–4038, 2007.

[12] Li, C., Wang, Z., Wang, P. I., Peles, Y., Koratkar, N., and Peterson, G. P.," Nanostructured copper interfaces for enhanced boiling," Small, Vol. 4, No. 8, pp.1084–1088, 2008.

[13] Chen, R., Lu, M. C., Srinivasan, V., Wang, Z., Cho, H. H. Cho, and Majumdar, A." Nanowires for enhanced boiling heat transfer," Nano Lett., Vol. 9, No. 2, pp. 548-553, 2009.

[14] Khoo, H. S., Tseng, F. G., "Spontaneous high-speed transport of subnanoliter water droplet on gradient nanotextured surfaces", Appl. Phys. Lett., Vol. 95, 063108, 2009.

[15] Chen, M. H., Hsu, T. H., Chuang, Y. J., Tseng, F. G., " Dual hierarchical biomimic superhydrophobic surface with three energy states", Appl. Phys. Lett., Vol. 95, 023702, 2009.

[16] Lin, C., Chen, C. J., Chieng, C. C., Tseng, F. G., "Dynamic effects of droplet impingement on nanotextured surface for enhanced spray cooling ," The 16th International Conference on Solid-State Sensors, Actuators and Microsystems (Transducers'11), Beijing, China, June5-9, pp.1252-1255,2011.

[17] Carey, V. P., "Liquid-vapor phase-change phenomena", Hemisphere: Washington, DC, 1992.

[18] Deng, T., Varanasi, K. K., Hsu, M., Bhate, N., Keimel, C., Stein, J., Blohm, M., "Nonwetting of impinging droplets on textured surfaces", Appl. Phys. Lett., Vol. 94, 133109, 2009.

[19] Mudawar I. and Estes, K. A., "Optimizing and predicting CHF in spray cooling of a square surface," J. Heat Transfer, Vol. 118, pp.672-679, 1996.

[20] Lin, C., Tseng, F. G., Kan, H. C., Chieng, C. C. , " Numerical studies on micropart self-alignment using surface tension forces," J. Microfluidic and Nanofluidic, Vol. 6, pp.63-75, 2009.

Chapter 9

Critical Heat Flux in Subcooled Flow Boiling of Water

Yuzhou Chen

Additional information is available at the end of the chapter

http://dx.doi.org/10.5772/52307

1. Introduction

Flow boiling has an extremely high heat transfer coefficient, and is applied in variety of practices. However, once the heat flux exceeds a certain high level the heated surface can no longer support continuous liquid contact, associated with substantial reduction in the heat transfer efficiency. It may result in a sudden rise of surface temperature in a heat flux controlled system, or a drastic decrease in power transferred in a temperature controlled system. This phenomenon is called the boiling crisis, and the maximum heat flux just before the boiling crisis is usually referred to as critical heat flux (CHF).

Depending on the flow regimes, two types of CHF are classified: (i) in subcooled or low quality region the CHF is characterized by the transition from nucleate boiling to film boiling, and it is termed as the departure from nucleate boiling (DNB); (ii) in higher quality region the CHF is characterized by the dryout of liquid film of annular flow. The DNB and dryout have substantially different mechanisms, and are generally cataloged as the first and the second kind of critical heat flux, respectively (Tong and Tang, 1997).

The CHF is an important subject to many engineering applications. Especially, in a nuclear reactor the occurrence of critical heat flux could lead to a failure of fuel element, and thus the CHF is a major limit for the reactor safety. During past five decades the CHF has been investigated extensively over the world theoretically and experimentally (IAEA-TECDOC-1203, 2001). A great number of empirical correlations and physical models have been proposed. In recent years, a Look-Up Table method (LUT) is widely accepted due to its advantages of higher accuracy, wider range of application, correct asymptotic trend and convenience for updating (Groeneveld, et al., 1996). This table is applied in the system code RELAP5 for reactor safety analysis. The LUT contains a tabulation of normalized CHF data of a uniformly heated tube of 8 mm in diameter at discrete local parameters of pressure, mass flux and quality. Several correction factors are incorporated for utilization of the LUT in other conditions to account for the effects of diameter, bundle, spacer, flux distribution, flow orientation, etc.. Unfortunately, there exists a scarcity of CHF data in low pressure/low flow/ subcooled region, as shown in Fig.1. Because of the extreme complexity of the phenomenon and the lack of adequate knowledge of the mechanisms, all these predictive methods are heavily relied on experimental data, and can not be extrapolated out of their ranges with confidence.

In China Institute of Atomic Energy (CIAE), in the past four decades a great number of CHF experimental data of subcooled boiling of flowing water were obtained in tubes and annuli at lower pressure with different diameter or gap to support the designs of research reactors, HFR and CARR, which were first put into operation in 1980 and 2011, respectively. In recent years the experiments were extended to the region of near-critical pressure with lower flow for the R&D of Supercritical Water-Cooled Reactor (SCWR). In these experiments the CHF (DNB) phenomena are studied with emphases on lower pressure and higher pressure with lower flow. The characteristics and parametric trends of the CHF are clarified, and the physical models are derived.

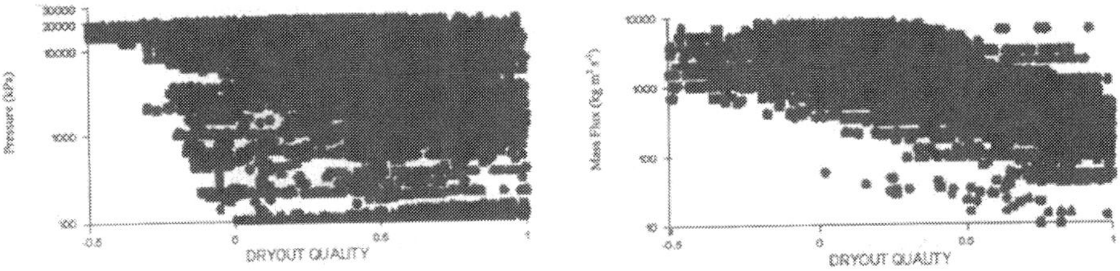

Figure 1. Range of database for 1996 CHF Look-up table (Groeneveld, et al., 1996)

2. Parametric trends

A comprehensive review on the subcooled flow boiling CHF was given by Berglest (1977). As concluded, the effect of heating length disappears when exceeding 200 mm and the subcooled boiling CHF depends only on local parameter ("local condition hypothesis"). The CHF increases with the local mass flux and subcooling increasing. The effect of pressure is more complicated: the CHF increases with pressure in low pressure region, is relatively constant over an intermediate range of pressure, and decreases in high pressure region. The effect of geometry on CHF is observed in many experiments, and is accounted by a factor $f = (D_e/8)^n$, in which D_e is the hydraulic diameter in mm. Different value of the exponent n is obtained from different experiments, ranging from -0.1 to -0.5. For applying the LUT in reactor safety analysis code RELAP5 the value of n = -0.5 is used. While for annulus or rectangular channels the effect of gap width is not observed in many experiments.

In Savannah River lab and Columbia University the subcooled CHF experiments were conducted with D_2O and H_2O coolant and aluminum and stainless-steel heaters (Knoebel et al., 1973). It was concluded that the CHF for D_2O is 16% greater than for H_2O at constant subcooling and velocity, and the CHF of aluminum heaters is a minimum of 20% greater than for stainless steel.

Some typical subcooled boiling CHF correlations are listed in Tab.1. Similar trends of the CHF with the flow rate and subcooling are represented in various correlations, but the degrees of these effects are different significantly. The present investigation has revealed that these effects are inter-dependent, associated with complicated parametrical trends for different regions of conditions.

Author	Correlation	Range of parameters
Knoebel et al. (1973)	$q_c = 0.485(1+0.169V)(1+0.124\Delta T_s)$ for H_2O and Stainless-steel annuli and rectangular channels	D_e=5.1-9.5mm,L/D=64, p=0.2-0.66MPa,ΔT_s=25 - 90K, V=3.9-18.3m/s
Mirshak et al. (1959)	$q_c = 1.51(1+0.12V)(1+0.00914\Delta T_s)(1+1.9p)$	D_e=5.3-11.7mm,p=0.17- 0.6 MPa,ΔT_s = 5-75K, V=1.5-13.7m/s
Zeigarnik (1994)	$q_c = A + B(G/1000)\Delta T_s$ A and B are given in curves in relation with D_e	D_e=3-14mm, L/D=50-62.5, p=0.5-3.0MPa,ΔT_s =20- 160K, G=4-25Mg/m²s
Sudo and Kaminaga (1993)	$q^*_{c,1} = 0.005\lvert G^*\rvert^{0.611}(1+5000\Delta T_0^*/\lvert G^*\rvert)$ for $G \ge G_1^*$ $q^*_{c,2} = (A/A_H)(\Delta h_{in}/h_{fg})G^*$ for $G_2^* \le G < G_1^*$ $q_{c,3}^* = 0.7(A/A_H)(W/\lambda)^{0.5}/(1+(\rho_g/\rho)^{0.25})^2$ for $G < G_2^*$ $q_c^* = q_c/[h_{fg}(\lambda g \rho_g(\rho - \rho_g))^{0.5}]$, $\lambda^* = [\sigma/(g(\rho - \rho_g))]^{0.5}$ $G^* = G/(\lambda g \rho_g(\rho - \rho_g))^{0.5}$, $\Delta T^* = C_{pl}\Delta T/h_{fg}$	P=0.1-4MPa, G=-25800 - +6250 kg/m²s, ΔT_s =1- 213K
Gambill (1963)	$q_c = q_{PB} + q_{FC}$ $q_{PB} = 0.15h_{fg}\rho_g\left(\sigma g(\rho_l - \rho_g)/\rho_g^2\right)^{0.25}$ $\left(1+(\rho_l/\rho_g)^{0.923}C_p\Delta T_s/(25h_{fg})\right)$ $q_{FC} = 0.023k/D_e \mathrm{Re}^{0.8}\mathrm{Pr}^{0.4}(57\ln 145p$ $-54p/(p+0.103) - V/1.22 - T_b)$	De=3-10mm,L/D_e=6.5-52, p=0.1-20 MPa, G= 0-5300kg/m²s
Tong and Celata (1993)	$q_c = Ch_{fg}\rho_g V/\mathrm{Re}^{0.5}$ $C = f(p)$	D_e=2-20mm,L/D_e=100-200, p=0.1-20MPa, G=1.3-40Mg/m²s, ΔT_s =50-150K
Gunther (1951)	$q_c = 0.072V^{0.5}\Delta T_s$	D_e=6.9mm,L/D_e=11, p=0.1-1.1MPa, ΔT_s=25-140K, V=1.5-12.2m/s
Tong (1967)	$q_c = 3.155\times\{(2.022 - 6.24\times10^{-2}p) + (0.1722 - 1.427\times10^{-2}p)$ $\times\exp[(18.177 - 0.5989p)x]\}\times[(0.1094 - 1.177x + 0.1275x\lvert x\rvert)\times$ $G/10^3 + 1.037]\times(1.157 - 0.869x)\times$ $[0.2664 + 0.8357\exp(-124.1D_e)]$ $\times[0.8258 + 0.3414\times10^{-6}(H_{fi} - H)]$ -- W-3	D_e=5.1-17.8mm, P=6.9-15.9Mpa, G=1356-6780kg/m²s X≤0.15 L=0.25 – 3.66m

Table 1. Subcooled flow boiling CHF correlations

3. Experiment

3.1. Technique

The onset of critical heat flux is characterized by a drastic increase in the wall temperature. For subcooled flow boiling of water with lower pressure the critical heat flux is higher and it could lead to a failure of the heated wall rapidly. This kind of CHF is also called as "fast burnout". In experiment the onset of CHF is usually detected by thermocouples for protection of the test section. However, the occurrence of CHF generally initiates from a small area (or a spot), and for a test section of larger size the CHF spot can not be expected exactly. This presents a challenge for prevention of the test section from burnout. For this condition the photocell has advantage for the detection of CHF.

In the present experiments the pressure and flow rate are kept at constant, and the CHF is approached by increasing slowly the water temperature or the power to test section. When the CHF occurs and the wall temperature exceeds about 500°C, the photocell produces an output, which switches off the power supply to test section. This technique is used by author for all the experiments in tubes and annuli.

3.2. Experimental results

3.2.1. Higher pressure CHF (Chen et al., 2011)

Experiment was performed in an uniformly heated inconel tube of 7.95 mm in diameter and 0.8 m in heating length, covering the ranges of pressure of p = 1.96 – 20.4 MPa, mass flux of G = 476 – 1653 kg/m²s, inlet subcooling of $\Delta T_{s,i}$ = 49 – 343 K, outlet subcooling of $\Delta T_{s,o}$ = 1 – 145 K and critical heat flux of q_C = 0.26 – 4.95 MW/m².

For the present low flow condition the CHF is related to the inlet condition, characterizing the mechanism of total power dominant. Totally, 193 data are obtained, and are formulated as the following empiric correlation,

$$q_C = cq_s$$

where q_s is the heat flux for the exit to reach the saturation temperature, evaluated by

$$q_s = \frac{(H_s - H_i)GD}{4L} \qquad (1)$$

and

$$c = Min\left[2350(1-0.0307p)(G(H_s - H_i))^{-0.35}, 1.0\right]$$

where p is the pressure, H_i and H_s the inlet enthalpy and saturation enthalpy, respectively, G the mass flux, D the diameter and L the heating length. Fig.2 shows the comparison of the prediction of Eq.(1) with the experimental data by plotting the ratio of $q_{CHF,c}/q_{CHF,M}$ versus P or G. The average deviation, AVG, is 0.75% and the standard deviation, RMS, is 5.34%.

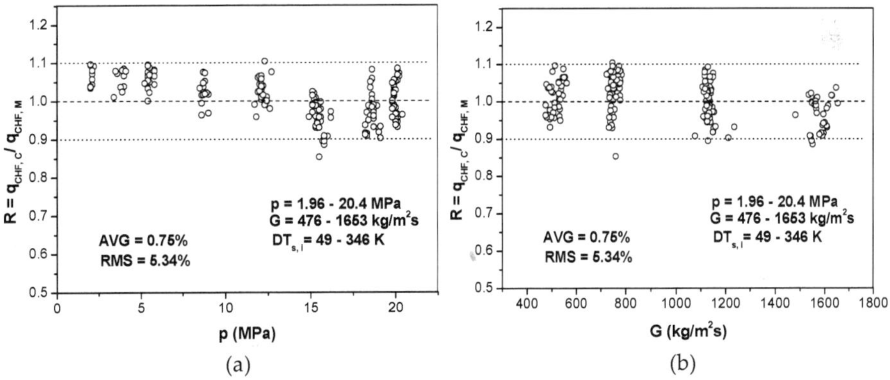

Figure 2. Comparison of experimental data of higher pressure with the prediction by Eq.(1)

The effects of mass flux, inlet subcooling and pressure are exemplified in Fig.3 and 4. The CHF decreases substantially with mass flux decreasing. As can be seen, at G > 1200 kg/m²s the data are close to the prediction of 96-CHF Look-Up Table (LUT), but at low mass flux the data are overpredicted significantly (Fig.3). For lower pressure the effects of inlet subcooling and mass flux are stronger than higher pressure, associated with complicated trend of the CHF with pressure. For G > 1200 kg/m²s higher CHF corresponds to lower pressure, especially in high subcooling region (Fig.4(a)). For G= 700 – 1200 kg/m²s the results of $p < 16$ MPa are not different appreciably for different pressures. For G< 600 kg/m²s, in high subcooling region lower CHF corresponds to higher pressure, while in low subcooling region lower CHF is attained at lower pressure (Fig.4 (b))

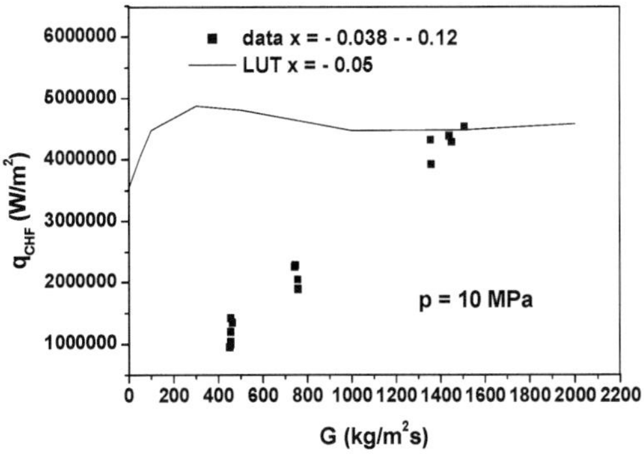

Figure 3. Effect of mass flux on the CHF

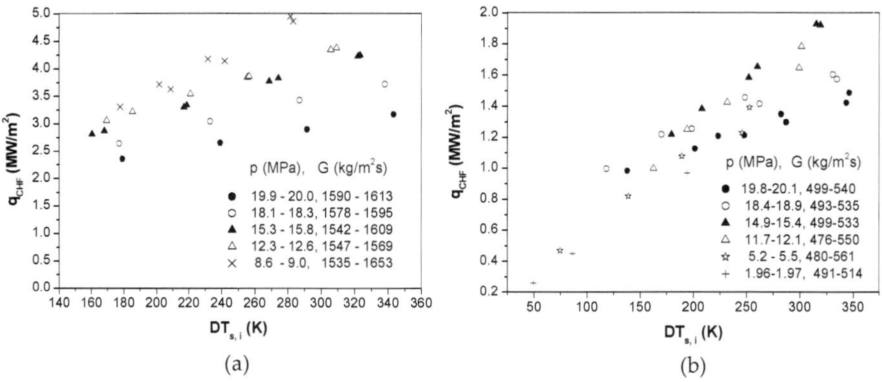

Figure 4. Variations of CHF with inlet subcooling for different mass fluxes and pressures

3.2.2. *Lower pressure CHF*

3.2.2.1. *CHF in tubes*

Medium and high subcooling (Chen et al., 2000)

Experimental data were obtained in uniformly heated stainless-steel tubes of 5.17, 8.05, 10.0 and 16.0 mm in diameter, covering the ranges of pressure p = 0.13 – 1.92 MPa, velocity V = 1.47 – 22.32 m/s, as listed in Tab.2.

Fig.5 and 6 exemplify the variations of critical heat flux with local velocity and subcooling for different pressures. The CHF increases strongly with the velocity increasing (Fig.5). For subcooling higher than about 30 K, the CHF exhibits an approximately linear increase with increase of subcooling. In medium subcooling region the CHF is not different greatly between different pressures. For p < 0.3 MPa the trend of CHF with subcooling is steeper than that of higher pressure, so that in low subcooling region lower CHF is obtained at lower pressure (Fig.6). The CHF also exhibits a general increase trend with the decrease of diameter, and at lower velocity this effect appears stronger than higher velocity.

No.	Diameter D (mm)	Length L (mm)	Pressure P (MPa)	Velocity V (m/s)	Subcooling ΔT_s (K)	Number of data
1	5.17	255	0.13-1.78	3.26-22.32	6.2-89.6	62
2	8.05	383, 396	0.14-1.92	1.85-16.05	8.3 – 88.2	65
3	10.0	295, 400	0.15-1.66	3.39-9.26	30.3-89.5	53
4	16.0	295, 390	0.19-1.29	1.47-13.7	36.7-108.7	56

Table 2. The Experimental conditions for CHF in tubes

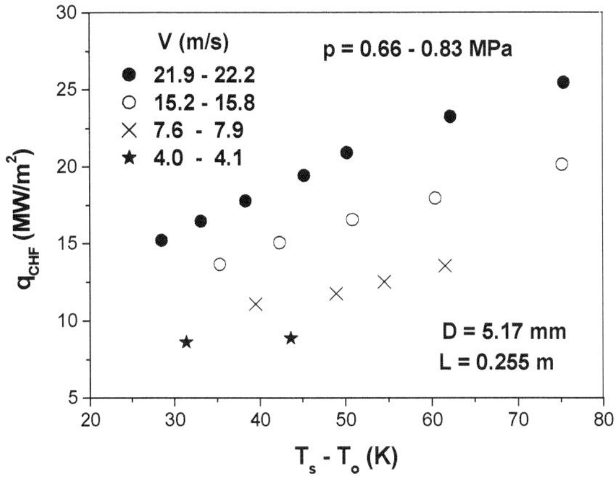

Figure 5. Effect of velocity on the CHF

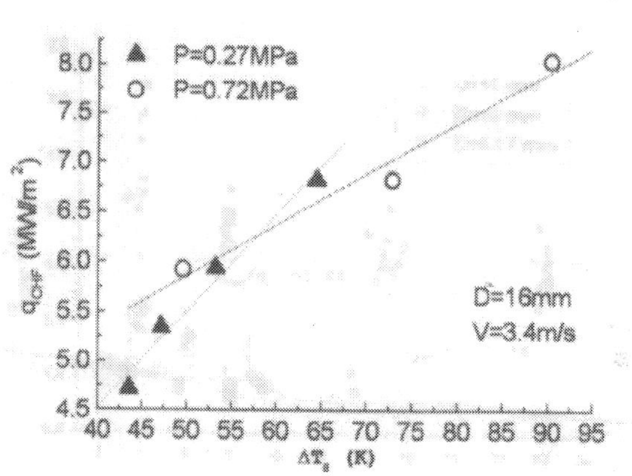

Figure 6. Effect of subcooling on the CHF

224 CHF data of subcooling higher than 35 K are formulated by the following empiric correlation with local parameters of p, V, ΔT_s and D:

$$q_{CHF} = 0.109\times10^6 \times(1+0.104V)\times(15P+\Delta T_s^{\,1-0.1p})\times(D/8.0)^{-0.35-0.05/\ln V} \tag{2}$$

where the pressure P is in MPa, velocity V in m/s, subcooling ΔT_s in K and diameter D in mm. Eq.(2) predicts the experimental data with AVG of 0.83% and RMS of 7.2%, as shown in Fig.7.

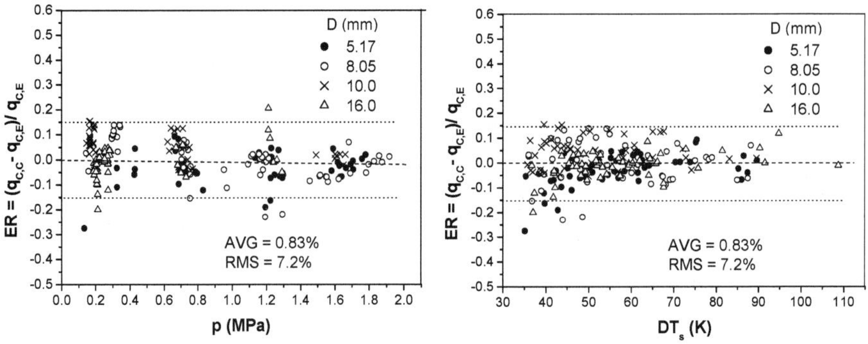

Figure 7. Comparison of experimental data of p<1.9 MPa in tubes with the prediction by Eq.(2)

Low subcooling (Chen et al., 2005).

The experiment was performed in a stainless-steel tube of D=15.9 mm with emphasis on the CHF characteristic in low subcooling region. The conditions cover the ranges of pressure of 0.2 - 1.7 MPa and velocity of 2.2 - 13.2 m/s. Fig.8 exemplifies the variations of CHF with subcooling. For p > 1.0 MPa, the CHF decreases with ΔT_s decreasing monotonously. For p < 0.3 MPa, however, this trend breaks at a certain low value of subcooling, and it turns to increase at further low subcooling. The subcooling at the minimum CHF varies from 13 to 30 K with lower value corresponding to lower velocity.

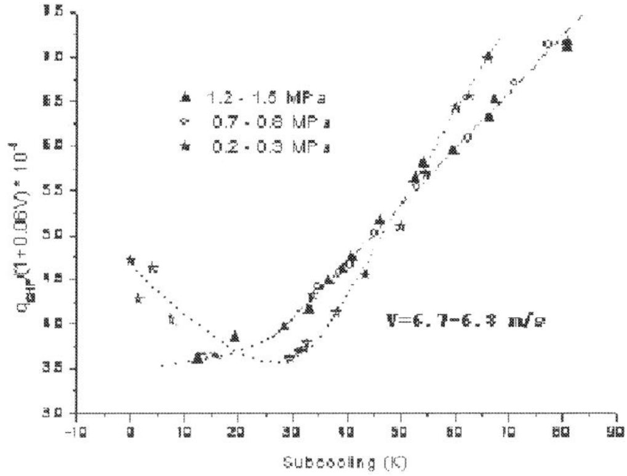

Figure 8. The variation of CHF with subcooling for different pressures

This behavior was also observed in the author's experiment in annuli and other similar experiments (Zeigarnik, 1994, Knoeble et al., 1973). It can be attributed to the onset of net vapor generation (NVG) or the onset of significant voiding (OSV). For p < 0.3 MPa, the vapor density is very small, so that in NVG regime the volumetric flow rate increases essentially, associated with a considerable increase in the liquid velocity. Therefore, the bubbles generated on the heated surface are more likely to enter into the liquid core, resulting in higher CHF. At high pressure the vapor density is much higher, and hence the CHF behavior could not vary distinctly in low subcooling region.

The NVG or OSV was generally identified by a sharper increase in the pressure drop. The following type of empiric correlation was derived

$$q'' = \eta \Delta T_s V^n$$

with the values of n ranging from 0.5 to 1.0. (Siman-Tov, et al. 1995). For the present experimental condition most correlations predict the OSV at subcooling of below 40 K.

Fig.9 shows the variations of pressure drop with the increase of exit temperature. As seen, before the OSV the pressure drop increases slowly with the temperature increasing. After the OSV this trend varies distinctly. Especially at low pressure it is much steeper than that at higher pressure.

In a reactor core the fuel elements are located in parallel channels. Therefore, at low pressure with low subcooling the OSV could result in flow instability or flow excursion (FE), characterized by a continuous decrease of flow rate in a channel. This could eventually lead to a failure of fuel element. Therefore, for reactor safety the limit of OSV is taken as a cr1terion in combination with the limit of minimum ratio of DNB

3.2.2.2. CHF in annuli

Single-side heating (Chen et al., 2004)

Experiments were performed in 8 annuli made of stainless-steel tubes with single-side heating. The diameter of outer wall is 16, 32 or 70mm and gap width is 2 to 4mm. Great majority of CHF data were obtained at outer wall and less at inner wall, covering the range of pressure of 0.17 - 1.8MPa, mass flux of 1300 to 18200 kg/m²s, outlet subcooling of 27-105 K and critical heat flux of 2.0 - 18.1 MW/m². The experimental conditions are listed in Tab.3.

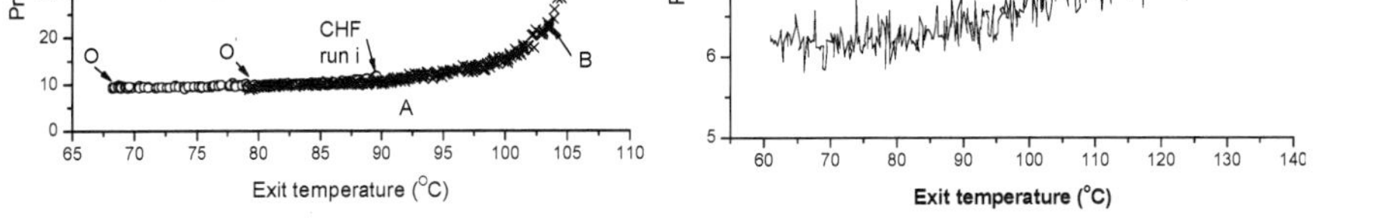

(1) p = 0.2 MPa (2) p = 1.0 MPa

Figure 9. The variations of pressure drop with exit temperature for different pressures

No	Diameters D_2/D_1 (mm)	Length L (mm)	Pressure P (MPa)	Mass flux G (Mg/m²s)	subcooling ΔT_s (K)	Number of data
1	16/12	260-400	0.17-1.31	2.8-18.2	34-89	67*
2	32/28	300	0.24-1.21	2.8-12.4	35-98	46**+27*
3	32/27.6	300	0.3-1.1	3.5-12.1	29-84	230*
4	32/26.8	275-400	0.28-1.1	2.9-13.0	37-102	67*
5	32/26	300	0.36-0.85	3.6-12.7	37-70	47*
6	32/24	300	0.18-1.8	1.3-8.0	27-105	61*
7	70/66	255	0.25-0.4	2.9-7.0	26-54	15*
8	70/65	255	0.25-0.37	3.2-5.7	34-49	18*

* outer-wall
** inner-wall

Table 3. Experimental conditions for CHF in annuli with single-side heating

The velocity and subcooling have predominant effects on the CHF. Under most conditions of interest the effect of pressure is not appreciable. The variation of gap width from 2.0 to 4.0mm does not make an appreciable effect on the CHF. This can probably be explained as follows: larger gap size associates with lower heat transfer coefficient to the liquid core, as suggested by the turbulent convection heat transfer correlation, on the other hand larger gap size corresponds to higher Reynolds number, which is benefit for the departure of bubbles from heated surface. For the present conditions these two opposite factors would be comparative, making the CHF not sensitive to the gap size. These results are consistent with many similar experiments (see Tab.1).

In the present experiment the effect of curvature of heated surface is not observed. This is understandable, because the curvature of heated surface is small, compared to the bubble, and it could not have a noticeable effect on the bubble behavior.

For the sake of simplicity in engineering applications, the effects of pressure, diameter and gap width are ignored, and the experimental data are formulated as the following empiric correlation

$$q_c = 1.21 \times 10^6 \times V^{0.5}\left(1 + 0.03\Delta T_S\right) \tag{3}$$

The comparison of calculation of Eq.(3) with the experimental data is shown in Fig.10. The AVG and RMS are -0.01 and 0.083, respectively.

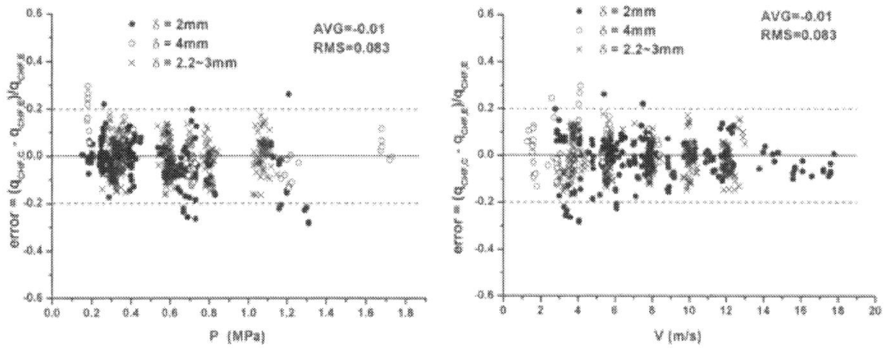

(a) (b)

Figure 10. Comparison of the experimental data in annuli of single-side heating with the predictions of Eq.(3)

Bilateral-side heating (Chen, et al., 1996)

Experiment was performed in a stainless-steel annulus of D_1=28 mm and D_2=35.7 mm and heating length of 160 mm with bilateral-side heating. The critical heat flux data were obtained at the outer wall. The conditions cover the range of pressure of 0.31 – 0.39 MPa, velocity of 2.55 – 8.12 mm, subcooling of 49.1 – 76.6 K and the ratio of inner to outer-wall heat fluxes of q_1/q_2 of 0 – 0.94.

For convenience in comparison, the result of outer-side heating is formulated by

$$q_{c,0} = 0.96 \times 10^6 \times V^{0.43} \times (1 + 0.057\Delta T_{s,0})$$

as shown in Fig.11.

The experimental results of bilateral-side heating are displayed in Fig.12 by plotting the ratio of $q_c/q_{c,0}$ against q_1/q_2 (1 and 2 denote the inner-wall and outer-wall, respectively). As seen, the critical heat flux exhibits an increase trend with q_1/q_2 increasing. When q_1/q_2 closes to 1.0 the CHF is increased by 15 – 20%. It can be attributed to the variation of temperature profile in liquid core, which results in an increase in condensation efficiency of the bubbly layer by subcooled liquid core, that is similar with that in single-phase convection heat transfer. This effect can be clarified further by the model analysis latter in paragraph 4.3.

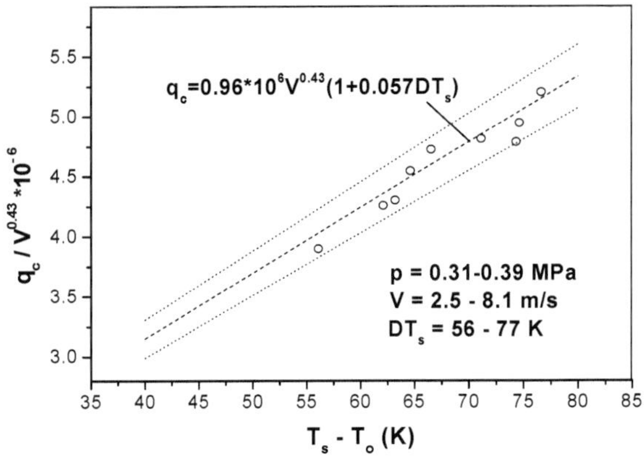

Figure 11. The CHF results of outer-wall heating

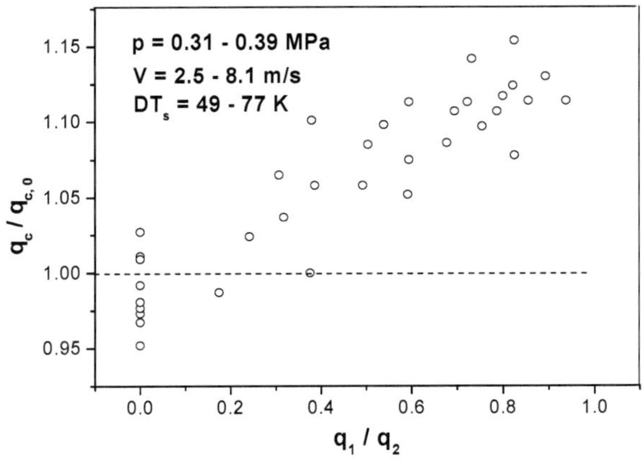

Figure 12. The ratio of $q_c/q_{c,0}$ versus q_1/q_2

3.2.3. *Transient CHF* (Chen et al., 2005)

Critical heat flux is more likely to occur under off-normal or accident conditions, in which a transient would experiences of flow rate, pressure and/or power. Many studies have focused on the transient CHF (Leung, 1978, Chang, et al., 1989, Iwamura, et al., 1987, 1994, Celata, et al., 1991, Weisman, 1993, Moon, et al., 2002). In higher quality region the experimental results revealed the inadequacy of the steady-state CHF correlation for transient conditions. While in subcooled and low quality region, the effect of transient on the CHF was found not appreciable. In general, the transient CHF has not been studied adequately for wider range of condition, and in evaluation of nuclear reactor safety the CHF for transient conditions is predicted with the correlations derived at steady-state conditions (IAEA-TECDOC-1203, 2001).

In the present study an experiment of flow-reduction transient CHF was performed in a stainless-steel tube of 15.9 mm in diameter, covering the range of pressure of 0.2 - 1.4 MPa, initial velocity of 4.5 - 13.5 m/s, and the initial inlet subcooling of 80 – 160 K. The flow rate was reduced linearly as

$$\dot{m} = \dot{m}_0 (1 - kt)$$

where $\dot{m}$ and $\dot{m}_0$ are the instantaneous and initial flow rate, respectively, t is the time, and k is the flow reduction rate, ranging from 0.0075 to 0.24 1/s.

The experimental results are shown in Fig.13, in which the P, V and ΔT_s are the instantaneous values. For P > 1.0 MPa the effect of flow transient on CHF appears not prominent. For P < 0.3 MPa, in high subcooling region the effect of transient is not appreciable, while for subcooling lower than about 50 K the result departs from the trend of steady-state distinctly, and higher CHF is attained at higher flow reduction rate.

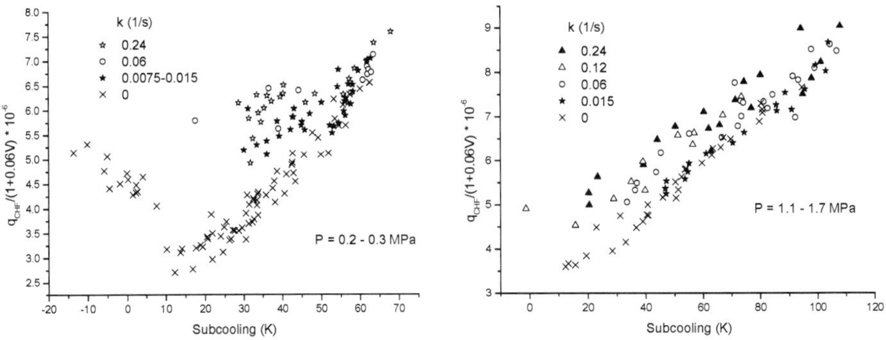

(a) p< 0.3 MPa (b) p >1.1 MPa

Figure 13. The experimental results under flow transient condition

Different effect on the CHF observed at different conditions can probably be explained by different mechanisms of the CHF. For low subcooling the CHF is induced by a limit of enthalpy of bubbly layer, while for high subcooling the CHF is induced by a limit of bubbly-layer condensation by the subcooled liquid core (Thorogerson et al., 1974). Compared to the steady-state, at a flow-reduction transient with constant heat flux and constant subcooling the enthalpy in the bubbly layer is higher and the temperature in the liquid core is lower. At high subcooling the thickness of bubbly layer is small, thus for a constant CHF the subcooling would not be different greatly from the steady-state. While at low subcooling the thickness of bubbly layer is larger, and a constant CHF would occur at a higher subcooling than that of steady-state. It would lead to the premature of the OSV, associated with prominent effect on the CHF, as observed at low pressure.

Fig 14 shows the variation of pressure drop before the onset of CHF in flow-reduction transients of p =0.23 MPa and k = 0.24 with heat flux of q = 890, 775 and 750 W/cm² for run iii, iv and v respectively. In run iii, the pressure drop exhibits a monotonous decrease with flow rate decrease, until the onset of CHF. While in run iv and v, before the onset of CHF the decrease of pressure drop is followed by a sharp increase. In these three runs the CHF are higher than those of steady-state by about 5%, 18% and 36%. The critical subcoolings are 54.7, 46.5 and 37 K respectively, all of which are higher than the values of steady-state, evidencing the premature of OSV.

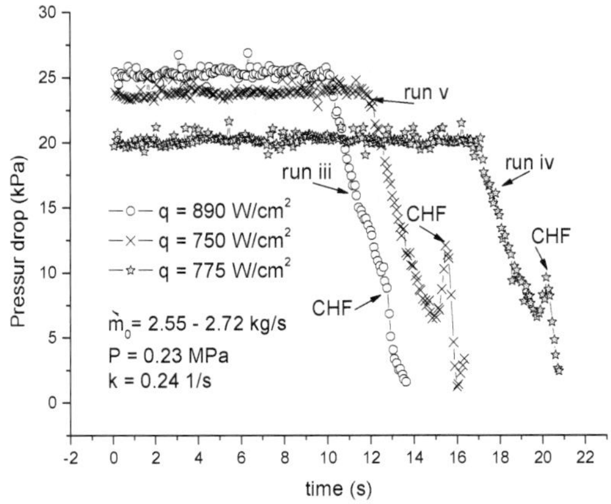

Figure 14. Variation of pressure drop before the onset of CHF

4. Physical model

4.1. Mechanism

In subcooled flow boiling the enthalpy of bubbly layer is determined by the heat transfer from the interface of bubbly layer to the liquid core, and the excessive bubble crowding serves as a thermal shield, leading to the onset of CHF. For the DNB with low subcooling or saturated condition, the critical enthalpy models were proposed by Weisman and Pei (1983) and Tong (1968), in which the heat transfer coefficient from the interface of bubbly layer to liquid core was estimated by correlations derived from their experiments. For the DNB with high subcooling, the liquid sublayer dryout models were proposed by Katto (1992), Lee and Mudawar (1983) and Celata et al. (1994), in which the bubble diameter, the thickness of liquid sublayer and the length of vapor blanket were determinant for the sublayer dryout.

For higher subcooling the major part of heat from wall is transferred to the liquid core and a minor part for increase of the enthalpy of bubbly layer. Therefore, the characteristic of bubbly layer is primarily controlled by the heat transfer from the bubbly layer to liquid core. This heat transfer is closely relative to the turbulence near the edge of bubbly layer, and is sensitive to the distance from the wall. The increase in thickness of bubbly layer has positive effect on the CHF due to increase in the heat transfer efficiency to liquid core, but negative effect due to increase in the thermal resistance of the bubbly layer. The balance of these two factors gives a critical value of the thickness. Therefore, in the liquid sublayer dryout model the thickness of bubbly layer is a determinant factor.

4.2. Model for tube (Chen et al., 2011)

The model is based on the mechanism of liquid sublayer dryout by modifying the Celata's model for the thickness of bubbly layer to cover both high and low subcooling region.

Bubbly layer

At high flow and high subcooling the minimum thickness of bubbly layer is determined by the size of a bubble, while at low subcooling it could be larger due to bubble crowding. For the present experimental condition the following expression on the thickness of bubbly layer is attempted,

$$\delta = k_1 D_B (1 + k_2 e^{-k_3 \mathrm{Pr} Q}) \tag{4}$$

where the factor k_1 , k_2 and k_3 are the constants, Pr is the Prandtl number, Q is a parameter group (see Eq.(8)). D_B is the bubble or vapor blanket equivalent diameter, evaluated by Staub correlation (1968), as

$$D_B = \frac{32\sigma f(\beta)\rho_l}{fG^2}$$

where σ is the surface tension, ρ_l the liquid density, G the mass flux, $f(\beta)$ is a function with parameter of contact angle and recommended as $f(\beta) = 0.02\text{-}0.03$. In the present model it is taken as

$$f(\beta) = 0.03 \text{ for } p \le 10MPa$$

and

$$f(\beta) = 0.03(1 - 0.055(p - 10)) \text{ for } p > 10MPa$$

where p is the pressure in MPa.

The friction factor, f , is calculated by Colebrook-White equation combined with Levy's rough surface model (1967), as

$$\frac{1}{\sqrt{f}} = 1.14 - 2.0\log(\frac{\varepsilon}{D} + \frac{9.35}{\mathrm{Re}\sqrt{f}})$$

where D is the tube diameter, Re the Reynolds number, ε is the surface roughness, accounted by $\varepsilon = 0.75 D_B$.

Liquid core

The velocity distribution in the liquid core is represented by Karman's universal law, as in Celata's model,

$$U^+ = y^+ \quad for\ 0 \le y^+ < 5$$

$$U^+ = 5.0 \ln y^+ - 3.05 \quad for\ 5 \le y^+ < 30$$

$$U^+ = 2.5 \ln y^+ + 5.5 \quad for\ y^+ \ge 30$$

with

$$U^+ = \frac{U}{U_\tau} \quad y^+ = \frac{yU_\tau \rho_l}{\mu_l}$$

and

$$U_\tau = \left(\frac{\tau_w}{\rho_l} \right)^{0.5}$$

where U is the liquid velocity, y the distance from the wall, μ_l the liquid viscosity and ρ_l the liquid density, U_τ the friction velocity, and τ_w is the wall shear stress, evaluated by

$$\tau_w = \frac{fG^2}{8\rho_l}$$

The temperature distribution in the liquid core is as follows (Martinelli, 1947) ,

$$T_0 - T = Q \Pr y^+ \quad for\ 0 \le y^+ < 5 \tag{5}$$

$$T_0 - T = 5Q \left\{ \Pr + \ln \left[1 + \Pr\left(\frac{y^+}{5} - 1\right) \right] \right\} \quad for\ 5 \le y^+ < 30 \tag{6}$$

$$T_0 - T = 5Q \left[\Pr + \ln(1 + 5Pr) + 0.5 \ln\left(\frac{y^+}{30}\right) \right] \quad \text{for } y^+ \ge 30 \tag{7}$$

with

$$Q = \frac{q}{\rho_l C_{pL} U_\tau} \tag{8}$$

Equations (5) to (7) are assumed valid in the region of $\delta \le y \le r$, and the T_0 is a referent value, which is determined by $T = T_s$ at $y = \delta$.

Calculation of critical heat flux

The local enthalpy, H, is calculated by

$$H\dot{m} = H_C(\dot{m} - \dot{m}_{B,g} - \dot{m}_{B,l}) + H_g \dot{m}_{B,g} + H_l \dot{m}_{B,l} \tag{9}$$

where $\dot{m}$ is the total flow rate, $\dot{m}_{B,g}$ and $\dot{m}_{B,l}$ are the vapor and liquid flow rate in the bubbly layer, respectively, H_g and H_l are the vapor and liquid enthalpy, and H_C is the enthalpy of liquid core.

The $\dot{m}$, $\dot{m}_{B,g}$ and $\dot{m}_{B,l}$ are evaluated by

$$\dot{m} = \frac{\pi D^2}{4} G \tag{10}$$

$$\dot{m}_{B,g} = \pi(D-\delta)\delta\alpha_B\rho_g\overline{U_B} \tag{11}$$

and

$$\dot{m}_{B,L} = \pi(D-\delta)\delta(1-\alpha_B)\rho_l\overline{U_B} \tag{12}$$

where α_B is the void fraction in the bubbly layer, and it is taken as α_B = 0.9, $\overline{U_B}$ is the average velocity of bubbly layer, estimated by

$$\overline{U_B} = 0.5U_{y=\delta}$$

H_C is calculated at the average temperature from the edge of bubbly layer to the center of tube, T_C, which is calculated by

$$T_C = \frac{\int_\delta^r TU(r-y)dy}{\int_\delta^r U(r-y)dy}$$

where r is the radius of tube, and δ is the distance from wall at which the temperature is equal to the saturation value.

The exit enthalpy, H, is evaluated from the heat balance equation, as

$$H = H_i + \frac{4qL}{GD} \tag{13}$$

where H_i is the inlet enthalpy, and L the heating length.

Calculation is started with a test heat flux q ($q < q_s$), and the δ, $\dot{m}_{B,g}$, $\dot{m}_{B,l}$ and T_C are calculated by Eq. (4), (10), (11) and (12). Then, the H is calculated by Eq.(9) and compared to that calculated by Eq.(13). The result of CHF is obtained through an iterative process.

To get the calculations better fit to the experimental data, the constants in Eq. (4) are as: k_1 = 0.75, k_2 = 1000, and k_3 = 1.0. At low subcooling q_C is close to q_s, and not sensitive to the δ, thus the maximum value of δ is simply set as 0.1D.

The experimental data in tubes are calculated by the present model. The comparison is shown in Fig.15 and 16 for p > 2 MPa and < 2 MPa, respectively.

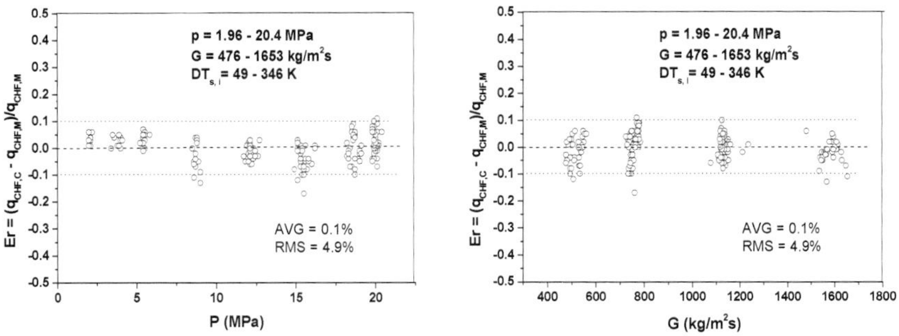

(a) (b)

Figure 15. Comparison of the calculations of present model with the experimental results in tube for p > 2 MPa

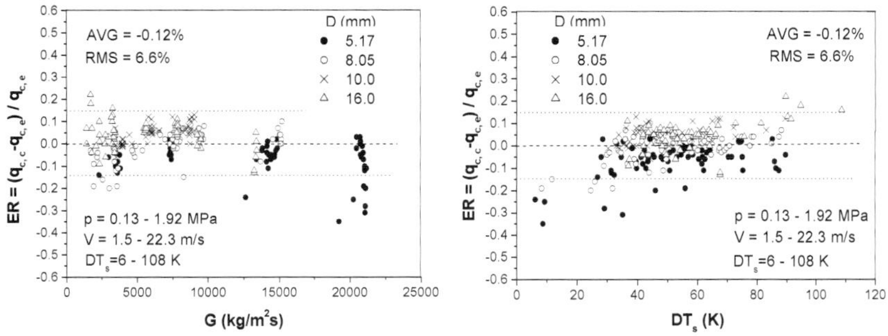

(a) (b)

Figure 16. Comparison of the calculations of present model with the experimental results in tubes for p < 2 MPa

4.3. Model for annulus (Chen et al., 1996)

Fig.17 shows schematically the profiles of velocity and temperature in the liquid core of an annulus. Some assumptions are made as follows:

- Each wall is heated uniformly, and the flow and heat transfer conditions are fully developed;
- At the edge of liquid bubbly layer the heat flux is equal to that of outer-wall
- In liquid core the properties are evaluated at the bulk temperature.

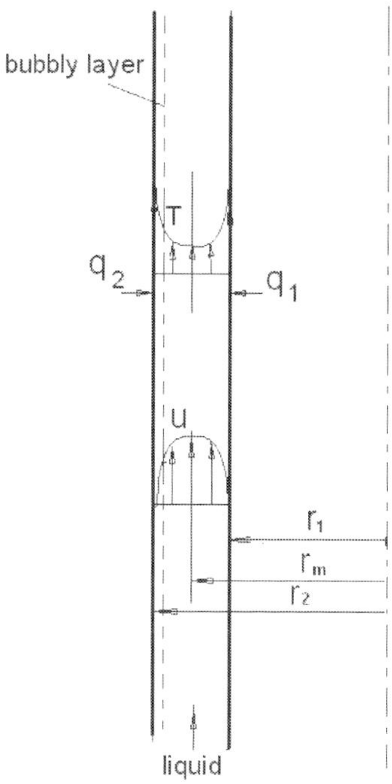

Figure 17. The profiles of velocity and temperature in the flow

In liquid core the energy balance equation is written as

$$u\frac{\partial t}{\partial x}=\frac{1}{r}\frac{\partial}{\partial r}\left[r\left(\frac{k_l}{\rho_l C_{pl}}+\frac{1}{\mathrm{Pr}_t}\varepsilon_m\right)\frac{\partial t}{\partial r}\right] \tag{14}$$

with

$$\frac{\partial t}{\partial x}=\frac{\partial \bar{t}}{\partial x}=\frac{2(q_1 r_1+q_2 r_2)}{\rho_l C_{pl}\bar{u}(r_2^2-r_1^2)} \tag{15}$$

For fully developed flow with uniform heating, the $\partial t / \partial x$ is a constant. Integrating Eq.(14) gives

$$\frac{\partial t}{\partial r}=\frac{1}{r(1+\frac{Pr}{\Pr_t}\frac{\varepsilon_m}{\nu_l})}\left[\frac{2(q_1r_1+q_2r_2)}{k_l(r_2^2-r_1^2)\bar{u}}\int_{r_1}^{r}urdr-\frac{q_1r_1}{k_l}\right] \tag{16}$$

The average bulk temperature is approximated by

$$\bar{t}=\frac{2}{\bar{u}(r_2^2-r_1^2)}\int_{r_1}^{r_i}turdr$$

Integrating Eq.(16) by parts gives

$$t_i-\bar{t}=\frac{2}{\bar{u}(r_2^2-r_1^2)}\int_{t_1}^{t_i}\int_{r_1}^{r}turdrdt \tag{17}$$

where t_i is the temperature at the outer edge of bubbly layer. For low pressure with higher subcooling the enthalpy of bubbly layer is negligible small.

By assuming t_i equal to the saturation temperature, Eq.(17) is approximated as the subcooling ΔT_s. Introducing R=r/r2 and $U = u/\bar{u}$, combining of Eq.(16) with Eq.(17) gives

$$\Delta T_s=\frac{2}{1-R_1^2}\int_{R_1}^{R_i}\frac{1}{(1+\frac{\Pr}{\Pr_t}\frac{\varepsilon_m}{\nu_l})R}\left[\frac{2(q_1r_1+q_2r_2)}{k_l(1-R_1^2)}(\int_{R_1}^{R}URdR)^2-\frac{q_1r_1}{k_l}\int_{R_1}^{R}URdR\right]dR$$

The velocity distribution is assumed in power law, as

$$\frac{u}{u_m}=\left[\frac{r_j-r}{r_j-r_m}\right]^{1/5} \qquad \begin{array}{l} \text{j=1 for } r_1<r<r_m \\ \text{j=2 } \textit{for } r_m<r<r_2 \end{array}$$

where r1 and r2 are the radius of inner and outer wall, respectively, and rm is the radius of maximum velocity u_m, calculated from

$$\tau^*=\frac{\tau_2}{\tau_1}=\frac{(1-R_m^2)R_1}{R_m^2-R_1^2}$$

where τ_1 and τ_2 are the shear stress at the inner and outer wall, respectively.

The friction factor is estimated by (Xu et al., 1979)

$$f=93\left(\frac{q}{h_{fg}\rho_g\bar{u}}\right)^{0.7}\left(\frac{h_{fg}\rho_g}{C_{pl}\rho_l\Delta T_s}\right)^{0.4}\left(\frac{\rho_g}{\rho_l}\right)^{0.2}f_0$$

with

$$f_0 = \frac{0.3164}{\text{Re}^{0.25}}$$

The momentum eddy diffusivity is evaluated by (Levy, 1967)

$$\frac{\varepsilon_m}{\nu_l} = \frac{1}{15}\left|(r_2 - r_m)\right|\frac{u^*}{\nu_l}\left(1-\eta^2\right)\left(1+2\eta^2\right)$$

with

$$\eta = (r_m - r)/(r_m - r_1)$$

and

$$u^* = \left(\frac{\tau_w}{\rho_l}\right)^{0.5}$$

The turbulent Prandtl number, Pr_t, is taken as $\text{Pr}_t = 1/1.2$.

Both the experimental data of 8 annuli with single-side heating and the annulus with bilateral-side heating are predicted by this model. The comparison is shown in Fig.18 to 20.

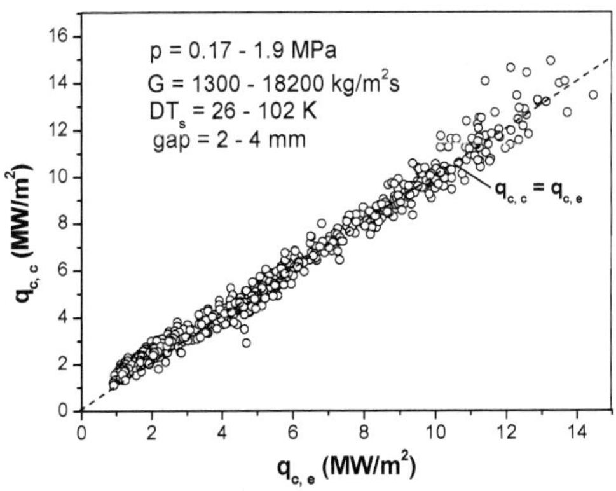

Figure 18. Comparison of the prediction of present model with the experimental data of 8 annuli

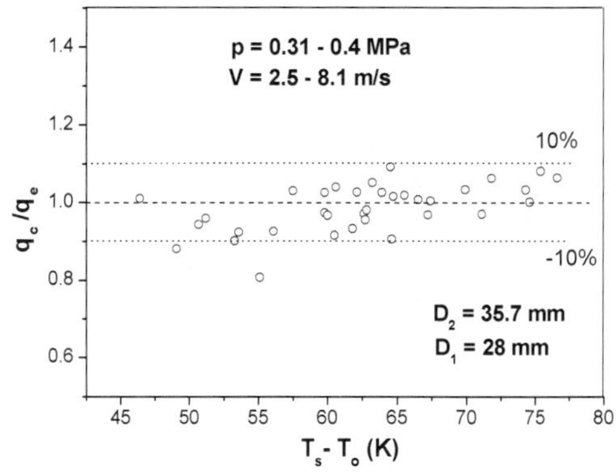

Figure 19. Comparison of the model prediction with experimental data of bilateral heating

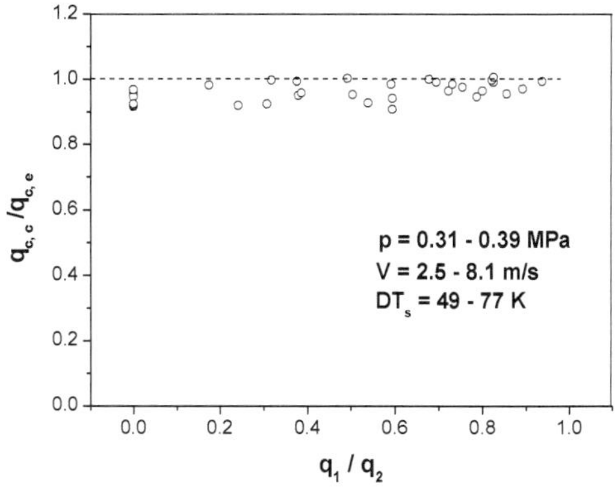

Figure 20. Effect of bilateral-side heating on the CHF by model prediction

5. Summary

Critical heat flux is an important subject to many applications. Especially for nuclear reactors, it is a major limit for the safety. The CHF has been investigated extensively over the world, and various prediction methods have been available. Unfortunately, there exists a scarcity of experimental data in certain regions. Because of extreme complexity of the phenomenon and the lack of adequate knowledge of the mechanisms, all these predictive methods are heavily relied on the experimental data, and can not be extrapolated out of the range with confidence.

In the present lab a great number of critical heat flux data of subcooled water have been obtained in tubes and annuli with different diameter and gap size over wide range of parameters with emphasis on lower pressure and higher pressure with low flow. The results fill the gap of database and the knowledge of the phenomenon.

The velocity and subcooling are the predominant parameters for the CHF. At lower pressure these effects are stronger. At the pressure below 0.3 MPa, when the subcooling decreases below a certain low value the CHF behavior varies substantially as a result of significant voiding. The effect of geometry is related to the pressure, subcooling and velocity. All these effects are inter-dependent, and are hardly to be represented in a single correlation for wide range of conditions. In the present study two models have been developed for the subcooled flow boiling CHF in circular tube and annulus, based on the mechanisms of CHF and the present experimental data. They will be validated and improved for extended range of conditions.

Nomenclature

A	flow area
A_H	heated area
C_p	specific heat
D	diameter
ΔT_s	subcooling
f	friction factor
G	mass flux
h	heat transfer coefficient
h_{fg}	latent heat
H	enthalpy
k	thermal conductivity
L	heating length
p	pressure
q	heat flux
r	radius
T,t	temperature
U u	velocity in axial direction
V	average liquid velocity at CHF
x	quality
Nu	Nusselt number
Pr	Prandtl number
Pr_t	turbulent Prandtl numbe
Re	Reynolds number
W	channel width
y	distance from the wall
ρ	density

σ	surface tension
ε	surface roughness
ε_m	momentum eddy diffusivity
μ	dynamic viscosity
ν	kinematic viscosity
τ_w	wall shear stress
α_B	void fraction
λ	critical wavelength

Subscript

B	bubble
c	critical, calculatiom
l	water
g	steam
m	maximum velocity
M	measurement
s	saturation
1	inner-wall
2	outer-wall

Author details

Yuzhou Chen
China Institute of Atomic Energy, China

6. References

Berglest, A. E. (1977), Burnout in Boiling Heat Transfer, Part II: Subcooled and Low-Quality Forced-Convection System, *Nuclear Safety*, Vol. 18, N2, pp 154 - 167

Chang, S. H., Lee, K. W. and Groeneveld, D. C. (1989), Transient-Effects Modelling of Critical Heat Flux, *Nucl. Eng. Design*, Vol. 113, PP 51 - 57

Celata, G. P., Cumo, M., D'annibale, F., Farello, G. E. and Mariani, A. (1991), CHF Behavior During Pressure, Power and/or Flow Rate Simultaneous Variations, *Int. J. Heat Mass Transfer*, 34(3). PP 723 - 738

Celata, G. P., Cuma, M., Mariani, A. Simosini, M. and Zummo, G. (1994), Rationalization of Existing Mechanistic Models for the Prediction of Water Subcooled Flow Boiling Critical Heat Flux, *Int. J. Heat and Mass Transfer* Vol. 37, Suppl. 1, PP 347 - 360

Chen, Y., Zhou, R. and Chen, H. (1996), Critical Heat Flux for Subcooled Flow in an Annulus with Bilateral Heating, in *Heat Transfer and Technology 1996*, Bu-xuan Wang, Higher Education. Press, ISBN 7-04-005894-4, pp 376 - 381

Chen, Y, Zhou. R., Hao L. and Chen, H. (1997), Critical Heat Flux with Subcooled Boiling of Water at Low Pressure, *Proc. 8th Int. Topical Meeting on Nuclear Reactor Thermal-Hydraulics*, 1997, Japan, Vol. 2, PP 958 - 964

Chen, Y., Zhang, H., Guo F. and Hao L. (2000), Subcooled Flow Boiling CHF in Tubes with Different Diameters, Presented at *Boiling 2000*, Alaska

Chen, Y., Zou, L. and Yang, C. (2004), Subcooled Critical Heat Flux in Annuli at Lower Pressure, *Proc. NUTHOS-5*, Japan

Chen, Y., Yang, C. and Mao, Y. (2005), An Experimental Study of Subcooled Flow Boiling Critical Heat Flux of Water under Steady-State and Flow-Transient Conditions at Lower Pressure, *Proc. 11th Int. Topical Meeting on Nuclear Reactor Thermal-Hydraulics (NURETH-11)*, Avignon, France

Chen, Y., Yang, C., Zhao, M., Bi, K., Du, K. and Zhang, S. (2011), Subcooled Boiling Critical Heat Flux of Water Flowing Upward in a Tube for Lower Flow and Pressure up to 20 MPa, *Proc. 14th Int. Topical Meeting on Nuclear Reactor Thermal Hydraulics (NURETH-14)*, LN-620

Gambill, W. R. (1963), Generalized Prediction of Burnout Heat Flux for Flowing Subcooled Wetting Liquids, *Chem. Eng. Prog. Sump. Series*, Vol.59, No.41

Groeneveld, D. C., et al. (1996), The 1996 Look-up Table for Critical Heat Flux in Tubes, *Nuclear Eng. Design* Vol. 163, PP 1 - 23

Gunther, F. C. (1951), Photographic Study of Surface Boiling Heat Transfer to Water with Forced Convection, *Trans. ASME*, Vol.73, PP 115

Iwamura, T. (1987), Transient Burnout Under Rapid Flow Reduction Condition, *J. Nucl. Sci. and Tech.* Vol.24(10), PP 811 - 820

Iwamura, T., Watanabe, H. and Murao, Y. (1994), Critical Heat Flux Experiments under Steady-state and Transient Conditions and Visualization of CHF Phenomenon with Neutron Radiography, *Nuclear Engineering and Design* Vol. 149, PP 195 - 206

Katto, Y. (1992), A Prediction Model of Subcooled Water Flow Boiling CHF for Pressures in the region 0.1-20.0 MPa, *Int. J. Heat ans Mass Transfer*, Vol. 35, PP 1115 - 1123.

Klausner, J. F., Mei, R., Bernhard, D. M. and Zeng, L. Z. (1993), Vapor Bubble Departure in Forced Convection Boiling, *Int. J. Heat Mass Transfer*, Vol. 36(3) PP 651 - 662.

Knoebel, D.H., Harris, S.D. and Crain, B. (1973), Forced Convection Subcooled Critical Heat Flux, D_2O and H_2O Coolant with Aluminum and Stainless-Steel Heaters, *DP - 1306*

Levy, S. (1967), Turbulent Flow in an Annulus, *Trans. ASME, C, J. Heat Transfer*, Vol. 89 (1), PP 25 - 31

Lee, C. H. and Mudawar, I. (1983) A Mechanistic Critical Heat Flux Model for Subcooled Flow Boiling Based on Local Bulk Flow Conditions", Int. J. Multiphase Flow, Vol. 14. PP 711 - 728

Leung, J. C. M. (1978), "CHF under Transient Conditions: a Literature survey," *NURETH/CP-0056*

Martinelli, R. C. (1947) "Heat Transfer to Molter Metals", Trans. ASME Vol. 69, pp 947 – 951

Mirshak, S., Durant, W. S. and Towell, R.H. (1959), Heat Flux at Burnout, *DP - 355*

Moon, S. K., Chun, S. Y., Choi, K. Y. and Baek, W. P. (2002), Transient Critical Heat Flux Under Flow Coastdown in a Vertical Annulus With Non-Uniform Heat Flux Distribution, *J. Korea Nuclear Science*, Vol. 34(4) PP 382 - 395

Siman-Tov, M., Felde, D. K., McDuffee, J. L. and Yoder. Jr. G. L. (1995), Static Flow-instability in Subcooled Flow Boiling in Wide Rectangular Parallel Channels, *2nd Int. Conf. on Multi-phase Flow*, Japan

Staub, F. W. (1968), "The Void Fraction in Subcooled Boiling – Prediction of the initial Point of Net Vapor Generation, *J. Heat Transfer*, Vol. 90, PP 151 - 157

Sudo, Y. and Kaminaga, M. (1993), A New CHF Correlation Scheme Proposed for Vertical Rectangular Channels Heated From Both Sides in Nuclear Research Reactors. *ASME, C.* Vol. 115, PP 426 - 434

Thermohydraulic relationships for advanced water cooled reactors (2001), *IAEA-TECDOC-1203*

Thorogerson, E. J., Knoebel, D. H., and Gibbons, J. H. (1974), "A Model to Predict Convective Subcooled Critical Heat Flux," Trans. ASME, J. Heat Transfer, 96, 79-82

Tong, L. S. (1968), Boundary Layer Analysis of the Flow Boiling Crisis, *Int. J. Heat Mass Transfer*, Vol. 11, PP 1208 – 1211

Tong, L. S. (1967), Prediction of Departure from Nucleate Boiling for An Axially Non-Uniform Heat Distribution, Journal Nuclear Energy, Vol. 21

Tong, L. S. and Tang, Y. S. (1997), *Boiling Heat Transfer and Two-Phase Flow*, Taylor & Francis Pub., ISBN 1-56032-485-6

Weisman, J. and Pei, B. S. (1983), Prediction of Critical Heat Flux in Flow Boiling at low qualities, *Int. J. Heat Mass Transfer*, Vol. 26, PP 1463 – 1477

Weisman, J. (1995), A Phenomenological Explanation of the Relationship Between Steady State and Transient CHF at Subcooled or low Quality Conditions, *Nucl. Eng. and Des.* Vol. 158, PP 157 - 160

Xu, G. at al. (1979), Heat Transfer and Pressure Drop in an Annulus with Subcooled Boiling Water, Analysis and Experiment of Reactor Thermohydraulics, Atomic Energy Press, PP 412 – 417 (in Chinese)

Zeigarnik, Yu. A. (1994), Critical Heat Flux with Boiling of Subcooled Water in Rectangular Channels with One Sided Supply of Heat, *Thermal Engineering*, Vol. 28 PP 40

Chapter 10

Condensate Drop Movement by Surface Temperature Gradient on Heat Transfer Surface in Marangoni Dropwise Condensation

Yoshio Utaka and Zhihao Chen

Additional information is available at the end of the chapter

http://dx.doi.org/10.5772/51830

1. Introduction

Marangoni dropwise condensation occurs in the condensation of a binary vapor mixture of a positive system, in which the surface tension of the mixture has a negative gradient with the mass fraction of the more volatile component, such as water–ethanol and water–ammonium mixtures. Thick condensate areas have higher liquid surface temperatures than thin areas; therefore, the surface tension flow is induced toward the peak of the condensate from thinner areas as a result of the vapor–liquid equilibrium and the variation in the surface tension in the binary vapor condensation of a positive system. This phenomenon differs essentially from so-called dropwise condensation on a hydrophobic surface, because there is a continuous thin liquid film between condensate drops and condensation occurs on a hydrophilic surface. This phenomenon was first reported by Mirkovich and Missen [1] in 1961 for a binary mixture of organic vapors. Ford and Missen [2] demonstrated that the criterion for instability of a condensate liquid film is $d\sigma/db > 0$, where b denotes the condensate film thickness. Fujii et al. [3] conducted an experimental investigation of the condensation of water–ethanol mixtures on a horizontal tube and observed several different condensation modes dependent on the concentration. Morrison and Deans measured the heat transfer characteristics of a water–ammonium vapor mixture and found that it exhibited enhanced heat transfer [4].

In recent years, Utaka and co-workers conducted research [5–9] on the dominant factors (surface subcooling, vapor mass fraction, and vapor velocity) in determining the condensation modes and heat transfer characteristics of Marangoni condensation. The major results on the heat transfer characteristics of Marangoni condensation were summarized in reference 9. Heat transfer was significantly enhanced for a low mass fraction of ethanol in a water–ethanol mixture. Murase et al. [10] studied Marangoni condensation of steam–ethanol mixtures using a horizontal condenser tube, and the results exhibited similar trends to those obtained by Utaka and Wang [7] for vertical surfaces.

The mechanisms of Marangoni condensation have also been studied. Hijikata et al. [11] presented a theoretical drop growth mechanism for Marangoni dropwise condensation. That is, the Marangoni effect occurs due to the surface tension difference which plays a more important role than the surface tension. Akiyama et al. [12] performed a 2-dimensional numerical simulation of the condensation of water-ethanol vapor on a horizontal heat transfer surface and found a 2 K temperature difference between the condensate film area and the crest of the condensate drop. Marangoni flow occurs in a condensate liquid and is driven by the surface tension gradient induced by the temperature difference. Utaka et al. [13] investigated the effect of the initial drop distance, which is the average distance between initially formed drops grown from a thin flat condensate film that forms immediately after a drop departs. They clarified that the initial drop distance is closely related to the heat transfer characteristics of Marangoni condensation. Furthermore, Utaka and Nishikawa [14] measured the thickness of condensate films on the tracks of departing drops and between drops using the laser extinction method, in which the proportion of laser light absorbed by the condensate liquid is dependent on the liquid thickness. The condensate film was approximately 1 μm thick and was strongly dependent on the initial drop distance and the heat transfer characteristics.

Marangoni condensation occurs due to the instability of Marangoni force acting on the condensate film. Condensate drops move spontaneously without any external forces when a bulk temperature gradient is applied to a horizontal heat transfer surface, only due to the imbalance of the surface tension distribution around the drops. This kind of phenomena could also occur in a low-gravity environment. This implies that condensate drops can be moved by applying a bulk surface temperature to a heat transfer surface. It is thus possible to remove a thick liquid film and large condensate drops by exploiting this spontaneous movement of condensate drops. A highly efficient heat exchanger could then be realized. Moreover, since non-uniform temperature distributions are often generated in heat exchangers, it is essential to clarify the heat transfer and condensate movement characteristics in Marangoni condensation when there is a temperature distribution on the heat transfer surface. It is also considered that the circulation of condensate driven by surface tension flow could be utilized in some heat transfer devices (e.g., a wickless heat pipe). Utaka and Kamiyama [15] examined the effect of the bulk surface tension gradient on condensate drop movement when a steady bulk temperature gradient was applied to horizontal and inclined heat transfer surfaces in the condensation of a water–ethanol vapor mixture. The condensate drops moved from the low-temperature side to the high-temperature side. The drop velocity increased with the surface tension gradient on the condensing surface and was independent of the drop size. Chen and Utaka [16] investigated the mechanisms and characteristics of drop movement on a horizontal condensing surface with a bulk temperature gradient for Marangoni dropwise condensation of a water–ethanol vapor mixture. In particular, experimental observations and measurements on the dominant factors affecting condensate drop movement were conducted, such as 1) bulk surface tension gradient, and 2) initial drop distance (adopted as a parameter for the Marangoni force and the condensate drop shape). The velocity of condensate drop movement was determined to correlate well with both the surface tension gradient and the initial drop distance.

In this chapter, the characteristics and mechanisms of condensate drop movement driven by a surface tension gradient in Marangoni dropwise condensation are summarized on the basis of the presented researches.

2. Marangoni dropwise condensation

2.1. Heat transfer characteristics in Marangoni dropwise condensation

In the condensation of a binary vapor mixture, such as water–ethanol vapor, the Marangoni force (indicated by the arrows in Fig. 1) pulls the condensate liquid from the periphery toward the peak along the surface of a condensate drop, whereby dropwise condensation occurs. The Marangoni force here is the driving force for condensate flow, which is considered to be caused by a surface tension difference on the condensate surface, based on the vapor–liquid equilibrium and the variation in the surface tension for a water–ethanol liquid mixture (see Fig. 2). This kind of phenomenon is referred to as 'Marangoni dropwise condensation'.

Utaka and Terachi [6] and Utaka and Wang [7] reported that significantly enhanced heat transfer could be realized by decreasing the thermal resistance of the condensate liquid in the Marangoni dropwise condensation of a water–ethanol vapor mixture. Utaka and Terachi [6] measured the condensation characteristics and clarified that surface subcooling is one of the dominant factors that determines the condensate and heat transfer characteristics of Marangoni condensation. More accurate measurements of a wider range of ethanol mass fraction and surface subcooling were conducted by Utaka and Wang [7], some of the results of which are shown in Fig. 3 and Fig. 4. Figure 3 shows the heat transfer coefficient of Marangoni condensation for water–ethanol vapor mixtures with various ethanol mass fractions. Figure 4 shows the variation in the ratio of the peak heat transfer coefficient of the mixture vapor to that of pure steam. These two figures indicate that the condensation heat transfer is significantly enhanced by the addition of an extremely small amount of ethanol and the heat transfer coefficient of the vapor mixture is approximately 8 times higher than that of pure steam.

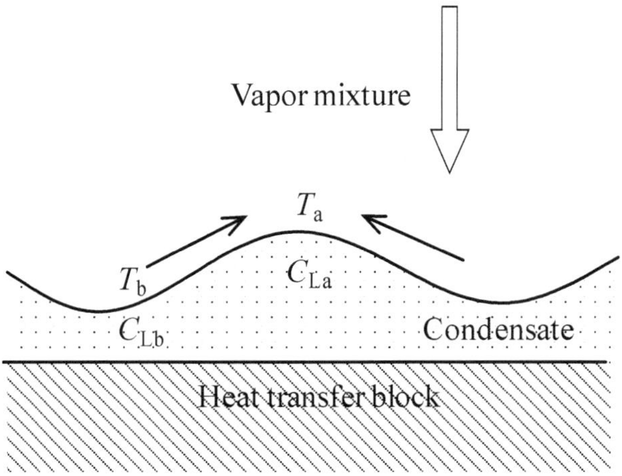

Figure 1. Mechanism for Marangoni dropwise condensation

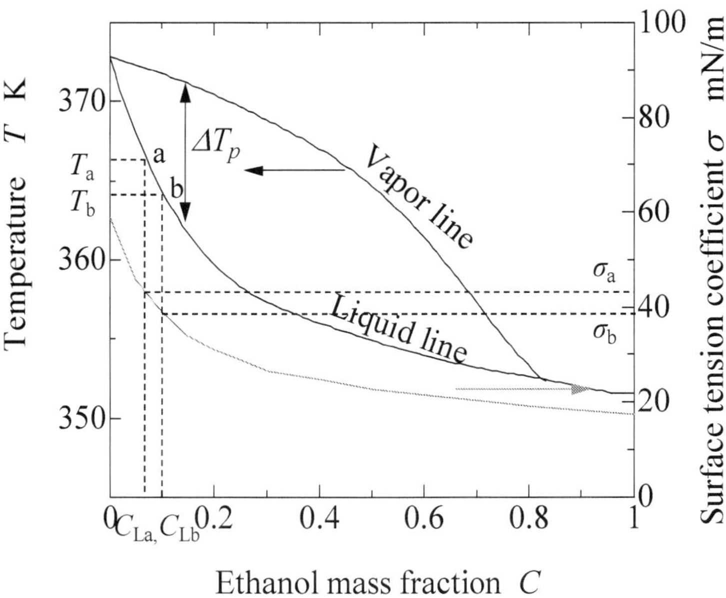

Figure 2. Vapor–liquid equilibrium and variation of surface tension coefficient for water–ethanol mixture

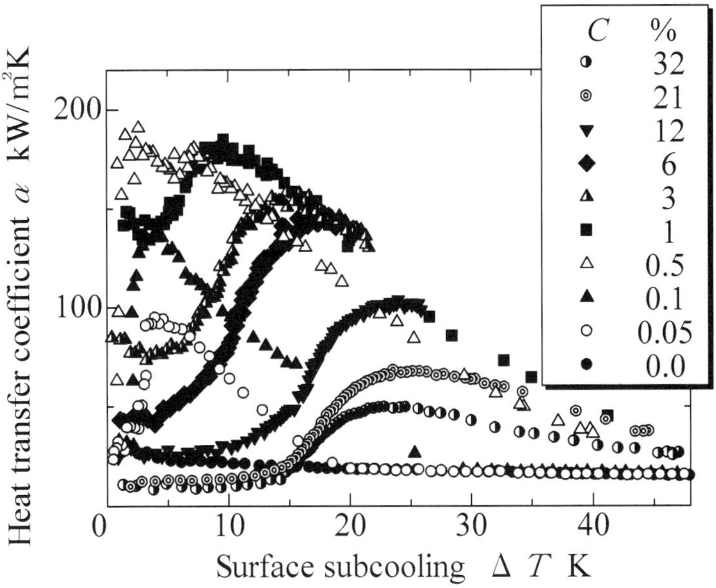

Figure 3. Condensation characteristics of water–ethanol vapor mixtures with various ethanol mass fractions

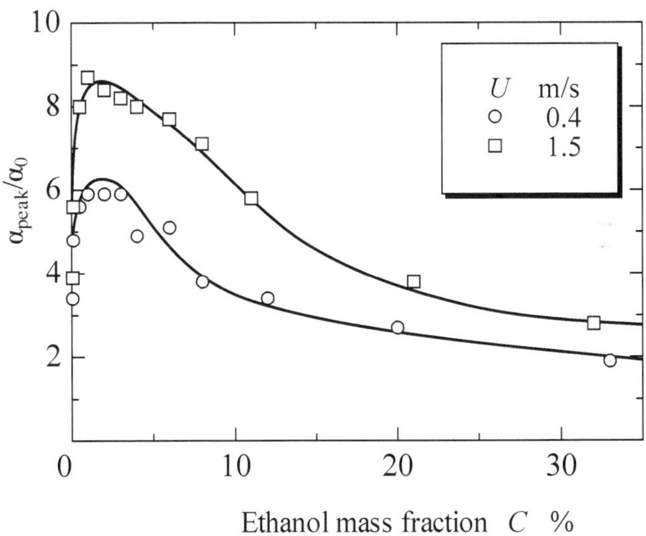

Figure 4. Variation in the ratio of the peak heat transfer coefficient of the vapor mixture to that of pure steam

2.2. Relations among initial drop distance, Marangoni force and shape (angle) of condensate drop

In Marangoni dropwise condensation, small condensate drops initially form from a smooth and thin liquid film adjacent to the periphery of a large condensate drop after its departure. Figure 5 shows time-series microscopic images of the formation of small condensate drops. These initially formed drops are called initial drops and the average distance between the centers of the initial drops is defined as the initial drop distance.

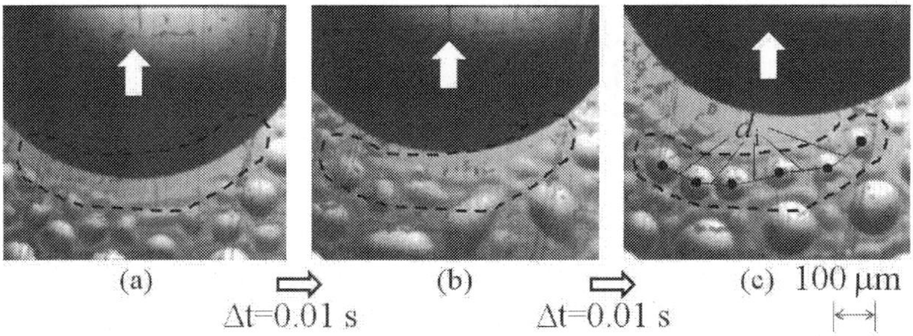

Figure 5. Initial drop formation process with initial drop distance (d_i) indicated

Condensate drops form in Marangoni dropwise condensation due to the Marangoni force acting on the surface of the condensate liquid. Therefore, the Marangoni force is considered to be closely related to the heat transfer mechanisms of Marangoni dropwise condensation. It is also reasonable that the formation of initial drops and the initial drop distance is determined by the strength of the Marangoni force. Thus, a close correlation exists between the initial drop distance and Marangoni force. Consequently, the initial drop distance is adopted as an important parameter of the heat transfer characteristics and mechanisms of Marangoni condensation in the studies of Utaka et al. [13] and Utaka and Nishikawa [14]. Figures 6(a) and (b) show respective plots of the heat transfer coefficient and the initial drop distance as a function of surface subcooling based on data measured by Utaka et al. [13]. The initial drop distances have U-shaped curves with minima that correspond to distances in the range 30–150 μm, depending on the surface subcooling and the mass fraction of ethanol. Surface subcooling at the minimum initial drop distances coincides with that at the maximum heat transfer coefficient. Utaka and Nishikawa [14] investigated the relationship between the liquid film thickness and the initial drop distance (Fig. 7) for a water–ethanol mixture using the laser extinction method. A condensate liquid film of approximately 1 μm thickness remained after sweeping by departing drops and between condensate drops. The minimum condensate film thickness decreased with initial drop distance for surface subcooling lower than the maximum heat transfer point, even when the condensation rate increased.

These two studies demonstrated that there is a close relationship between the heat transfer coefficient and characteristic parameters such as the initial drop distance and the minimum condensate thickness. In the surface subcooling region near the maximum heat transfer coefficient, the initial drop distance and minimum film thickness tend to assume minimum values as a result of the driving force being a maximum, due to the surface tension gradient on the condensate surface. Thus, when the initial drop distance decreases, heat transfer is enhanced by thinning of the condensate film that could result in a reduction in the thermal resistance of the condensate. In addition, the condensate drop shape changes with increasing Marangoni force and the condensate film becomes thinner, even when condensation rate increases. This implies that the drop height increases as the drops approach hemispherical shapes due to an increase in the Marangoni force. The correlation among the Marangoni force, initial drop distance and shape (angle) of condensate drops, as

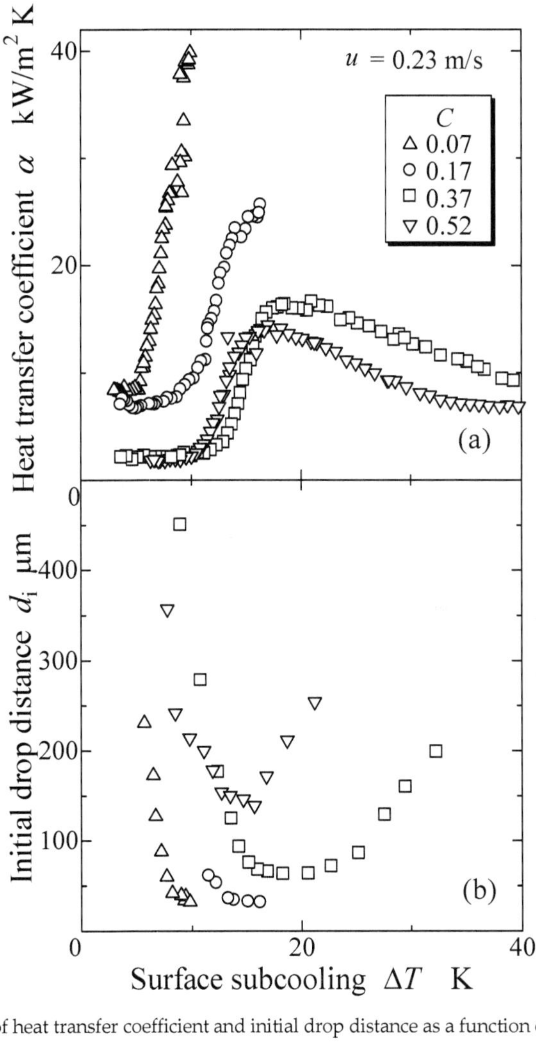

Figure 6. Variation of heat transfer coefficient and initial drop distance as a function of surface subcooling

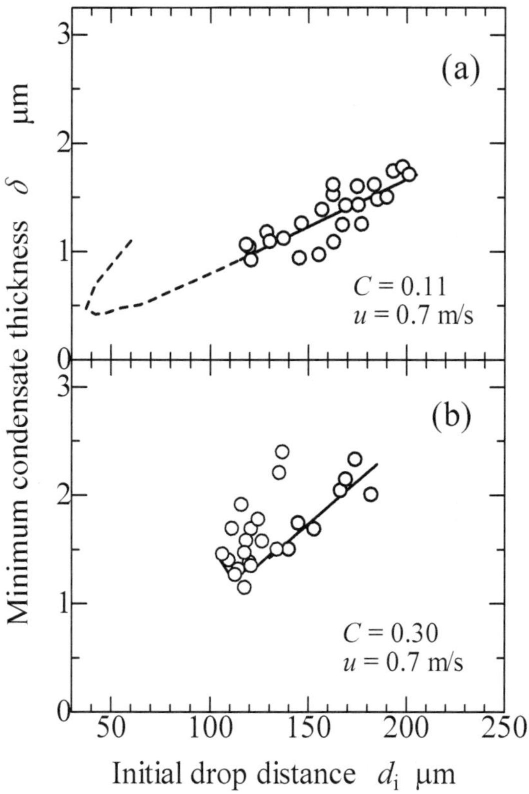

Figure 7. Variation of minimum condensate film thickness as a function of initial drop distance

shown in Fig. 8, could be inferred on the basis of these experimental results. For certain mass fractions of ethanol, the experimental condition of surface subcooling determines the strength of the Marangoni force, and thus the initial drop distance and shape of the condensate drop are also determined. Therefore, if any one of the three factors is known, the two other factors can be determined based on the corresponding correlations. The qualitative correlations are inferred from the experimental results. The quantitative correlations were experimentally studied and are introduced in the following section.

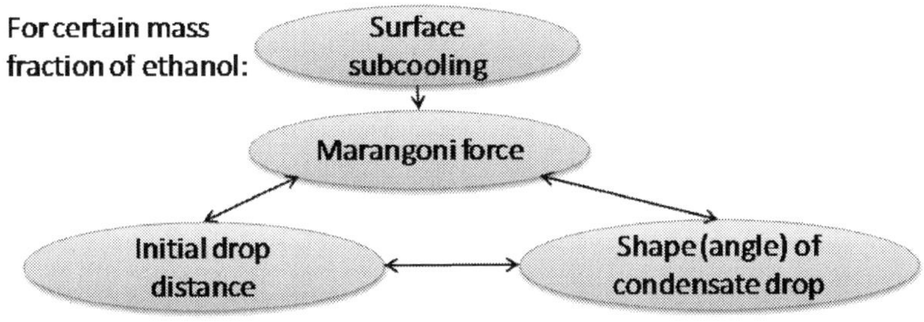

Figure 8. Correlation between Marangoni force, initial drop distance and condensate drop shape

3. Spontaneous movement of condensate drops in Marangoni dropwise condensation

When a bulk temperature gradient is applied to a horizontal heat transfer surface or in a low-gravity environment under a Marangoni condensation field, condensate drops move spontaneously without external forces. The reason for condensate drop movement is considered to be as follows. A Marangoni force (F_H or F_L in Fig. 9, the letter 'H' represents the high temperature side and 'L' represents the high temperature side.) is induced by the difference in surface tension on the condensate surface in Marangoni dropwise condensation. The condensate near the periphery of a condensate drop is pulled along the condensate liquid surface toward the peak of the drop. A reactive force against the surface tension flow caused by the Marangoni force is induced at the drop periphery. When there is no bulk temperature gradient, the reactive force is uniform around the drop periphery and averages out over time, so that the drop does not move to a large extent. In contrast, when a bulk temperature gradient is applied to a horizontal heat transfer surface, the horizontal component of the reactive force (F_{HX} or F_{LX} in Fig. 9, the letter 'X' represents the horizontal component.) around a condensate drop becomes nonuniform, as shown in Fig. 9. Consequently, condensate drops move spontaneously on the heat transfer surface without external forces.

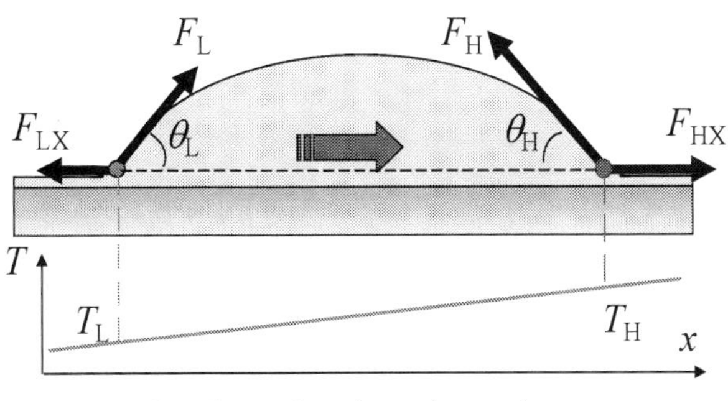

Figure 9. Schematic diagram of the driving force for condensate drop movement

It is considered that the imbalance of the reactive force is determined by the bulk surface tension gradient of the condensate liquid. Consequently, the velocity of condensate drop movement is considered to be affected by the bulk surface tension gradient. The bulk surface tension gradient is calculated from the surface tension difference, which corresponds to the time-averaged surface temperature distribution of the extremely thin liquid film covering the heat transfer surface. The horizontal component of the Marangoni force depends on the overall magnitude of the Marangoni force and the shape (angle) of the condensate drop. Therefore, it is conceivable that the movement of the condensate drop is also determined by these two factors. Based on the correlation between the Marangoni force, initial drop distance and the angle of the condensate drop shown in Fig. 8, it follows that condensate drop movement is also affected by these three factors. Utaka and co-workers [15, 16] have focused on these factors and carried out several experimental and numerical studies on the characteristics and mechanisms of condensate drop movement in Marangoni dropwise condensation.

3.1. Experimental apparatus

Figure 10 shows a schematic of a typical experimental system. A vapor mixture is generated by electrically heating a water–ethanol mixture with a certain mass fraction in a vapor generator. The vapor is partially condensed on a heat transfer block, and is almost completely condensed in an auxiliary condenser after passing through the condensing chamber. The vapor pressure is maintained close to atmospheric pressure by a small opening to the atmosphere between the auxiliary condenser and condensate receiver. The condensate is fed back into the vapor generator after deaeration to remove non-condensable gases dissolved in the condensate. In addition, the vapor is made to flow in the opposite direction to the condensate drop movement to distinguish the driving force of drop movement from the shear force of the vapor flow. Figure 11 shows a schematic of the condensing chamber, where the condensate drop behavior is observed through front and side windows.

Figure 12 shows a schematic of the heat transfer block, which was made of brass with a surface area of 20×20 mm^2 that was positioned horizontally for the experiments. A triangular cross-section of constantan, which has low thermal conductivity, was soldered onto the cooling surface of the heat transfer block. This allowed a bulk temperature gradient to be applied to the heat transfer surface by uniformly cooling the constantan surface with multiple water jet spray. Temperature was measured using thermocouples located inside the heat transfer block, and the surface temperature distribution was determined by two-dimensional extrapolation. The heat transfer surface was coated with titanium dioxide to make it hydrophilic to distinguish it from dropwise condensation on a hydrophobic surface. Experiments were conducted continuously using quasi-steady-state measurements, in which the temperature of the cooling water was changed very slowly.

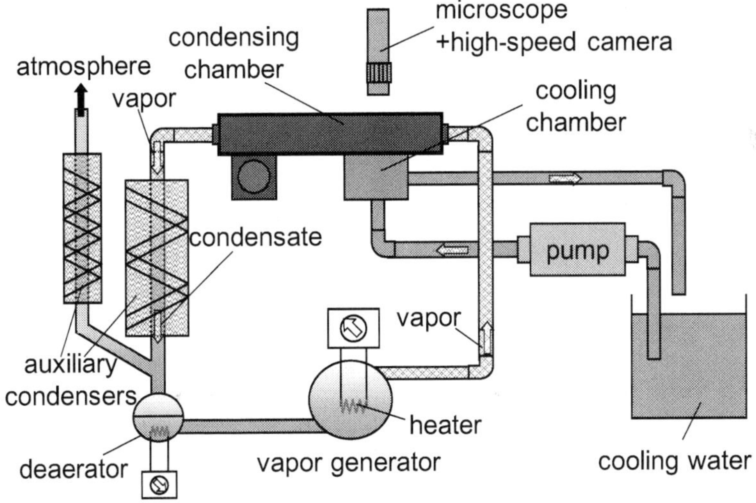

Figure 10. Schematic of a typical experimental apparatus setup

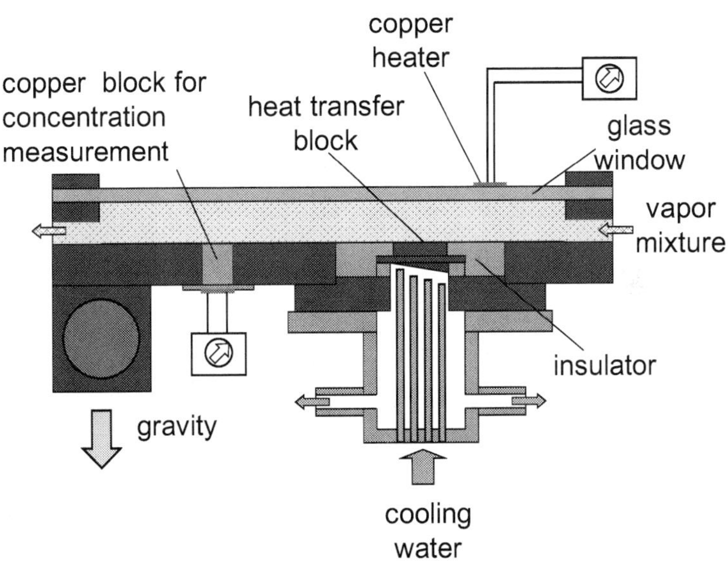

Figure 11. Schematic of the condensing chamber

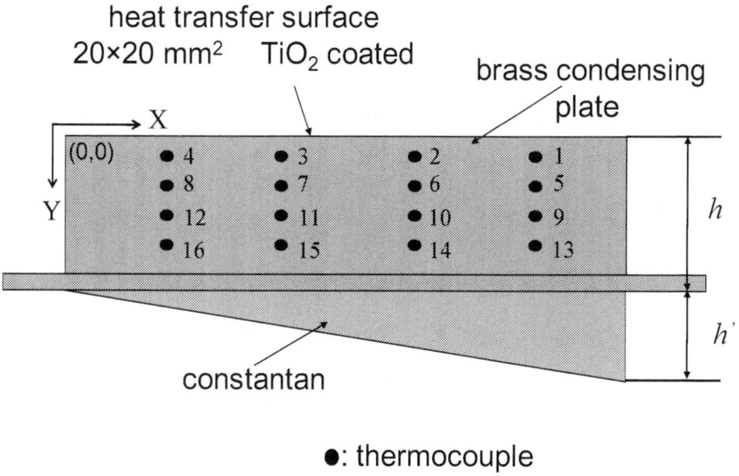

Figure 12. Schematic of the heat transfer block

3.2. Variations of condensate drop shape, initial drop distance and heat transfer coefficient against surface subcooling

To confirm the correlations among the Marangoni force, initial drop distance and shape of condensate inferred in section 2.2, experimental studies were conducted to investigate the quantitative relations. The angle of the condensate drop, initial drop distance and heat transfer coefficient were experimentally measured and the relations are discussed. A profile image of a condensate drop taken through the side view window is shown in Fig. 13. Vapor flows from the right side to the left side, which is the high-temperature side of the heat transfer surface, as does the condensate drop. The angle between the surface of the condensate drop and the heat transfer surface near the drop base as shown in the profile image is defined as the angle of the condensate drop. The angle in the direction of forward movement is the advancing angle θ_a, and that at the opposite side is the receding angle, θ_r. Since similar tendencies in the variation of advancing and receding angles were observed, an average value of the advancing and receding angle for single condensate drops was calculated. The variation of average condensate drop angle, initial drop distance and the heat transfer coefficient as a function of surface subcooling are shown in Fig. 14.

Several tendencies are evident in Fig. 14. For each mass fraction of ethanol, as with the previous results, the heat transfer coefficient increases and the initial drop distance decreases with increasing surface subcooling. In addition, the average condensate drop angle increases with increasing surface subcooling. This indicates that the decrease in the initial drop distance corresponds to an increase in the angle of the condensate drop. The maximum average angle of a condensate drop was approximately 35-45°, which is slightly smaller than the typical contact angle of a condensate drop on a hydrophobic surface. Moreover, for the same surface subcooling, a higher heat transfer coefficient and smaller initial drop distance were realized for a smaller mass fraction of ethanol.

The experimental results indicate that greater surface subcooling or lower mass fraction of ethanol gives a smaller initial drop distance, and the average angle of the condensate drop is larger due to the stronger Marangoni force. Therefore, it was confirmed that the three main factors have quantitative correlations. In addition, there was a large amount of scatter in the data for the condensate drop angle, which is caused by frequent coalescence when the drops are moving or by variation of the temperature distribution on the heat transfer surface. This scatter is considered to be an essential characteristic of Marangoni dropwise condensation. Therefore, in the three important factors, the Marangoni force cannot be measured and the condensate drop angle has large amount of scatter. In contrast with the other two factors, initial drops grow from a thin flat condensate film that appears immediately after a drop departs, the state before the initial drops form is relatively stable, and thus, the measurement of the initial drop distance has good repeatability. Therefore, it is appropriate to adopt the initial drop distance as the dominant parameter of Marangoni dropwise condensation that represents the Marangoni force and the shape of a condensate drop.

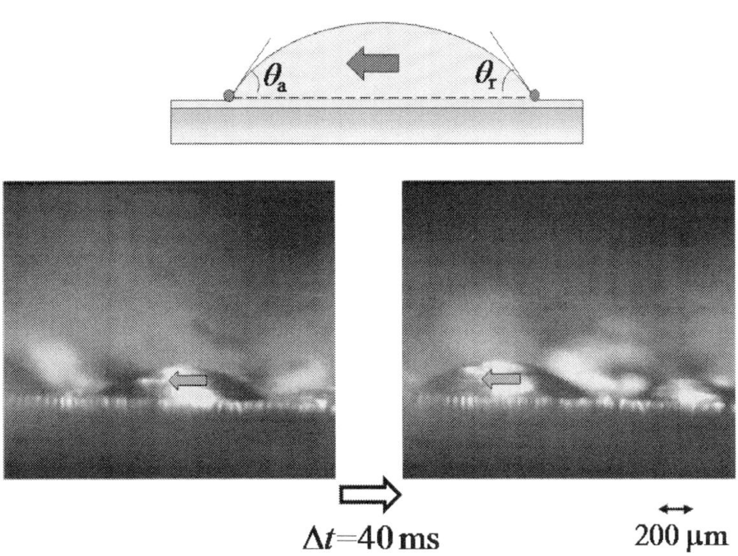

Figure 13. Profile image of condensate drop shape

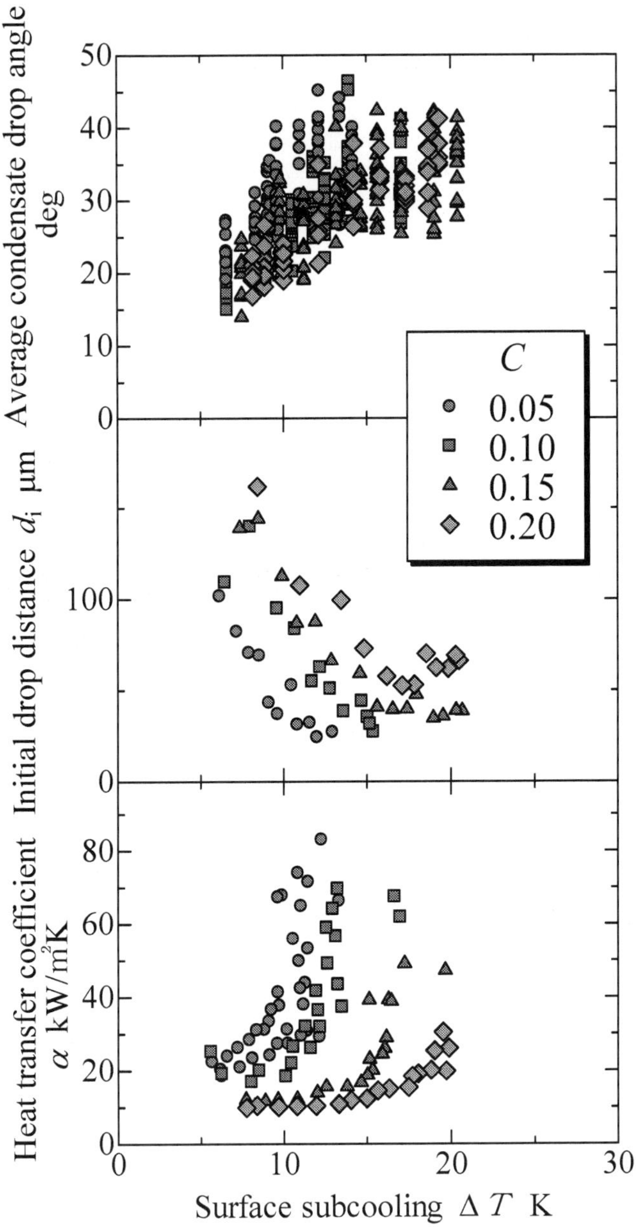

Figure 14. Variation of condensate drop angle, initial drop distance and heat transfer coefficient as a function of surface subcooling for various mass fractions of ethanol

3.3. Effect of bulk surface tension gradient on velocity of drop movement

Utaka and Kamiyama [15] examined the effect of the bulk surface tension gradient on condensate drop movement when a steady bulk temperature gradient was applied to horizontal and inclined heat transfer surfaces during the condensation of a water–ethanol vapor mixture. Figure 15 shows images of condensate drop movement on the horizontal heat transfer surface. The upper part of the image is the high-temperature side and the lower part is the low-temperature side. The condensate drops move from the low-temperature side to the high-temperature side. The variations of condensate drop velocity are shown for various ethanol mass fractions as a function of the bulk surface tension gradient in Fig. 16. The drop velocity increased with increasing surface tension gradient on the condensing surface and was independent of the drop size. Moreover, although there is a large scatter in the drop velocities due to frequent coalescence of the condensate drops, qualitatively similar tendencies of drop velocity were shown.

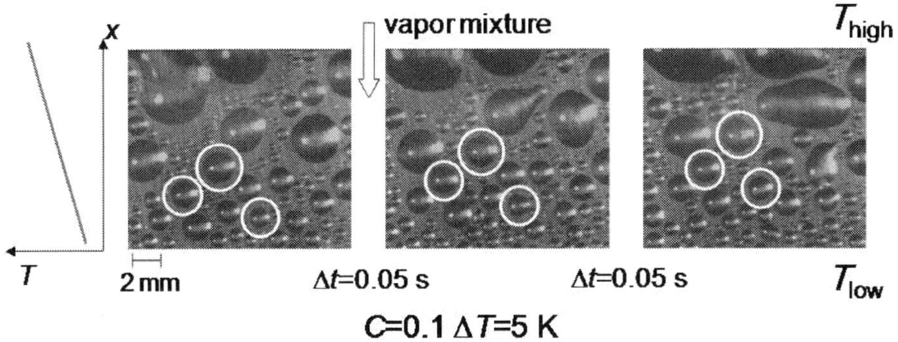

Figure 15. Appearance of condensate drop movement

3.4. Effect of initial drop distance on velocity of drop movement

Chen and Utaka [16] investigated the affects of the Marangoni force and the condensate drop angle on the velocity of condensate drop movement. As discussed in section 3.2, the initial drop distance was adopted as the dominant parameter representing the Marangoni force and angle of condensate drop. The variations of velocity of drop movement as functions of the initial drop distance and bulk surface tension gradient are shown in Figs. 17 and 18, respectively. The drop velocities vary significantly, so that an average velocity of all condensate drops was adopted for each set of conditions.

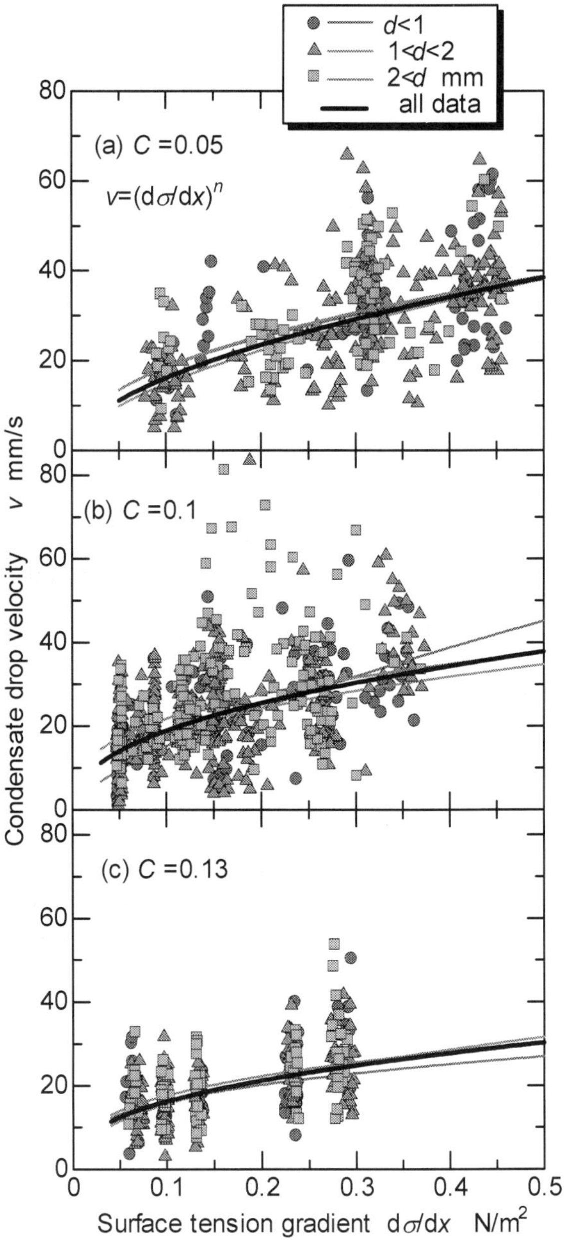

Figure 16. Variation of condensate drop velocity against bulk surface tension gradient

Figures 17 show plots of the average drop velocity as a function of the initial drop distance for ethanol mass fractions *C*, of 0.1, 0.15, 0.2, and 0.25 for six different bulk surface tension gradients. All experiments were performed in the surface subcooling range for which the heat transfer coefficient is less than its maximum value. Distinct trends between the average drop velocity and the initial drop distance are observed in Fig. 17. Qualitatively similar tendencies are evident and the curves have maximum values for each ethanol mass fraction and each surface tension gradient. When the initial drop distance decreases, the average drop velocity initially increases and then decreases after reaching a maximum value at almost the same surface tension gradient for all ethanol mass fractions. For example, for C = 0.15 and a bulk surface tension gradient of 0.2–0.3 N/m^2, the average drop velocity increases from v = 16 mm/s to a maximum value of v = 21 mm/s when the initial drop distance decreases from d_i = 190 to 130 μm. The average velocity then tended to decrease to v = 0 mm/s as d_i decreased to 25 μm. While the changes in drop velocity were gradual at relatively high bulk surface tension gradients, the drop velocities over the entire range of initial drop distances decreased significantly over a smaller range of bulk surface tension gradients for all ethanol mass fractions. Although it is not surprising that the driving force is approximately 0 at low surface tension gradients in the ranges –0.05 to 0 N/m^2 and 0–0.05 N/m^2 for all mass fractions, it is notable that the driving force is also very small for small d_i in the initial drop distance range of 30–40 μm, even at high surface tension gradients.

Figure 18 shows the effect of the bulk surface tension gradient on the drop velocity for four initial drop distances and for ethanol mass fractions *C*, of 0.1, 0.15, 0.2, and 0.25. Data with similar initial drop distances as those in Fig. 17 were selected and were plotted together and fitted with lines that pass through the origin. The drop velocity increases linearly with increasing bulk surface tension gradient for each initial drop distance range. Furthermore, the rate of increase in the drop velocity with the bulk surface tension gradient increases with increasing initial drop distance in the lower ranges of initial drop distance up to the peak average drop velocity shown in Fig. 17. Similar increasing rate were obtained in the larger ranges of initial drop distance for each mass fraction. The effect of the surface tension gradient on the drop velocity became stronger when the initial drop distance approached values that give rise to the maximum velocities shown in Fig. 17. For example, when the bulk surface tension gradient is –0.05 N/m^2, the velocity of condensate drops is around 0 mm/s when the initial drop distances are in the range of 30–35 μm. The velocity then increases with increasing bulk surface tension gradient; the velocity is 10 mm/s at a bulk surface tension gradient of 0.34 N/m^2. For larger initial drop distances, the increase in the drop velocity as a function of bulk surface tension gradient becomes more rapid when the initial drop distance was in the range of 60–80 μm. This increase becomes much more rapid and the velocity increases from 3.6 to 18.9 mm/s in the initial drop distance range of 155–255 μm, when the corresponding bulk surface tension gradient is increased from 0.09 to 0.28 N/m^2. For comparatively large initial drop distances in the ranges of 155–215 μm and 200–270 μm, the variations in the drop velocity as a function of the bulk surface tension gradient were similar, as shown in Fig. 18(b).

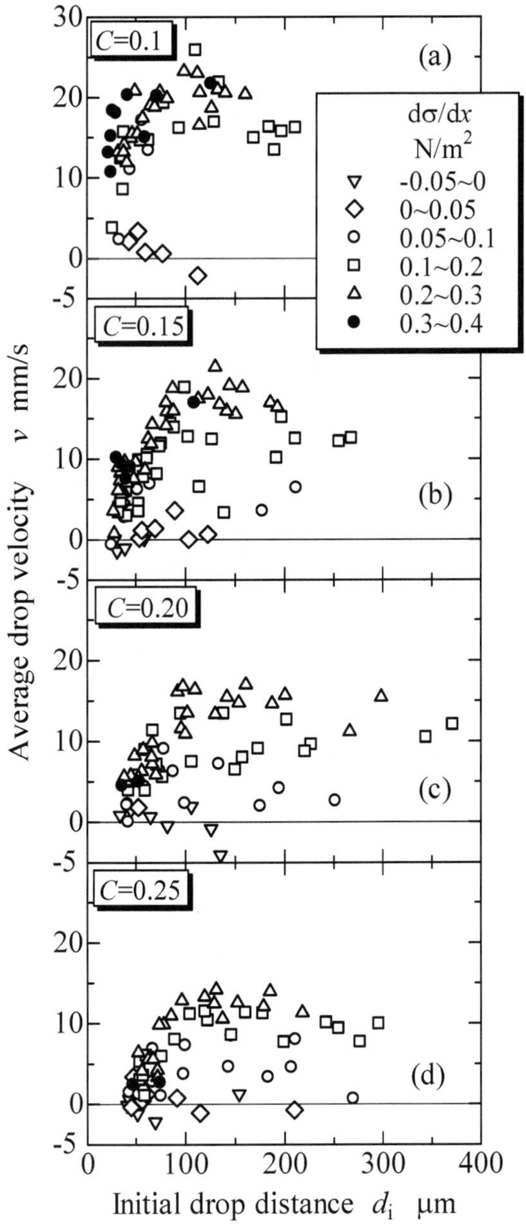

Figure 17. Variation of average drop velocity as a function of initial drop distance for various ethanol mass fractions and bulk surface tension gradients

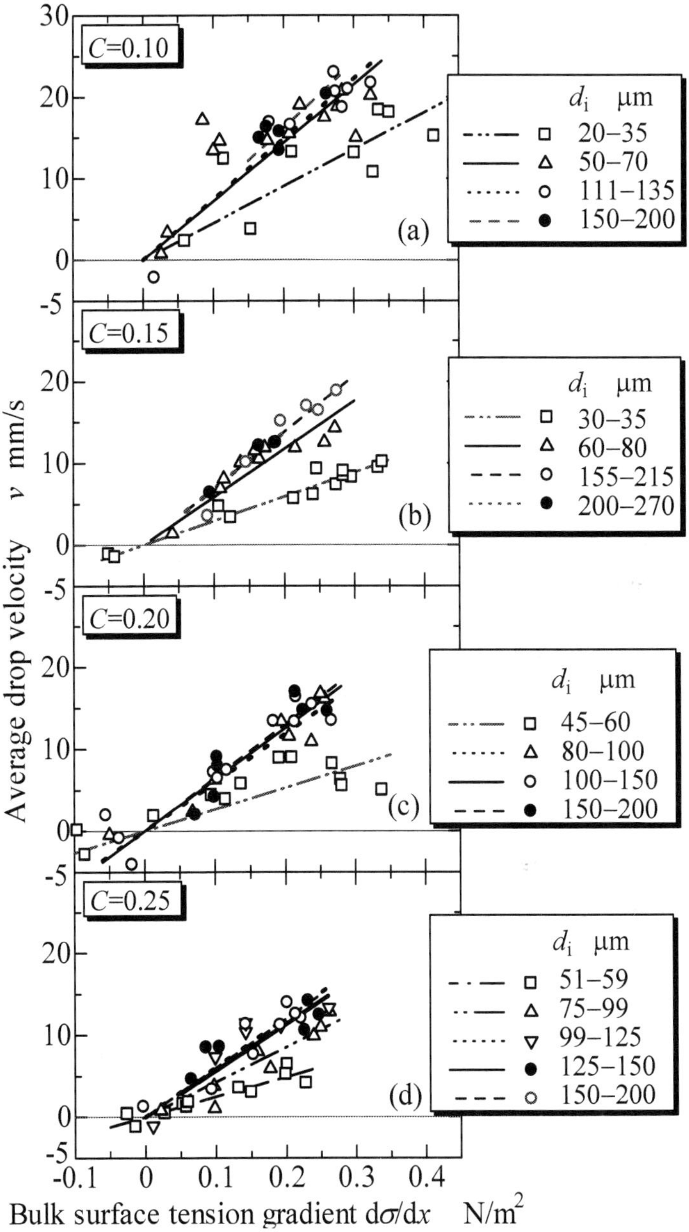

Figure 18. Variation of average drop velocity as a function of bulk surface tension gradient for various ethanol mass fractions and initial drop distances

3.5. Mechanisms of condensate drop movement

Experimental studies on the characteristics (effects of several parameters) of drop movement under a bulk temperature gradient on a heat transfer surface have been conducted for Marangoni dropwise condensation of water-ethanol vapor. However, the essential factor relating to Marangoni dropwise condensation and the condensate drop movement, the Marangoni force, cannot be experimentally measured. Therefore, to better understand the relationship between the Marangoni force and condensate drop movement, numerical simulation of the spontaneous movement of condensate drops was conducted using the volume of fluid (VOF) method. In this section, the 3-dimensional phenomenon of condensate drop movement was simulated using a 2-dimensional calculation in the domain presented in Fig. 19; therefore, only qualitative discussion is presented in this section.

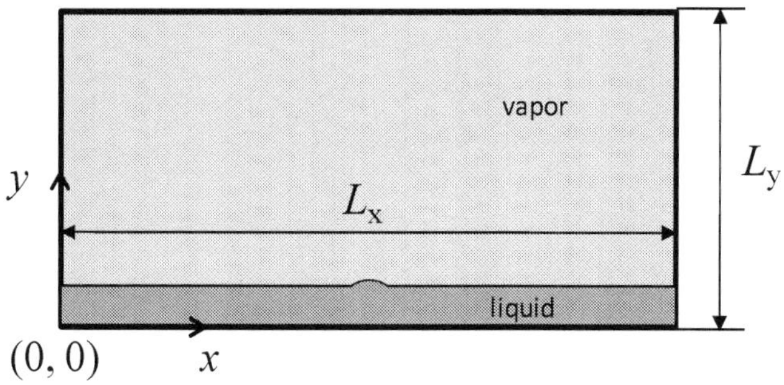

Figure 19. Calculation domain used for the numerical simulation

3.5.1. *Governing equations*

3.5.1.1. *Liquid phase*

In the calculation, the liquid phase was treated as incompressible. The continuity, momentum, and energy equations were solved.

$$\frac{\partial u_{\mathrm{L}}}{\partial x}+\frac{\partial v_{\mathrm{L}}}{\partial y}=0 \tag{1}$$

$$\frac{\partial u_{\mathrm{L}}}{\partial t}+u_{L}\frac{\partial u_{\mathrm{L}}}{\partial x}+v_{L}\frac{\partial u_{\mathrm{L}}}{\partial y}=g_{x}-\frac{1}{\rho_{\mathrm{L}}}\frac{\partial P}{\partial x}+\upsilon_{\mathrm{L}}\left(\frac{\partial^{2}u_{\mathrm{L}}}{\partial x^{2}}+\frac{\partial^{2}u_{\mathrm{L}}}{\partial y^{2}}\right) \tag{2}$$

$$\frac{\partial v_{\mathrm{L}}}{\partial t}+u_{\mathrm{L}}\frac{\partial v_{\mathrm{L}}}{\partial x}+v_{\mathrm{L}}\frac{\partial v_{\mathrm{L}}}{\partial y}=g_{y}-\frac{1}{\rho_{\mathrm{L}}}\frac{\partial P}{\partial y}+\upsilon_{\mathrm{L}}\left(\frac{\partial^{2}v_{\mathrm{L}}}{\partial x^{2}}+\frac{\partial^{2}v_{\mathrm{L}}}{\partial y^{2}}\right) \tag{3}$$

$$\frac{\partial T_{\mathrm{L}}}{\partial t}+u_{\mathrm{L}}\frac{\partial T_{\mathrm{L}}}{\partial x}+v_{\mathrm{L}}\frac{\partial T_{\mathrm{L}}}{\partial y}=\frac{\lambda_{\mathrm{L}}}{\rho_{\mathrm{L}}C_{\mathrm{P,L}}}\left(\frac{\partial^{2}T_{\mathrm{L}}}{\partial x^{2}}+\frac{\partial^{2}T_{\mathrm{L}}}{\partial y^{2}}\right) \tag{4}$$

$$\frac{\partial C_L}{\partial t} + u_L \frac{\partial C_L}{\partial x} + v_L \frac{\partial C_L}{\partial y} = D_L \left(\frac{\partial^2 C_L}{\partial x^2} + \frac{\partial^2 C_L}{\partial y^2} \right) \quad (5)$$

$$\frac{\partial F}{\partial t} + \frac{\partial u_L F}{\partial x} + \frac{\partial v_L F}{\partial y} = 0. \quad (6)$$

3.5.1.2. Vapor phase

In the vapor phase, the vapor velocity in the y-direction was determined by the condensation rate at the vapor-liquid interface, and the vapor velocity in the x-direction was ignored. The diffusion equation was solved based on the assigned vapor velocity. After the results of diffusion equation were obtained, the temperature of the vapor mixture was calculated from the mass fraction based on the vapor line in the relation of vapor-liquid equilibrium.

$$\rho_V v_V = \dot{m} \quad (7)$$

$$\frac{\partial C_V}{\partial t} + v_V \frac{\partial C_V}{\partial y} = D_V \left(\frac{\partial^2 C_V}{\partial x^2} + \frac{\partial^2 C_V}{\partial y^2} \right) \quad (8)$$

$$T_V = f(C_V). \quad (9)$$

3.5.1.3. Vapor-liquid interface

The velocity distribution at the vapor-liquid interface is calculated by taking account of the effect of the surface tension gradient (stress balance at the vapor-liquid interface). This was used as the velocity boundary condition of the liquid phase. The temperature and mass fraction of ethanol at the interface were calculated based on the vapor-liquid equilibrium, energy balance and mass balance.

Condensation rate:

$$\rho_V v_V = \dot{m} \quad (10)$$

Energy balance:

$$-\lambda_L \frac{\partial T_L}{\partial y} = \Delta h_V \dot{m} \quad (11)$$

Mass balance:

$$\dot{m}_E = \rho_V D_V \frac{\partial C_V}{\partial y} + C_V \dot{m} \quad (12)$$

Mass fraction of liquid in the cell at the interface:

$$C_L = \frac{\dot{m}_E}{\dot{m}} \quad (13)$$

Vapor-liquid equilibrium:

$$C_{\mathrm{V}} = f\left(T_{\mathrm{surf}}\right) \tag{14}$$

$$C_{\mathrm{L}} = f\left(T_{\mathrm{surf}}\right) \tag{15}$$

Relationship between the surface tension coefficient and concentration of the liquid:

$$\sigma = f\left(C_{\mathrm{L}}\right) \tag{16}$$

Increase of F caused by condensation:

$$\frac{\partial F}{\partial t} = \frac{1}{\rho_{\mathrm{L}}}\frac{\dot{m}}{\Delta y} \tag{17}$$

Stress balance at the vapor-liquid interface:

$$\left(P_{\mathrm{V}} - P_{\mathrm{L}} + \frac{\sigma}{R}\right)n_i = \left(-\tau_{\mathrm{L},ij}\right)n_k + \frac{\partial \sigma}{\partial x_i} \tag{18}$$

3.5.2. Boundary and initial conditions

Considering the real phenomenon and the computation time, the calculation is conducted in a relatively small region of 600×200 μm² (Fig. 19). The boundary conditions are summarized in Table 1. The boundary at y=0 is set as the condensing surface, and $y=L_y$ is the steady temperature/concentration boundary. In addition, the boundary condition at x=0, L_x is the free inlet/outlet flow. A thin (1.5 μm) liquid film is set initially on the condensing surface. A tiny protuberance is also given in the center of the initial liquid film as a disturbance. During the calculation, a certain mass fraction of ethanol vapor and the corresponding vapor line temperature are assigned to the boundary of $y=L_y$, and the temperature gradient (the right side is set as the high-temperature side) was assigned directly to the condensing surface of y=0. During the calculation, a constant temperature was initially assigned to the condensing surface. After the temperature/concentration distribution in the calculation region became close to that for the actual phenomenon with elapse of time, the temperature gradient was applied to the condensing surface.

Furthermore, the basic equations were discretized using a staggered grid. The convective term was approximated by a 1st–order upwind difference and the diffusion term by 2nd–order central difference. Pressure was calculated implicitly. Other variables such as velocity, temperature and mass fraction were calculated explicitly. The velocity field in the calculation region was calculated using the SOLA method. In addition, the variations of F at the vapor-liquid interface were calculated using the donor-acceptor method.

$y = L_y$	$\partial u/\partial x = 0\ ,\ v = 0$	$T = T_0$	$C = C_0$
$y = 0$	$u = 0\ ,\ v = 0$	$T = T_w(x)$	$\partial C/\partial y = 0$
$x = 0$	$\partial u/\partial x = 0\ ,\ \partial v/\partial y = 0$	$\partial T/\partial x = 0$	$\partial C/\partial x = 0$
$x = L_x$	$\partial u/\partial x = 0\ ,\ \partial v/\partial y = 0$	$\partial T/\partial x = 0$	$\partial C/\partial x = 0$

Table 1. Summary of the boundary conditions employed for the numerical simulation

3.5.3. *Calculation results and discussion*

3.5.3.1. *Variation of liquid film and shape of condensate drop*

The calculation results for an ethanol mass fraction of $C = 0.09$ are shown in Figs. 20–23. The variation in the thickness of the condensate film and the form of the free surface are shown in Fig. 20. The condensate film became thicker over time. Several condensate drops formed and became larger on the condensing surface including the spot where the initial disturbance was located.

3.5.3.2. *Angle of the condensate drop*

Figure 21 shows a comparison of condensate drops forming on condensing surfaces with different surface subcooling (ΔT=6 and 10 K). Condensate drops with similar diameters were selected and the shape of the drops was compared. Developed condensate drops were investigated to avoid the influence of initial conditions. The condensate drop was higher for larger surface subcooling (ΔT = 10 K). Thus, the angle of the condensate drop becomes larger when the surface subcooling is larger, which is in agreement with the experimental results and indicates the condensate drop angle becomes larger because of the stronger Marangoni force.

3.5.3.3. *Driving force of drop movement*

To investigate the driving force of condensate drop movement, the momentum of condensate liquid pulled into a condensate drop by the Marangoni force around the periphery was calculated. In the two-dimensional simulation, the momenta on the high and low-temperature sides of a condensate drop were calculated. The qualitative relation between the drop movement and the imbalance of momentum in the horizontal direction is discussed. In addition, the momentum was calculated at the position where the condensate film around the periphery of a condensate drop is the thinnest (in the valley around the base of a condensate drop).

The experimental results obtained so far indicate that there is a large amount of scatter in the velocities and angles of condensate drops, due to the coalescence of drops or unstable temperature distributions near the periphery of drops. Similar to the experimental results, it is considered that the calculation results also vary significantly around the average values in the numerical simulation. Thus, to avoid the influence of adjacent condensate drops, the condensate drop formed in the vicinity of the center of a condensing surface (Fig. 20) was selected. Because it was considered that the characteristics of a relatively isolated condensate drop in the numerical simulation is nearly equal to that of condensate drops in the experiments.

The aspects of growth and movement of a condensate drop after the temperature gradient was applied are shown in Fig. 22 for C=0.09 and ΔT =10 K. The crests of the condensate drops are indicated by open circles. In addition, the horizontal component of momentum at the high/low-temperature side of the condensate drop and the temperature difference of the liquid surface between the high- (right side) and low- (left side) temperature sides of the condensing surface (boundary condition) are shown in Figs. 23(a) and (b), respectively. The surface temperature of the condensate on the high-temperature side of the condensate drop is higher than that of the low-temperature side, and the horizontal momentum of the condensate liquid is larger on the high-temperature side than that on the low-temperature side.

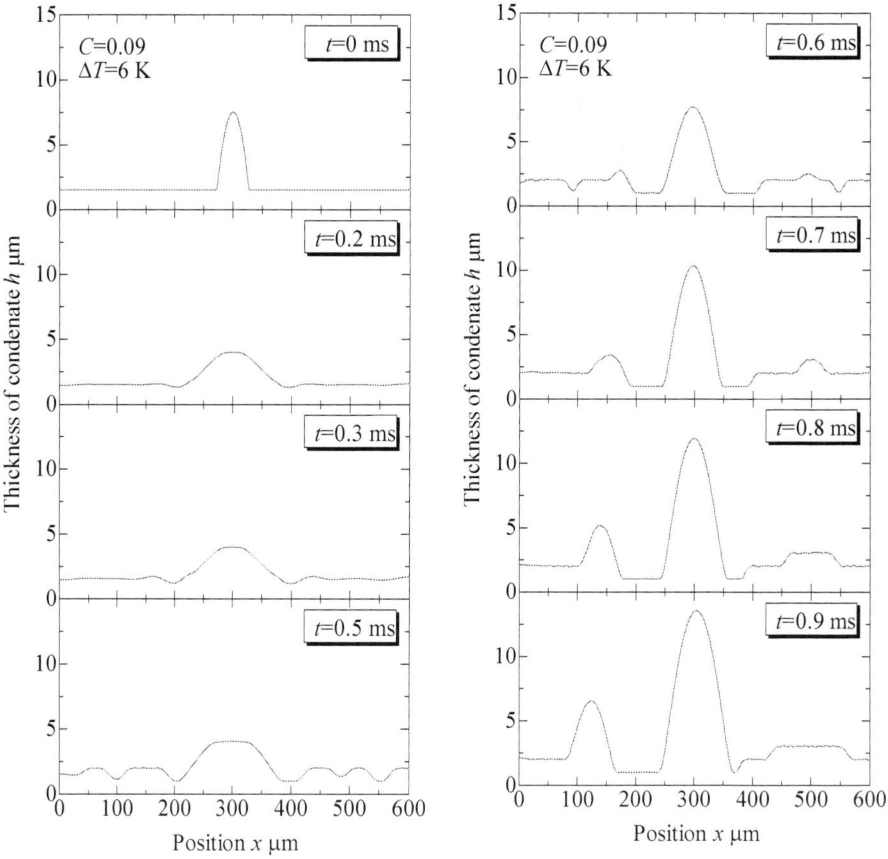

Figure 20. Variation of condensate liquid film thickness over time

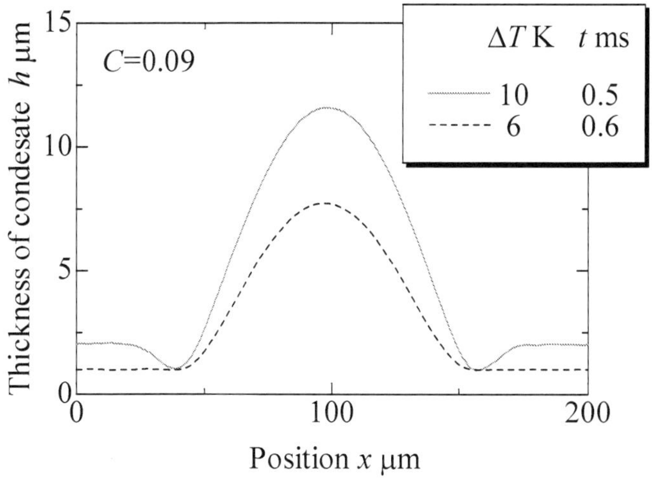

Figure 21. Comparison of condensate drop shape for different subcooling temperatures

These results correspond to those given in Fig. 22, where the growing condensate drop moves towards the high-temperature side. In conclusion, the condensate drop movement is in the direction of the side with the larger momentum of condensate liquid being pulled into the condensate drop by the Marangoni force. Thus, it could be inferred that an imbalance of the horizontal component of the Marangoni force is the driving force for condensate drop movement.

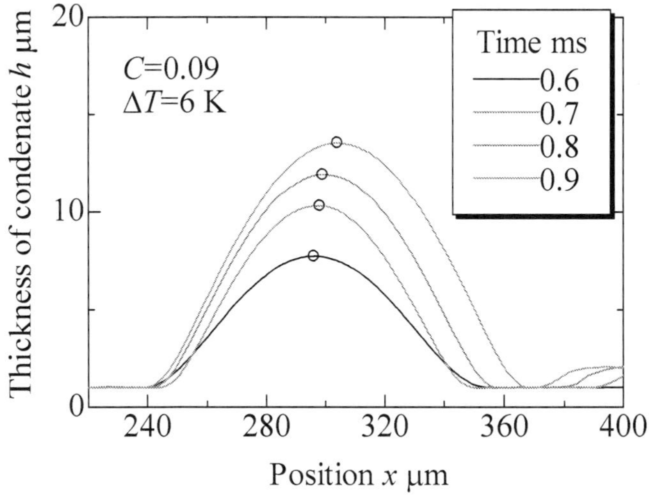

Figure 22. Growth and movement of a condensate drop over time

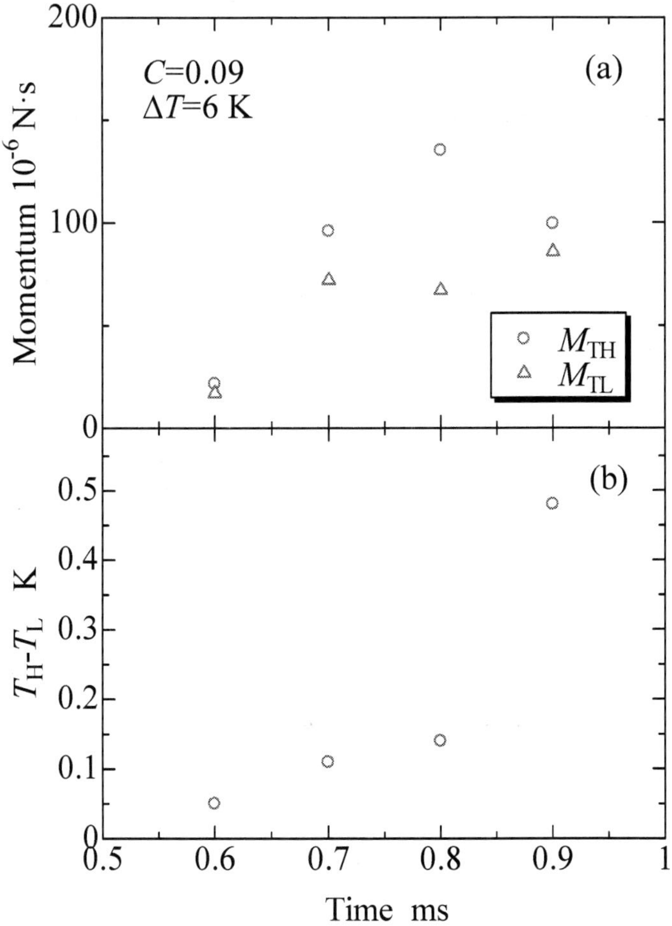

Figure 23. (a) Horizontal momentum of a condensate liquid driven into a condensate drop, and (b) surface temperature difference between the low- and high-temperature side of a condensate drop

4. Conclusion

Condensate drops move spontaneously on a heat transfer surface with a bulk temperature gradient in Marangoni dropwise condensation. It is conceivable that the velocity of a condensate drop is determined by the bulk surface tension gradient, Marangoni force, and the shape of the condensate drop. With a focus on these three factors, experimental and numerical studies were conducted on the spontaneous movement of condensate drops in the Marangoni condensation of a water–ethanol mixture. The results are summarized as follows:

1. Condensate drops move from the low-temperature to high-temperature side of a heat transfer surface. The velocity of condensate drop movement increases with the bulk surface tension gradient.
2. There are correlations among the Marangoni force, initial drop distance and angle of the condensate drop. It is appropriate to adopt the initial drop distance as a dominant parameter to express the characteristics of drop velocity.
3. When the initial drop distance decreases, the average drop velocity initially increases and then decreases after reaching a maximum at almost the same surface tension gradient. The average drop velocity increases linearly with bulk surface tension gradient for each initial drop distance range. The rate of increase in the drop velocity increases with the increasing initial drop distance.
4. Condensate drop movement is directed toward the side with a larger momentum of condensate liquid being pulled into the condensate drop by Marangoni force. It could be inferred that an imbalance of the horizontal component of Marangoni force is the driving force for condensate drop movement.

Nomenclature

C		Mass fraction of ethanol vapor
C_P	[J/kg·K]	Specific heat at constant pressure
D	[m^2/s]	Diffusion coefficient
d	[mm]	Diameter of condensate drop
d_i	[μm]	Initial drop distance
F		VOF function, Force
g	[m/s^2]	Gravity acceleration
Δh	[kJ/kg]	Latent heat
M	[N·s]	Momentum
$\dot{m}$	[kg/m^2·s]	Mass flux of condensation
P	[kPa]	Pressure
T	[K]	Temperature

ΔT	[K]	Surface subcooling
t	[s]	Time
U	[m/s]	Velocity of vapor mixture
u	[m/s]	Horizontal velocity
v	[mm/s], [m/s]	Velocity of drop movement, Vertical velocity
x	[m]	Cartesian axis direction
y	[m]	Cartesian axis direction

Greek characters

α	[kW/m^{2}k]	Heat transfer coefficient
θ	[°]	Angle of condensate drop
σ	[mN/m]	Surface tension coefficient
δ	[μm]	Minimum condensate thickness
λ	[W/m·K]	Thermal conductivity
υ	[m^2/s]	Kinematic viscosity
ρ	[kg/m^3]	Density
τ	[Pa]	Shear stress

Subscripts

a	Advancing angle
r	Receding angle
surf	Vapor-liquid interface
E	Ethanol
L	Liquid phase, low-temperature side
V	Vapor phase
H	High-temperature side
x	Horizontal

Author details

Yoshio Utaka and Zhihao Chen
Division of Systems Research, Faculty of Engineering, Yokohama National University, Japan

5. References

[1] Mirkovich, V.V. and Missen, R.W. (1961). Non-Filmwise Condensation of Binary Vapor of Miscible Liquids, *Can. J. Chem. Eng.*, Vol. 39, pp. 86–87.

[2] Ford, J.D. and Missen, R.W., (1968). On the Conditions for Stability of Falling Films Subject to Surface Tension Disturbances; the Condensation of Binary Vapor, *Can. J. Chem. Eng.*, Vol. 48, pp. 309–312.

[3] Fujii, T., Osa, N. and Koyama, S., (1993). Free Convective Condensation of Binary Vapor Mixtures on a Smooth Horizontal Tube: Condensing Mode and Heat Transfer Coefficient of Condensate, *Proc. US Engineering Foundation Conference on Condensation and Condenser Design*, St. Augustine, Florida, ASME, pp. 171–182.

[4] Morrison, J.N.A. and Deans, J., (1997). Augmentation of Steam Condensation Heat Transfer by Addition of Ammonia, *Int. J. Heat Mass Transfer*, Vol. 40, pp. 765–772.

[5] Utaka, Y. and Terachi, N., (1995). Measurement of Condensation Characteristic Curves for Binary Mixture of Steam and Ethanol Vapor, *Heat Transfer-Japanese Research*, Vol. 24, pp. 57–67.

[6] Utaka, Y. and Terachi, N., (1995). Study on Condensation Heat Transfer for Steam–Ethanol Vapor Mixture (Relation between Condensation Characteristic Curve and Modes of Condensate), *Trans. Jpn. Soc. Mech. Eng.*, Series B, Vol. 61, No. 588, pp. 3059–3065.

[7] Utaka, Y. and Wang, S., (2004). Characteristic Curves and the Promotion Effect of Ethanol Addition on Steam Condensation Heat Transfer, *Int. J. Heat Mass Transfer*, Vol. 47, pp. 4507–4516.

[8] Utaka, Y. and Kobayashi, H., (2003). Effect of Vapor Velocity on Condensation Heat Transfer for Water–Ethanol Binary Vapor Mixture, *Proceedings of 6th ASME-JSME Thermal Engineering Conference*.

[9] Utaka, Y., (2011). Marangoni Condensation Heat Transfer. In: Belmiloudi, A., editor. Experimental Investigations and Industrial Systems, Heat Transfer-Theoretical Analysis, InTech. ISBN 978-953-307-226-5. pp. 327-350.

[10] Murase, T., Wang, H.S. and Rose, J.W., (2007). Marangoni condensation of steam-ethanol mixtures on a horizontal tube, *Int. J. Heat Mass Transfer*, Vol. 50, pp. 3774–3779.

[11] Hijikata, K., Fukasaku, Y. and Nakabeppu, O., (1996). Theoretical and Experimental Studies on the Pseudo-Dropwise Condensation of a Binary Vapor Mixture, *J. Heat Transfer*, Vol. 118, pp. 140–147.

[12] Akiyama, H., Nagasaki, T. and Ito, Y., (2001). A Study on the Mechanism of Dropwise Condensation in Water-Ethanol Vapor Mixture, *Thermal Science & Engineering*, Vol. 9, No. 6, pp. 19-27.

[13] Utaka, Y., Kenmotsu, T. and Yokoyama, S., (1998). Study on Marangoni Condensation (Measurement and Observation for Water and Ethanol Vapor Mixture), *Proceedings of 11th International Heat Transfer Conference*, Vol. 6, pp. 397–402.

[14] Utaka, Y. and Nishikawa, T., (2003). Measurement of Condensate Film Thickness for Solutal Marangoni Condensation Applying Laser Extinction Method, *J. Enhanc. Heat Transf.*, Vol. 10, No. 1, pp. 119–129.

[15] Utaka, Y. and Kamiyama, T., (2008). Condensate Drop Movement in Marangoni Condensation by Applying Bulk Temperature Gradient on Heat Transfer Surface, *Heat Transfer–Asian Research*, Vol. 37, No. 7, pp. 387-397.

[16] Chen, Z. and Utaka, Y., (2011). Characteristics of condensate drop movement with application of bulk surface temperature gradient in Marangoni dropwise condensation, *Int. J. Heat Mass Transfer*, Vol. 54, pp. 5049–5059.

Section 4

Two Phase Flow, Heat Generation and Removal

Chapter 11

Two-Phase Flow

M.M. Awad

Additional information is available at the end of the chapter

http://dx.doi.org/10.5772/76201

1. Introduction

A phase is defined as one of the states of the matter. It can be a solid, a liquid, or a gas. Multiphase flow is the simultaneous flow of several phases. The study of multiphase flow is very important in energy-related industries and applications. The simplest case of multiphase flow is two-phase flow. Two-phase flow can be solid-liquid flow, liquid-liquid flow, gas-solid flow, and gas-liquid flow. Examples of solid-liquid flow include flow of corpuscles in the plasma, flow of mud, flow of liquid with suspended solids such as slurries, motion of liquid in aquifers. The flow of two immiscible liquids like oil and water, which is very important in oil recovery processes, is an example of liquid-liquid flow. The injection of water into the oil flowing in the pipeline reduces the resistance to flow and the pressure gradient. Thus, there is no need for large pumping units. Immiscible liquid-liquid flow has other industrial applications such as dispersive flows, liquid extraction processes, and co-extrusion flows. In dispersive flows, liquids can be dispersed into droplets by injecting a liquid through an orifice or a nozzle into another continuous liquid. The injected liquid may drip or may form a long jet at the nozzle depending upon the flow rate ratio of the injected liquid and the continuous liquid. If the flow rate ratio is small, the injected liquid may drip continuously at the nozzle outlet. For higher flow rate ratio, the injected liquid forms a continuous jet at the end of the nozzle. In other applications, the injected liquid could be dispersed as tiny droplets into another liquid to form an emulsion. In liquid extraction processes, solutes dissolved in a liquid solution are separated by contact with another immiscible liquid. Polymer processing industry is an instance of co-extrusion flow where the products are required to manifest a steady interface to obtain superior mechanical properties. Examples of gas-solid flow include fluidized bed, and transport of powdered cement, grains, metal powders, ores, coal, and so on using pneumatic conveying. The main advantages in pneumatic conveying over other systems like conveyor belt are the continuous operation, the relative flexibility of the pipeline location to avoid obstructions or to save space, and the capability to tap the pipeline at any location to remove some or all powder.

Sometimes, the term two-component is used to describe flows in which the phases do not consist of the same chemical substance. Steam-water flow found in nuclear power plants and other power systems is an example of two-phase single-component flow. Argon-water is an instance of two-phase two-component flow. Air-water is an example of two-phase multi component flow. Actually, the terms two-component flow and two-phase flow are often used rather loosely in the literature to mean liquid-gas flow and liquid-vapor flow respectively. The engineers developed the terminology rather than the chemists. However, there is little danger of ambiguity.

2. Basic definitions and terminology

The total mass flow rate ($\dot{m}$) (in kg per second) is the sum of the mass flow rate of liquid phase ($\dot{m}_l$) and the mass flow rate of gas phase ($\dot{m}_g$).

$$\dot{m} = \dot{m}_l + \dot{m}_g \tag{1}$$

The total volumetric flow rate ($\dot{Q}$) (in cubic meter per second) is the sum of the volumetric flow rate of liquid phase ($\dot{Q}_l$) and the volumetric flow rate of gas phase ($\dot{Q}_g$).

$$\dot{Q} = \dot{Q}_l + \dot{Q}_g \tag{2}$$

The volumetric flow rate of liquid phase ($\dot{Q}_l$) is related to the mass flow rate of liquid phase ($\dot{m}_l$) as follows:

$$\dot{Q}_l = \frac{\dot{m}_l}{\rho_l} \tag{3}$$

The volumetric flow rate of gas phase ($\dot{Q}_g$) is related to the mass flow rate of gas phase ($\dot{m}_g$) as follows:

$$\dot{Q}_g = \frac{\dot{m}_g}{\rho_g} \tag{4}$$

The total mass flux of the flow *(G)* is defined the total mass flow rate ($\dot{m}$) divided by the pipe cross-sectional area *(A)*.

$$G = \frac{\dot{m}}{A} \tag{5}$$

The quality (dryness fraction) *(x)* is defined as the ratio of the mass flow rate of gas phase ($\dot{m}_g$) to the total mass flow rate ($\dot{m}$).

$$x = \frac{\dot{m}_g}{\dot{m}} = \frac{\dot{m}_g}{\dot{m}_l + \dot{m}_g} \tag{6}$$

The volumetric quality (β) is defined as the ratio of the volumetric flow rate of gas phase ($\dot{Q}_g$) to the total volumetric flow rate ($\dot{Q}$).

$$\beta = \frac{\dot{Q}_g}{\dot{Q}} = \frac{\dot{Q}_g}{\dot{Q}_l + \dot{Q}_g} \tag{7}$$

The volumetric quality (β) can be related to the mass quality (x) as follows:

$$\beta = \frac{x v_g}{x v_g + (1-x) v_l} = \frac{1}{1 + \left(\frac{1-x}{x}\right)\left(\frac{\rho_g}{\rho_l}\right)} \tag{8}$$

The void fraction (α) is defined as the ratio of the pipe cross-sectional area (or volume) occupied by the gas phase to the pipe cross-sectional area (or volume).

$$\alpha = \frac{\dot{A}_g}{\dot{A}} = \frac{\dot{A}_g}{\dot{A}_l + \dot{A}_g} \tag{9}$$

The superficial velocity of liquid phase flow (U_l) is the velocity if the liquid is flowing alone in the pipe. It is defined as the volumetric flow rate of liquid phase ($\dot{Q}_l$) divided by the pipe cross-sectional area (A).

$$U_l = \frac{\dot{Q}_l}{A} \tag{10}$$

The superficial velocity of gas phase flow (U_g) is the velocity if the gas is flowing alone in the pipe. It is defined as the volumetric flow rate of gas phase ($\dot{Q}_g$) divided by the pipe cross-sectional area (A).

$$U_g = \frac{\dot{Q}_g}{A} \tag{11}$$

The mixture velocity of flow (U_m) is defined as the total volumetric flow rate ($\dot{Q}$) divided by the pipe cross-sectional area (A).

$$U_m = \frac{\dot{Q}}{A} \tag{12}$$

The mixture velocity of flow (U_m) (in meter per second) can also be expressed in terms of the superficial velocity of liquid phase flow (U_l) and the superficial velocity of gas phase flow (U_g) as follows:

$$U_m = U_l + U_g \tag{13}$$

The average velocity of liquid phase flow (u_l) is defined as the volumetric flow rate of liquid phase ($\dot{Q}_l$) divided by the pipe cross-sectional area occupied by the liquid phase flow (A_l).

$$u_l = \frac{\dot{Q}_l}{A_l} = \frac{\dot{Q}_l}{(1-\alpha)A} = \frac{U_l}{(1-\alpha)} \tag{14}$$

The average velocity of gas phase flow (u_g) is defined as the volumetric flow rate of gas phase ($\dot{Q}_g$) divided by the pipe cross-sectional area occupied by the gas phase flow (A_g).

$$u_g = \frac{\dot{Q}_g}{A_g} = \frac{\dot{Q}_g}{\alpha A} = \frac{U_g}{\alpha} \tag{15}$$

In order to characterize a two-phase flow, the slip ratio (S) is frequently used instead of void fraction. The slip ratio is defined as the ratio of the average velocity of gas phase flow (u_g) to the average velocity of liquid phase flow (u_l). The void fraction (α) can be related to the slip ratio (S) as follows:

$$S = \frac{u_g}{u_l} = \frac{\dot{Q}_g / A\ \alpha}{\dot{Q}_l / A\ (1-\alpha)} = \frac{\dot{Q}_g (1-\alpha)}{\dot{Q}_l\, \alpha} \tag{16}$$

$$S = \frac{u_g}{u_l} = \frac{G\ x / A\ \alpha\ \rho_g}{G\ (1-x) / A\ (1-\alpha)\ \rho_l} = \frac{\rho_l\ x\ (1-\alpha)}{\rho_g\ (1-x)\ \alpha} \tag{17}$$

Equations (16) and (17) can be rewritten in the form:

$$\alpha = \frac{\dot{Q}_g}{S\ \dot{Q}_l + \dot{Q}_g} \tag{18}$$

$$\alpha = \frac{1}{1 + S\left(\frac{1-x}{x}\right)\left(\frac{\rho_g}{\rho_l}\right)} \tag{19}$$

It is obvious from Eqs. (7), and (18) or from Eqs. (8), and (19) that the volumetric quality (β) is equivalent to the void fraction (α) when the slip ratio (S) is 1. The void fraction (α) is called the homogeneous void fraction (α_m) when the slip ratio (S) is 1. This means that $\beta = \alpha_m$. When (ρ_l/ρ_g) is large, the void fraction based on the homogeneous model (α_m) increases very rapidly once the mass quality (x) increases even slightly above zero. The prediction of the void fraction using the homogeneous model is reasonably accurate only for bubble and mist flows since the entrained phase travels at nearly the same velocity as the continuous phase. Also, when (ρ_l/ρ_g) approaches 1 (i.e. near the critical state), the void fraction based on the homogeneous model (α_m) approaches the mass quality (x) and the homogeneous model is applicable at this case.

2.1. Dimensionless parameters

Dimensionless groups are useful in arriving at key basic relations among system variables that are valid for various fluids under various operating conditions. Dimensionless groups can be divided into two types: (a) Dimensionless groups based on empirical considerations, and (b) Dimensionless groups based on fundamental considerations. The first type has been derived empirically, often on the basis of experimental data. This type has been proposed in literature on the basis of extensive data analysis. The extension to other systems requires rigorous validation, often requiring modifications of constants or exponents. The convection number (*Co*), and the boiling number (K_f) are examples of this type. Although the Lockhart-Martinelli parameter (*X*) is derived from fundamental considerations of the gas and the liquid phase friction pressure gradients, it is used extensively as an empirical dimensionless group in correlating experimental results on pressure drop, void fraction, as well as heat transfer coefficients.

On the other hand, fundamental considerations of the governing forces and their mutual interactions lead to the second type that provides important insight into the physical phenomena. The Capillary number (*Ca*), and the Weber number (*We*) are examples of this type.

It should be noted that using of dimensionless groups is important in obtaining some correlations for different parameters in two-phase flow. For example, Kutateladze (1948) combined the critical heat flux (CHF) with other parameters through dimensional analysis to obtain a dimensionless group. Also, Stephan and Abdelsalam (1980) utilized eight dimensionless groups in developing a comprehensive correlation for saturated pool boiling heat transfer.

Also, the dimensionless groups are used in obtaining some correlations for two-phase frictional pressure drop such as Friedel (1979), Lombardi and Ceresa (1978), Bonfanti et al. (1979), and Lombardi and Carsana (1992).

Moreover, a dimensional analysis can be used to resolve the equations of electrohydrodynamics (EHD), in spite of their complexity, in two-phase flow. The two dimensionless EHD numbers that will result from the analysis of the electric body force are the EHD number or conductive Rayleigh number and the Masuda number or dielectric Rayleigh number (Cotton et al., 2000, Chang and Watson, 1994, and Cotton et al., 2005).

The use of traditional dimensionless numbers in two-phase flow is very limited in correlating data sets (Kleinstreuer, 2003). However, a large number of dimensionless groups found in literature to represent two phase-flow data into more convenient forms. Examples of these dimensionless groups are discussed below.

Archimedes Number (*Ar*)

The Archimedes number (*Ar*) is defined as

$$Ar = \frac{\rho_l(\rho_l - \rho_g)gd^3}{\mu_l^2} \tag{20}$$

And represents the ratio of gravitational force to viscous force. It is used to determine the motion of fluids due to density differences (ρ_l-ρ_g).

Quan (2011) related the Archimedes number (Ar) to the inverse viscosity number (N_f) as follows:

$$N_f = Ar^{1/2} = \frac{\sqrt{\rho_l(\rho_l - \rho_g)gd^3}}{\mu_l} \tag{21}$$

Recently, Hayashi et al. (2010 and 2011) used the inverse viscosity number (N_f) in the study of terminal velocity of a Taylor drop in a vertical pipe.

Atwood Ratio (At)

The Atwood ratio (At) is defined as

$$At = \frac{\rho_l - \rho_g}{\rho_l + \rho_g} \tag{22}$$

The important consideration that one must remember is the Atwood ratio (At) and the effect of the gravitational potential field, Froude number (Fr) on causing a drift or allowing a relative velocity to exist between the phases. If these differences are large, then one should use a separated flow model. For instance, for air-water flows at ambient pressure, the density ratio (ρ_l/ρ_g) is ~1000 while the Atwood ratio (At) is ~1. As a result, a separated flow model may be dictated. On the other hand, when the density ratio (ρ_l/ρ_g) approaches 1, a homogenous model becomes more appropriate for wide range of applications.

Bond Number (Bo)

The Bond number (Bo) is defined as:

$$Bo = \frac{gd^2(\rho_l - \rho_g)}{4\sigma} \tag{23}$$

And represents the ratio of gravitational (buoyancy) and capillary force scales. The length scale used in its definition is the pipe radius. The Bond number (Bo) is used in droplet atomization and spray applications. The gravitational force can be neglected in most cases of liquid-gas two-phase flow in microchannels because $Bo \ll 1$. As a result, the other forces like surface tension force, the gas inertia and the viscous shear force exerted by the liquid phase are found to be the most critical forces in the formation of two-phase flow patterns.

In addition, Li and Wu (2010) analyzed the experimental results of adiabatic two-phase pressure drop in micro/mini channels for both multi and single-channel configurations from collected database of 769 data points, covering 12 fluids, for a wide range of operational conditions and channel dimensions. The researchers observed a particular trend with the Bond number (*Bo*) that distinguished the data in three ranges, indicating the relative importance of surface tension. When $1.5 \leq Bo$, in the region dominated by surface tension, inertia and viscous forces could be ignored. When $1.5 < Bo \leq 11$, surface tension, inertia force, and viscous force were all important in the micro/mini-channels. However, when $11 < Bo$, the surface tension effect could be neglected.

Recently, Li and Wu (2010) obtained generalized adiabatic pressure drop correlations in evaporative micro/mini-channels. The researchers observed a particular trend with the Bond number (*Bo*) that distinguished the entire database into three ranges: $Bo < 0.1$, $0.1 \leq Bo$ and $BoRe_l^{0.5} \leq 200$, and $BoRe_l^{0.5} > 200$. Using the Bond number, they established improved correlations of adiabatic two-phase pressure drop for small Bond number regions. The newly proposed correlations could predict the database well for the region where $BoRe_l^{0.5} \leq 200$.

Bodenstein Number (*Bod*)

The Bodenstein number (*Bod*) is defined as follows:

$$Bod = \frac{U_b d}{D} \tag{24}$$

And represents the ratio of the product of the bubble velocity and the microchannel diameter to the mass diffusivity. For example, Salman et al. (2004) developed numerical model for the study of axial mass transfer in gas–liquid Taylor flow at low values of this dimensionless group. The researchers found that their model was suitable for $Bod < 500$. Also, for $Bod > 10$, their model could be approximated by a simple analytical expression.

Capillary Number (*Ca*)

The Capillary number (*Ca*) is defined as:

$$Ca = \frac{\mu_l U}{\sigma} \tag{25}$$

And is a measure of the relative importance of viscous forces and capillary forces. Frequently, it arises in the analysis of flows containing liquid drops or plugs. In the case of liquid plugs in a capillary tube, the Capillary number (*Ca*) can be viewed as a measure of the scaled axial viscous drag force and the capillary or wetting force. The Capillary number (*Ca*) is useful in analyzing the bubble removal process. For two-phase flow in microchannels, *Ca* is expected to play a critical role because both the surface tension and the viscous forces are important in microchannel flows.

This dimensionless group is used in flow pattern maps. For example, Suo and Griffith (1964) used the Capillary number (Ca) as a vertical axis in their flow pattern maps. The researchers gave a transition from slug flow to churn flow by $CaRe^2 = 2.8\times10^5$ that agreed more or less with aeration of the slugs at the development of turbulence.

In addition, Taha and Cui (2006a) showed that in CFD modeling of slug flow inside square capillaries at low Ca, both the front and rear ends of the bubbles were nearly spherical. With increasing Ca, the convex bubble end inverted gradually to concave. As the Ca increased, the bubble became longer and more cylindrical. At higher Ca numbers, they had cylindrical bubbles.

The Capillary number (Ca) controls principally the liquid film thickness (δ) surrounds the gas phase in gas-liquid two-phase plug flows or the immiscible liquid phase in liquid-liquid two-phase plug flows. In the literature, the is a number of well known models for the film thickness in a gas-liquid Taylor flow such as Fairbrother and Stubbs (1935), Marchessault and Mason (1960), Bretherton (1961), Taylor (1961), Irandoust and Andersson (1989), Bico and Quere (2000), and Aussillous and Quere (2000). Kreutzer et al. (2005a, 2005b) reviewed a number of correlations for liquid film thickness available in the literature. Moreover, Angeli and Gavriilidis (2008) reviewed additional relationships for the liquid film thickness. Recently, Howard et al. (2011) studied Prandtl and capillary effects on heat transfer performance within laminar liquid–gas slug flows. The researchers focused on understanding the mechanisms leading to enhanced heat transfer and the effect of using various Prandtl number fluids, leading to variations in Capillary number. They found that varying Prandtl and Capillary numbers caused notable effects in the transition region between entrance and fully developed flows.

For liquid-liquid immiscible flows, Grimes et al. (2007) investigated the validity of the Bretherton (1961) and Taylor (1961) laws through an extensive experimental program in which a number of potential carrier fluids were used to segment aqueous droplets over a range of flow rates. The researchers observed that there were significant discrepancies between measured film thicknesses and those predicted by the Bretherton (1961) and Taylor (1961) laws, and that when plotted against capillary number, film thickness data for the fluids collapsed onto separate curves. By multiplying the capillary number (Ca) by the ratio of the liquid plug viscosity (μ_p) to the liquid film viscosity (μ_l), the data for the different fluids collapsed onto a single curve with very little scatter.

Table 1 shows different equations for dimensionless film thickness (δ/R).

It should be noted that most of the expressions available in the literature are correlating the dimensionless liquid film thickness (δ/R) against the Capillary number (Ca). Recently, Han and Shikazono (2009a, 2009b) measured the local liquid film thickness in microchannels by laser confocal method. For larger Capillary numbers ($Ca > 0.02$), inertial effects must be considered and hence the researchers suggested an empirical correlation of the dimensionless bubble diameter by considering capillary number (Ca) and Weber number (We). The Han and Shikazono (2009a) correlation was

$$\cdot\frac{\delta}{R}=\begin{cases}\dfrac{1.34\ Ca^{2/3}}{13{+}1\ 3\quad Ca^{2/3}+0.504Ca^{0.672}Re^{0.5890}-0.352We^{0.629}} & Re<2000\\[2ex] \dfrac{212\left(\dfrac{\mu^2}{\rho\sigma d}\right)^{2/3}}{149\quad\left(\dfrac{\mu^2}{\rho\sigma d}\right)^{2/3}7773\quad\left(\dfrac{\mu^2}{\rho\sigma d}\right)^{0.672}0\text{-}500\quad 0\left(\dfrac{\mu^2}{\rho\sigma d}\right)^{0.629}} & Re>2000\end{cases} \tag{26}$$

Researcher	δ/R	Notes
Fairbrother and Stubbs (1935)	$\frac{\delta}{R}=0.5Ca^{1/2}$	$5\times10^{-5}\le Ca\le 3\times10^{-1}$
Marchessault and Mason (1960)	$\frac{\delta}{R}=\left(0.89-\frac{0.05}{U_g^{1/2}}\right)Ca^{1/2}$	$7\times10^{-6}\le Ca\le 2\times10^{-4}$ U_g in cm/s
Bretherton (1961)	$\frac{\delta}{R}=1.34Ca^{2/3}$	$10^{-3}\le Ca\le 10^{-2}$
Irandoust and Andersson (1989)	$\frac{\delta}{R}=0.36[1-\exp(-3.08(Ca^{0.54}))]$	$9.5\times10^{-4}\le Ca\le 1.9$
Bico and Quere (2000)	$\frac{\delta}{R}=1.34(2Ca)^{2/3}$	Bretherton (1961) is corrected by a factor of $2^{2/3}$ for $\mu_c>\mu_d$
Aussillous and Quere (2000)	$\frac{\delta}{R}=\frac{1.34Ca^{2/3}}{1+2.5(1.34Ca^{2/3})}$	$10^{-3}\le Ca\le 1.4$ approaches Bretherton (1961) for $Ca\to0$
Grimes et al. (2007)	$\frac{\delta}{R}=5Ca^{2/3}$	$10^{-5}\le Ca\le 10^{-1}$ $Ca=\mu_p U/\sigma$

Table 1. Different Equations for Dimensionless Film Thickness (δ/R).

In fact, the Weber number includes the capillary number (Ca) and Reynolds number (Re) (Sobieszuk et al., 2010). Therefore, the term ($\mu^2/\rho\,\sigma d$) in the second equation of Eq. (26) for $Re>2000$ is equal to (Ca^2/We) or (Ca/Re). As the capillary number approached zero ($Ca\to0$), the first equation of Eq. (26) for $Re<2000$ should follow Bretherton's theory (1961), so the coefficient in the numerator was taken as 1.34. The other coefficients were obtained by least linear square method from their experimental data. If Reynolds number became larger than 2000, liquid film thickness remained constant due to the flow transition from laminar to turbulent. As a result, liquid film thickness was fixed to the value at $Re = 2000$. The second equation of Eq. (26) for $Re>2000$ could be obtained from the first equation by substituting $Re = 2000$. Capillary number (Ca) and Weber number (We) should be also replaced with the values when Reynolds number = 2000. The first equations of Eq. (26) were replaced as follows:

$$Ca=Re\times\left(\frac{\mu^2}{\rho\sigma d}\right)=2000\times\left(\frac{\mu^2}{\rho\sigma d}\right)\quad(Re=2000) \tag{27}$$

$$We = \text{Re} \ . \ Ca = 2000^2 \times \left(\frac{\mu^2}{\rho \sigma d} \right) \quad (\text{Re} = 2000) \tag{28}$$

In Eqs. (27) and (28), ($\mu^2/\rho\ \sigma\ d$) was a constant value if pipe diameter and fluid properties were fixed. The researchers mentioned that their correlation, Eq. (26), could predict δ within the range of ±15% accuracy.

In addition, Yun et al. (2010) used the Weber number (We) to correlate the maximum and minimum film thickness (δ_{max} and δ_{min}) because the maximum and the minimum film thickness could be evaluated approximately and calculated statistically from the shade boundaries of Taylor bubbles observed in the images. On the other hand, it was difficult to determine the mean film thickness from the 2-D optical images of slug flow due to the irregular shapes of liquid film around Taylor bubbles in rectangular microchannels. Their maximum and minimum film thickness (δ_{max} and δ_{min}) correlations were

$$\frac{\delta_{\max}}{R} = 0.78\ We^{0.09} \tag{29}$$

$$\frac{\delta_{\min}}{R} = 0.04\ We^{0.62} \tag{30}$$

Cahn number (Cn)

The Cahn number (Cn) is defined as:

$$Cn = \frac{\delta}{d} \tag{31}$$

And represents the ratio of the interface thickness (δ) and the tube diameter (d). For example, He et al. (2010) used this dimensionless group in their dimensionless governing equations for heat transfer modeling of gas–liquid slug flow without phase change in a micro tube.

Convection Number (Co)

The Convection number (Co) is a modified Lockhart-Martinelli parameter (X). It is defined as:

$$Co = \left(\frac{1-x}{x} \right)^{0.9} \left(\frac{\rho_g}{\rho_l} \right)^{0.5} \tag{32}$$

This dimensionless number was introduced by Shah (1982) in correlating flow boiling data. It was not based on any fundamental considerations. For example, based on more than 10

000 experimental data points for various fluids, including water, refrigerants, and cryogents, Kandlikar (1990) proposed a generalized heat transfer correlation for convective boiling in both vertical and horizontal tubes. One of the dimensionless numbers used in his correlation was the convection number (*Co*).

Courant Number (*Cou*)

A very important step in numerical simulation is transient time step sizing. The Courant number (*Cou*) is a dimensionless group that can be used to adjust the time step. It is defined as follows:

$$Cou = \frac{U\Delta t}{\Delta x} \tag{33}$$

And represents a comparison between the particle moving distance during the assumed time step and control volume dimension. A low *Cou* value means a small time step size (Δt) and consequently a large simulation time. On the other hand, a high *Cou* value leads to an unstable numeric approach. As a result, there is a need to optimize *Cou* using appropriate time step size (Δt). Furthermore, as the mesh becomes finer (Δx), the time step (Δt) should be decreased as well in order to hold *Cou* in its safe range. A typical time step (Δt) order of magnitude of 1×10^{-5} (s) or 1×10^{-6} (s) has been used by the researchers. For example, Cherlo et al. (2010) performed the three-dimensional simulation in their numerical investigations of two-phase (liquid–liquid) flow behavior in rectangular microchannels.

A wiser time step adjustment is using a variable time step by implementing a fixed Courant number (*Cou*) that is available in ANSYS Fluent. For example, Gupta et al. (2009) applied this technique in the CFD modeling of Taylor flow in microchannels. In this method, the time step (Δt) is being modified based on the critical cells size and local velocity components to hold the maximum Courant number (Cou_{max}) to a fixed value.

EHD Number (E_{hd})

The EHD number (E_{hd}) or conductive Rayleigh number is defined as:

$$E_{hd} = \frac{I_o L^3}{\rho_o v^2 \mu_c A} \tag{34}$$

E_M Number

The E_M is defined as:

$$E_M = \frac{h_{lg}}{\dfrac{\sigma d^2}{\rho_l d^3}} \tag{35}$$

And represents the ratio of two energies. The numerator of the term represents the latent heat of vaporization that can further be referred as latent energy per unit mass. The denominator of the term represents the surface tension energy per unit mass. Sabharwall et al. (2009) expressed this dimensionless number as the ratio of latent heat of vaporization to the capillary pressure and used it in phase-change thermosyphon and heat-pipe heat exchangers.

E_r Number

The E_r is defined as:

$$E_r = \frac{h_{lg}}{U_s^2} \tag{36}$$

And relates the ratio of thermal to kinetic energies. Thermal energy is the energy that is required by the fluid for phase change from the liquid to vapor state, and the square of the velocity represents the kinetic energy head. Sabharwall et al. (2009) expressed this dimensionless number as the ratio of latent heat of vaporization to the pressure drop across the heat pipes and thermosyphons and used it in phase-change thermosyphon and heat-pipe heat exchangers.

Eötvös Number (*Eo*)

The Eötvös number (*Eo*) is defined as:

$$Eo = \frac{gd^2(\rho_l - \rho_g)}{\sigma} \tag{37}$$

And represents the ratio of gravitational (buoyancy) and capillary force scales. The length scale used in its definition is the pipe diameter.

Brauner and Moalem-Maron (1992) identified the range of 'small diameters' conduits, regarding two-phase flow pattern transitions. The researchers used the Eötvös number (*Eo*) to characterize the surface tension dominance in the two-phase flow in microchannels. They took $Eo < (2\pi)^2$ as the criterion for the surface tension dominance.

Recently, Ullmann and Brauner (2007) reexamined the channel diameter effect on the flow regime transitions in mini channels and suggested that new mechanistic models be expressed in terms of the non-dimensional Eötvös number. In their definition of Eötvös number, they multiplied the surface tension (σ) in the denominator in Eq. (37) by the factor 8. The researchers suggested that in small Eötvös number systems (of the order of 0.04), the negligibly small bubble velocity, even in vertical systems, led to flow regimes resembling those obtained in conventional channels under microgravity conditions. They used the experimental flow regime data presented by Triplett et al. (1999) for air-water in 1.097 mm Pyrex pipe, corresponding to an Eötvös number of 0.021, to calibrate and determine the efficacy of their approach for small Eötvös number configurations.

Euler Number (*Eu*)

The Euler number (*Eu*) is often written in terms of pressure differences (Δp) and is defined as:

$$Eu = \frac{\Delta p}{\rho U^2} \tag{38}$$

And represents the ratio of pressure forces to inertial forces. It expresses the relationship between the pressure drop and the kinetic energy per volume, and is used to characterize losses in the flow, where a perfect frictionless flow corresponds to $Eu = 1$.

Fourier Number (*Fo*)

The Fourier number (*Fo*) is defined as:

$$Fo = \frac{\alpha t}{d^2} \tag{39}$$

And represents the ratio of the heat conduction rate to the rate of thermal energy storage.

When used in connection with mass transfer, the thermal diffusivity (α) is replaced by the mass diffusivity (D).

$$Fo = \frac{Dt}{d^2} \tag{40}$$

Using the above definition of the Fourier number (*Fo*) with the liquid film thickness (δ) as the characteristic length, Pigford (1941) analyzed in his Ph. D. thesis the transient mass transfer to a falling film in laminar flow. His analysis is most conveniently found on the book of Sherwood et al. (1975). In addition, van Baten and Krishna (2004) formulated a mass transfer model of penetration theory for the film for shorter unit cells (or higher velocities) using Eq. (40) with $d = \delta$.

Froude Number (*Fr*)

The Froude number (*Fr*) is defined as:

$$Fr = \frac{\rho U^2}{\rho g d} = \frac{U^2}{gd} \tag{41}$$

And represents a measure of inertial forces and gravitational forces. When $Fr < 1$, small surface waves can move upstream; when $Fr > 1$, they will be carried downstream; and when $Fr = 1$ (said to be the critical Froude number), the velocity of flow is equal to the velocity of surface waves.

Also, there is (Fr^*), which is defined as:

$$Fr^{*} = \frac{Fr}{At} = \left(\frac{\rho_l + \rho_g}{\rho_l - \rho_g} \right) \frac{U^2}{gd} \tag{42}$$

In addition, the Froude number (Fr) is frequently defined as $Fr = U/(gd)^{0.5}$.

Electric Froude Number (Fr_e)

The electric Froude number (Fr_e) is defined as:

$$\mathrm{Fr_e} = \frac{\mathrm{xG}}{(\rho_g[(\rho_l - \rho_g)d_0 g - f_e^{''}])^{1/2}} \tag{43}$$

This dimensionless group was given by Chang (1989) and Chang (1998).

Froude Rate (Ft)

The Froude rate (Ft) is defined as:

$$Ft = \left(\frac{x^3 G^2}{\rho_g^2 g d (1 - x)} \right)^{0.5} \tag{44}$$

And represents the ratio of the vapor kinetic energy to the energy required to lift the liquid phase around the tube. This parameter was derived by Hulburt and Newell (1997). Graham et al. (1999) obtained an expression for the void fraction in terms of the Lockhart-Martinelli parameter for turbulent-turbulent flow (X_{tt}) and the Froude rate (Ft). Also, Thome (2003) deduced the transition from annular flow (viscous forces predominate) to intermittent flow (gravitational forces predominate) with the aid of his maps using the combination of two parameters: the Froude rate (Ft), and the Lockhart-Martinelli parameter for turbulent-turbulent flow (X_{tt}). In addition, Wilson et al. (2003) obtained an expression for the void fraction in terms of the Lockhart-Martinelli parameter for turbulent-turbulent flow (X_{tt}) and the Froude rate (Ft) for smooth tube, 18° Helix, and 0° Helix, respectively.

Galileo Number (Ga)

The Galileo number (Ga) is defined as:

$$Ga = \frac{gd^3}{\nu^2} \tag{45}$$

And represents the ratio of gravitational and viscous force scales. The Galileo number (Ga) is an important number in two-phase gas-liquid flow in determining the motion of a bubble/droplet under the action of gravity in the gravity-driven viscous flow. For instance, Haraguchi et al. (1994) expressed the condensation heat transfer coefficient in terms of Nusselt number as a combination of forced convection condensation and gravity controlled convection condensation terms. They expressed the gravity controlled convection condensation term as a function of the Galileo number (Ga_l).

The modified Galileo number (Ga^*) is defined as:

$$Ga^* = \frac{\rho \sigma^3}{\mu^4 g} \tag{46}$$

And accounts for the influences of surface tension and viscosity. The modified Galileo number (Ga^*) is sometimes referred to as the "film number". For example, Hu and Jacobi (1996) reported experiments that explored viscous, surface tension, inertial, and gravitational effects on the falling-film mode transitions. Their study covered a variety of fluids including water, ethylene glycol, hydraulic oil, water/ethylene glycol mixture, and alcohol, tube diameters, tube pitches, flow rates and with/without concurrent gas flow. Based on their 1000 experimental observations, the researchers provided new flow classifications, a novel flow regime map, and unambiguous transition criteria for every of the mode transitions. Over the range of their experiments, they found that the mode transitions were relatively independent of geometric effect (tube diameter and spacing). In a simplified map neglecting hysteresis (transition with an increasing flow rate compared with that with a decreasing flow rate), the coordinates of their flow mode map were the film Reynolds number (Re_f) versus the modified Galileo number (Ga^*). The mixed mode zones of jet-sheet and droplet-jet were transition zones between the three dominant modes of sheet, jet, and droplet in which both modes were present. Their four flow transition expressions between these five zones were given by the film Reynolds number (Re_f) as a function of the modified Galileo number (Ga^*) (valid for passing through the transition in either direction). Their map was applicable to plain tubes for air velocities less than 15 m/s.

Graetz Number (Gz)

The Graetz number (Gz) is defined as:

$$Gr = \frac{d}{L} \mathrm{Re} \, \mathrm{Pr} \tag{47}$$

When used in connection with mass transfer the Prandtl number (Pr) is replaced by the Schmidt number (Sc) that expresses the ratio of the momentum diffusivity to the mass diffusivity (Kreutzer, 2003).

$$Gr = \frac{d}{L} \mathrm{Re} \, Sc \tag{48}$$

Dimensionless Vapor Mass Flux (J_g)

The dimensionless vapor mass flux (J_g) is defined as follows:

$$J_g = \frac{Gx}{\sqrt{dg\rho_g(\rho_l - \rho_g)}} \tag{49}$$

In two-phase flow, transition criteria for flow regimes are determined using the dimensionless vapor mass flux (J_g) and the Lockhart-Martinelli parameter for the turbulent-turbulent flow (X_{tt}). For example, Breber et al. (1980) used the dimensionless vapor mass flux (J_g) in the prediction of horizontal tube-side condensation of pure components using flow regime criteria. Also, Sardesai et al. (1981) used the dimensionless vapor mass flux (J_g) in the determination of flow regimes for condensation of a vapor inside a horizontal tube. In addition, Tandon et al. (1982, 1985) used the dimensionless vapor mass flux (J_g) in the prediction of flow patterns during condensation in a horizontal tube. Moreover, Cavallini et al. (2002) allowed an assessment of the limits of the two-phase flow structures of the condensation of refrigerants in channels with the aid of the dimensionless vapor mass flux (J_g), and the Lockhart-Martinelli parameter for the turbulent-turbulent flow (X_{tt}).

Jacob Number (*Ja*)

The Jacob number (*Ja*) is defined as:

$$Ja = \frac{c_p \Delta T}{h_{lg}} \tag{50}$$

And represents the ratio of the sensible heat for a given volume of liquid to heat or cool through the temperature difference (ΔT) in arriving to its saturation temperature, to the latent heat required in evaporating the same volume of vapor. It is used in film condensation and boiling. For instance, *Ja* may be used in studying the influences of liquid superheat prior to initiation of nucleation in microchannels. Also, it may be useful in studying the subcooled boiling conditions. The Jacob number (*Ja*) can be modified to produce the modified Jacob number (Ja^*) by multiplying it by the density ratio (ρ_l/ρ_g) (Yang et al., 2000).

$$Ja^* = \frac{\rho_l}{\rho_g} Ja = \frac{\rho_l c_p \Delta T}{\rho_g h_{lg}} \tag{51}$$

Yang et al. (2000) used both the Jacob number (*Ja*) and the modified Jacob number (Ja^*) in their study on bubble dynamics for pool nucleate boiling.

Recently, Charoensawan and Terdtoon (2007) modified the Jacob number (*Ja*) by adding the influence of filling ratio (*FR*) in their study on the thermal performance of horizontal closed-loop oscillating heat pipes.

$$Ja^* = \frac{FR}{1 - FR} Ja = \frac{FR c_p \Delta T}{(1 - FR) h_{lg}} \tag{52}$$

Boiling Number (K_f)

The Boiling number (K_f) is defined as:

$$K_f = \frac{q}{Gh_{lg}} \tag{53}$$

In this dimensionless number, heat flux (q) is non-dimensionalized with mass flux (G) and latent heat (h_{lg}). It is based on empirical considerations. It can be used in empirical treatment of flow boiling because it combines two important flow parameters, q and G. It is used as one of the parameters for correlating the flow boiling heat transfer in both macro-scale and micro-scale. For example, Lazarek and Black (1982) proposed the nondimensional correlation for the flow boiling Nusselt number for their heat transfer experiments on R-113 as a function of the all-liquid Reynolds number (Relo), and the Boiling number (Kf). Also, Tran et al. (1996) obtained a correlation for the heat transfer coefficient in their experiments on R-12 and R-113 as a function of the all-liquid Weber number (We_{lo}), the Boiling number (K_f), and the liquid to vapor density ratio (ρ_l/ρ_g) to account for variations in fluid properties

New Non-Dimensional Constants of Kandlikar (K_1, K_2, and K_3)

These three new nondimensional groups, K_1, K_2, and K_3 are relevant to flow boiling phenomenon in microchannels. K1 and K2 were derived by Kandlikar (2004). The new non-dimensional constant (K1) is defined as:

$$K_1 = \frac{\left(\frac{q}{h_{lg}}\right)^2 \frac{d}{\rho_g}}{\frac{G^2 d}{\rho_l}} = \left(\frac{q}{Gh_{lg}}\right)^2 \frac{\rho_l}{\rho_g} = K_f^2 \frac{\rho_l}{\rho_g} \tag{54}$$

And represents the ratio of the evaporation momentum force, and the inertia force. This dimensionless group includes the Boiling number (K_f) and the liquid to vapor density ratio (ρ_l/ρ_g). The Boiling number (Kf) alone does not represent the true influence of the evaporation momentum, and its coupling with the density ratio (ρl/ρg) is important in representing the evaporation momentum force. A higher value of K1 means that the numerator (i.e. the evaporation momentum forces) is dominant and is likely to alter the interface movement.

The new non-dimensional constant (K_2) is defined as:

$$K_2 = \frac{\left(\frac{q}{h_{lg}}\right)^2 \frac{d}{\rho_g}}{\sigma} = \left(\frac{q}{h_{lg}}\right)^2 \frac{d}{\rho_g \sigma} \tag{55}$$

And represents the ratio of the evaporation momentum force, and the surface tension force. Kandlikar (2004) mentioned that the contact angle was not included in K_2 although the actual force balance in a given situation might involve more complex dependence on contact angles and surface orientation. It should be recognized that the contact angles play an important role in bubble dynamics and contact line movement and need to be taken into account in a comprehensive analysis.

A higher value of K_2 means that the numerator (i.e. the evaporation momentum forces) is dominant and causes the interface to overcome the retaining surface tension force. Kandlikar (2001) used effectively the group K_2 in developing a model for the critical heat flux (CHF) in pool boiling. He replaced the characteristic dimension (*d*) with the departure bubble diameter.

Kandlikar (2004) mentioned that these two groups were able to represent some of the key flow boiling characteristics, including the critical heat flux (CHF). In his closing remarks, he mentioned that the usage of the new non-dimensional groups K_1 and K_2 in conjunction with the Weber number (*We*) and the Capillary number (*Ca*) was expected to provide a better tool for analyzing the experimental data and developing more representative models.

Awad (2012a) mentioned that similar to the combination of the nondimensional groups, $K_2K_1^{0.75}$ (Kandlikar, 2004) that used in representing the flow boiling CHF data by Vandervort et al. (1994), these two new nondimensional groups, K_1 and K_2, can be combined using Eqs. (54) and (55) as:

$$\frac{K_2}{K_1} = \frac{\left(\frac{q}{h_{lg}}\right)^2 \frac{d}{\rho_g \sigma}}{\left(\frac{q}{Gh_{lg}}\right)^2 \frac{\rho_l}{\rho_g}} = \frac{dG^2}{\rho_l \sigma} = We_{lo} \quad (56)$$

i.e. the ratio of K_2 and K_1 is equal to the all liquid Weber number (We_{lo}). As a result, it is enough to use the new non-dimensional groups K_1 and K_2 in conjunction with the Capillary number (*Ca*) only to provide a better tool for analyzing the experimental data and developing more representative models for heat transfer mechanisms during flow boiling in microchannels because $K_2/K_1 = We_{lo}$.

Moreover, it should be noted that the ratio of K_2 and K_1 (K_2/K_1) and the Capillary number (*Ca*) can be combined as (Awad, 2012a):

$$\frac{K_2/K_1}{Ca} = \frac{We_{lo}}{Ca} = \mathrm{Re}_{lo} \quad (57)$$

i.e. the ratio of (K_2/K_1) and *Ca* is equal to the all liquid Reynolds number (Re_{lo}).

Recently, K_3 was derived by Kandlikar (2012). The new non-dimensional constant (K_3) is defined as:

$$K_3 = \frac{\text{Evaporation Momentum Force}}{\text{Viscous Force}} \quad (58)$$

And represents the ratio of the evaporation momentum force, and the viscous force. Kandlikar (2012) mentioned that this nondimensional group K_3 had not been independently used yet, but it was relevant if the evaporation momentum and viscous forces were considered in a process. K_3 could also be represented as:

$$K_3 = K_1 \, Re = \frac{K_2}{Ca} \tag{59}$$

In his summary, Kandlikar (2012) mentioned that recognizing the evaporation momentum force as an important force during the boiling process opened up the possibilities of three new relevant nondimensional groups, K_1, K_2, and K_3. Any two of these groups could be represented by combining the third one with one of the other relevant nondimensional groups *Re*, *We*, and *Ca*.

Kapitza Number (*Ka*)

The Kapitza number (*Ka*) is defined as:

$$Ka = \frac{\mu^4 g}{\rho \sigma^3} \tag{60}$$

And accounts for the influences of viscosity and surface tension. Using Eqs. (46) and (60), it should be noted that the Kapitza number (*Ka*) is equal to the inverse of the modified Galileo number (Ga^*).

$$Ka = Ga^{*-1} \tag{61}$$

The Kapitza number (*Ka*) is used in wave on liquid film. For example, Mudawwar and El-Masri (1986) found that it was impossible to obtain universal correlations of the heat transfer coefficients for different fluids in terms of Reynolds and Prandtl numbers alone because the heat transfer data across freely-falling turbulent liquid films had a strong dependence on the Kapitza number (*Ka*) below Re = 10 000. On the other hand, the researchers recommended turbulent-film correlations based on *Re* and *Pr* similar to those used in conventional internal or external flows for higher Reynolds numbers ($Re > 15\,000$).

Knudsen Number (*Kn*)

At microscales, the no-slip boundary condition can be applied in many situations specially when there is a liquid flow inside microchannels. Deciding on slip or no-slip boundary condition is dependent on a dimensionless group that is called Knudsen number (*Kn*). It is defined as:

$$Kn = \frac{\lambda}{L} \tag{62}$$

And represents the ratio of the molecular mean free path length (λ) to a representative physical length scale such as the hydraulic diameter. However, a slip condition, Navier slip condition can be applied to avoid numerical clutches where there is a moving contact line. For example, Chen et al. (2009) applied this technique in their numerical study on the formation of Taylor bubbles in capillary tubes. More fundamental details can be found in Renardy et al. (2001) and Spelt (2005). The Knudsen number (Kn) is an important number in two-phase gas-liquid flow in determining the continuum approximation.

von Karman Number (Kr)

The von Karman number (Kr) is defined as:

$$Kr = \mathrm{Re}\, f^{1/2} \tag{63}$$

i.e. it is the product of the Reynolds number (Re) and the square root of the friction factor ($f^{1/2}$). It is does not contain the velocity (U), but it is determined from the pipe dimensions (d and L), fluid properties (ρ and μ), and pressure drop (Δp). Computation of the von Karman number (Kr) in problem in which the flow is the only variable to be determined saves a solution by trial and error. However, this trial and error is relatively simple and takes only a few steps. A plot in which f, $f^{1/2}$ or $f^{1/2}$ is plotted versus Kr with relative roughness (ε) as a third parameter can be used in order to avoid the trial and error solution.

On the other hand, Charoensawan and Terdtoon (2007) defined the von Karman number (Kr) as:

$$Kr = \mathrm{Re}^2 f \tag{64}$$

i.e. it is the product of the Reynolds number square (Re^2) and the friction factor (f). Charoensawan and Terdtoon (2007) found that one of the influence dimensionless groups on the thermal performance of horizontal closed-loop oscillating heat pipes was the von Karman number (Kr).

Kutateladze Number (Ku)

The Kutateladze number (Ku) is defined as:

$$Ku = U\rho^{1/2}[g\sigma(\rho_l - \rho_g)]^{-1/4} \tag{65}$$

And represents a balance between the dynamic head, surface tension, and gravitational force. For example, Kutateladze (1972) used the Kutateladze number (Ku) in the Kutateladze two-phase flow stability criterion, in which the inertia, buoyancy, and surface tension forces were balanced for the prediction of flooding limit of open two-phase systems.

Recently, Charoensawan and Terdtoon (2007) defined the Kutateladze number (Ku) as:

$$Ku = \frac{q}{h_{lg}\rho_g \left[\dfrac{g\sigma(\rho_l - \rho_g)}{\rho_g^2}\right]^{1/4}} \tag{66}$$

The researchers developed successfully the thermal performance correlation of a horizontal closed-loop oscillating heat pipe (HCLOHP) in the non-dimensional form of power function using the curve fitting. In their correlation, they presented the Kutateladze number (*Ku*) as a function of liquid Prandtl number (Pr_l), modified Jacob number (Ja^*) with adding the influence of filling ratio, Bond number (Bo), von Karman number (*Kr*), and k_c/k_a (the ratio of the thermal conductivities of the cooling fluid at the required temperature and the ambient air at 25°C). From their non-dimensional correlation, they concluded that *Ku* or dimensionless group representing the thermal performance of HCLOHP improved with increasing Pr_l, *Ka* and k_c/k_a and with decreasing Ja^* and *Bo*.

Laplace Number (*La*)

The Laplace number (*La*) is also known as the Suratman number (*Su*). It is defined as:

$$La = Su = \sqrt{\frac{\sigma}{gd^2(\rho_l - \rho_g)}} \tag{67}$$

And represents the ratio of capillary and gravitational (buoyancy) force scales. The length scale used in its definition is the pipe diameter.

In addition, the Laplace number (*La*) is known as the confinement number (*Co*). The threshold to confined bubble flow is one of the most widely-used criterions to distinguish between macro and microscale flow boiling. Following the classification by Kew and Cornwell (1997), channels are classified as micro-channels if $Co \geq 0.5$.

The confinement number (*Co*) can represent the ratio of capillary length (L_c) and the pipe diameter (Phan et al., 2011).

$$Co = \frac{L_c}{d} \tag{68}$$

$$L_c = \sqrt{\frac{\sigma}{g(\rho_l - \rho_g)}} \tag{69}$$

In Eq. (69), the capillary length (L_c) scales all the phenomena involving liquid-vapor interfaces, such as bubble growth and detachment, interface instabilities and oscillation wavelengths. Also, it is used as a characteristic boiling length to scale the heater size. As L_c increases with decreasing gravity, it is clear that bubbles will become "larger" and heaters will become "smaller" in low gravity, and vice versa. This should be accounted for in evaluating experimental results, as it is well known that boiling on "small" heaters has various features.

From Eqs. (23), (37), and (67), it is clear that three dimensionless numbers, *Bo*, *Eo*, and *La* are related as follows:

$$Eo = 4Bo = La^{-2} \tag{70}$$

When the pipe size is large, the Laplace constant (*La*) is a useful property length scale for multiphase flow calculations compared to the bubble diameter.

Also, it is defined as:

$$La = Su = \frac{\rho \sigma d}{\mu^2} \tag{71}$$

The above definition is used in flow pattern maps. For example, Jayawardena et al. (1997) proposed a flow pattern map to obtain flow transition boundaries for microgravity two-phase flows. The researchers used bubble-slug and slug-annular flow pattern transitions on the Suratman number (*Su*) of the system. The bubble-slug transition occurred at a transitional value of the gas to liquid Reynolds number ratio (Re_g/Re_l) that decreased with increasing Suratman number (*Su*) and increased with increasing Suratman number (*Su*) at slug-annular transitions for low Suratman number systems. For high Suratman number systems, the slug-annular transition occurred at a transitional value of the gas Reynolds number (Re_g) that increased with increasing Suratman number (*Su*).

In two-phase flow, presenting the Chisholm constant (*C*) as a function of the Laplace number (*La*) in order to represent the hydraulic diameter (d_h) in a dimensionless form overcame the main disadvantage in some correlations available in the open literature, which is the dimensional specification of d_h, as it is easy to miscalculate C if the proper dimensions are not used for d_h.

Lewis Number (*Le*)

The Lewis number (*Le*) is defined as:

$$Le = \frac{\alpha}{D} = \frac{Sc}{\mathrm{Pr}} \tag{72}$$

And represents the ratio of the thermal diffusivity to the mass diffusivity.

The Lewis number (*Le*) controls the relative thickness of the thermal and concentration boundary layers. When *Le* is small, this corresponds to the thickness of the concentration boundary layer is much bigger than the thermal boundary layer. On the other hand, when *Le* is high, this corresponds to the thickness of the concentration boundary layer is much smaller than the thermal boundary layer.

Lo Number

The *Lo* number is defined as:

$$Lo = \frac{G^2 d}{\rho_m \sigma}\left(\frac{\mu_g}{\mu_l}\right)^{0.5} \tag{73}$$

This dimensionless number is used first in CISE-DIF-2 correlation (Lombardi and Ceresa, 1978). In CISE-DIF-2 correlation, a remarkable analogy with single-phase correlations was evident if the Reynolds number (*Re*) was substituted by the *Lo* number. Also, this dimensionless number is used in CISE-DIF-3 correlation (Bonfanti et al., 1979). Later, this dimensionless number is used in CESNEF-2 correlation (Lombardi and Ceresa, 1992).

Masuda Number (M_a)

The Masuda number (M_a) or dielectric Rayleigh number is defined as:

$$M_a = \frac{\varepsilon_0 E_0^2 T_0 (\partial\varepsilon_s / \partial T)_\rho L^2}{2\rho_0 v^2} \tag{74}$$

Using the analogy to free convective flows, Cotton et al. (2005) mentioned that the combined effects of electric and forced convection must be considered when (E_{hd}/Re^2) ~ 1 and/or (M_a/Re^2) ~ 1. If the inequalities (E_{hd}/Re^2) << 1 or (M_a/Re^2) << 1 are satisfied, electric convection influences may be neglected, and conversely, if (E_{hd}/Re^2) >> 1 or (M_a/Re^2) >> 1, forced convection influences may be neglected. This is exactly analogous to buoyancy driven flow and a similar argument may be made by comparing the EHD numbers to the Grashof number in the absence of forced convection. This order of magnitude analysis helps determine the range and extent to which EHD may affect the flow and must be identified to determine the voltage levels required to induce the migration of the liquid in order to influence heat transfer. Based on dimensionless analysis (Chang and Watson, 1994) it is expected that E_{hd}/Re_l^2 ~ ≥ 0.1 is sufficient to define the minimum condition above which electric fields significantly affect the liquid flow.

HEM Mach Number (*Ma*)

The Homogeneous Equilibrium Model (HEM) Mach number (*Ma*) is defined as:

$$Ma = G^2 \left[\frac{(1-x)^2}{\rho_l^2 c_l^2} + \frac{x^2}{\rho_g^2 c_g^2} \right] \tag{75}$$

Morton Number (*Mo*)

The Morton number (*Mo*) is defined as:

$$Mo = \frac{g\mu_c^4(\rho_c - \rho_p)}{\rho_c^2 \sigma^3} \tag{76}$$

And uses together with the Eötvös number (*Eo*) to characterize the shape of bubbles or drops moving in a surrounding fluid or continuous phase. For example, Taha and Cui (2006b) found that in their CFD modeling of slug flow in vertical tubes by decreasing the

Morton number (*Mo*) under a constant value of Eötvös number (Eo), the curvature of the bubble nose increases and the bubble tail flattens that results in an increment of the liquid film thickness around the bubble. In addition, the curvature of bubble nose increases as *Eo* goes up.

For the case of $\rho_p << \rho_c$, the Morton number (*Mo*) can be simplified to

$$Mo = \frac{g\mu_c^4}{\rho_c \sigma^3} \tag{77}$$

Nusselt Number (*Nu*)

The Nusselt number (*Nu*) is defined as:

$$Nu = \frac{(Q/A)d}{k(T_w - T_m)} = \frac{hd}{k} \tag{78}$$

And can be viewed as either the dimensionless heat transfer rate or dimensionless heat transfer coefficient. It represents the ratio of convective to conductive heat transfer across (normal to) the boundary. $Nu \sim 1$, namely convection and conduction of similar magnitude, is characteristic of "slug flow" or laminar flow. A larger Nusselt number corresponds to more active convection, with turbulent flow typically in the range of 100–1000.

Ohnesorge Number (*Oh*)

The Ohnesorge number (*Oh*) is defined as:

$$Oh = \frac{\mu_l}{\sqrt{\sigma d \rho_l}} = \frac{\sqrt{We}}{\text{Re}} \approx \frac{\text{viscous forces}}{\sqrt{\text{inertia . surface tension}}} \tag{79}$$

And relates the viscous forces to inertial and surface tension forces. The combination of these three forces into one masks the individual effects of every force. Often, it is used to relate to free surface fluid dynamics like dispersion of liquids in gases and in spray technology (Ohnesorge, 1936, Lefebvre, 1989). Also, it is used in analyzing liquid droplets and droplet atomization processes. Larger Ohnesorge numbers indicate a greater effect of the viscosity.

Using Eqs. (67) and (79), it is clear that there is an inverse relationship, between the Laplace number (*La*) and the Ohnesorge number (*Oh*) as follows:

$$Oh = \frac{1}{\sqrt{La}} \tag{80}$$

Historically, it is more correct to use the Ohnesorge number (*Oh*), but often mathematically neater to use the Laplace number (*La*).

Peclet Number (*Pe*)

For diffusion of heat (thermal diffusion), the Peclet number is defined as:

$$Pe = \mathrm{Re}\,\mathrm{Pr} = \frac{Ud}{\alpha} \tag{81}$$

For diffusion of particles (mass diffusion), it is defined as:

$$Pe = \mathrm{Re}\,Sc = \frac{Ud}{D} \tag{82}$$

For example, Muradoglu et al. (2007) studied the influences of the Peclet number (*Pe*) on the axial mass transfer (dispersion) in the liquid slugs. The researchers found that "convection" and "molecular diffusion" control the axial dispersion for various Peclet numbers (*Pe*). They introduced three various regimes of *Pe*:

1. Convection-controlled regime when $Pe > 10^3$.
2. Diffusion-controlled regime when $Pe < 10^2$.
3. Transition regime when $10^2 \leq Pe \leq 10^3$.

Phase Change Number (*Ph*)

The phase change number (*Ph*) is defined as follows:

$$Ph = \frac{c_p(T_s - T_{w,i})}{h_{\mathrm{lg}}} \tag{83}$$

And represents the ratio of the enthalpy change due to the temperature difference between the saturation temperature and inner wall temperature to the latent heat of vaporization. For example, Haraguchi et al. (1994) expressed the condensation heat transfer coefficient in terms of Nusselt number as a combination of forced convection condensation and gravity controlled convection condensation terms. They expressed the gravity controlled convection condensation term as a function of the phase change number (Ph_l).

Prandtl Number (*Pr*)

The Prandtl number (Pr) is defined as:

$$\mathrm{Pr} = \frac{\nu}{\alpha} = \frac{\mu c_p}{k} \tag{84}$$

And represents a measure of the rate of momentum diffusion versus the rate of thermal diffusion. It should be noted that Prandtl number contains no such length scale in its definition and is dependent only on the fluid and the fluid state. As a result, Prandtl number is often found in property tables alongside other properties like viscosity and thermal conductivity. It is used to characterize heat transfer in fluids. Typical values for *Pr* are:

$$\begin{array}{ll} \Pr \ll 1 & \text{Liquid Metals} \\ \Pr \approx 1 & \text{Gases} \\ \Pr \gg 1 & \text{Viscous Liquids} \end{array} \tag{85}$$

Low *Pr* means that conductive transfer is strong while high *Pr* means that convective transfer is strong. For example, heat conduction is very effective compared to convection for mercury (i.e. thermal diffusivity is dominant). On the other hand, convection is very effective in transferring energy from an area, compared to pure conduction for engine oil (i.e. momentum diffusivity is dominant).

In heat transfer problems, the Prandtl number controls the relative thickness of the momentum and thermal boundary layers. When *Pr* is small, it means that the heat diffuses very quickly compared to the velocity (momentum). This means that for liquid metals the thickness of the thermal boundary layer is much bigger than the velocity boundary layer. On the other hand, when *Pr* is high, it means that the heat diffuses very slowly compared to the velocity (momentum). This means that for viscous liquids the thickness of the thermal boundary layer is much smaller than the velocity boundary layer.

As seen, for example, in the two definitions of Graetz number (*Gz*) and Peclet number (*Pe*), the mass transfer analog of the Prandtl number (*Pr*) is the Schmidt number (*Sc*).

Reynolds Number (*Re*)

The Reynolds number (*Re*) is defined as:

$$\mathrm{Re} = \frac{\rho U d}{\mu} \tag{86}$$

And is traditionally defined as the ratio of inertial to viscous force scales. Often, it is most used to determine whether the flow is laminar or turbulent.

Recently, Shannak (2009) analyzed the historical definition of dimensionless number as a ratio of the most important forces that acts in single-phase flow to be applicable for the multiphase flow. He presented new expressions for the multiphase flow like Reynolds number (*Re*) and the Froude number (*Fr*) as a function of the primary influencing parameters. Therefore, the presented extension for Reynolds number and Froude number in his study could be simply extrapolated and used as well as more extensive applied for all other dimensions numbers. Pressure drop, friction factors and flow maps of two- and multiphase flow could be simply presented and graphically showed as a function of such new defined numbers.

Ca/*Re* Number

The ratio of the Capillary number (*Ca*) and Reynolds number (*Re*) appears as a group in plug flows. It results in a group that is independent of flow velocity:

$$\frac{Ca}{\mathrm{Re}} = \frac{\mu^2}{\rho d \sigma} = La^{-1} = Su^{-1} \tag{87}$$

This dimensionless group in Eq. (87) is used in flow pattern maps. For example, Jayawardena et al. (1997) plotted (Re_g/Re_l) or Re_g versus (Re_l/Ca) (i.e. $(Ca/Re_l)^{-1} = Su$) in their flow pattern maps. Using this plot, the boundaries for a large set of experimental data, obtained using various fluids and geometries, could be accurately predicted.

Also, the Ca/Re number is associated with Taylor plug flows. When combined with dimensionless liquid slug length (L^*) = L_s/d, it provides a measure of the effect that plug characteristics have on pressure drop or fluid friction. Walsh et al. (2009) showed that when:

$$\begin{aligned} L^*\left(\frac{Ca}{\mathrm{Re}}\right)^{0.33} &\gg 1 \quad \text{Taylor Flow} \\ L^*\left(\frac{Ca}{\mathrm{Re}}\right)^{0.33} &\ll 1 \quad \text{Poiseuille Flow} \end{aligned} \tag{88}$$

Equivalent All Liquid Reynolds Number (Re_{eq})

The equivalent all liquid Reynolds number (Re_{eq}) is defined as:

$$\mathrm{Re}_{eq} = \frac{G_{eq} d}{\mu_l} \tag{89}$$

$$G_{eq} = G\left[(1-x) + x\left(\frac{\rho_l}{\rho_g}\right)^{0.5}\right] \tag{90}$$

This dimensionless number was proposed by Akers et al. (1959) and used, for example, in the empirical correlations of Yan and Lin (1998) and Ma et al. (2004) for the two-phase frictional pressure gradient.

Film Reynolds Number (Re_f)

The film Reynolds number (Re_f) is defined as:

$$\mathrm{Re}_f = \frac{2\Gamma}{\mu} \tag{91}$$

This dimensionless group is used as a vertical coordinates in the flow mode map of Hu and Jacobi (1996).

Particle Reynolds Number (Re_p)

The particle or relative Reynolds number (Re_p) is defined as:

$$\mathrm{Re}_p = \frac{\rho_c \left| U_c - U_p \right| d_p}{\mu_c} \tag{92}$$

The particle Reynolds number (Re_p) is an important parameter in many industrial applications with small droplets/bubbles in two-phase, two component flow because it determines whether the flow falls into the category of the Stokes flow or not. Also, this number is a benchmark to determine the appropriate drag coefficient (C_D).

If $Re_p \ll 1$, the two-phase flow would be termed Stokes flow. In the Stokes flow regime, viscous bubbles or drops remain spherical; regardless of the Eötvös number (Eo) value. Even at low particle Reynolds numbers, a wake is formed behind the sphere. This is a steady-state wake, which becomes stronger as the Reynolds number increases and the inertia of the flow around the bubbles/droplets overcomes the viscosity effects on the surface of the bubbles/droplets (Crowe, 2006).

Richardson Number (*Ri*)

The Richardson number (Ri) is defined as:

$$Ri = \frac{\Delta\rho g d}{\rho U^2} \tag{93}$$

And represents the ratio of buoyancy forces to inertial forces. It is clear that the Richardson number (Ri) can be obtained from combining the density ratio ($\Delta\rho/\rho$) with the Froude number (Fr).

Schmidt Number (*Sc*)

The Schmidt number (Sc) is defined as:

$$Sc = \frac{\nu}{D} \tag{94}$$

And represents the ratio of the momentum diffusivity to the mass diffusivity.

In mass transfer problems, the Schmidt number (Sc) controls the relative thickness of the momentum and concentration boundary layers. When Sc is small, it means that the mass diffuses very quickly compared to the velocity (momentum). This corresponds to the thickness of the concentration boundary layer is much bigger than the velocity boundary layer. On the other hand, when Sc is high, it means that the mass diffuses very slowly compared to the velocity (momentum). This corresponds to the thickness of the concentration boundary layer is much smaller than the velocity boundary layer.

As seen, for example, in the two definitions of Graetz number (Gz) and Peclet number (Pe), the heat transfer analog of the Schmidt number (Sc) is the Prandtl number (Pr).

Sherwood Number (*Sh*)

The Sherwood number (*Sh*) is also called the mass transfer Nusselt number) and is defined as:

$$Sh = \frac{Kd}{D} \tag{95}$$

And represents the ratio of convective to diffusive mass transport. It is used in mass-transfer operation.

For example, Kreutzer (2003) calculated the liquid–solid mass transfer with a finite-element method, arriving at different values than reported by Duda and Vrentas (1971). These results gave an expression for the length-averaged mass transfer from a circulating vortex to the wall, without a lubricating film in between:

$$Sh = \sqrt{40^2 \left[1 + 0.28 \left(\frac{\mathrm{L_{slug}}}{\mathrm{d}} \right)^{-4/3} \right]^2 + \left[90 + 104 \left(\frac{\mathrm{L_{slug}}}{\mathrm{d}} \right)^{-4/3} \right] \left[\frac{\mathrm{dRe\ Sc}}{\mathrm{L}} \right]} \tag{96}$$

Equation (96) is defined per unit slug volume and should be multiplied with the liquid holdup to obtain a mass transfer coefficient based on channel volume. In addition, Eq. (96) is only valid for the region in which the circulating vortex has at least circulated once.

Stanton Number (St)

In heat transfer problem, the Stanton number (*St*) is also called modified Nusselt number and is defined as:

$$St = \frac{h}{\rho U c_p} = \frac{Nu}{\mathrm{Re}\,\mathrm{Pr}} = \frac{Nu}{Pe} \tag{97}$$

And represents the ratio of the Nusselt number (*Nu*) to the Peclet number (*Pe*).

In mass transfer problem, the Stanton number (*St*) is also called modified Sherwood number and is defined as:

$$St = \frac{K}{U} = \frac{Sh}{\mathrm{Re}\,Sc} = \frac{Sh}{Pe} \tag{98}$$

And represents the ratio of the Sherwood number (*Nu*) to the Peclet number (*Pe*).

Stefan Number (*Ste*)

The Stefan number (*Ste*) is defined as:

$$\mathrm{Ste} = \frac{c_p \Delta T}{h_{sl}} \tag{99}$$

And represents the ratio of sensible heat to latent heat. It is used in melting and solidification.

Stokes Number (*Stk*)

The Stokes number (*Stk*) is defined as:

$$Stk = \frac{\tau_p}{\tau_c} = \frac{\rho_p d_p^2 / 18\mu_c}{d / U_c} \tag{100}$$

And represents the ratio of the particle momentum response time over a flow system time. The Stokes number (*Stk*) is a very important parameter in liquid-particle motion and particle dynamics, where particles are suspended in a fluid flow. Also, the Stokes number (*Stk*) can be further related to the slip ratio (*S*) as follows:

$$S \approx \frac{1}{1 + Stk} \tag{101}$$

From Eq. (101), it is clear that if the Stokes number (*Stk*) tends to be zero, there would be no-slip between the two phases (i.e. $S = 1$).

There are three kinds of situations can be observed for particles (bubbles/droplets) suspended in fluid, namely: Case i) If $Stk \ll 1$, the response time of the particles (τ_p) is much less than the characteristic time associated with the flow field (τ_c). In this case, the particles will have ample time to respond to changes in flow velocity. Case ii) $Stk \sim 0$, where the two phases are in thermodynamic or velocity equilibrium. Case c) if $Stk \gg 1$, the response time of the particles (τ_p) is much higher than the characteristic time associated with the flow field (τ_c). In this case, the particle will have essentially no time to respond to the fluid velocity changes and the particle velocity will be little affected by fluid velocity change (Crowe, 2006).

Strouhal Number (*Str*)

The Strouhal number (*Str*) is defined as:

$$Str = \frac{fd}{U} \tag{102}$$

In their study of the inclination effects on wave characteristics in annular gas–liquid flows, Al-Sarkhi et al. (2012) correlated reasonably the Strouhal number (*Str*) for all inclination angles as a function of the modified Lockhart-Martinelli parameter (X^*).

Weber Number (*We*)

The Weber number (*We*) is defined as:

$$We = \frac{\rho U^2 d}{\sigma} \tag{103}$$

And represents a measure of inertial forces to interfacial forces. The Weber number (We) is useful in analyzing the formation of droplets and bubbles. If the surface tension of the fluid decreases, bubbles/droplets will have the tendency to decrease because of higher momentum transfer between the phases.

This dimensionless group is used in flow pattern maps. Thus, the influence of pipe diameter on the flow regimes is well accounted for through the use of the Weber number dimensionless groups. For example, Zhao and Rezkallah (1993) and updated later with new literature data by Rezkallah (1995, 1996) showed that three different regimes at a microgravity environment (g = 0.0981 m/s^2; on average) might be identified: (1) a surface tension dominated regime with bubbly and slug flow, (2) an inertia dominated regime with annular flow and (3) a transitional regime in between with frothy slug-annular flow. Then, the boundary between the regimes was determined by the Weber number (We_g) that was based on gas properties and gas superficial velocity. Roughly, the surface tension dominated regime (regime 1) was delimited by $We_g < 1$ and the inertial regime (regime 2) was delimited by $We_g > 20$. Rezkallah (1996) mentioned that the experimental data could be better predicted using the mapping coordinates; We_g, and We_l that were based on the actual gas and liquid velocities rather than the superficial ones. The transition from bubble/slug type flows to transitional flow was shown to occur at a constant value of We_g (based on the actual gas velocity) of about 2, while the transition from frothy slug-annular type flows to fully-developed annular flow was shown to take place at $We_g = 20$.

In addition, Akbar et al. (2003) found an important resemblance between two-phase flow in microchannels and in common large channels at microgravity. In both system types the surface tension, inertia, and the viscosity are important factors, while buoyancy is suppressed. As a result, the researchers used for microchannels two-phase flow regime maps that had previously been developed for microgravity with the mapping coordinates; We_g, and We_l that were based on the superficial gas and liquid velocities.

The Weber number (We) can be related to the Reynolds number (Re), Eötvös number (Eo), and Morton Number (Mo) as:

$$We = \mathrm{Re}^2 \left(\frac{Mo}{Eo} \right)^{-0.5} \tag{104}$$

Also, the Weber number (We) can be expressed by using a combination of the Froude number (Fr), Morton number (Mo), and Reynolds number (Re) as:

$$We = (Fr Mo \mathrm{Re}^4)^{1/3} \tag{105}$$

In addition, the Weber number (We) can be expressed by using a combination of the Capillary number (Ca), and Reynolds number (Re) as (Sobieszuk et al., 2010):

$$We = Ca\,\mathrm{Re} \tag{106}$$

From Eq. (106), it is clear that the *Ca/Re* number can be expressed as:

$$\frac{Ca}{\mathrm{Re}} = La^{-1} = Su^{-1} = \frac{We\,/\,\mathrm{Re}}{\mathrm{Re}} = \frac{We}{\mathrm{Re}^2} \tag{107}$$

It should be noted that the length scale of the dispersed phase (bubble diameter) can be used as the characteristic length instead of the pipe diameter (d) in these dimensionless groups depending on the specific application.

Two-Phase Flow Frictional Multiplier

The two-phase flow frictional multiplier is defined as the ratio of the two-phase flow frictional pressure gradient to some reference single-phase flow frictional pressure gradient, usually based on one of the components flowing by itself. The reference phase can be either the liquid phase pressure gradient that results in ϕ_l^2 or the gas phase pressure gradient that results in ϕ_g^2.

In some models, the reference pressure gradient is based on the total mass flow as either a liquid or gas. If the reference phase pressure gradient is based on the liquid phase properties and the total mass flow, this results in ϕ_{lo}^2 or if the reference phase pressure gradient is based on the gas phase properties and the total mass flow, this results in ϕ_{go}^2. The definitions of different two-phase frictional multipliers will be presented later in Table 4.

Finally, the Lockhart-Martinelli parameter (X) is defined as:

$$X = \frac{\phi_g}{\phi_l} \tag{108}$$

From Eq. (108), it should be noted that $\phi_g = \phi_l$ when $X = 1$ although some correlations available in the literature such as Goto et al. correlation (2001) of their obtained data for all kinds of the refrigerant and the tube during the evaporation and the condensation do not satisfy this condition (Awad, 2007b).

For the definition of two-phase frictional multipliers and Lockhart-Martinelli parameter (X) in liquid-liquid flow, the liquid with higher density is used similar to the liquid phase in gas-liquid two-phase flow while the liquid with lower density is used similar to the gas phase in gas-liquid two-phase flow (Awad and Butt (2009a)).

Muzychka and Awad (2010) mentioned that the Lockhart-Martinelli parameter (X) can be viewed as a reference scale that defines the extent to which the two-phase flow frictional pressure drop is characterized, i.e. dominated by liquid phase or dominated by gas phase:

$$\begin{array}{ll} X \ll 1 & \text{Gas Flow} \\ X \approx 1 & \text{Two-Phase Flow} \\ X \gg 1 & \text{Liquid Flow} \end{array} \tag{109}$$

The Lockhart-Martinelli parameter for turbulent-turbulent flow (X_{tt}) can be related to the Convection number (Co) as:

$$X_{tt} = Co\left(\frac{\mu_l}{\mu_g}\right)^{0.125} \tag{110}$$

From Eq. (110), it should be noted that the Lockhart-Martinelli parameter for turbulent-turbulent flow (X_{tt}) is equal to the Convection number (Co) at the critical state.

The modified Lockhart-Martinelli parameter (X^*) is defined as:

$$X^* = \sqrt{\frac{\rho_g}{\rho_l}}\frac{\dot{m}_l}{\dot{m}_g} = \sqrt{\frac{\rho_l}{\rho_g}}\frac{U_l}{U_g} = \frac{Fr_l}{Fr_g} \tag{111}$$

$$Fr_l = \sqrt{\frac{\rho_l U_l^2}{(\rho_l - \rho_g) g d \cos\theta}} \tag{112}$$

$$Fr_g = \sqrt{\frac{\rho_g U_g^2}{(\rho_l - \rho_g) g d \cos\theta}} \tag{113}$$

For example, Al-Sarkhi et al. (2012) studied inclination effects on wave characteristics in annular gas–liquid flows. The researchers proposed correlations for wave celerity, frequency, and liquid film Reynolds number as a function of the modified Lockhart-Martinelli parameter (X^*).

Two-Phase Heat Transfer Multiplier (*E*)

The two-phase heat transfer multiplier (E) is defined as follows:

$$\mathrm{E} = \frac{\mathrm{h_{tp,r,o}}}{\mathrm{h_{l,r,o}}} \tag{114}$$

And represents the ratio of two-phase heat transfer coefficient of refrigerant–oil mixture to liquid-phase heat transfer coefficient of refrigerant–oil mixture. It is defined to quantify and analyze the oil influence on two-phase heat transfer performance.

Besides the two-phase heat transfer multiplier (E), there is also the enhanced factor (EF). It is defined as:

$$\mathrm{EF} = \frac{\mathrm{h_{tp,r,o}}}{\mathrm{h_{tp,r}}} \tag{115}$$

And represents the ratio of heat transfer coefficient of refrigerant–oil mixture to that of pure refrigerant. It is generally used to address oil effect on heat transfer.

Wei et al. (2007) mentioned that the measured data of heat transfer coefficient of refrigerant–oil mixtures can be normalized by using (a) the two-phase heat transfer coefficient of pure refrigerant, and (b) the liquid-phase heat transfer coefficient of refrigerant–oil mixture.

For more dimensionless groups, the reader can see (Catchpole and Fulford, 1966) and (Fulford and Catchpole, 1968). In the first paper, Catchpole and Fulford (1966) compiled 210 dimensionless groups. Their constituent variables provided guide for solution of design and development problems where dimensional analysis was utilized as practical tool. In the second paper, Fulford and Catchpole (1968) compiled and tabulated 75 dimensionless groups, published in literature since authors' extensive compilation in March 1966, for interpolation into previous listing. The researchers gave table of alphabetical list of new groups with serial No., name, symbol, definition, significance, field of use, and reference. Also, they gave tables for identifying dimensionless groups with parameter, symbol, dimensions, exponent and groups.

2.2. Scale analysis

Scale analysis is the art of examining the magnitude order of terms appearing in the governing equations. The objective of scale analysis is to use the basic principles in order to produce the magnitude order. For example, in two-phase flow, Kandlikar (2010a) presented a scaling analysis to identify the relative effects of various forces on the boiling process at microscale. There were five major forces that come into play. There were inertia, surface tension, shear, gravity, and evaporation momentum forces. Also, Kandlikar (2010b) applied scale analysis to identify the relevant forces leading to the critical heat flux (CHF) condition during saturated flow boiling in microchannels and minichannels. Using these forces (the evaporation momentum, surface tension, shear, inertia, and gravity forces), the researcher developed a local parameter model to predict the flow boiling CHF.

3. Flow patterns in two-phase flow

Flow patterns in two-phase flow depend on different flow parameters, including the physical properties of fluids (the density of the gas and liquid phases (ρ_g and ρ_l), the viscosity of the gas and liquid phases (μ_g and μ_l), and the surface tension (σ)), the flow rate of the gas and liquid phases ($\dot{Q}_g$ and $\dot{Q}_l$), as well as the geometrical dimensions of the flow system. For example, Weisman et al. (1979) obtained extensive new data on the transitions between two-phase flow patterns during co-current gas liquid flow in horizontal lines. The researchers varied fluid properties in a systematic manner to determine the effects of liquid viscosity, liquid density, interfacial tension and gas density. Line sizes varied from 1.2 to 5 cm for most of the tests. They supplemented visual observations by an analysis of pressure drop fluctuations and hence their present data were believed to be less subjective than most past observations. They compared the transition data from their present tests, as well as available literature data, to the most frequently used transition line correlations. In almost all cases serious deficiencies were observed. Revised dimensionless correlations that fit their present data, and those previously available, were presented.

Physically, the formation of specific flow patterns is governed by the competition of different forces in the system such as inertia, viscous, gravitational, and surface tension forces. Flow patterns in two-phase flow at both horizontal and vertical orientations are discussed below.

4. Flow pattern maps

Flow pattern maps are an attempt, on a two-dimensional graph, to separate the space into areas corresponding to the different flow patterns. Simple flow pattern maps use the same axes for all flow patterns and transitions while complex flow pattern maps use different axes for different transitions. Flow pattern maps exist for both horizontal and vertical flow.

4.1. Flow pattern map in a horizontal two-phase flow

Flow Patterns maps are constructed of liquid superficial velocity (U_l) versus gas superficial velocity (U_g). In these maps, experimentally determined flow patterns are plotted with different markers, and the boundaries, i.e., the transitions of one flow pattern to the other, are plotted by lines. Which flow pattern actually occurs in a system depends on many factors such as the gas and liquid properties (ρ_g, μ_g, ρ_l, μ_l, σ), pipe geometry (at least d) and gas and liquid superficial velocities (U_g, U_l). The number of relevant dimensionless groups is large, and most experimental flow maps in the literature are applicable only to the specific systems in which they were obtained. Most of the transitions depend on a disturbance to grow, and the amplitude of the disturbances introduced has a profound influence on the flow map.

The Baker map is an example of flow pattern map for horizontal flow in a pipe. Figure 1 shows Baker flow pattern map for horizontal flow in a pipe. This map was first suggested by Baker (1954), and was subsequently modified by Scott (1964). The axes are defined in terms of G_g/λ and $G_l\,\psi$, where

$$G_g = \frac{\dot{m}_g}{A} \tag{116}$$

$$G_l = \frac{\dot{m}_l}{A} \tag{117}$$

$$\lambda = \left(\frac{\rho_g}{\rho_{air}} \frac{\rho_l}{\rho_{water}} \right)^{\frac{1}{2}} \tag{118}$$

$$\psi = \frac{\sigma_{water}}{\sigma}\left(\frac{\mu_l}{\mu_{water}}\left[\frac{\rho_{water}}{\rho_l}\right]^2\right)^{\frac{1}{3}} \tag{119}$$

The dimensionless parameters λ and ψ, were introduced to account for variations in the density, surface tension, and dynamic viscosity of the flowing media. These parameters are functions of the fluid properties normalized with respect to the properties of water and air at standard conditions. Both λ and ψ reduce to 1 for water/air mixtures at standard conditions. The Baker map is reasonably well for water/air and oil/gas mixtures in small diameter (< 0.05 m) pipes.

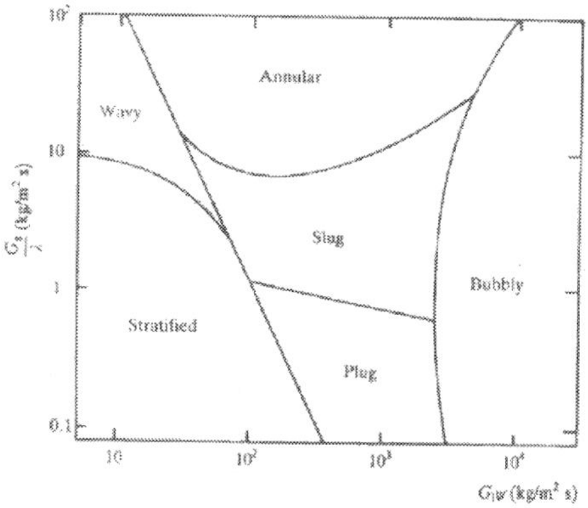

Figure 1. Baker Flow Pattern Map for Horizontal Flow in a Pipe (Whalley, 1987).

The Taitel and Dukler (1976) map is the most widely used flow pattern map for horizontal two-phase flow. This map is based on a semi-theoretical method, and it is computationally more difficult to use than other flow maps. The horizontal coordinate of the Taitel and Dukler (1976) map is the Lockhart-Martinelli parameter (X), Eq. (108). The vertical coordinates of the Taitel and Dukler (1976) map are K on the left hand side and T or F on the right hand side. They are defined as follows:

$$F = \sqrt{\frac{\rho_g}{\rho_l - \rho_g}}\frac{U_g}{\sqrt{dg\cos\theta}} \tag{120}$$

$$K = \left[\frac{\rho_g U_g^2}{(\rho_l - \rho_g)dg\cos\theta}\frac{dU_l}{v_l}\right]^{1/2} \tag{121}$$

$$T = \left[\frac{(dp/dz)_{f,l}}{(\rho_l - \rho_g)g\cos\theta}\right]^{1/2} \tag{122}$$

It should be noted that determination of the flow regime using Taitel and Dukler's (1976) map requires $(dp/dz)_{f,l}$ and $(dp/dz)_{f,g}$ that should be determined using appropriate flow models. Taitel and Dukler (1976) concluded that the different transitions were controlled by the grouping tabulated below:

i. Stratified to annular – *X, F*.
ii. Stratified to intermittent – *X, F*.
iii. Intermittent to bubble – *X, T*.
iv. Stratified smooth to Stratified wavy – *X, K*.

It should be noted that the Taitel and Dukler (1976) map was obtained for adiabatic two-phase flow; however, the transition boundaries between various flow regimes depend on the heat flux. Nevertheless, this flow map is often used to determine the flow patterns for evaporation and condensation inside pipes, for which external heating or cooling is required. As with any extrapolation, application of this flow map to forced convective boiling or condensation inside a pipe may not yield reliable results. Taitel (1990) presented a good review about flow pattern transition in two-phase flow.

4.2. Flow pattern map in a vertical two-phase flow

The Hewitt and Roberts (1969) map is an example of flow pattern map for vertical flow in a pipe. Figure 2 shows Hewitt and Roberts flow pattern map for vertical upflow in a pipe. Since the axes are defined in terms of G_g/ρ_g and G_l/ρ_l (phase momentum flux). So all the transitions are assumed to depend on the phase momentum fluxes. Wispy annular flow is a sub-category of annular flow that occurs at high mass flux when the entrained drops are said to appear as wisps or elongated droplets. The Hewitt and Roberts map is reasonably well for all water/air and water/steam systems over a range of pressures in small diameter pipes. For both horizontal and vertical maps, it should be noted also that the transitions between adjacent flow patterns do not occur suddenly but over a range of flow rates. So, the lines should really be replaced by rather broad transition bands.

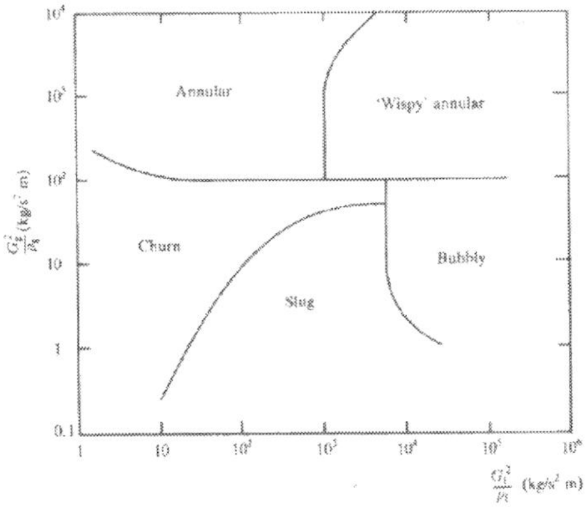

Figure 2. Hewitt and Roberts Flow Pattern Map for Vertical Upflow in a Pipe (Whalley, 1987).

5. Pressure drop in two-phase flow

The pressure drop, which is the change of fluid pressure occurring as a two-phase flow passes through the system. The pressure drop is very important parameter in the design of both adiabatic systems and systems with phase change, like boilers and condensers. In natural circulation systems, the pressure drop dictates the circulation rate, and hence the other system parameters. In forced circulation systems, the pressure drop governs the pumping requirement.

In addition, the pressure drop is very important in pipelines because co-current flow of liquid and vapor (gas) create design and operational problems due to formation of different types of two-phase flow patterns. Estimation of pressure drop in these cases helps the piping designer in reaching an optimum line size and a better piping system design.

Not only accurate prediction of pressure drop is extremely important when designing both horizontal and vertical two-phase flow systems, but also it is extremely important when designing inclined two-phase flow systems like directional wells or hilly terrain pipelines. Pipe inclination has an appreciable effect on flow patterns, slippage between phases and energy transfer between phases. There is no method for performing these calculations, which is accurate for all flow conditions. Historically, pressure drop in inclined flow has often been calculated using horizontal or vertical two-phase flow correlations. This is often satisfactory if the pipe inclination is very near to the horizontal case or the vertical case. However, this may not be the case in many applications.

The total measured pressure drop in two-phase flow (Δp) consists of three contributions. The first contribution is the frictional pressure drop (Δp_f). The second contribution is the acceleration pressure drop (Δp_a). The third contribution is the gravitational pressure drop (Δp_{grav}).

$$\Delta p = \Delta p_f + \Delta p_a + \Delta p_{grav} \tag{123}$$

The acceleration pressure drop (Δp_a) can be neglected in the adiabatic flow. For flow in a horizontal pipe, the gravitational pressure drop (Δp_{grav}) is zero. Thus, the total measured pressure drop (Δp) in the adiabatic experiments in horizontal pipes comes from the frictional pressure drop (Δp_f) only. Two-phase pressure drop can be measured for gas–liquid adiabatic flow or it can also be measured for vapor–liquid nonadiabatic, boiling or condensing flow. Laboratory measurements tend to be made with adiabatic gas–liquid flow, for instance, air–water flow, rather than vapor–liquid flow with phase change because of ease and low cost of operation.

The frictional pressure drop results from the shear stress between the flowing fluid and the pipe wall and also from the shear stress between the liquid and gas phases. To compute the frictional component of pressure drop, either the two-phase friction factor or the two-phase frictional multiplier must be known. It is necessary to know the void fraction (the ratio of gas flow rate to total flow rate) to compute the acceleration, and gravitational components of pressure drop (ASHRAE, 1993).

The acceleration component of pressure drop (Δp_a) reflects the change in kinetic energy of the flow. Assuming the vapor and liquid velocities to be uniform in each phase, the acceleration component of pressure drop can be obtained by the application of a simplified momentum equation in the form:

$$\Delta p_a = G^2 \left\{ \left[\frac{(1-x_o)^2}{\rho_l(1-\alpha_o)} + \frac{x_o^2}{\rho_g \alpha_o} \right] - \left[\frac{(1-x_i)^2}{\rho_l(1-\alpha_i)} + \frac{x_i^2}{\rho_g \alpha_i} \right] \right\} \tag{124}$$

The gravitational component of pressure gradient can be expressed in terms of the void fraction as follows:

$$\left(\frac{dp}{dz}\right)_{grav} = g[\alpha \rho_g + (1-\alpha)\rho_l]\sin\theta \tag{125}$$

Using Eq. (125) and knowing that $\alpha_m = \beta$, the gravitational component of pressure gradient based on the homogeneous model can be expressed as follows:

$$\left(\frac{dp}{dz}\right)_{grav,m} = \frac{g\sin\theta}{\dfrac{x}{\rho_g} + \dfrac{1-x}{\rho_l}} \tag{126}$$

5.1. Two-phase frictional multiplier

The two-phase frictional pressure drop (Δp_f) can be expressed in terms of two-phase frictional multiplier. This representation method is often useful for calculation and comparison needs. For example, the two-phase frictional pressure drop (Δp_f) can be expressed in terms of the single-phase frictional pressure drop for the total flow considered as liquid $(\Delta p_{f,lo})$ using two-phase frictional multiplier for total flow assumed liquid in the pipe (ϕ_{lo}^2). The single-phase frictional pressure drop for the total flow considered as liquid is computed from the total mass flux (G) and the physical properties of the liquid. The concept of all-liquid frictional pressure drop is useful because it allows the correlation to be tied into single-phase results at one end and eliminates any ambiguity about the physical properties to use, especially viscosity. Moreover, the all-liquid frictional pressure drop is chosen over the all-gas frictional pressure drop, because the liquid density generally does not vary in a problem, while the gas density changes with pressure. Also, the correlation of frictional pressure drop in terms of the parameter (ϕ_{lo}^2) is more convenient for boiling and condensation problems than (ϕ_l^2). The parameter (ϕ_{lo}^2) was first introduced by Martinelli and Nelson (1948). Table 2 shows definitions of different two-phase frictional multipliers.

The relationship between two-phase frictional multiplier for all flow as liquid (ϕ_{lo}^2) and two-phase frictional multiplier for liquid fraction only (ϕ_l^2) is:

$$\phi_{lo}^2 = \phi_l^2 (1-x)^{2-n} \tag{127}$$

Two-Phase Frictional Multiplier	Mass Flux	Density	**Reynolds Number**	Symbol
All flow as liquid	$G_l + G_g$	ρ_l	$(G_l + G_g)d/\mu_l$	ϕ_{lo}^2
Liquid fraction only	G_l	ρ_l	$G_l d/\mu_l$	ϕ_l^2
Gas fraction only	G_g	ρ_g	$G_g d/\mu_g$	ϕ_g^2
All flow as gas	$G_l + G_g$	ρ_g	$(G_l + G_g)d/\mu_g$	ϕ_{go}^2

Table 2. Definitions of Different Two-Phase Frictional Multipliers.

Also, the relationship between two-phase frictional multiplier for all flow as gas (ϕ_{go}^2) and two-phase frictional multiplier for gas fraction only (ϕ_g^2) is:

$$\phi_{go}^2 = \phi_g^2 x^{2-n} \tag{128}$$

In addition, the relationship between two-phase frictional multiplier for all flow as gas (ϕ_{go}^2) and two-phase frictional multiplier for all flow as liquid (ϕ_{lo}^2) can be obtained using Eqs. (108), (127), and (128) as follows:

$$\phi_{go}^2 = \phi_{lo}^2 X^2 \left(\frac{x}{1-x} \right)^{2-n} \tag{129}$$

In Eqs. (127-129), $n = 1$ for laminar-laminar flow while $n = 0.25$ for turbulent-turbulent flow.

Moreover, the relationship between two-phase frictional multiplier for all flow as liquid (ϕ_{lo}^2) and two-phase frictional multiplier for all flow as gas (ϕ_{go}^2) can be related to physical property coefficient (Γ) introduced by Chisholm (1973) as follows:

$$\frac{\phi_{lo}^2}{\phi_{go}^2} = \frac{\Delta p_{f,go}}{\Delta p_{f,lo}} = \Gamma^2 \tag{130}$$

5.2. Some forms of dimensionless two-phase frictional pressure drop (Δp_f^*)

Keilin et al. (1969) expressed two-phase frictional pressure drop (Δp_f) in a dimensionless form as follows:

$$\Delta p_f^* = \frac{\Delta p_f}{x \Delta p_{f,go} + (1-x) \Delta p_{f,lo}} \tag{131}$$

The above expression satisfies the following limiting conditions:

$$at\ x = 0, Dp_f = Dp_{f,lo}\ and\ Dp_f^* = 0; \qquad at\ x = 1, Dp_f = Dp_{f,go}\ and\ Dp_f^* = 1; \tag{132}$$

The dimensionless two-phase frictional pressure drop (Δp_f^*) can be expressed as a function of two-phase frictional multipliers as follows:

$$\Delta p_f^* = \frac{\phi_{lo}^2}{1 - x + x\left(\phi_{lo}^2 / \phi_{go}^2\right)} \tag{133}$$

For turbulent-turbulent flow, and using the Blasius equation (1913) to define the friction factor, Eq. (133) can be expressed as follows:

$$\Delta p_f^* = \frac{\left[1 - x + x\left(\frac{\rho_l}{\rho_g}\right)\right]\left[1 - x + x\left(\frac{\mu_l}{\mu_g}\right)\right]^{-0.25}}{1 - x + x\left(\frac{\rho_l}{\rho_g}\right)\left(\frac{\mu_g}{\mu_l}\right)^{0.25}} \tag{134}$$

Borishansky et al. (1973) expressed two-phase frictional pressure drop (Δp_f) in a dimensionless form as follows:

$$\Delta p_f^* = \frac{\Delta p_f - \Delta p_{f,lo}}{\Delta p_{f,go} - \Delta p_{f,lo}} \tag{135}$$

The above expression satisfies the following limiting conditions:

$$\text{at x } = 0, Dp_f = Dp_{f,lo} \; and \; Dp_f^* = 0; \qquad at \; x = 1, Dp_f = Dp_{f,go} \; and \; Dp_f^* = 1; \tag{136}$$

The dimensionless two-phase frictional pressure drop (Δp_f^*) can be expressed as a function of two-phase frictional multipliers as follows:

$$\Delta p_f^* = \frac{\phi_{lo}^2 - 1}{\left(\phi_{lo}^2 / \phi_{go}^2\right) - 1} \tag{137}$$

For turbulent-turbulent flow, and using the Blasius equation (1913) to define the friction factor, Eq. (137) can be expressed as follows:

$$\Delta p_f^* = \frac{\left[1 - x + x\left(\frac{\rho_l}{\rho_g}\right)\right]\left[1 - x + x\left(\frac{\mu_l}{\mu_g}\right)\right]^{-0.25} - 1}{\left(\frac{\rho_l}{\rho_g}\right)\left(\frac{\mu_g}{\mu_l}\right)^{0.25} - 1} \tag{138}$$

6. Methods of analysis

Two-phase flows obey all of the basic equations of fluid mechanics (continuity equation, momentum equation, and energy equation). However, the equations for two-phase flows are more complicated than those of single-phase flows. The techniques for analyzing one-dimensional two-phase flows include correlations, the phenomenological models, simple analytical model, and other methods such as integral analysis, differential analysis, computational fluid dynamics (CFD), and artificial neural network (ANN).

6.1. Correlations

The basic procedure used in predicting the frictional pressure drop in two-phase flow is developing a general correlation based on statistical evaluation of the data. The main disadvantage of this procedure is the difficulty in deciding on a method of properly weighing the fit in each flow pattern. For example, it is difficult to decide whether a correlation giving a poor fit with stratified flow and a good fit with annular flow is a better correlation than one giving a fair fit for both kinds of flow. Although the researchers that deal with two-phase flow problems still continue to use general correlations, alternate procedures must be developed to improve the ability to predict the pressure drop. In addition, correlations fitted to data banks that contain measurements with a number of liquid-gas combinations for different flow conditions and pipe diameters often have the disadvantage of containing a large number of constants and of being inconvenient in use. The correlation developed by Bandel (1973), is an example of this type of correlations.

The prediction of frictional pressure drop in two-phase flow can also be achieved by empirical correlations. Correlating the experimental data in terms of chosen variables is a convenient way of obtaining design equations with a minimum of analytical work. There are a considerable number of empirical correlations for the prediction of frictional pressure drop in two-phase flow. Although the empirical correlations require a minimum of knowledge of the system characteristics, they are limited by the range of data available for correlation construction. Most of these empirical correlations can be used beyond the range of the data from that they were constructed but with poor reliability (Dukler et al., 1964). Also, deviations of several hundred percent between predicted and measured values may be found for conditions outside the range of the original data from that these correlations were derived (Dukler et al., 1964).

The prediction of void fraction in two-phase flow can also be achieved by empirical correlations. There are a considerable number of empirical correlations for the prediction of void fraction. The empirical correlations are usually presented in terms of the slip ratio (S).

6.2. Phenomenological models

The phenomenological models can be developed based on the interfacial structure. Including phenomenon specific information like interfacial shear stress and slug frequency is used to obtain a complete picture of the flow. To reduce the dependence on empirical data, modeling on a theoretical basis is used. However, some empiricism is still required. The prediction of pressure gradient, void fraction, and the heat transfer coefficient simultaneously means that the phenomenological model is now preferred. For design purposes, the phenomenological models are often brought together within a framework provided by a flow pattern map such as Taitel, and Dukler (1976) flow pattern map. These flow pattern-based phenomenological models take into consideration the interfacial structure and the phase velocity distribution for every individual two-phase flow pattern. Quiben and Thome (2007) mentioned that the flow pattern-based phenomenological models are able to provide a more accurate prediction of the frictional pressure drop than the homogeneous and separated flow models. For example, the researchers developed an empirical correlation for the interfacial friction factor (f_i) in annular flow by considering the effects of the liquid film thickness, interfacial wave, viscosity ratio of the gas and liquid phases and the liquid inertia. Their correlation was

$$f_i = 0.67\left(\frac{\delta}{d}\right)^{1.2}\left[\frac{(\rho_l-\rho_g)g\delta^2}{\sigma}\right]^{-1.4}\left(\frac{\mu_g}{\mu_l}\right)^{0.08}\left(\frac{\rho_l U_l^2 d}{\sigma}\right)^{-0.034} \tag{139}$$

In Eq. (139), the expression $((\rho_l-\rho_g)g\delta^2/\sigma)$ came from manipulation of the Helmholtz instability equation using δ as the scaling factor for the most dangerous wavelength for the formation of interfacial waves. The Quiben and Thome (2007) model predicted a maximum in the frictional pressure drop but no explicit value of x_{maz} was proposed. According to their theory, the maximum occurred either in the annular flow regime or at the annular-to-dryout transition or at the annular-to-mist flow transition depending on the conditions.

Recently, Revellin and Haberschill (2009) presented an alternative approach for the prediction of frictional pressure drop during flow boiling of refrigerants in horizontal tubes over the entire range of vapor quality. The researchers developed an explicit expression (never proposed before) for the vapor quality corresponding to the maximum pressure drop (x_{maz}). This maximum was obtained by a flow regime analysis. Based on this maximum and on the pressure drop for liquid and vapor, they developed a simple linear function for predicting the frictional pressure drop. They mentioned that their method presented the best accuracy and predicts almost 86% of the data within a ± 30% error band. Their method did not include any new empirical parameters and could be used for a wide range of experimental conditions. Furthermore, the experimental data were also segregated into flow regimes and compared to every individual prediction method. The linear approach presented the best statistics for every flow regime.

For two-phase flow in microchannels, phenomenological models were developed primarily for the slug and annular flow patterns due to their dominance on the two-phase flow maps as well as their direct engineering relevance (Kreutzer et al., 2005a, Gunther and Jensen, 2006, Angeli, and Gavriilidis, 2008). For example, Kreutzer et al. (2005b) developed a phenomenological model for pressure drop in gas-liquid plug flow that includes the effects of plug length (L_s), Reynolds number (*Re*) and Capillary number (*Ca*). Later, Walsh et al. (2009) improved this model by considering additional data and the data of Kreutzer et al. (2005b). The proposed model of Walsh et al. (2009) is the asymptotic superposition of single-phase Poiseuille flow and an empirically derived result for Taylor flow. Their model takes the form:

$$f\,\mathrm{Re} = 16 + \frac{1.92}{L^*}\left(\frac{\mathrm{Re}}{Ca}\right)^{0.33} \tag{140}$$

Where $L^* = L_s/d$ is the dimensionless liquid plug length and the Reynolds number (*Re*) is computed based on the velocity and properties of the liquid slug. The transition from Taylor flow to Poiseuille flow occurs when:

$$L^*\left(\frac{Ca}{\mathrm{Re}}\right)^{0.33} > 1 \tag{141}$$

Below this critical value, significant contribution to the total pressure drop occurs due to the Laplace pressure contributions from the bubble train.

Walsh et al. (2009) mentioned that their plug flow model was found to have accuracy of 4.4% rms when compared with the data.

As shown in Fig. 3, the author suggests that the plug flow model of Walsh et al. (2009), Eq. (140), can be extended in order to calculate total pressure drop in two phase slug/bubble flows in mini scale capillaries for non-circular shapes as follows:

$$f\mathrm{Re} = \begin{cases} 16\left[1+\frac{0.12}{L^*}\left(\frac{\mathrm{Re}}{Ca}\right)^{0.33}\right] & circular \\ 24\left[1+\frac{0.12}{L^*}\left(\frac{\mathrm{Re}}{Ca}\right)^{0.33}\right] & parallel\ plates \\ 14.23\left[1+\frac{0.12}{L^*}\left(\frac{\mathrm{Re}}{Ca}\right)^{0.33}\right] & square \end{cases} \tag{142}$$

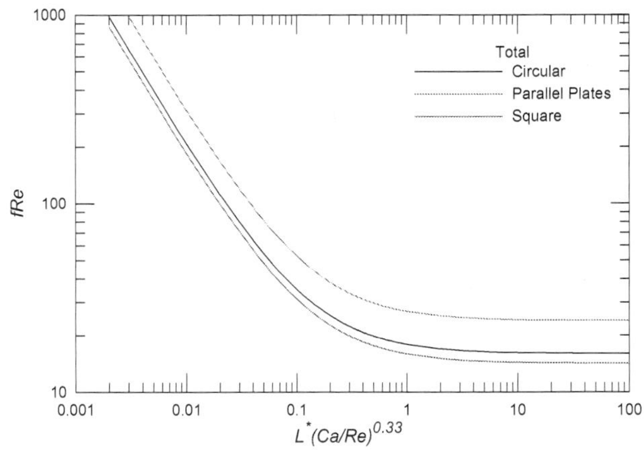

Figure 3. The Extension of the Plug Flow Model of Walsh et al. (2009), Eq. (140), for Total Pressure Drop in Two Phase Slug/Bubble Flows in Mini Scale Capillaries for Different Shapes (Circular, Parallel Plates, Square).

Moreover, instead of 16, 24, and 14.23 for circular, parallel plates, and square, respectively, the Shah and London (1978) relation for (*fRe*) for laminar flow forced convection in rectangular ducts as a function of the aspect ratio (*AR*):

$$\mathrm{fRe} = 24(1 - 1.3553AR + 1.9467AR^2 - 1.7012AR^3 + 0.9564AR^4 - 0.2537AR^5) \tag{143}$$

Equation (142) still needs verifications using experimental/numerical data to check if the constant multiplied by term $(Re/Ca)^{0.33}/L^*$ is 0.12 or not for non-circular shapes. Dividing both sides of Eq. (142) by the Poiseuille flow limit, which is 16, 24, and 14.23 for circular, parallel plates, and square, respectively, we obtain

$$\frac{f\mathrm{Re}}{f\mathrm{Re}_{Poise}} = \begin{cases} 1 + \dfrac{0.12}{L^*}\left(\dfrac{\mathrm{Re}}{Ca}\right)^{0.33} & \text{circular} \\ 1 + \dfrac{0.12}{L^*}\left(\dfrac{\mathrm{Re}}{Ca}\right)^{0.33} & \text{parallel plates} \\ 1 + \dfrac{0.12}{L^*}\left(\dfrac{\mathrm{Re}}{Ca}\right)^{0.33} & \text{square} \end{cases} \tag{144}$$

It is clear from Eq. (144) that the normalized variable (fRe/fRe_{Poise}) has the same value for different shapes such as circular, parallel plates, and square.

Recently, Talimi et al. (2012) reviewed numerical studies on the hydrodynamic and heat transfer characteristics of two-phase flows in small tubes and channels. These flows were non-boiling gas–liquid and liquid–liquid slug flows. Their review began with some general notes and important details of numerical simulation setups. Then, they categorized the review into two groups of studies: circular and non-circular channels. Various aspects like slug formation, slug shape, flow pattern, pressure drop and heat transfer were of interest.

The prediction of void fraction in two-phase flow can also be achieved by using models for specific flow regimes. The Taitel and Dukler (1976) model is an example for this type of model.

It should be noted that the precision and accuracy of phenomenological models are equal to those of empirical methods, while the probability density function is less sensitive to changes in fluid system (Tribbe and Müller-Steinhagen, 2000).

6.3. Simple analytical models

Simple analytical models are quite successful method for organizing the experimental results and for predicting the design parameters. Simple analytical models take no account of the details of the flow. Examples of simple analytical models include the homogeneous flow model, the separated flow model, and the drift flux model.

6.3.1. The homogeneous flow model

The homogeneous flow model provides the simplest technique for analyzing two-phase (or multiphase) flows. In the homogeneous model, both liquid and vapor phases move at the same velocity (slip ratio = 1). Consequently, the homogeneous model has also been called the zero slip model. The homogeneous model considers the two-phase flow as a single-phase flow having average fluid properties depending on quality. Thus, the frictional pressure drop is calculated by assuming a constant friction coefficient between the inlet and outlet sections of the pipe.

Using the homogeneous modeling approach, the frictional pressure gradient can be calculated using formulas from single-phase flow theory using mixture properties (ρ_m and μ_m). For flow in pipes and channels, it can be obtained using the familiar equations:

$$f = \left(\frac{d}{2\rho U^2}\right)\left(\frac{dp}{dz}\right)_f \tag{145}$$

6.3.1.1. Simple friction models

Since the homogeneous flow model is founded on using single-phase flow models with appropriate mixture models for ρ_m and μ_m, some useful results for laminar, turbulent, and transition flows in circular and non-circular shapes are provided. The models given below are for the Fanning friction factor that is related to the Darcy friction factor by means of $f = f_D/4$.

6.3.1.1.1. Hagen-Poiseuille model

For $Re_{dh} < 2300$, the Hagen-Poiseuille flow (White, 2005) gives:

$$f\,\mathrm{Re}_{d_h} = \begin{cases} 16 & circular \\ 24 & parallel\ plates \\ 14.23 & square \end{cases} \tag{146}$$

Moreover, instead of 16, 24, and 14.23 for circular, parallel plates, and square, respectively, Eq. (146) that represents the Shah and London (1978) relation for (fRe) for laminar flow forced convection in rectangular ducts as a function of the aspect ratio (AR) can be used.

For two-phase flow, Awad and Muzychka (2007, 2010b) used the Hagen-Poiseuille flow to represent the Fanning friction factor based on laminar-laminar flow assumption.

6.3.1.1.2. Blasius model

For turbulent flow, the value of the Fanning friction factor cannot be predicted from the theory alone, but it must be determined experimentally. Dimensional analysis shows that the Fanning friction factor is a function of the Reynolds number (Re_{dh}) and relative roughness (ε/d_h). For turbulent flow in smooth pipes, Blasius (1913) obtained the relationship between the Fanning friction factor (f) and the Reynolds number (Re_{dh}) as

$$f = \frac{0.079}{\mathrm{Re}_{d_h}^{0.25}} \quad 3\,000 < \mathrm{Re}_{d_h} < 100\,000 \tag{147}$$

For two-phase flow, Awad and Muzychka (2005a) used the Blasius equation to represent the Fanning friction factor for obtaining the lower bound of two-phase frictional pressure gradient based on turbulent-turbulent flow assumption.

6.3.1.1.3. Drew et al. model

Drew et al. (1932) obtained a relationship between the Fanning friction factor (f) and the Reynolds number (Re_{dh}) with a deviation of ±5% using their own experimental data and those of other investigators on smooth pipes. Their relationship was

$$f = 0.0014 + \frac{0.125}{\mathrm{Re}_{d_h}^{0.32}} \quad 3\,000 < \mathrm{Re}_{d_h} < 3\,000\,000 \tag{148}$$

6.3.1.1.4. Petukhov model

Petukhov (1970) developed a single correlation that encompassed the large Reynolds number range for friction factor in turbulent pipe flow with variable physical properties. His correlation was

$$f = 0.25(0.790 \ln \mathrm{Re}_{d_h} - 2.0)^{-2} \quad 3\,000 \le \mathrm{Re}_{d_h} \le 5000\,000 \tag{149}$$

6.3.1.1.5. Glielinski model

Glielinski (1976) derived an equation for the calculation of heat and mass transfer coefficients in the case of pipe and channel flow, taking into account the experimental data for high Reynolds and Prandtl numbers. His equation was valid for the transition region and for the range of fully developed turbulent flows. His equation for the friction factor was:

$$f = \frac{0.25}{(1.82\ \mathrm{lnRe} - 1.64)^2} \tag{150}$$

Also, Glielinski (1976) used his friction factor equation to calculate the Nusselt number (Nu) as follows:

$$Nu = \frac{(f/2)\ (\mathrm{Re} - 1000)\ \mathrm{Pr}}{1 + 12.7\ (f/2)^{1/2}\ (\mathrm{Pr}^{2/3} - 1)} \tag{151}$$

This heat transfer correlation was valid for $0.5 < Pr < 10^6$ and $Re > 3\ 000$. Later, Gnielinski (1999) provided an alternative approach to the Nusselt numbers prediction in the transition region based on a complex linear interpolation between $Re = 2\ 300$ and 10 000.

6.3.1.1.6. Swamee and Jain model

Swamee and Jain (1976) proposed an alternate form of the turbulent friction model developed by Colebrook (1939). It allows the influence of pipe or channel wall roughness to be considered. Swamee and Jain (1976) reported that it provided accuracy within ±1.5% when compared with the Colebrook model. The Swamee and Jain model is:

$$f = \frac{1}{16\left[\log\left(\frac{\varepsilon / d_h}{3.7} + \frac{5.74}{\mathrm{Re}_{d_h}^{0.9}}\right)\right]^2} \tag{152}$$

6.3.1.1.7. Churchill model

Since the definitive picture of the transition process and the transition mechanism are still unclear, the laminar to turbulent transition region should be considered a metastable and complicated region. The transition region is a varying mixture of various transport mechanisms and the mixed degree relies on the Reynolds number value and other conditions. In this study, Churchill (1977) developed a model to define the Fanning friction factor. In his model, he developed a correlation of the Moody chart (1944). His correlation spanned the entire range of laminar, transition, and turbulent flow in pipes. The Churchill model equations that define the Fanning friction factor were

$$f = 2\left[\left(\frac{8}{\mathrm{Re}_{d_h}}\right)^{12} + \frac{1}{(a+b)^{3/2}}\right]^{1/12} \tag{153}$$

$$a = \left[2.457 \ln \frac{1}{(7/\mathrm{Re}_{d_h})^{0.9} + (0.27\varepsilon / d_h)}\right]^{16} \tag{154}$$

$$b = \left(\frac{37530}{\mathrm{Re}_{d_h}}\right)^{16} \tag{155}$$

The Churchill model can be used due to its simplicity and accuracy of prediction. It is preferable since it encompasses all Reynolds numbers and includes roughness effects in the turbulent regime. Also, when a computer is used, the Churchill model equations are more recommended than the Blasius equations to define the Fanning friction factor (Chisholm, 1983). The Churchill model may be extended to non-circular shapes, by introducing the more general Poiseuille constant (*Po*). The factor of eight appearing above in Eq. (153) can be replaced by $fRe_{dh}/2$.

For two-phase flow, Awad and Muzychka (2004a) presented a simple two-phase frictional multiplier calculation method using the Churchill model to define the Fanning friction factor to take into account the effect of the mass flux on ϕ_{lo}^2. Also, Awad and Muzychka (2004b) used the Churchill model to calculate the single-phase friction factors (f_l and f_g). These friction factors were used to calculate single-phase frictional pressure gradients for liquid and gas flowing alone. Two-phase frictional pressure gradient was then expressed using a nonlinear superposition of the asymptotic single-phase frictional pressure gradients for liquid and gas flowing alone. Moreover, Awad and Muzychka (2008, 2010b) calculated the friction factor for the case of minichannels and microchannels using the Churchill model to allow for prediction over the full range of laminar-transition-turbulent regions.

6.3.1.1.8. Phillips model

Phillips (1987) was not aware of published results of the developing turbulent flow average friction factor for $x/d_h > 20$. Therefore, he decided to re-integrate the curves of the local friction factor. He obtained the following equation for the average turbulent friction factor:

$$f = A\,\mathrm{Re}^{B} \tag{156}$$

$$A = 0.09290 + \frac{1.01612}{x/d_h} \tag{157}$$

$$B = -0.26800 - \frac{0.31930}{x/d_h} \tag{158}$$

Equation (156) applies for circular pipes. In order to obtain the friction factor for rectangular ducts, *Re* is replaced by Re^*as follows:

$$\mathrm{Re}^* = \frac{\rho U d_{el}}{\mu} \tag{159}$$

$$d_{el} = \left[\left(\frac{2}{3}\right) + \left(\frac{11}{24AR}\right)\left(2 - \frac{1}{AR}\right)\right] d_h \tag{160}$$

Recently, Ong and Thome (2011) found that the single-phase friction factor for turbulent flow in small horizontal circular channels compared well with the correlation by Philips (1987).

6.3.1.1.9. García et al. model

García et al. (2003) took data from 2 435 gas-liquid flow experiments in horizontal pipelines, including new data for heavy oil. The definition of the Fanning friction factor for gas–liquid flow used in their study is based on the mixture velocity and density. Their universal (independent of flow type) composite (for all Reynolds number) correlation for gas-liquid Fanning friction factor (FFUC) was

$$f_m = 0.0925\,\mathrm{Re}_m^{-0.2534} + \frac{13.98\,\mathrm{Re}_m^{-0.9501} - 0.0925\,\mathrm{Re}_m^{-0.2534}}{\left(1 + \left(\frac{\mathrm{Re}_m}{293}\right)^{4.864}\right)^{0.1972}} \tag{161}$$

$$\mathrm{Re}_m = \frac{U_m d}{\nu_l} \tag{162}$$

$$U_m = U_l + U_g \tag{163}$$

The standard deviation of the correlated friction factor from the measured value was estimated to be 29.05% of the measured value. They claimed that the above correlation was a best guess for the pressure gradient when the flow type was unknown or different flow types were encountered in one line.

It should be noted that García et al. (2003) definition of the mixture Reynolds number is not suitable at high values of the dryness fraction. For example, for single-phase gas flow of air-water mixture at atmospheric conditions, García et al. (2003) definition gives $Re_m = 14.9Re_g$ instead of $Re_m = Re_g$.

6.3.1.1.10. Fang et al. model

Fang et al. (2011) evaluated the existing single-phase friction factor correlations. Also, the researchers obtained new correlations of single-phase friction factor for turbulent pipe flow. For turbulent flow in smooth pipes, they proposed the following correlation:

$$\mathrm{f} = 0.0625\left[\log\left(\frac{150.39}{\mathrm{Re}^{0.98865}} - \frac{152.66}{\mathrm{Re}}\right)\right]^{-2} \tag{164}$$

In the range of Re = 3000–10^8, their new correlation had the mean absolute relative error (MARE) of 0.022%. For turbulent flow in both smooth and rough pipes, they proposed the following correlation:

$$\mathrm{f} = 0.4325\left[\ln\left(0.234\varepsilon^{1.1007} - \frac{60.525}{\mathrm{Re}^{1.1105}} + \frac{56.291}{\mathrm{Re}^{1.0712}}\right)\right]^{-2} \tag{165}$$

In the range of Re = 3000–108 and ε = 0.0–0.05, the new correlation had the MARE of 0.16%.

6.3.1.2. Effective density models

For the homogeneous model, the density of two-phase gas-liquid flow (ρ_m) can be expressed as follows:

$$\rho_m = \left(\frac{x}{\rho_g} + \frac{1-x}{\rho_l}\right)^{-1} \tag{166}$$

Equation (166) can be derived knowing that the density is equal to the reciprocal of the specific volume and using thermodynamics relationship for the specific volume

$$v_m = (1-x)v_l + xv_g \tag{167}$$

Equation (166) can also be obtained based on the volume averaged value as follows:

$$\rho_m = \alpha_m \rho_g + (1-\alpha_m)\rho_l = \left(\frac{x}{\rho_g} + \frac{1-x}{\rho_l}\right)^{-1} \tag{168}$$

Equation (166) satisfies the following limiting conditions between (ρ_m) and mass quality (x):

$$\left.\begin{aligned} x=0, \quad \rho_m=\rho_l \\ x=1, \quad \rho_m=\rho_g \end{aligned}\right\} \tag{169}$$

There are other definitions of two-phase density (ρ_m). For example, Dukler et al. (1964) defined two-phase density (ρ_m) as follows:

$$\rho_m=\rho_l\frac{(1-\alpha_m)^2}{H_l}\alpha_m+\rho_g\frac{\alpha_m^2}{1-H_l} \tag{170}$$

Also, Oliemans (1976) defined two-phase density (ρ_m) as follows:

$$\rho_m=\frac{\rho_l(1-\alpha_m)+\rho_g(1-H_l)}{(1-\alpha_m)+(1-H_l)} \tag{171}$$

In addition, Ouyang (1998) defined two-phase density (ρ_m) as follows:

$$\rho_m=\rho_l H_l+\rho_g(1-H_l) \tag{172}$$

6.3.1.3. Effective viscosity models

In the homogeneous model, the mixture viscosity for two-phase flows (μ_m) has received much attention in literature. There are some common expressions for the viscosity of two-phase gas-liquid flow (μ_m). The expressions available for the two-phase gas-liquid viscosity are mostly of an empirical nature as a function of mass quality (x). The liquid and gas are presumed to be uniformly mixed due to the homogeneous flow. The possible definitions for the viscosity of two-phase gas-liquid flow (μ_m) can be divided into two groups. In the first group, the form of the expression between (μ_m) and mass quality (x) satisfies the following important limiting conditions:

$$\left.\begin{aligned} x=0, \quad \mu_m=\mu_l \\ x=1, \quad \mu_m=\mu_g \end{aligned}\right\} \tag{173}$$

In the second group, the form of the expression between (μ_m) and mass quality (x) does not satisfy the limiting conditions of Eq. (173). In gas-liquid two-phase flows the most commonly used formulas are summarized below in Table 3.

In Table 3, it should be noted the following:

i. Awad and Muzychka (2008) Definition 4 is based on the Arithmetic Mean (AM) for Awad and Muzychka (2008) Definition 1 and Awad and Muzychka (2008) Definition 2.
ii. Muzychka et al. (2011) Definition 1 is based on the Geometric Mean (GM) for Awad and Muzychka (2008) Definition 1 and Awad and Muzychka (2008) Definition 2.
iii. Muzychka et al. (2011) Definition 2 is based on the Harmonic Mean (HM) for Awad and Muzychka (2008) Definition 1 and Awad and Muzychka (2008) Definition 2.

Researcher	Model
Arrhenius (1887)	$\mu_m = \mu_l^{1-\alpha_m} \mu_g^{\alpha_m}$
Bingham (1906)	$\mu_m = \left(\frac{1-\alpha_m}{\mu_l} + \frac{\alpha_m}{\mu_g} \right)^{-1}$
MacAdams et al. (1942)	$\mu_m = \left(\frac{x}{\mu_g} + \frac{1-x}{\mu_l} \right)^{-1}$
Davidson et al. (1943)	$\mu_m = \mu_l \left[1 + x \left(\frac{\rho_l}{\rho_g} - 1 \right) \right]$
Vermeulen et al. (1955)	$\mu_m = \frac{\mu_l}{\alpha_m} \left[1 + \left(\frac{1.5\mu_g (1-\alpha_m)}{\mu_l + \mu_g} \right) \right]$
Akers et al. (1959)	$\mu_m = \mu_l \left[(1-x) + x \left(\frac{\rho_l}{\rho_g} \right)^{0.5} \right]^{-1}$
Hoogendoorn (1959)	$\mu_m = \mu_l^{H_l} \mu_g^{1-H_l}$
Cicchitti et al. (1960)	$\mu_m = x\mu_g + (1-x)\mu_l$
Bankoff (1960)	$\mu_m = H_l \mu_l + (1-H_l)\mu_g$
Owen (1961)	$\mu_m = \mu_l$
Dukler et al. (1964)	$\mu_m = \rho_m \left[x \frac{\mu_g}{\rho_g} + (1-x) \frac{\mu_l}{\rho_l} \right]$
Oliemans (1976)	$\mu_m = \frac{\mu_l (1-\alpha_m) + \mu_g (1-H_l)}{(1-\alpha_m) + (1-H_l)}$
Beattie and Whalley (1982)	$\mu_m = \mu_l (1-\alpha_m)(1+2.5\alpha_m) + \mu_g \alpha_m$ $= \mu_l - 2.5\mu_l \left(\frac{x\rho_l}{x\rho_l + (1-x)\rho_g} \right)^2 + \left(\frac{x\rho_l (1.5\mu_l + \mu_g)}{x\rho_l + (1-x)\rho_g} \right)$
Lin et al. (1991)	$\mu_m = \frac{\mu_l \mu_g}{\mu_g + x^{1.4}(\mu_l - \mu_g)}$
Fourar and Bories (1995)	$\mu_m = \rho_m \left(\sqrt{x \frac{\mu_g}{\rho_m}} + \sqrt{(1-x) \frac{\mu_l}{\rho_l}} \right)^2$
García et al. (2003, 2007)	$\mu_m = \mu_l \left(\frac{\rho_m}{\rho_l} \right) = \frac{\mu_l \rho_g}{x\rho_l + (1-x)\rho_g}$

Researcher	Model
Awad and Muzychka (2008) Definition 1	$\mu_m = \mu_l \frac{2\mu_l + \mu_g - 2(\mu_l - \mu_g)x}{2\mu_l + \mu_g + (\mu_l - \mu_g)x}$
Awad and Muzychka (2008) Definition 2	$\mu_m = \mu_g \frac{2\mu_g + \mu_l - 2(\mu_g - \mu_l)(1-x)}{2\mu_g + \mu_l + (\mu_g - \mu_l)(1-x)}$
Awad and Muzychka (2008) Definition 3	$(1-x)\frac{\mu_l - \mu_m}{\mu_l + 2\mu_m} + x\frac{\mu_g - \mu_m}{\mu_g + 2\mu_m} = 0$
Awad and Muzychka (2008) Definition 4	$\mu_m = \left[\frac{\mu_l}{2}\frac{2\mu_l + \mu_g - 2(\mu_l - \mu_g)x}{2\mu_l + \mu_g + (\mu_l - \mu_g)x} + \frac{\mu_g}{2}\frac{2\mu_g + \mu_l - 2(\mu_g - \mu_l)(1-x)}{2\mu_g + \mu_l + (\mu_g - \mu_l)(1-x)}\right]$
Muzychka et al. (2011) Definition 1	$\mu_m = \left[\mu_l \frac{2\mu_l + \mu_g - 2(\mu_l - \mu_g)x}{2\mu_l + \mu_g + (\mu_l - \mu_g)x} * \mu_g \frac{2\mu_g + \mu_l - 2(\mu_g - \mu_l)(1-x)}{2\mu_g + \mu_l + (\mu_g - \mu_l)(1-x)}\right]^{0.5}$
Muzychka et al. (2011) Definition 2	$\mu_m = \left[2\mu_l \frac{2\mu_l + \mu_g - 2(\mu_l - \mu_g)x}{2\mu_l + \mu_g + (\mu_l - \mu_g)x} * \mu_g \frac{2\mu_g + \mu_l - 2(\mu_g - \mu_l)(1-x)}{2\mu_g + \mu_l + (\mu_g - \mu_l)(1-x)}\right] / \left[\mu_l \frac{2\mu_l + \mu_g - 2(\mu_l - \mu_g)x}{2\mu_l + \mu_g + (\mu_l - \mu_g)x} + \mu_g \frac{2\mu_g + \mu_l - 2(\mu_g - \mu_l)(1-x)}{2\mu_g + \mu_l + (\mu_g - \mu_l)(1-x)}\right]$

Table 3. The Most Commonly Used Formulas of the Mixture Viscosity in Gas-Liquid Two-Phase Flow.

The relationships between the Arithmetic Mean (*AM*), the Geometric Mean (*GM*), and the Harmonic Mean (*HM*) are as:

$$GM^2 = 2.AM.HM \tag{174}$$

$$HM < GM < AM \quad 0 < x < 1 \tag{175}$$

Agrawal et al. (2011) investigated recently new definitions of two-phase viscosity, based on its analogy with thermal conductivity of porous media, for transcritical capillary tube flow, with CO_2 as the refrigerant. The researchers computed friction factor and pressure gradient quantities based on the proposed two-phase viscosity model using homogeneous modeling approach. They assessed the proposed new models based on test results in the form of temperature profile and mass flow rate in a chosen capillary tube. They showed that all the proposed models of two-phase viscosity models showed a good agreement with the existing models like McAdams et al. (1942), Cicchitti et al. (1960), etc. They found that the effect of the viscosity model to be insignificant unlike to other conventional refrigerants in capillary tube flow.

Banasiak and Hafner (2011) presented a one-dimensional mathematical model of the R744 two-phase ejector for expansion work recovery. The researchers computed friction factor and pressure gradient quantities based on the proposed two-phase viscosity model using homogeneous modeling approach. They approximated the two-phase viscosity according to the Effective Medium Theory. This formulation was originally derived for the averaged thermal conductivity and successfully tested by Awad and Muzychka (2008) for the average viscosity of vapor-liquid mixtures for different refrigerants. They predicted the friction factor (f) using the Churchill model (1977).

In liquid-liquid two-phase flows, Taylor (1932) presented the effective viscosity for a dilute emulsion of two immiscible incompressible Newtonian fluids by

$$\mu_m = \mu_c \left(1 + 2.5\alpha \frac{\mu_d + 0.4\mu_c}{\mu_d + \mu_c} \right) = \mu_c \left(1 + \alpha \frac{1 + 2.5(\mu_d / \mu_c)}{1 + (\mu_d / \mu_c)} \right) \tag{176}$$

If the viscosity of the dispersed phase (μ_d) is much lower than the continuous liquid (μ_c), like when water is mixed with silicone oil, the value of (μ_d/μ_c) would be much smaller than 1. Hence, Eq. (176) can be simplified as

$$\mu_m = \mu_c(1 + \alpha) \tag{177}$$

If the viscosity of the dispersed phase (μ_d) is much higher than the continuous liquid (μ_c), the value of (μ_d/μ_c) would be much greater than 1. Hence, Eq. (176) can be simplified as

$$\mu_m = \mu_c(1 + 2.5\alpha) \tag{178}$$

Equation (178) is the well known Einstein model (1906, 1911). It is frequently used in prediction of nano fluid viscosity.

Instead of Eq. (178) being a first order equation in α, can be written as a virial series,

$$\mu_m = \mu_c(1 + K_1\alpha + K_2\alpha^2 + K_3\alpha^3 + ...) \tag{179}$$

Where K_1, K_2, K_3, ….. are constants. For example, K_1 = 2.5, K_2 = -11.01 and K_3 = 52.62 in the Cengel (1967) definition for viscosity of liquid-liquid dispersions in laminar and turbulent flow. For more different definitions of the viscosity of emulsion, the reader can see Chapter 3: Physical Properties of Emulsions in the book by Becher (2001).

For different definitions of the viscosity of solid-liquid two-phase flow that are commonly used in the nanofluid applications, the reader can see, for example, Table 2: Models for effective viscosity in Wang and Mujumdar (2008a). Also, Wang and Mujumdar (2008b) reported that there were limited rheological studies in the literature in comparison with the experimental studies on thermal conductivity of nanofluids.

Similar to the idea of bounds on two-phase flow developed by Awad and Muzychka (2005a, 2005b, and 2007), these different definitions of two-phase viscosity can be used for bounding the data in an envelope using the homogeneous model. For example, Cicchitti et al. (1960), represents the upper bound while Dukler et al. (1964), represents the lower bound in gas-liquid two-phase flow. Using the different definitions of a certain property such as thermal conductivity in bounding the data is available in the open literature. For instance, Carson et al. (2005) supported the use of different definitions as thermal conductivity bounds by experimental data from the literature.

The homogeneous flow modeling approach can be used for the case of bubbly flows with appropriate mixture models for density and viscosity in order to obtain good predictive results. For example, this approach has been examined by Awad and Muzychka (2008), Cioncolini et al. (2009), and Li and Wu (2010) for both microscale and macroscale flows.

The homogenous flow modeling approach using the different mixture models reported earlier, typically provides an accuracy within 15% rms, (Awad and Muzychka (2008), Cioncolini et al. (2009), and Li and Wu (2010)).

In the two-phase homogeneous model, the selection of a suitable definition of two-phase viscosity is inevitable as the Reynolds number would require this as an input to calculate the friction factor. It is possible, as argued by Collier and Thome (1994), that the failure of establishing an accepted definition is that the dependence of the friction factor on two-phase viscosity is small.

The opinion of the author is that which definition of two-phase viscosity to use depends much on the two-phase flow regime and less on the physical structure of the two-phase viscosity itself. As a matter of fact, till today some water-tube boiler design methods still use single-phase water viscosity in the homogeneous model with good accuracy. This could be explained by the high mass flux and mass quality always below 0.1.

6.3.2. The separated flow model

In the separated model, two-phase flow is considered to be divided into liquid and vapor streams. Hence, the separated model has been referred to as the slip flow model. The separated model was originated from the classical work of Lockhart and Martinelli (1949) that was followed by Martinelli and Nelson (1948). The Lockhart-Martinelli method is one of the best and simplest procedures for calculating two-phase flow pressure drop and hold up. One of the biggest advantages of the Lockhart-Martinelli method is that it can be used for all flow patterns. However, relatively low accuracy must be accepted for this flexibility. The separated model is popular in the power plant industry. Also, the separated model is relevant for the prediction of pressure drop in heat pump systems and evaporators in refrigeration. The success of the separated model is due to the basic assumptions in the model are closely met by the flow patterns observed in the major portion of the evaporators.

The separated flow model may be developed with different degrees of complexity. In the simplest situation, only one parameter, like velocity, is allowed to differ for the two phases while conservation equations are only written for the combined flow. In the most sophisticated situation, separate equations of continuity, momentum, and energy are written for each phase and these six equations are solved simultaneously, together with rate equations which describe how the phases interact with each other and with the walls of the pipe. Correlations or simplifying assumptions are introduced when the number of variables to be determined is greater than the available number of equations.

For void fraction, the separated model is used by both analytical and semi-empirical methods. In the analytical theories, some quantities like the momentum or the kinetic energy is minimized to obtain the slip ratio *(S)*. The momentum flux model and the Zivi model (1964) are two examples of this technique, where the slip ratio *(S)* equals $(\rho_l/\rho_g)^{1/2}$ and $(\rho_l/\rho_g)^{1/3}$.

For two-phase flow modeling in microchannels and minichannels, it should be noted that the literature review on this topic can be found in tabular form in a number of textbooks such as Celata (2004), Kandlikar et al. (2006), Crowe (2006), Ghiaasiaan (2008), and Yarin et al. (2009).

For two-phase frictional pressure gradient, a number of models have been developed with varying the sophistication degrees. These models are all reviewed in this section in a chorological order starting from the oldest to the newest.

6.3.2.1. Lockhart-Martinelli model

Lockhart and Martinelli (1949) presented data for the simultaneous flow of air and liquids including benzene, kerosene, water, and different types of oils in pipes varying in diameter from 0.0586 in. to 1.017 in. There were four types of isothermal two-phase, two-component flow. In the first type, flow of both the liquid and the gas were turbulent. In the second type, flow of the liquid was viscous and flow of the gas was turbulent. In the third type, flow of the liquid was turbulent and flow of the gas was viscous. In the fourth type, flow of both the liquid and the gas were viscous. The data used by Lockhart and Martinelli consisted of experimental results obtained from a number of sources as detailed in their original paper and covered 810 data sets including 191 data sets that are for inclined and vertical pipes and 619 data sets for horizontal flow (Cui and Chen (2010)).

Lockhart and Martinelli (1949) correlated the pressure drop resulting from these different flow mechanisms by means of the Lockhart-Martinelli parameter *(X)*. The Lockhart-Martinelli parameter *(X)* was defined as:

$$X^2 = \frac{(dp/dz)_{f,l}}{(dp/dz)_{f,g}} \tag{180}$$

In addition, they expressed the two-phase frictional pressure drop in terms of factors, which multiplied single-phase drops. These multipliers were given by:

$$\phi_l^2 = \frac{(dp/dz)_f}{(dp/dz)_{f,l}} \tag{181}$$

$$\phi_g^2 = \frac{(dp/dz)_f}{(dp/dz)_{f,g}} \tag{182}$$

Using the generalized Blasius form of the Fanning friction factor, the frictional component single-phase pressure gradient could be expressed as

$$\left(\frac{dp}{dz}\right)_{f,l} = \frac{2C_l\mu_l^n U_l^{2-n}\rho_l^{1-n}}{d^{1+n}} \tag{183}$$

$$\left(\frac{dp}{dz}\right)_{f,g} = \frac{2C_g\mu_g^n U_g^{2-n}\rho_g^{1-n}}{d^{1+n}} \tag{184}$$

Values of the exponent (n) and the constants C_l and C_g for different flow conditions are given in Table 4.

	turbulent-turbulent	**laminar-turbulent**	**turbulent-laminar**	**laminar-laminar**
n	0.2	1.0	0.2	1.0
C_l	0.046	16	0.046	16
C_g	0.046	0.046	16	16
Re_l	> 2000	< 1000	> 2000	< 1000
Re_g	> 2000	> 2000	< 1000	< 1000

Table 4. Values of the Exponent (n) and the Constants C_l and C_g for Different Flow Conditions.

Also, they presented the relationship of ϕ_l and ϕ_g to X in graphical forms. They proposed tentative criteria for the transition of the flow from one type to another. Equations to calculate the parameter (X) under different flow conditions are given in Table 5.

Flow Condition	X
turbulent-turbulent	$X_{tt}^2 = \left(\frac{1-x}{x}\right)^{1.8}\left(\frac{\rho_g}{\rho_l}\right)\left(\frac{\mu_l}{\mu_g}\right)^{0.2}$
laminar-turbulent	$X_{lt}^2 = \mathrm{Re}_g^{-0.8}\left(\frac{C_l}{C_g}\right)\left(\frac{1-x}{x}\right)\left(\frac{\rho_g}{\rho_l}\right)\left(\frac{\mu_l}{\mu_g}\right)$
turbulent- laminar	$X_{tl}^2 = \mathrm{Re}_l^{0.8}\left(\frac{C_l}{C_g}\right)\left(\frac{1-x}{x}\right)\left(\frac{\rho_g}{\rho_l}\right)\left(\frac{\mu_l}{\mu_g}\right)$
laminar-laminar	$X_{ll}^2 = \left(\frac{1-x}{x}\right)\left(\frac{\rho_g}{\rho_l}\right)\left(\frac{\mu_l}{\mu_g}\right)$

Table 5. Equations to Calculate the Parameter (X) under Different Flow Conditions.

It should be noted that Lockhart and Martinelli (1949) only presented the graphs for ϕ_g versus X for the t-t, l-t and l-l flow mechanisms of liquid-gas, and the graph of t-l flow mechanisms of liquid-gas (the third type) was not given.

Recently, Cui and Chen (2010) used 619 data sets for horizontal flow to recalculate the original data of Lockhart-Martinelli following the procedures of Lockhart-Martinelli. Once the researchers separated the data into the four flow mechanisms based on the superficial Reynolds number of the gas phase (Re_g) and liquid phase (Re_l) respectively, the corresponding values of X, ϕ_g and ϕ_l were calculated, and the data points were plotted on the ϕ_g-X diagram. They compared these data points with the four Lockhart-Martinelli correlation curves respectively. They commented that there was no mention of how the correlation curves were developed from the data points and there was also no evidence of any statistical analysis in the original Lockhart-Martinelli paper. It appeared that the curves were drawn by following the general trend of the data points. Furthermore, from the original graph of the correlation curves given in the original Lockhart-Martinelli paper, it was noted that the middle and some of the right-hand portions of the curves were shown as "solid lines" while the left-hand portion of the curves were drawn as "dashed lines". It was obvious that the "dashed lines" were not supported by data points and were extrapolations. They mentioned that computers and numerical analysis were not so readily accessible when the Lockhart-Martinelli paper was published in 1949. With the help of modern computers, the goodness of fit of data to empirical correlations could be analyzed and new empirical curves that better fit the existing data points might be obtained using the non-linear least squares method. The t-l curve had a percentage error significantly lower than for the other curves. However, this did not necessarily mean that the t-l curve was the best-fitted correlation because there were only nine data points associated with this curve. Also, these data points were in a very narrow range of $10 < X < 100$ while the empirical correlation given was for the range $0.01 < X < 100$. The t-t, l-t and l-l curves had similar but larger values of percentage error compared with the t-l curve.

Moreover, Cui and Chen (2010) re-categorized the Lockhart-Martinelli data according to flow pattern. In order to re-categorize the Lockhart-Martinelli data according to flow pattern, the researchers needed to make use of the Mandhane-Gregory-Aziz (1974) flow pattern map because the original Lockhart-Martinelli data had no information on flow patterns. Having calculated the superficial velocities of the gas phase (U_g) and liquid phase (U_l) respectively, the Lockhart-Martinelli data were plotted as scatter points on the Mandhane-Gregory-Aziz (1974) flow pattern map with the X-axis "U_g" was the superficial velocity of the gas phase, while the Y-axis "U_l" was the superficial velocity of the liquid phase. It was clear that the data used by Lockhart-Martinelli fell into five categories in terms of flow patterns: A, Annular flow; B, Bubbly flow; W, Wave flow; S, Slug flow and Str, Stratified flow. There were no data in the D, Dispersed flow region. They observed that the majority of the data fell within the annular, slug and wavy flow patterns. A few points fell within the stratified flow and the bubbly flow patterns. Also, in every flow pattern, the distribution of data points based on the four flow mechanisms of t-t, t-l, l-t and l-l flow was presented. After all the data had been re-categorized according to flow pattern, Cui and Chen (2010) compared the new data groups with the Lockhart-Martinelli curves. Again, the "Mean Absolute Percentage Error" , which referred to the vertical distance between the data point and the curve expressed as a percentage deviation from the curve, was used for making the comparison. The t-t curve was the best correlation for the annular (13.4% error), bubbly (9.0% error) and slug (15.8% error) data used by Lockhart-Martinelli. The wavy data showed an error greater than about 20% when compared with any one of the Lockhart-

Martinelli curves, while the stratified data was best represented by the l-l curve with an average error of 14.3%. As a result, when the data were categorized according to flow patterns, none of the four curves (t-t, t-l, l-t, and l-l) provided improved correlation, but with the exception of the bubbly flow data that showed an averaged error of 9.7%. It should be noted that the bubbly data points were located at large X values where the four ϕ_g-X curves tended to merge.

Although the Lockhart-Martinelli correlation related to the adiabatic flow of low pressure air-liquid mixtures, they purposely presented the information in a generalized form to enable the application of the model to single component systems, and, in particular, to steam-water mixtures. Their empirical correlations were shown to be as reliable as any annular flow pressure drop correlation (Collier and Thome, 1994). The Lockhart-Martinelli model (1949) is probably the most well known method, commonly used in refrigeration and wet steam calculations. The disadvantage of this method was its limit to small-diameter pipes and low pressures because many applications of two-phase flow fell beyond these limits.

Since Lockhart and Martinelli published their paper on two-phase or two-component flows in 1949 to define the methodology for presenting two-phase flow data in non-boiling and boiling flows, their paper has received nearly 1000 citations in journal papers alone is a testament to its contribution to the field of two-phase flow.

6.3.2.2. Turner model

In his Ph. D. thesis, Turner (1966) developed the separate-cylinder model by assuming that the two-phase flow, without interaction, in two horizontal separate cylinders and that that the areas of the cross sections of these cylinders added up to the cross-sectional area of the actual pipe. The liquid and gas phases flow at the same flow rate through separate cylinders. The pressure gradient in each of the imagined cylinders was assumed to be equal, and its value was taken to be equal to the two-phase frictional pressure gradient in the actual flow. For this reason, the separate-cylinder model was not valid for gas-liquid slug flow, which gave rise to large pressure fluctuations. The pressure gradient was due to frictional effects only, and was calculated from single-phase flow theory. The separate-cylinder model resembled Lockhart and Martinelli correlation (1949) but had the advantage that it could be pursued to an analytical conclusion. The results of his analysis were

$$\left(\frac{1}{\phi_l^2}\right)^{1/n} + \left(\frac{1}{\phi_g^2}\right)^{1/n} = 1 \tag{185}$$

The values of n were dependent on whether the liquid and gas phases were laminar or turbulent flow. The different values of n are given in Table 6.

In Table 6, it should be noted the following for turbulent flow (analyzed on a basis of friction factor):

i. $n = 2.375$ for $f_l = 0.079/Re_l^{0.25}$ and $f_g = 0.079/Re_g^{0.25}$.

ii. $n = 2.4$ for $f_l = 0.046/Re_l^{0.2}$ and $f_g = 0.046/Re_g^{0.2}$.

iii. $n = 2.5$ for f_l = constant (i.e. not function of Re_l) and f_g = constant (i.e. not function of Re_g).

Flow Type	n
Laminar Flow	2
Turbulent Flow (analyzed on a basis of friction factor)	2.375-2.5
Turbulent Flows (calculated on a mixing-length basis)	2.5-3.5
Turbulent-Turbulent Regime	4
All Flow Regimes	3.5

Table 6. Values of Exponent *(n)* for Different Flow Types.

In the case of the two mixed flow regimes, Awad (2007a) mentioned in his Ph. D. thesis that the generalization of the Turner method could lead to the following implicit expressions:

$$\phi_l^2 = \left[1 + (\phi_l^2)^{(3/38)} \left(\frac{1}{X^2} \right)^{1/2.375} \right]^2 \tag{186}$$

for the laminar liquid-turbulent gas case ($f_l = 16/Re_l$ and $f_g = 0.079/Re_g^{0.25}$), and

$$\phi_l^2 = \left[1 + (\phi_l^2)^{(-3/38)} \left(\frac{1}{X^2} \right)^{0.5} \right]^{2.375} \tag{187}$$

for the turbulent liquid-laminar gas case ($f_l = 0.079/Re_l^{0.25}$ and $f_g = 16/Re_g$). Equations (186) and (187) can be solved numerically.

Also, Muzychka and Awad (2010) mentioned that the values of n in Eq. (185) for the case of the two mixed flow regimes were n = 2.05 for the turbulent liquid-laminar gas case and n = 2.10 for the laminar liquid-turbulent gas case.

Wallis (1969) mentioned in his book that there is no rationale for the good agreement between the analytical results the separate-cylinder model and the empirical results of Lockhart and Martinelli (1949). In spite of this statement, the method is still widely accepted because of its simplicity.

6.3.2.3. Chisholm model

In the following year after Turner (1966) proposed the separate cylinders model in his Ph. D. thesis, Chisholm (1967) proposed a more rigorous analysis that was an extension of the Lockhart-Martinelli model, except that a semi-empirical closure was adopted. Chisholm's rationale for his study was the fact that the Lockhart-Martinelli model failed to produce suitable equations for predicting the two-phase frictional pressure gradient, given that the empirical curves were only presented in graphical and tabular form. In spite of Chisholm's claims, he developed his approach in much the same manner as the Lockhart-Martinelli model. The researcher developed equations in terms of the Lockhart-Martinelli correlating groups for the friction pressure drop during the flow of gas-liquid or vapor-liquid mixtures in pipes. His theoretical development was different from previous treatments in the method of allowing for the interfacial shear force between the phases. Also, he avoided some of the anomalies occurring in previous "lumped flow". He gave simplified equations for use in engineering design. His equations were

$$\phi_l^2 = 1 + \frac{C}{X} + \frac{1}{X^2} \tag{188}$$

$$\phi_g^2 = 1 + CX + X^2 \tag{189}$$

The values of C were dependent on whether the liquid and gas phases were laminar or turbulent flow. The values of C were restricted to mixtures with gas-liquid density ratios corresponding to air-water mixtures at atmospheric pressure. The different values of C are given in Table 7.

Liquid	**Gas**	*C*
Turbulent	Turbulent	20
Laminar	Turbulent	12
Turbulent	Laminar	10
Laminar	Laminar	5

Table 7. Values of Chisholm Constant (C) for Different Flow Types.

He compared his predicted values using these values of C and his equation with the Lockhart-Martinelli values. He obtained good agreement with the Lockhart-Martinelli empirical curves.

The meaning of the Chisholm constant (C) can be easily seen if we multiply both sides of Eq. (188) by $(dp/dz)_{f,l}$ or both sides of Eq. (189) by $(dp/dz)_{f,g}$ to obtain:

$$\left(\frac{dp}{dz}\right)_{f,tp} = \left(\frac{dp}{dz}\right)_{f,l} + \underbrace{C\left[\left(\frac{dp}{dz}\right)_{f,l}\left(\frac{dp}{dz}\right)_{f,g}\right]^{0.5}}_{interfacial} + \left(\frac{dp}{dz}\right)_{f,g} \tag{190}$$

The physical meaning of Eq. (190) is that the two-phase frictional pressure gradient is the sum of three components: the frictional pressure of liquid-phase alone, the interfacial contribution to the total two-phase frictional pressure gradient, and the frictional pressure of gas-phase alone. As a result, we may now write

$$\left(\frac{dp}{dz}\right)_{f,i} = \underbrace{C\left[\left(\frac{dp}{dz}\right)_{f,l}\left(\frac{dp}{dz}\right)_{f,g}\right]^{0.5}}_{interfacial} \tag{191}$$

The means that the constant C in Chisholm's model can be viewed as a weighting factor for the geometric mean (GM) of the single-phase liquid and gas only pressure gradients.

The Chisholm parameter (C) is a measure of two-phase interactions. The larger the value, the greater the interaction, hence the Lockhart-Martinelli parameter (X) can involve ll, tl, lt, and tt regimes. It just causes the data to shift outwards on the Lockhart-Martinelli plot.

The Chisholm constant (C) can be derived analytically for a number of special cases. For instance, Whalley (1996) obtained for a homogeneous flow having constant friction factor:

$$C=\left[\left(\frac{\rho_l}{\rho_g}\right)^{0.5}+\left(\frac{\rho_g}{\rho_l}\right)^{0.5}\right] \tag{192}$$

that for an air-water combination gives $C \approx 28.6$ that is in good agreement with Chisholm's value for turbulent-turbulent flows. Also, Whalley (1996) shows that for laminar and turbulent flows with no interaction between phases the values of $C \approx 2$ and $C \approx 3.66$ are obtained, respectively.

In addition, Awad and Muzychka (2007, 2010b) mentioned that a value of $C = 0$ can be used as a lower bound for two-phase frictional pressure gradient in minichannels and microchannels. The physical meaning of the lower bound ($C = 0$) is that the two-phase frictional pressure gradient is merely the sum of the frictional pressure of liquid phase alone and the frictional pressure of gas phase alone:

$$\left(\frac{dp}{dz}\right)_{f,tp}=\left(\frac{dp}{dz}\right)_{f,l}+\left(\frac{dp}{dz}\right)_{f,g} \tag{193}$$

This means there is no contribution to the pressure gradient through phase interaction. The above result can also be obtained using the asymptotic model for two-phase frictional pressure gradient (Awad and Muzychka (2004b)) with linear superposition. Further, using the homogeneous model with the Dukler et al. (1964) definition of two-phase viscosity for laminar-laminar flow leads to the same result as Eq. (193).The value of $C = 0$ is also in agreement with recent models in microchannel flows such as (Mishima and Hibiki correlation (1996) and English and Kandlikar correlation (2006)) that implies that as $d_h \rightarrow 0$, $C \rightarrow 0$. The only disadvantage in these mentioned correlations is the dimensional specification of d_h, as it is easy to miscalculate C if the proper dimensions are not used for d_h. Other researchers such as Zhang et al. (2010) overcame this disadvantage by representing the hydraulic diameter (d_h) in a dimensionless form using the Laplace number (La).

Moreover, if a laminar plug flow is assumed, a value of $C = 0$ can be easily derived that implies that the total pressure gradient is just the sum of the component pressure gradients based on plug length and component flow rate. This is a reasonable approximation provided that plug lengths are longer than fifteen diameters (Walsh et al., 2009).

In his Ph. D. thesis, Awad (2007a) reviewed additional extended Chisholm type models.

6.3.2.4. *Hemeida-Sumait model*

The Lockhart-Martinelli (1949) correlation in its present form cannot be used to study a large set of data because it requires the use of charts and hence cannot be simulated numerically. As a result, Hemeida and Sumait (1988) developed a correlation between Lockhart and Martinelli parameters ϕ and X for a two-phase pressure drop in pipelines using the Statistical Analysis System (SAS). To calculate the parameter ϕ as a function of X using SAS software, their equation was

$$\phi = \exp\left[2.303a + bLn(X) + \frac{c}{2.30}(LnX)^2\right] \tag{194}$$

Where a, b, and c were constants. They selected the values of the constants a, b, and c according to the type of fluid and flow mechanisms (Table 8).

Parameter	a	b	c
$\phi_{g,ll}$	0.4625	0.5058	0.1551
$\phi_{g,lt}$	0.5673	0.4874	0.1312
$\phi_{g,tl}$	0.5694	0.4982	0.1255
$\phi_{g,tt}$	0.6354	0.4810	0.1135
$\phi_{l,ll}$	0.4048	0.4269	0.1841
$\phi_{l,lt}$	0.5532	-0.4754	0.1481
$\phi_{l,tl}$	0.5665	-0.4586	0.1413
$\phi_{l,tt}$	0.6162	-0.5063	0.124

Table 8. Values of a, b, and c for Different Flow Mechanisms.

In Table 8, the first subscript refers to whether the liquid is laminar or turbulent while the second subscript refers to whether the gas is laminar or turbulent. Equation (194) enabled the development of a computer program for the analysis of data using the Lockhart-Martinelli (1949) correlation. Using this program, they analyzed field data from Saudi flow lines. The results showed that the improved Lockhart-Martinelli correlation predicted accurately the downstream pressure in flow lines with an average percent difference of 5.1 and standard deviation of 9.6%.

It should be noted that the Hemeida-Sumait (1988) model is not famous in the literature like other models such as the Chisholm (1967) model although it gave an accurate prediction of two-phase frictional pressure gradient.

6.3.2.5. *Modified Turner model*

Awad and Muzychka (2004b) arrived at the same simple form as the empirical Turner (1966) model, but with a different physical approach. Rather than model the fluid as two distinct fluid streams flowing in separate pipes, the researchers proposed that the two- phase frictional pressure gradient could be predicted using a nonlinear superposition of the component pressure gradient that would arise from every stream flowing alone in the same pipe, through application of the Churchill-Usagi (1972) asymptotic correlation method. This form was asymptotically correct for either phase as the mass quality varied from $0 < x < 1$. Moreover, rather than approach the Lockhart-Martinelli parameter (X) from the point of view of the four flow regimes using simple friction models, they proposed using the Churchill (1977) model for the friction factor in smooth and rough pipes for all values of the Reynolds number. In this way, the proposed model was more general and contained only one empirical coefficient, the Churchill-Usagi blending parameter. The resulting model takes the form:

$$\left(\frac{dp}{dz}\right)_f = \left[\left(\frac{dp}{dz}\right)_{f,l}^p + \left(\frac{dp}{dz}\right)_{f,g}^p\right]^{1/p} \tag{195}$$

or when written as a two-phase frictional liquid multiplier:

$$\phi_l^2 = \left[1 + \left(\frac{1}{X^2}\right)^p\right]^{1/p} \tag{196}$$

or when written as a two-phase frictional gas multiplier:

$$\phi_g^2 = [1 + (X^2)^p]^{1/p} \tag{197}$$

which are the same equations from the Turner approach, when $p = 1/n$. The main exception is that the values of p were developed for different flow regimes using the Churchill friction model to calculate X.

The principal advantages of the above approach over the Turner (1966) method are twofold. First, all four Lockhart-Martinelli flow regimes can be handled with ease because the Turner (1966) method leads to implicit relationships for the two mixed regimes. Second, since the friction model used is only a function of Reynolds number and roughness, broader applications involving rough pipes can be easily modeled. Using Eqs. (196) and (197), Awad (2007b) found that $p \approx 0.307$ for large tubes and $p \approx 0.5$ for microchannels, minichannels, and capillaries. The modified Turner model is also a one parameter correlating scheme. Recently, Awad and Butt (2009a, 2009b, and 2009c) have shown that the asymptotic method works well for petroleum industry applications for liquid-liquid flows, flows through fractured media, and flows through porous media. Moreover, Awad and Muzychka (2010a) have shown that the asymptotic method works well for two-phase gas-liquid flow at microgravity conditions.

Approximate equivalence between Eq. (188) and Eq. (196) (or Eq. (189) and Eq. (197)) can be found when $p = 0.36$, 0.3, 0.285, and 0.245 when $C = 5$, 10, 12, and 20, respectively. This yields differences of 3-9% rms. The special case of $p = 1$ leads to a linear superposition of the component pressure gradients that corresponds to $C = 0$. This limiting case is only valid for plug flows when plug length to diameter ratios exceed 15 (Walsh et al., 2009).

6.3.2.6. *Modified Chisholm models*

Finally, in a recent series of studies by Saisorn and Wongwises (2008, 2009, and 2010), correlation was proposed having the form:

$$\phi_l^2 = 1 + \frac{6.627}{X^{0.761}} \tag{198}$$

for experimental data for slug flow, throat-annular flow, churn flow, and annular-rivulet flow, Saisorn and Wongwises (2008), and

$$\phi_l^2 = 1 + \frac{2.844}{X^{1.666}} \tag{199}$$

for experimental data for annular flow, liquid unstable annular alternating flow (LUAAF), and liquid/annular alternating flow (LAAF), Saisorn and Wongwises (2009). These correlations neglect the $1/X^2$ term that represents the limit of primarily gas flow in the Lockhart-Martinelli (1949) formulation. Neglecting this term ignores this important limiting case, which is an essential contribution in the Lockhart-Matrinelli modeling approach. As a result, at low values of X, the proposed correlations undershoot the trend of the data, limiting their use in the low X range. Thus, a more appropriate and generalized form of the above correlations should be:

$$\phi_l^2 = 1 + \frac{A}{X^m} + \frac{1}{X^2} \tag{200}$$

or

$$\phi_g^2 = 1 + AX^m + X^2 \tag{201}$$

These formulations, Eqs. (200) and (201), can be considered extended Chisholm type models. They will be utilized in the next section as a means of modeling the two-phase flow interfacial pressure gradient.

6.3.3. *Interfacial pressure gradient*

Gas-liquid two-phase flow will be examined from the point of view of interfacial pressure gradient. Recognizing that in a Lockhart-Martinelli reduction scheme, single-phase flow characteristics must be exhibited in a limiting sense, they will be subtracted from the experimental data being considered to illustrate some benefits of using the one and two parameter models.

The two-phase frictional pressure gradient can be defined as a linear combination of three pressure gradients. These are the single-phase liquid, single-phase gas, and interfacial pressure gradient. The rationale for such a choice lies in the definition of the Lockhart-Martinelli approach, whereby, one obtains single-phase gas flow for small values of the

Lockhart-Martinelli parameter (X) and single-phase liquid flow for large values of the Lockhart-Martinelli parameter (X). While in the transitional region between $0.01 < X < 100$, interfacial effects result in a large spread of data depending upon flow regime.

Beginning with

$$\left(\frac{dp}{dz}\right)_{f,tp} = \left(\frac{dp}{dz}\right)_{f,l} + \left(\frac{dp}{dz}\right)_{f,i} + \left(\frac{dp}{dz}\right)_{f,g} \quad (202)$$

Rearranging Eq. (202), we obtain

$$\left(\frac{dp}{dz}\right)_{f,i} = \left(\frac{dp}{dz}\right)_{f,tp} - \left(\frac{dp}{dz}\right)_{f,l} - \left(\frac{dp}{dz}\right)_{f,g} \quad (203)$$

Dividing both sides of Eq. (203) by the single-phase liquid frictional pressure gradient, we obtain

$$\phi_{1,i}^2 = \left(\frac{dp}{dz}\right)_{f,i} / \left(\frac{dp}{dz}\right)_{f,l} = \varphi_1^2 - 1 - \frac{1}{X^2} \quad (204)$$

On the other hand, dividing both sides of Eq. (203) by the single-phase gas frictional pressure gradient, we obtain

$$\phi_{g,i}^2 = \left(\frac{dp}{dz}\right)_{f,i} / \left(\frac{dp}{dz}\right)_{f,g} = \varphi_g^2 - X^2 - 1 \quad (205)$$

Where $\phi_{l,i}^2$ and $\phi_{g,i}^2$ are two-phase frictional multiplier for the interfacial pressure gradient. This can be viewed as an extended form of the Chisholm model, where the interfacial contribution is what is to be modeled. The data defined using Eqs. (204) and (205) may then be modeled using one, two, or multi-parameter forms. We discuss these approaches below.

It should be noted that this analysis is useful to show that $\phi_{g,i}$ does not exist at high values of X_{tt} for some correlations available in the literature such as the ϕ_g correlation of Ding et al. (2009) to predict the pressure drop of R410A–oil mixtures in microfin tubes, the ϕ_g correlation of Hu et al. (2008) to predict the pressure drop of R410A/POE oil mixture in micro-fin tubes, and the ϕ_g correlation of Hu et al. (2009) to predict the pressure drop of R410A/oil mixture in smooth tubes because $\phi^2_{g,i}$ has negative values at high values of X_{tt} (Awad, 2010a, Awad, 2010b, and Awad, 2011). Also, it this analysis is useful to show that $\phi_{l,i}$ does not exist at certain values of X_{tt} for some correlations available in the literature like the ϕ_l correlation of Changhong et al. (2005) to predict the pressure drop in two vertical narrow annuli (Awad, 2012b).

6.3.3.1. One parameter models

Comparison with the Chisholm (1967) formulation gives:

$$\phi_{l,i}^{2} = \frac{C}{X} \tag{206}$$

for the liquid multiplier formulation, or

$$\phi_{g,i}^{2} = CX \tag{207}$$

for the gas multiplier formulation.

This represents a simple one parameter model, whereby closure can be found with comparison with experimental data. Also, the simple asymptotic form of Eqs. (196) or (197) represents a one parameter model. If the interfacial effects can be modeled by Chisholm's proposed model or Eqs. (196) or (197), then all of the reduced data should show trends indicated by Eqs. (206) or (207). However, if data do not scale according to Eqs. (206) or (207), i.e. a slope of negative one for the liquid multiplier formulation or positive one for the gas multiplier formulation, then a two parameter model is likely required.

6.3.3.2. *Two parameter models*

Muzychka and Awad (2010) extended Eqs. (206) and (207) to develop a simple two parameter power law model such that:

$$\phi_{l,i}^{2} = \frac{A}{X^{m}} \tag{208}$$

or

$$\phi_{g,i}^{2} = AX^{m} \tag{209}$$

leading to Eqs. (200) or (201).

These forms have the advantage that experimental data for a particular flow regime can be fit to the simple power law after removal of the single-phase pressure contributions (Muzychka and Awad (2010)). Also, the advantage of the A and m model over the Chisholm model (1967) is the Chisholm model (1967) is destined to fail as they do not scale with X properly when data deviate from the -1 and +1 slope. For example, this two parameter power law model can be use for the analysis of stratified flow data separated into different categories (t–t, l–t and l–l) in Cui and Chen (2010) for their study on a re-examination of the data of Lockhart-Martinelli. The researchers used the 619 data sets for horizontal flow. Their 619 data sets were classified based on the flow patterns as follows: 191 data sets for Annular flow, 277 data sets for Slug flow, 94 data sets for Wavy flow, 32 data sets for Bubbly flow, and 25 data sets for Stratified flow.

The analysis is presented here for the stratified flow data because it has only 25 data points (the lowest number of data points for the different flow patterns: annular (191), slug (277), wavy (94), bubbly (32), and stratified (25)). The interfacial component ($\phi_{g,i}$) for stratified flow data of Lockhart-Martinelli is calculated as follows:

$$\phi_{g,i} = (\phi_g^2 - X^2 - 1)^{0.5} \tag{210}$$

Using this analysis, the interfacial component for the high pink triangle at the right hand side of stratified flow data separated into different categories (t–t, v–t and v–v) (Cui and Chen (2010)) does not exist. This is because (ϕ_g^2-X^2-1) < 0 for this point so that the square root of a negative value does not exist. This means that there is an error in the measurement in one data point for the stratified flow at lt flow mechanisms of liquid-gas. In this analysis, the data points of tt, lt, and ll flow mechanisms of liquid-gas were fit with only one line instead of three different lines for each flow mechanism of liquid-gas (tt, lt, and ll) because tt has only one data point. As shown in Fig. 4, the fit equation was:

$$\phi_{g,i} = 2.1X^{0.678} \tag{211}$$

However, drawing a different line of the interfacial component for the stratified flow for each flow mechanism of liquid-gas (tt, lt, and ll) will be more accurate. This analysis can be also done for other flow patterns: annular (191), slug (277), wavy (94), and bubbly (32).

It should be noted that the nonexistence of the interfacial component for some data sets for any flow patterns: annular (191), slug (277), wavy (94), bubbly (32), and stratified (25)) means that there is an error in the measurement of some data sets of Lockhart-Martinelli although their paper has received nearly 1000 citations in journal papers.

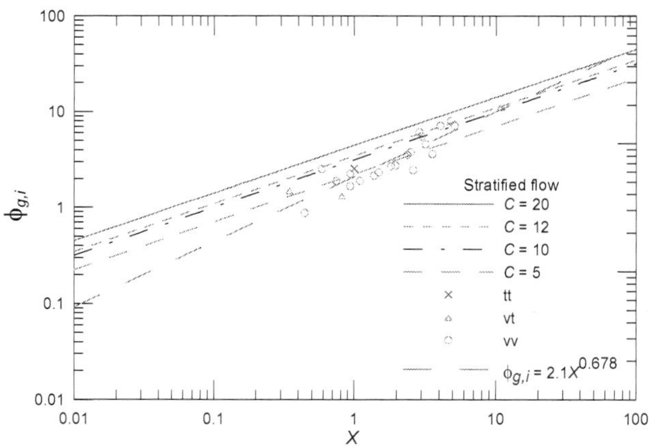

Figure 4. Analysis of Stratified Flow Data Separated into Different Categories (t–t, v–t and v–v) Using Two Parameter Power Law Model (Muzychka and Awad (2010)).

6.3.3.3. Multi-parameter models

Multi-parameter models may be developed using both the Chisholm model and the modified Chisholm models, by correlating the constants *C*, *A*, and *m* with other dimensionless parameters. For example, Sun and Mishima (2009) adopted an approach that led to the development of *C* in the laminar flow region as a function of the following dimensionless parameters: the Laplace constant (La), and the liquid Reynolds number (Re_l). Also, Venkatesan et al. (2011) adopted an approach that led to the development of C in circular tubes with d = 0.6, 1.2, 1.7, 2.6 and 3.4 mm using air and water as a function of the following dimensionless parameters: Weber number(We), superficial liquid Reynolds number (Re_l), and superficial gas Reynolds number (Re_g). In addition, Kawahara et al. (2011) used their two-phase frictional pressure drop data in a rectangular microchannel with a T-junction type gas-liquid mixer to correlate the Chisholm constant (C) as a function of the following dimensionless parameters: Bond number(Bo), superficial liquid Reynolds number (Re_l), and superficial gas Weber number (We_g). But care must be taken because even with the introduction of additional variables, increased accuracy will not necessarily be obtained.

6.3.4. *The drift flux model*

The drift flux model is a type of separated flow model. In the drift flux model, attention is focused on the relative motion rather than on the motion of the individual phases. The drift flux model was developed by Wallis (1969). The drift flux model has widespread application to bubble flow and plug flow. The drift flux model is not particularly suitable to a flow such as annular flow that has two characteristic velocities in one phase: the liquid film velocity and the liquid drop velocity. However, the drift flux model has been used for annular flows, but with no particular success.

The drift flux model is the fifth example of the existing void fraction models. The Rouhani and Axelsson (1970) model is an instance for this type of model. In the drift-flux model, the void fraction (α) is a function of the gas superficial velocity (U_g), the total superficial velocity (U), the phase distribution parameter (C_o), and the mean drift velocity (u_{gj}) that includes the effect of the relative velocity between the phases. The form of the drift-flux model is

$$\alpha = \frac{U_g}{C_o U + u_{gj}} \tag{212}$$

The drift-flux correlations often present procedures to compute C_o and u_{gj}. Since the expressions of C_o and u_{gj} are usually functions of the void fraction (α), the predictions of the void fraction (α) are calculated using method of solving of non-linear equation.

6.3.5. *Two-fluid model*

This model is known as the two-fluid model designating two phases or components. This model is an advanced predictive tool for liquid-gas two-phase flow in engineering applications. It is based on the mass, momentum and energy balance equations for every phase (Ishii, 1987). In this model, every phase or component is treated as a separate fluid with its own set of governing balance equations. In general, every phase has its own velocity, temperature and pressure. This approach enables the prediction of important non-equilibrium phenomena of two-phase flow like the velocity difference between liquid and gas phase. This prediction is important for two-phase flows in large shell sides of steam generators and kettle reboilers, where even different gas and liquid velocity directions exist.

6.4. Other methods

There are other methods of analysis like integral analysis, differential analysis, computational fluid dynamics (CFD), and artificial neural network (ANN).

6.4.1. *Integral analysis*

In a one-dimensional integral analysis, the form of certain functions which describe, for instance, the velocity or concentration distribution in a pipe is assumed first. Then, these functions are made to satisfy appropriate boundary conditions and the basic equations of fluid mechanics (continuity equation, momentum equation, and energy equation) in integral form. Single-phase boundary layers are analyzed using similar techniques.

6.4.2. *Differential analysis*

The velocity and concentration fields are deduced from suitable differential equations. Usually, the equations are written for time-average quantities, like in single-phase theories of turbulence.

6.4.3. *Computational Fluid Dynamics (CFD)*

Two-phase flows are encountered in a wide range of industrial and natural situations. Due to their complexity such flows have been investigated only analytically and experimentally. New computing facilities provide the flexibility to construct computational models that are easily adapted to a wide variety of physical conditions without constructing a large-scale prototype or expensive test rigs. But there is an inherent uncertainty in the numerical predictions due to stability, convergence and accuracy. The importance of a well-placed mesh is highlighted in the modeling of two-phase flows in horizontal pipelines (Lun et al., 1996).

Also, with the increasing interest in multiphase flow in microchannels and advancement in interface capturing techniques, there have recently been a number of attempts to apply computational fluid dynamics (CFD) to model Taylor flow such as van Baten and Krishna (2004), Taha and Cui (2006a, 2006b) and Gupta et al. (2009). The CFD package, Fluent was used in these numerical studies of CFD modeling of Taylor flow.

In addition, Liu et al. (2011) developed recently a new two-fluid two-component computational fluid dynamics (CFD) model to simulate vertical upward two-phase annular flow. The researchers utilized the two-phase VOF scheme to model the roll wave flow, and described the gas core by a two-component phase consisting of liquid droplets and gas phase. They took into account the entrainment and deposition processes by source terms of the governing equations. Unlike the previous models, their newly developed model included the influence of liquid roll waves directly determined from the CFD code that was able to provide more detailed and, the most important, more self-standing information for both the gas core flow and the film flow as well as their interactions. They compared predicted results with experimental data, and achieved a good agreement.

6.4.4. *Artificial Neural Network (ANN)*

In recent years, artificial neural network (ANN) has been universally used in many applications related to engineering and science. ANN has the advantage of self-learning and self-organization. ANN can employ the prior acquired knowledge to respond to the new information rapidly and automatically. When the traditional methods are difficult to be carried out or sometimes the specific models of mathematical physics will not be thoroughly existing, the neural network will be considered as a very good tool to tackle these time-consuming and complex nonlinear relations because neural network has the excellent characteristics of parallel processing, calculating for complex computation and self-learning. The development of any ANN model involves three basic steps. First, the generation of data required for training. Second, the training and testing of the ANN model using the information about the inputs to predict the values of the output. Third, the evaluation of the ANN configuration that leads to the selection of an optimal configuration that produces the best results based on some preset measures. The optimum ANN model is also validated using a larger dataset. In the area of two-phase flow, the applications of the ANN include the prediction of pressure drop (Osman and Aggour, 2002), identifying flow regimes (Selli and Seleghim, 2007), predicting liquid holdup (Osman, 2004) and (Shippen and Scott, 2004), and the determination of condensation heat transfer characteristics during downward annular flow of R134a inside a vertical smooth tube (Balcilar et. al., 2011).

7. Summary and conclusions

This chapter aims to introduce the reader to the modeling of two-phase flow in general, liquid-gas flow in particular, and the prediction of frictional pressure gradient specifically. Different modeling techniques were presented for two-phase flow. Recent developments in theory and practice are discussed. The reader of this chapter is encouraged to pursue the associated journal and text references for additional theory not covered, especially the state of the art and review articles because they contain much useful information pertaining to the topics of interest. Given the rapid growth in the research topic of two-phase flow, new models and further understanding in areas like nano fluids will likely be achieved in the near future. Although, for most design and research applications, the topics covered in this chapter represent the state of the art.

Author details

M.M. Awad

Mechanical Power Engineering Department, Faculty of Engineering, Mansoura University, Egypt

Acknowledgement

The author acknowledges his Ph. D. supervisor, Prof. Yuri S. Muzychka, who introduced him to the possibilities of analytical modeling during his Ph. D. thesis. Also, the author gratefully acknowledges ASME International Petroleum Technology Institute (IPTI) scholarship awarded to him in 2005 and 2006. In addition, the author wants to thank the Editor, Prof. M. Salim Newaz Kazi, for inviting him to prepare this chapter.

Nomenclature

A	area, m^2
A	constant in the modified Chisholm model
A	Phillips parameter
a	Churchill parameter
a	constant in Hemeida and Sumait (1988) correlation
AM	Arithmetic mean
AR	Aspect ratio
Ar	Archimedes number
At	Atwood ratio
B	Phillips parameter
b	Churchill parameter
b	constant in Hemeida and Sumait (1988) correlation
Bo	Bond number
Bod	Bodenstein number
C	Chisholm constant
C	constant
c	constant in Hemeida and Sumait (1988) correlation
c	sound speed, m/s
C_D	drag coefficient
C_o	the phase distribution parameter
c_p	constant-pressure specific heat, J/kg.K
Ca	Capillary number
Cn	Cahn number
Co	Confinement number
Co	Convection number
Cou	Courant number
D	mass diffusivity, m^2/s
d	pipe diameter, m
E	electric field strength, V/m
E	two-phase heat transfer multiplier
E_{hd}	EHD number or conductive Rayleigh number
E_M	dimensionless number
E_r	dimensionless number
EF	Enhanced factor
Eo	Eötvös number
Eu	Euler number
F	parameter in the Taitel and Dukler (1976) map
f	Fanning friction factor
f	wave frequency, Hz

f''_e electric force density, N/m^2
Fo Fourier number
FR filling ratio
Fr Froude number
Fr_e Electric Froude number
Fr^* ratio of Froude number to Atwood ratio
G mass flux, kg/m^2.s
g gravitational acceleration, m/s^2
Ga Galileo number
Ga^* modified Galileo number
GM Geometric mean
Gz Graetz number
h heat transfer coefficient, W/m^2.K
H_l liquid holdup fraction
h_{lg} latent heat of voporization, J/kg
h_{sl} latent heat of melting, J/kg
HM Harmonic mean
I current, A
J_g dimensionless vapor mass flux
Ja Jacob number
Ja^* modified Jacob number
K mass transfer coefficient, m/s
K parameter in the Taitel and Dukler (1976) map
k thermal conductivity, W/m.K
K_1, K_2, K_3 constants in the Cengel (1967) definition for viscosity
K_1, K_2, K_3 new non-dimensional constant of Kandlikar
K_f Boiling number
Ka Kapitza number
Kn Knudsen number
Kr von Karman number
Ku Kutateladze number
L characteristic length, m
L length, m
L_c capillary length, m
L_s liquid plug length, m
L^* dimensionless liquid plug length = L_s/d
La Laplace number
Le Lewis number
Lo dimensionless number
m exponent in the modified Chisholm model
$\dot{m}$ mass flow rate, kg/s
M_a Masuda number or dielectric Rayleigh number
Ma Homogeneous Equilibrium Model (HEM) Mach number
Mo Morton number

n	Blasius index
n	exponent
N_f	inverse viscosity number
Nu	Nusselt number
Oh	Ohnesorge number
p	fitting parameter
dp/dz	pressure gradient, Pa/m
Δp	pressure drop, Pa
Δp_f^*	dimensionless frictional pressure drop, Pa
Pe	Peclet number
Ph	phase change number
Po	Poiseuille constant
Pr	Prandtl number
Q	heat transfer rate, W
$\dot{Q}$	volumetric flow rate, m^3/s
q	heat flux, W/m^2
R	pipe radius, m
Re	Reynolds number
Re_f	film Reynolds number
Re_p	particle Reynolds number
Re^*	laminar equivalent Reynolds number
Ri	Richardson number
S	slip ratio
Sc	Schmidt number
Sh	Sherwood number
St	Stanton number
Stk	Stokes number
Str	Strouhal number
Su	Suratman number
T	parameter in the Taitel and Dukler (1976) map
T	temperature, K
ΔT	temperature difference, K
U	superficial velocity, m/s
u_{gj}	mean drift velocity, m/s
v	specific volume, m^3/kg
X	Lockhart-Martinelli parameter
x	distance in x-direction, m
x	mass quality
X^*	modified Lockhart-Martinelli parameter

Greek

α	concentration
α	thermal diffusivity, m^s/s

α	void fraction
β	volumetric quality
Δt	time step size, s
Δx	mesh becomes finer, m
δ	liquid film thickness, m
ε_s	dielectric constant ($\varepsilon_s = \varepsilon/\varepsilon_0$)
ε	permittivity, N/V^2
ε	permittivity of free space ($\varepsilon_0 = 8.854 \times 10^{-12}$ N/V^2)
ε	pipe roughness, m
ρ	density, kg/m^3
μ	dynamic viscosity, kg/m.s
μ_c	ion mobility, m^2/Vs
ϕ_g^2	two-phase frictional multiplier for gas alone flow
ϕ_{go}^2	two-phase frictional multiplier for total flow assumed gas
ϕ_l^2	two-phase frictional multiplier for liquid alone flow
ϕ_{lo}^2	two-phase frictional multiplier for total flow assumed liquid
λ	dimensionless parameter used in Baker flow pattern map
λ	molecular mean free path length, m
ν	kinematic viscosity, m^2/s
ψ	dimensionless parameter used in Baker flow pattern map
σ	surface tension, N/m
τ_c	characteristic flow system time, s
τ_p	particle momentum response time, s
θ	inclination angle to the horizontal
Γ	physical property coefficient
Γ	total liquid mass flow rate on both sides of the tube per unit length of tube

Subscripts

0	vacuum or reference
a	acceleration
air	air
b	bubble
c	continuous phase
D	Darcy
d	dispersed phase
d_h	hydraulic diameter
eq	equivalent
f	frictional
g	gas
go	gas only (all flow as gas)
h	hydraulic
i	inner or inlet
i	interfacial

l	liquid
le	laminar equivalent
ll	laminar liquid-laminar gas flow type
lo	liquid only (all flow as liquid)
lt	laminar liquid- turbulent gas flow type
m	homogeneous mixture
m	mean
max	maximum
min	minimum
o	outer or outlet
o	oil
p	particle
p	plug
Poise	Poiseuille flow
r	refrigerant
s	saturation
s	sound
slug	slug
tl	turbulent liquid-laminar gas flow type
tp	two-phase
tt	turbulent liquid-turbulent gas flow type
w	wall
water	water

8. References

Agrawal, N., Bhattacharyya, S., and Nanda, P., 2011, Flow Characteristics of Capillary Tube with CO_2 Transcritical Refrigerant Using New Viscosity Models for Homogeneous Two-Phase Flow, International Journal of Low-Carbon Technologies, 6 (4), pp. 243-248.

Akbar, M. K., Plummer, D. A., and Ghiaasiaan, S. M., 2003, On Gas-Liquid Two-Phase Flow Regimes in Microchannels, International Journal of Multiphase Flow, 29 (5) pp. 855-865.

Akers, W. W., Deans, H. A., and Crosser, O. K., 1959, Condensation Heat Transfer within Horizontal Tubes, Chemical Engineering Progress Symposium Series, 55 (29), pp. 171–176.

Al-Sarkhi, A., Sarica, C., and Magrini, K., 2012, Inclination Effects on Wave Characteristics in Annular Gas–Liquid Flows, AIChE Journal, 58 (4), pp. 1018-1029.

Angeli, P., and Gavriilidis, A., 2008, Hydrodynamics of Taylor Flow in Small Channels: A Review, Proceedings of the Institution of Mechanical Engineers, Part C: Journal of Mechanical Engineering Science, 222 (5), pp. 737-751.

Arrhenius, S., 1887, On the Internal Friction of Solutions in Water, Zeitschrift für Physikalische Chemie (Leipzig), 1, pp. 285-298.

ASHRAE, 1993, Handbook of Fundamentals, ASHRAE, Atlanta, GA, Chap. 4.

Aussillous, P., and Quere, D., 2000, Quick Deposition of a Fluid on the Wall of a Tube, Physics of Fluids, 12 (10), pp. 2367-2371.

Awad, M. M., 2007a, Two-Phase Flow Modeling in Circular Pipes, Ph.D. Thesis, Memorial University of Newfoundland, St. John's, NL, Canada.

Awad, M. M., 2007b, Comments on Condensation and evaporation heat transfer of R410A inside internally grooved horizontal tubes by M. Goto, N. Inoue and N. Ishiwatari, International Journal of Refrigeration, 30 (8), pp. 1466.

Awad, M. M., 2010a, Comments on Experimental Investigation and Correlation of Two-Phase Frictional Pressure Drop of R410A-Oil Mixture Flow Boiling in a 5 mm Microfin Tube Int. J. Refrigeration 32/1 (2009) 150-161, by Ding, G., Hu, H., Huang, X., Deng, B., and Gao, Y., International Journal of Refrigeration, 33 (1), pp. 205-206.

Awad, M. M., 2010b, Comments on Measurement and Correlation of Frictional Two-Phase Pressure Drop of R410A/POE Oil Mixture Flow Boiling in a 7 mm Straight Micro-Fin Tube by H.-t. Hu, G.-l. Ding, and K.-j. Wang, Applied Thermal Engineering, 30 (2-3), pp. 260-261.

Awad, M. M., 2011, Comments on "Pressure drop during horizontal flow boiling of R410A/oil mixture in 5 mm and 3 mm smooth tubes" by H-t Hu, G-l Ding, X-c Huang, B. Deng, and Y-f Gao, Applied Thermal Engineering, 31 (16), pp. 3629-3630.

Awad, M. M., 2012a, Discussion: Heat Transfer Mechanisms During Flow Boiling in Microchannels (Kandlikar, S. G., 2004, ASME Journal of Heat Transfer, 126 (2), pp. 8-16), ASME Journal of Heat Transfer, 134 (1), Article No. (015501).

Awad, M. M., 2012b, Comments on "Two-phase flow and boiling heat transfer in two vertical narrow annuli", Nuclear Engineering and Design, 245, pp. 241-242.

Awad, M. M., and Butt, S. D., 2009a, A Robust Asymptotically Based Modeling Approach for Two-Phase Liquid-Liquid Flow in Pipes, ASME 28th International Conference on Offshore Mechanics and Arctic Engineering (OMAE2009), Session: Petroleum Technology, OMAE2009-79072, Honolulu, Hawaii, USA, May 31-June 5, 2009.

Awad, M. M., and Butt, S. D., 2009b, A Robust Asymptotically Based Modeling Approach for Two-Phase Gas-Liquid Flow in Fractures, 12th International Conference on Fracture (ICF12), Session: Oil and Gas Production and Distribution, ICF2009-646, Ottawa, Canada, July 12-17, 2009.

Awad, M. M., and Butt, S. D., 2009c, A Robust Asymptotically Based Modeling Approach for Two-Phase Flow in Porous Media, ASME Journal of Heat Transfer, 131 (10), Article (101014) (The Special Issue of JHT on Recent Advances in Porous Media Transport), Also presented at ASME 27th International Conference on Offshore Mechanics and Arctic Engineering (OMAE2008), Session: Offshore Technology, Petroleum Technology II, OMAE2008-57792, Estoril, Portugal, June 15-20, 2008.

Awad, M. M., and Muzychka, Y. S., 2004a, A Simple Two-Phase Frictional Multiplier Calculation Method, Proceedings of IPC2004, International Pipeline Conference, Track: 3. Design & Construction, Session: System Design/Hydraulics, IPC04-0721, Vol. 1, pp. 475-483, Calgary, Alberta, October 4-8, 2004.

Awad, M. M., and Muzychka, Y. S., 2004b, A Simple Asymptotic Compact Model for Two-Phase Frictional Pressure Gradient in Horizontal Pipes, Proceedings of IMECE 2004, Session: FE-8 A Gen. Pap.: Multiphase Flows - Experiments and Theory, IMECE2004-61410, Anaheim, California, November 13-19, 2004.

Awad, M. M., and Muzychka, Y. S., 2005a, Bounds on Two-Phase Flow. Part I. Frictional Pressure Gradient in Circular Pipes, Proceedings of IMECE 2005, Session: FED-11 B Numerical Simulations and Theoretical Developments for Multiphase Flows-I, IMECE2005-81493, Orlando, Florida, November 5-11, 2005.

Awad, M. M., and Muzychka, Y. S., 2005b, Bounds on Two-Phase Flow. Part II. Void Fraction in Circular Pipes, Proceedings of IMECE 2005, Session: FED-11 B Numerical Simulations and Theoretical Developments for Multiphase Flows-I, IMECE2005-81543, Orlando, Florida, November 5-11, 2005.

Awad, M. M. and Muzychka, Y. S., 2007, Bounds on Two-Phase Frictional Pressure Gradient in Minichannels and Microchannels, Heat Transfer Engineering, 28 (8-9), pp. 720-729. Also presented at The 4th International Conference on Nanochannels, Microchannels and Minichannels (ICNMM 2006), Session: Two-Phase Flow, Numerical and Analytical Modeling, ICNMM2006-96174, Stokes Research Institute, University of Limerick, Ireland, June 19-21, 2006.

Awad, M. M. and Muzychka, Y. S., 2008, Effective Property Models for Homogeneous Two Phase Flows, Experimental and Thermal Fluid Science, 33 (1), pp. 106-113.

Awad, M. M., and Muzychka, Y. S., 2010a, Review and Modeling of Two-Phase Frictional Pressure Gradient at Microgravity Conditions, ASME 2010 3rd Joint US-European Fluids Engineering Summer Meeting and 8th International Conference on Nanochannels, Microchannels, and Minichannels (FEDSM2010-ICNMM2010), Symposium 1-14 4th International Symposium on Flow Applications in Aerospace, FEDSM2010-ICNMM2010-30876, Montreal, Canada, August 1-5, 2010.

Awad, M. M., and Muzychka, Y. S., 2010b, Two-Phase Flow Modeling in Microchannels and Minichannels, Heat Transfer Engineering, 31 (13), pp. 1023-1033. Also presented at The 6th International Conference on Nanochannels, Microchannels and Minichannels (ICNMM2008), Session: Two-Phase Flow, Modeling and Analysis of Two-Phase Flow, ICNMM2008-62134, Technische Universitaet of Darmstadt, Darmstadt, Germany, June 23-25, 2008.

Baker, O., 1954, Simultaneous Flow of Oil and Gas, Oil and Gas Journal, 53, pp.185-195.

Balcilar, M., Dalkilic, A. S., and Wongwises, S., 2011, Artificial Neural Network Techniques for the Determination of Condensation Heat Transfer Characteristics during Downward Annular Flow of R134a inside a Vertical Smooth Tube, International Communications in Heat and Mass Transfer 38 (1), pp. 75-84.

Banasiak, K., and Hafner, A., 2011, 1D Computational Model of a Two-Phase R744 Ejector for Expansion Work Recovery, International Journal of Thermal Sciences, 50 (11), pp. 2235-2247.

Bandel, J., 1973, Druckverlust und Wärmeübergang bei der Verdampfung siedender Kältemittel im durchströmten waagerechten Rohr, Doctoral Dissertation, Universität Karlsruhe.

Bankoff, S. G., 1960, A Variable Density Single-Fluid Model for Two-Phase Flow with Particular Reference to Steam-Water Flow, Journal of Heat Transfer, 82 (4), pp. 265-272.

Beattie, D. R. H., and Whalley, P. B., 1982, A Simple Two-Phase Frictional Pressure Drop Calculation Method, International Journal of Multiphase Flow, 8 (1), pp. 83-87.

Becher, P., 2001, Emulsions: Theory and Practice, 3rd edition, Oxford University Press, New York, NY.

Bico, J., and Quere, D., 2000, Liquid Trains in a Tube, Europhysics Letters, 51 (5), pp. 546-550.

Blasius, H., 1913, Das Ähnlichkeitsgesetz bei Reibungsvorgängen in Flüssikeiten, Forsch. Gebiete Ingenieurw., 131.

Bonfanti, F., Ceresa, I., and Lombardi, C., 1979, Two-Phase Pressure Drops in the Low Flowrate Region, Energia Nucleare, 26 (10), pp. 481-492.

Borishansky, V. M., Paleev, I. I., Agafonova, F. A., Andreevsky, A. A., Fokin, B. S., Lavrentiev, M. E., Malyus-Malitsky, K. P., Fromzel V. N., and Danilova, G. P., 1973, Some Problems of Heat Transfer and Hydraulics in Two-Phase Flows, International Journal of Heat and Mass Transfer, 16 (6), pp. 1073-1085.

Brauner, N., and Moalem-Maron, D., 1992, Identification of the Range of 'Small Diameters' Conduits, Regarding Two-Phase Flow Pattern Transitions, International Communications in Heat and Mass Transfer, 19 (1), pp. 29-39.

Breber, G., Palen, J., and Taborek, J., 1980, Prediction of Horizontal Tube-Side Condensation of Pure Components Using Flow Regime Criteria, ASME Journal of Heat Transfer, 102 (3), pp. 471-476.

Bretherton, F. P., 1961, The Motion of Long Bubbles in Tubes, Journal of Fluid Mechanics, 10 (2), pp. 166-188.

Carson, J. K., Lovatt, S. J., Tanner, D. J., and Cleland, A. C., 2005, Thermal Conductivity Bounds for Isotropic, Porous Materials, International Journal of Heat and Mass Transfer, 48 (11), pp. 2150-2158.

Catchpole, J. P., and Fulford, G. D., 1966, Dimensionless Groups, Industrial and Engineering Chemistry, 58 (3), pp. 46-60.

Cavallini, A., Censi, G., Del Col, D., Doretti, L., Longo, G. A., and Rossetto, L., 2002, Condensation of Halogenated Refrigerants inside Smooth Tubes, HVAC and R Research, 8 (4), pp. 429-451.

Celata, G. P., 2004, Heat Transfer and Fluid Flow in Microchannels, Begell House, Redding, CT.

Cengel, J., 1967, Viscosity of Liquid-Liquid Dispersions in Laminar and Turbulent Flow, PhD Dissertation Thesis, Oregon State University.

Chang, J. S., 1989, Stratified Gas–Liquid Two-Phase Electrohydrodynamics in Horizontal Pipe Flow, IEEE Transactions on Industrial Applications, 25 (2), pp. 241–247.

Chang, J. S., 1998, Two-Phase Flow in Electrohydrodynamics, in: Castellanos, A., (Ed.), Part V, Electrohdyrodynamics, International Centre for Mechanical Sciences Courses and Lectures No. 380, Springer, New York.

Chang., J. S., and Watson, A., 1994, Electromagnetic Hydrodynamics, IEEE Transactions on Dielectrics and Electrical Insulation, 1 (5), pp. 871-895.

Changhong, P., Yun, G., Suizheng, Q., Dounan, J., and Changhua, N., 2005. Two-Phase Flow and Boiling Heat Transfer in Two Vertical Narrow Annuli, Nuclear Engineering and Design, 235 (16), pp. 1737–1747.

Charoensawan, P., and Terdtoon, P., 2007, Thermal Performance Correlation of Horizontal Closed-Loop Oscillating Heat Pipes, 9th Electronics Packaging Technology Conference (EPTC 2007), pp. 906-909, 10-12 December 2007, Grand Copthorne Waterfront Hotel, Singapore.

Chen, Y., Kulenovic, R., and Mertz, R., 2009, Numerical Study on the Formation of Taylor Bubbles in Capillary Tubes, International Journal of Thermal Sciences, 48 (2), pp. 234-242. Also presented at Proceedings of the 5th International Conference on Nanochannels, Microchannels and Minichannels (ICNMM2007), ICNMM2007-30182, pp. 939-946, June 18-20, 2007, Puebla, Mexico.

Cherlo, S. K. R., Kariveti, S., and Pushpavanam, S., 2010, Experimental and Numerical Investigations of Two-Phase (Liquid–Liquid) Flow Behavior in Rectangular Microchannels, Industrial and Engineering Chemistry Research, 49 (2), pp. 893-899.

Chisholm, D., 1967, A Theoretical Basis for the Lockhart-Martinelli Correlation for Two-Phase Flow, International Journal of Heat and Mass Transfer, 10 (12), pp. 1767-1778.

Chisholm, D., 1973, Pressure Gradients due to Friction during the Flow of Evaporating Two-Phase Mixtures in Smooth Tubes and Channels, International Journal of Heat and Mass Transfer, 16 (2), pp. 347-358.

Chisholm, D., 1983, Two-Phase Flow in Pipelines and Heat Exchangers, George Godwin in Association with Institution of Chemical Engineers, London.

Churchill, S. W., 1977, Friction Factor Equation Spans all Fluid Flow Regimes, Chemical Engineering, 84 (24), pp. 91-92.

Churchill, S. W. and Usagi, R., 1972, A General Expression for the Correlation of Rates of Transfer and Other Phenomena, American Institute of Chemical Engineers Journal, 18 (6), pp. 1121-1128.

Cicchitti, A., Lombaradi, C., Silversti, M., Soldaini, G., and Zavattarlli, R., 1960, Two-Phase Cooling Experiments- Pressure Drop, Heat Transfer, and Burnout Measurements, Energia Nucleare, 7 (6), pp. 407-425.

Cioncolini, A., Thome, J. R., and Lombardi, C., 2009, Unified Macro-to-Microscale Method to Predict Two-Phase Frictional Pressure Drops of Annular Flows, International Journal of Multiphase Flow, 35 (12), pp. 1138-1148.

Colebrook, C. F., 1939, Turbulent Flow in Pipes, with Particular Reference to the Transition between the Smooth and Rough Pipe Laws, J. Inst. Civ. Eng. Lond., 11, pp. 133-156.

Collier, J. G. and Thome, J. R., 1994, Convective Boiling and Condensation (3rd Edn), Claredon Press, Oxford.

Cotton, J., Robinson, A. J., Shoukri, M., and Chang, J. S., 2005, A Two-Phase Flow Pattern Map for Annular Channels under a DC Applied Voltage and the Application to Electrohydrodynamic Convective Boiling Analysis, International Journal of Heat and Mass Transfer, 48 (25-26), pp. 5563-5579.

Cotton, J. S., Shoukri, M., Chang, J. S., and Smith-Pollard, T., 2000, Electrohydrodynamic (EHD) Flow and Convective Boiling Augmentation in Single-Component Horizontal Annular Channels, 2000 International Mechanical Engineering Congress and Exposition, HTD-366-4, pp. 177–184.

Crowe, C. T., 2006, Multiphase Flow Handbook, CRC: Taylor & Francis, Boca Raton, FL.

Cui, X., and Chen, J. J. J., 2010, A Re-Examination of the Data of Lockhart-Martinelli, International Journal of Multiphase Flow, 36 (10), pp. 836-846.

Davidson, W. F., Hardie, P. H., Humphreys, C. G. R., Markson, A. A., Mumford, A. R., and Ravese, T., 1943, Studies of Heat Transmission Through Boiler Tubing at Pressures from 500-3300 Lbs, Trans. ASME, 65 (6), pp. 553-591.

Ding, G., Hu, H., Huang, X., Deng, B., and Gao, Y., 2009, Experimental Investigation and Correlation of Two-Phase Frictional Pressure Drop of R410A–Oil Mixture Flow Boiling in a 5 mm Microfin Tube, International Journal of Refrigeration, 32 (1), pp. 150-161.

Drew, T. B., Koo, E. C., and McAdams, W. H., 1932, The Friction Factor for Clean Round Pipe, Trans. AIChE, 28, pp. 56.

Duda, J. L., and Vrentas, J. S., 1971, Heat Transfer in a Cylindrical Cavity, Journal of Fluid Mechanics, 45, pp. 261–279.

Dukler, A. E., Moye Wicks and Cleveland, R. G., 1964, Frictional Pressure Drop in Two-Phase Flow. Part A: A Comparison of Existing Correlations for Pressure Loss and Holdup, and Part B: An Approach through Similarity Analysis AIChE Journal, 10 (1), pp. 38-51.

Einstein, A., 1906, Eine neue Bestimmung der Moleküldimensionen (A New Determination of Molecular Dimensions), Annalen der Physik (ser. 4), 19, pp. 289-306.

Einstein, A., 1911, Berichtigung zu meiner Arbeit: Eine neue Bestimmung der Moleküldimensionen (Correction to My Paper: A New Determination of Molecular Dimensions), Annalen der Physik (ser. 4), 34, pp. 591-592.

English, N. J., and Kandlikar, S. G., 2006, An Experimental Investigation into the Effect of Surfactants on Air–Water Two-Phase Flow in Minichannels, Heat Transfer Engineering, 27 (4), pp. 99-109. Also presented at The 3rd International Conference on Microchannels and Minichannels (ICMM2005), ICMM2005-75110, Toronto, Ontario, Canada, June 13–15, 2005.

Fairbrother, F., and Stubbs, A. E., 1935, Studies in Electro-Endosmosis. Part VI. The Bubble-Tube Method of Measurement, Journal of the Chemical Society (Resumed), pp. 527-529.

Fang, X., Xu, Y., and Zhou, Z., 2011, New Correlations of Single-Phase Friction Factor for Turbulent Pipe Flow and Evaluation of Existing Single-Phase Friction Factor Correlations, Nuclear Engineering and Design, 241 (3), pp. 897-902.

Fourar, M. and Bories, S., 1995, Experimental Study of Air-Water Two-Phase Flow Through a Fracture (Narrow Channel), International Journal of Multiphase Flow, 21 (4), pp. 621-637.

Friedel, L., 1979, Dimensionless Relationship for The Friction Pressure Drop in Pipes during Two-Phase Flow of Water and of R 12, Verfahrenstechnik, 13 (4), pp. 241-246.

Fulford, G. D., and Catchpole, J. P., 1968, Dimensionless Groups, Industrial and Engineering Chemistry, 60 (3), pp. 71-78.

García, F., García, R., Padrino, J. C., Mata, C., Trallero J. L., and Joseph, D. D., 2003, Power Law and Composite Power Law Friction Factor Correlations for Laminar and Turbulent Gas-Liquid Flow in Horizontal Pipelines, International Journal of Multiphase Flow, 29 (10), pp. 1605-1624.

García, F., García, J. M., García, R., and Joseph, D. D., 2007, Friction Factor Improved Correlations for Laminar and Turbulent Gas-Liquid Flow in Horizontal Pipelines, International Journal of Multiphase Flow, 33 (12), pp. 1320-1336.

Ghiaasiaan, S. M., 2008, Two-Phase Flow, Boiling and Condensation in Conventional and Miniature Systems, Cambridge University Press, New York.

Glielinski, V., 1976, New Equations for Heat and Mass Transfer in Turbulent Pipe and Channel Flow, International Chemical Engineering, 16 (2), pp. 359–367.

Gnielinski, V., 1999, Single-Phase Convective Heat Transfer: Forced Convection in Ducts, Heat Exchanger Design Updates, Heat Exchanger Design Handbook, Begell House, New York, NY, Chapter 5.

Goto, M., Inoue, N., and Ishiwatari, N., 2001, Condensation and Evaporation Heat Transfer of R410A inside Internally Grooved Horizontal Tubes, International Journal of Refrigeration, 24 (7), pp. 628-638.

Graham, D. M., Kopke, H. P., Wilson, M. J., Yashar, D. A., Chato, J. C. and Newell, T. A., 1999, An Investigation of Void Fraction in the Stratified/Annular Flow Regions in Smooth Horizontal Tubes, ACRC TR-144, Air Conditioning and Refrigeration Center, University of Illinois at Urbana-Champaign.

Grimes, R., King, C., and Walsh, E., 2007, Film Thickness for Two Phase Flow in a Microchannel, Advances and Applications in Fluid Mechanics, 2 (1), pp. 59-70.

Gunther, A., and Jensen, K. F., 2006, Multiphase Microfluidics: From Flow Characteristics to Chemical and Materials Synthesis, Lab on a Chip, 6 (12), pp. 1487-1503.

Gupta, R., Fletcher, D. F., and Haynes, B. S., 2009, On the CFD Modelling of Taylor Flow in Microchannels, Chemical Engineering Science, 64 (12), pp. 2941-2950.

Han, Y., and Shikazono, N., 2009a, Measurement of the Liquid Film Thickness in Micro Tube Slug Flow, International Journal of Heat and Fluid Flow, 30 (5), pp. 842–853.

Han, Y., and Shikazono, N., 2009b, Measurement of the Liquid Film Thickness in Micro Square Channel, International Journal of Multiphase Flow, 35 (10), pp. 896-903.

Haraguchi, H., Koyama, S., and Fujii, T., 1994, Condensation of Refrigerants HCF C 22, HFC 134a and HCFC 123 in a Horizontal Smooth Tube (2nd Report, Proposals of Empirical Expressions for the Local Heat Transfer Coefficient), Transactions of the JSME, Part B, 60 (574), pp. 2117-2124.

Hayashi, K., Kurimoto, R., and Tomiyama, A., 2010, Dimensional Analysis of Terminal Velocity of Taylor Bubble in a Vertical Pipe, Multiphase Science and Technology, 22 (3), pp. 197-210.

Hayashi, K., Kurimoto, R., and Tomiyama, A., 2011, Terminal Velocity of a Taylor Drop in a Vertical Pipe, International Journal of Multiphase Flow, 37 (3), pp. 241-251.

He, Q., Hasegawa, Y., and Kasagi, N., 2010, Heat Transfer Modelling of Gas–Liquid Slug Flow without Phase Change in a Micro Tube, International Journal of Heat and Fluid Flow, 31 (1), pp. 126–136.

Hemeida, A., and Sumait, F., 1988, Improving the Lockhart and Martinelli Two-Phase Flow Correlation by SAS, Journal of Engineering Sciences, King Saud University, 14 (2), pp. 423-435.

Hewitt, G. F., and Roberts, D. N., 1969, Studies of Two-Phase Flow Patterns by Simultaneous Flash and X-Ray Photography, AERE-M2159.

Hoogendoorn, C. J., 1959, Gas-Liquid Flow in Horizontal Pipes, Chemical Engineering Science, 9, pp. 205-217.

Howard, J. A., Walsh, P. A., and Walsh, E. J., 2011, Prandtl and Capillary Effects on Heat Transfer Performance within Laminar Liquid–Gas Slug Flows, International Journal of Heat and Mass Transfer, 54 (21-22), pp. 4752-4761.

Hu, H. -t., Ding, G. -l., and Wang, K. -j., 2008, Measurement and Correlation of Frictional Twophase Pressure Drop of R410A/POE Oil Mixture Flow Boiling in a 7 mm Straight Micro-Fin Tube, Applied Thermal Engineering, 28 (11-12), pp. 1272-1283.

Hu, H. -t., Ding, G. -l., Huang, X. –c., Deng, B., and Gao, Y. –f., 2009, Pressure Drop During Horizontal Flow Boiling of R410A/Oil Mixture in 5 mm and 3 mm Smooth Tubes, Applied Thermal Engineering, 29 (16), pp. 3353-3365.

Hu, X., and Jacobi, A. M., 1996, The Intertube Falling Film: Part 1—Flow Characteristics, Mode Transitions, and Hysteresis, ASME Journal of Heat Transfer, 118 (3), pp. 616-625.

Hulburt, E. T. and Newell, T. A., 1997, Modeling of the Evaporation and Condensation of Zeotropic Refrigerants Mixtures in Horizontal Annular Flow, ACRC TR-129, Air Conditioning and Refrigeration Center, University of Illinois at Urbana-Champaign.

Irandoust, S., and Andersson, B., 1989, Liquid Film in Taylor Flow through a Capillary, Industrial & Engineering Chemistry Research, 28 (11), pp. 1684-1688.

Ishii, M., 1987, Two-Fluid Model for Two Phase Flow, Multiphase Science and Technology, 5 (1), pp. 1-63.

Jayawardena, S. S., Balakotaiah, V., and Witte, L., 1997, Pattern Transition Maps for Microgravity Two-Phase Flow, AIChE Journal, 43 (6), pp. 1637-1640.

Kandlikar, S. G., 1990, A General Correlation for Saturated Two-Phase Flow Boiling Heat Transfer inside Horizontal and Vertical Tubes, ASME Journal of Heat Transfer 112, (1) pp. 219–228.

Kandlikar, S. G., 2001, A Theoretical Model to Predict Pool Boiling CHF Incorporating Effects of Contact Angle and Orientation, ASME Journal of Heat Transfer, 123 (12), pp. 1071–1079.

Kandlikar, S. G., 2004, Heat Transfer Mechanisms During Flow Boiling in Microchannels, ASME Journal of Heat Transfer, 126 (2), pp. 8-16.

Kandlikar, S. G., 2010a, Scale Effects on Flow Boiling Heat Transfer in Microchannels: A Fundamental Perspective, International Journal of Thermal Sciences, 49 (7), pp. 1073-1085.

Kandlikar, S. G., 2010b, A Scale Analysis Based Theoretical Force Balance Model for Critical Heat Flux (CHF) During Saturated Flow Boiling in Microchannels and Minichannels, ASME Journal of Heat Transfer, 132 (8), Article No. (081501).

Kandlikar, S. G., 2012, Closure to Discussion of `Heat Transfer Mechanisms During Flow Boiling in Microchannels (2012, ASME J. Heat Transfer, 134, p. 015501), ASME Journal of Heat Transfer, 134 (1), Article No. (015502).

Kandlikar, S. G., Garimella, S., Li, D., Colin, S., and King, M. R., 2006, Heat Transfer and Fluid Flow in Minichannels and Microchannels, Elsevier, Oxford, UK.

Kawahara, A., Sadatomi, M., Nei, K., and Matsuo, H., 2011, Characteristics of Two-Phase Flows in a Rectangular Microchannel with a T-Junction Type Gas-Liquid Mixer, Heat Transfer Engineering, 32 (7-8), pp. 585-594.

Keilin, V. E., Klimenko, E. Yu., and Kovalev, I. A., 1969, Device for Measuring Pressure Drop and Heat Transfer in Two-Phase Helium Flow, Cryogenics, 9 (2), pp. 36-38.

Kew, P., and Cornwell, K., 1997, Correlations for the Prediction of Boiling Heat Transfer in Small-Diameter Channels, Applied Thermal Engineering, 17 (8-10), pp. 705–715.

Kleinstreuer, C., 2003, Two-Phase Flow: Theory and Applications, Taylor & Francis, New York, NY.

Kreutzer, M. T., 2003. Hydrodynamics of Taylor Flow in Capillaries and Monoliths Channels, Doctoral dissertation. Delft University of Technology, Delft, The Netherlands.

Kreutzer, M. T., Kapteijn, F., Moulijin, J. A., and Heiszwolf, J. J., 2005a, Multiphase Monolith Reactors: Chemical Reaction Engineering of Segmented Flow in Microchannels, Chemical Engineering Science, 60 (22), pp. 5895–5916.

Kreutzer, M. T., Kapteijn, F., Moulijin, J. A., Kleijn, C. R., and Heiszwolf, J. J., 2005b, Inertial and Interfacial Effects on Pressure Drop of Taylor Flow in Capillaries, AIChE Journal, 51 (9), pp. 2428-2440.

Kutateladze, 1948, On the Transition to Film Boiling under Natural Convection, Kotloturbostroenie, 3, pp. 10–12.

Kutateladze, S. S., 1972, Elements of Hydrodynamics of Gas-Liquid Systems, Fluid Mechanics – Soviet Research, 1, pp. 29-50.

Lazarek, G. M., and Black, S. H., 1982, Evaporative Heat Transfer, Pressure Drop and Critical Heat Flux in a Small Vertical Tube with R-113, International Journal of Heat and Mass Transfer, 25 (7), pp. 945–960.

Lefebvre, A. H., 1989,. Atomization and Sprays. Hemisphere Publishing Corp., New York and Washington, D. C.

Li, W., and Wu, Z., 2010, A General Correlation for Adiabatic Two Phase Flow Pressure Drop in Micro/Mini-Channels, International Journal of Heat and Mass Transfer, 53 (13-14), pp. 2732-2739.

Li, W., and Wu, Z., 2011, Generalized Adiabatic Pressure Drop Correlations in Evaporative Micro/Mini-Channels, Experimental Thermal and Fluid Science, 35 (6,) pp. 866–872.

Lin, S., Kwok, C. C. K., Li, R. Y., Chen, Z. H., and Chen, Z. Y., 1991, Local Frictional Pressure Drop during Vaporization for R-12 through Capillary Tubes, International Journal of Multiphase Flow, 17 (1), pp. 95-102.

Liu, Y., Cui, J., and Li, W. Z., 2011, A Two-Phase, Two-Component Model for Vertical Upward Gas–Liquid Annular Flow, International Journal of Heat and Fluid Flow, 32 (4), pp. 796–804.

Lockhart, R. W., and Martinelli, R. C., 1949, Proposed Correlation of Data for Isothermal Two-Phase, Two-Component Flow in Pipes, Chemical Engineering Progress Symposium Series, 45 (1), pp. 39-48.

Lombardi, C., and Ceresa, I., 1978, A Generalized Pressure Drop Correlation in Two-Phase Flow, Energia Nucleare, 25 (4), pp. 181-198.

Lombardi, C., and Carsana, C. G., 1992, Dimensionless Pressure Drop Correlation for Two-Phase Mixtures Flowing Upflow in Vertical Ducts Covering Wide Parameter Ranges, Heat and Technology, 10 (1-2), pp. 125-141.

Lun, I., Calay, R. K., and Holdo, A. E., 1996, Modelling Two-Phase Flows Using CFD, Applied Energy, 53 (3), pp. 299-314.

Ma, X., Briggs, A., and Rose, J. W., 2004, Heat Transfer and Pressure Drop Characteristics for Condensation of R113 in a Vertical Micro-Finned Tube with Wire Insert, International Communications in Heat and Mass Transfer, 31, pp. 619-627.

Mandhane, J. M., Gregory, G. A., and Aziz, K., 1974, A Flow Pattern Map of Gas-Liquid Flow in Horizontal Pipes, International Journal of Multiphase Flow, 1 (4), pp. 537-553.

Marchessault, R. N., and Mason, S. G., 1960, Flow of Entrapped Bubbles through a Capillary, Industrial & Engineering Chemistry, 52 (1), pp. 79-84.

Martinelli, R. C., and Nelson, D. B., 1948, Prediction of Pressure Drop during Forced-Circulation Boiling of Water, Trans. ASME, 70 (6), pp. 695-702.

McAdams, W. H., Woods, W. K. and Heroman, L. C., 1942, Vaporization inside Horizontal Tubes. II -Benzene-Oil Mixtures, Trans. ASME, 64 (3), pp. 193-200.

Mishima, K., and Hibiki, T., 1996, Some Characteristics of Air-Water Two-Phase Flow in Small Diameter Vertical Tubes, International Journal of Multiphase Flow, 22 (4), pp. 703-712.

Moody, L. F., 1944, Friction Factors for Pipe Flow, Trans. ASME, 66 (8), pp. 671-677.

Mudawwar, I. A., and El-Masri, M. A., 1986, Momentum and Heat Transfer across Freely-Falling Turbulent Liquid Films, International Journal of Multiphase Flow 12 (5), pp. 771-790.

Muradoglu, M., Gunther, A., and Stone, H. A., 2007, A Computational Study of Axial Dispersion in Segmented Gas–Liquid Flow, Physics of Fluids, 19 (7), Article No. (072109).

Muzychka, Y. S., Walsh, E., Walsh, P., and Egan, V., 2011, Non-boiling Two Phase Flow in Microchannels, in Microfluidics and Nanofluidics Handbook: Chemistry, Physics, and Life Science Principles, Editors: Mitra, S. K., and Chakraborty, S., CRC Press Taylor & Francis Group, Boca Raton, FL.

Ohnesorge, W., 1936, Formation of Drops by Nozzles and the Breakup of Liquid Jets, Zeitschrift für Angewandte Mathematik und Mechanik (ZAMM) (Applied Mathematics and Mechanics) 16, pp. 355–358.

Oliemans, R., 1976, Two Phase Flow in Gas-Transmission Pipelines, ASME paper 76-Pet-25, presented at Petroleum Division ASME meeting, Mexico, September 19-24, 1976.

Ong, C. L., and Thome, J. R., 2011, Experimental Adiabatic Two-Phase Pressure Drops of R134a, R236fa and R245fa in Small Horizontal Circular Channels, Proceedings of the ASME/JSME 2011 8th Thermal Engineering Joint Conference (AJTEC2011), AJTEC2011-44010, March 13-17, 2011, Honolulu, Hawaii, USA.

Osman, E. A., 2004, Artificial Neural Network Models for Identifying Flow Regimes and Predicting Liquid Holdup in Horizontal Multiphase Flow, SPE Production and Facilities, 19 (1), pp. 33-40.

Osman, E. A., and Aggour, M. A., 2002, Artificial Neural Network Model for Accurate Prediction of Pressure Drop in Horizontal and Near-Horizontal-Multiphase Flow, Petroleum Science and Technology, 20 (1-2), pp. 1-15.

Ouyang, L., 1998, Single Phase and Multiphase Fluid Flow in Horizontal Wells, PhD Dissertation Thesis, Department of Petroleum Engineering, School of Earth Sciences, Stanford University, CA.

Owens, W. L., 1961, Two-Phase Pressure Gradient, ASME International Developments in Heat Transfer, Part II, pp. 363-368.

Petukhov, B. S., 1970, Heat Transfer and Friction in Turbulent Pipe Flow with Variable Physical Properties, Advances in Heat Transfer, 6, pp. 503-564.

Phan, H. T., Caney, N., Marty, P., Colasson, S., and Gavillet, J., 2011, Flow Boiling of Water in a Minichannel: The Effects of Surface Wettability on Two-Phase Pressure Drop, Applied Thermal Engineering, 31 (11-12), pp. 1894-1905.

Phillips, R. J., 1987, Forced Convection, Liquid Cooled, Microchannel Heat Sinks, Master's Thesis, Department of Mechanical Engineering, Massachusetts Institute of Technology, Cambridge, MA.

Pigford, R. L., 1941, Counter-Diffusion in a Wetted Wall Column, Ph. D. Dissertation, The University of Illinois/Urbana, IL.

Quan, S. P., 2011, Co-Current Flow Effects on a Rising Taylor Bubble, International Journal of Multiphase Flow, 37 (8), pp. 888–897.

Quiben, J. M., and Thome, J. R., 2007, Flow Pattern Based Two-Phase Frictional Pressure Drop Model for Horizontal Tubes. Part II: New Phenomenological Model, International Journal of Heat and Fluid Flow, 28 (5), pp. 1060-1072.

Renardy, M., Renardy, Y., and Li, J., 2001, Numerical Simulation of Moving Contact Line Problems Using a Volume-of-Fluid Method, Journal of Computational Physics, 171 (1), pp. 243-263.

Revellin, R., and Haberschill, P., 2009, Prediction of Frictional Pressure Drop During Flow Boiling of Refrigerants in Horizontal Tubes: Comparison to an Experimental Database, International Journal of Refrigeration, 32 (3) pp. 487–497.

Rezkallah, K. S., 1995, Recent Progress in the Studies of Two-Phase Flow at Microgravity Conditions, Journal of Advances in Space Research, 16, pp. 123-132.

Rezkallah, K. S., 1996, Weber Number Based Flow-Pattern Maps for Liquid-Gas Flows at Microgravity, International Journal of Multiphase Flow, 22 (6), pp. 1265–1270.

Rouhani S. Z., and Axelsson, E., 1970, Calculation of Volume Void Fraction in the Subcooled and Quality Region, International Journal of Heat and Mass Transfer, 13 (2), pp. 383-393.

Sabharwall, P., Utgikar, V., and Gunnerson, F., 2009, Dimensionless Numbers in Phase-Change Thermosyphon and Heat-Pipe Heat Exchangers, Nuclear Technology, 167 (2), pp. 325-332.

Saisorn, S., and Wongwises, S., 2008, Flow Pattern, Void Fraction and Pressure Drop of Two-Phase Air-Water Flow in a Horizontal Circular Micro-Channel, Experimental Thermal and Fluid Science, 32 (3), pp. 748-760.

Saisorn, S., and Wongwises, S., 2009, An Experimental Investigation of Two-Phase Air-Water Flow Through a Horizontal Circular Micro-Channel, Experimental Thermal and Fluid Science, 33 (2), pp. 306-315.

Saisorn, S., and Wongwises, S., 2010, The Effects of Channel Diameter on Flow Pattern, Void Fraction and Pressure Drop of Two-Phase Air-Water Flow in Circular Micro-Channels, Experimental Thermal and Fluid Science, 34 (4), pp. 454-462.

Salman, W., Gavriilidis, A., and Angeli, P., 2004, A Model for Predicting Axial Mixing During Gas–Liquid Taylor Flow in Microchannels at Low Bodenstein Numbers, Chemical Engineering Journal, 101 (1-3), pp. 391-396.

Sardesai, R. G., Owen, R. G., and Pulling, D. J., 1981, Flow Regimes for Condensation of a Vapour Inside a Horizontal Tube, Chemical Engineering Science, 36 (7), pp. 1173-1180.

Scott, D. S., 1964, Properties of Co-Current Gas- Liquid Flow, Advances in Chemical Engineering, 4, pp. 199-277.

Selli, M. F., and Seleghim, P., Jr., 2007, Online Identification of Horizontal Two-Phase Flow Regimes Through Gabor Transform and Neural Network Processing, Heat Transfer Engineering, 28 (6), pp. 541–548.

Shah, M. M., 1982, Chart Correlation for Saturated Boiling Heat Transfer: Equations and Further Study, ASHRAE Trans., 88, Part I, pp. 185-196.

Shannak, B., 2009, Dimensionless Numbers for Two-Phase and multiphase flow. In: International Conference on Applications and Design in Mechanical Engineering (ICADME), Penang, Malaysia, 11–13 October, 2009.

Sherwood, T. K., Pigford, R. L., and Wilke, C. R., 1975, Mass Transfer, Mc-Graw Hill, New York, NY, USA.

Shippen, M. E., and Scott, S. L., 2004, A Neural Network Model for Prediction of Liquid Holdup in Two-Phase Horizontal Flow, SPE Production and Facilities, 19 (2), pp 67-76.

Sobieszuk, P., Cygański, P., and Pohorecki, R., 2010, Bubble Lengths in the Gas–Liquid Taylor Flow in Microchannels, Chemical Engineering Research and Design, 88 (3), pp. 263-269.

Spelt, P. D. M., 2005, A Level-Set Approach for Simulations of Flows with Multiple Moving Contact Lines with Hysteresis, Journal of Computational Physics, 207 (2), pp. 389-404.

Stephan, K., and Abdelsalam, M., 1980, Heat Transfer Correlation for Natural Convection Boiling, International Journal of Heat and Mass Transfer, 23 (1), pp. 73–87.

Suo, M., and Griffith, P., 1964, Two Phase Flow in Capillary Tubes, ASME Journal of Basic Engineering, 86 (3), pp. 576–582.

Swamee, P. K., and Jain, A. K., 1976, Explicit Equations for Pipe Flow Problems, Journal of the Hydraulics Divsion - ASCE, 102 (5), pp. 657-664.

Taha, T., and Cui, Z. F., 2006a, CFD Modelling of Slug Flow inside Square Capillaries, Chemical Engineering Science 61 (2), pp. 665-675.

Taha, T., and Cui, Z. F., 2006b, CFD Modelling of Slug Flow in Vertical Tubes, Chemical Engineering Science, 61 (2), pp. 676-687.

Taitel, Y., 1990, Flow Pattern Transition in Two-Phase Flow, Proceedings of 9th International Heat Transfer Conference (IHTC9), Jerusalem, Vol. 1, pp. 237-254.

Taitel, Y., and Dukler, A. E., 1976, A Model for Predicting Flow Regime Transitions in Horizontal and Near Horizontal Gas–Liquid Flow, AIChE Journal, 22 (1), pp. 47–55.

Talimi, V., Muzychka, Y. S., and Kocabiyik, S., 2012, A Review on Numerical Studies of Slug Flow Hydrodynamics and Heat Transfer in Microtubes and Microchannels, International Journal of Multiphase Flow, 39, pp. 88-104.

Tandon, T. N., Varma, H. K., and Gupta. C. P., 1982, A New Flow Regime Map for Condensation Inside Horizontal Tubes, ASME Journal of Heat Transfer, 104 (4), pp. 763-768.

Tandon, T. N., Varma, H. K., and Gupta. C. P., 1985, Prediction of Flow Patterns During Condensation of Binary Mixtures in a Horizontal Tube, ASME Journal of Heat Transfer, 107 (2), pp. 424-430.

Taylor, G. I., 1932, The Viscosity of a Fluid Containing Small Drops of Another Fluid, Proceedings of the Royal Society of London, Series A, 138 (834), pp. 41-48.

Taylor, G. I., 1961, Deposition of a Viscous Fluid on the Wall of a Tube, Journal of Fluid Mechanics, 10 (2), pp. 161-165.

Thome, J. R., 2003, On Recent Advances in Modeling of Two-Phase Flow and Heat Transfer, Heat Transfer Engineering, 24 (6), pp. 46–59.

Tran, T. N., Wambsganss, M. W., and France, D. M., 1996, Small Circular- and Rectangular-Channel Boiling with Two Refrigerants, International Journal of Multiphase Flow, 22 (3), pp. 485-498.

Tribbe, C., and Müller-Steinhagen, H. M., 2000, An Evaluation of the Performance of Phenomenological Models for Predicting Pressure Gradient during Gas-Liquid Flow in Horizontal Pipelines, International Journal of Multiphase Flow, 26 (6), pp. 1019-1036.

Triplett, K. A., Ghiaasiaan, S. M., Abdel-Khalik, S. I., and Sadowski, D. L., 1999, Gas-Liquid Two-Phase Flow in Micro-Channels, Part 1: Two-Phase Flow Pattern, International Journal of Multiphase Flow 25 (3), pp. 377-394.

Turner, J. M., 1966, Annular Two-Phase Flow, Ph.D. Thesis, Dartmouth College, Hanover, NH, USA.

Ullmann, A., and Brauner, N., 2007, The Prediction of Flow Pattern Maps in Mini Channels, Multiphase Science and Technology, 19 (1), pp. 49-73.

van Baten, J. M., and Krishna, R., 2004, CFD Simulations of Mass Transfer from Taylor Bubbles Rising in Circular Capillaries, Chemical Engineering Science, 59 (12), pp. 2535–2545.

Vandervort, C. L., Bergles, A. E., and Jensen, M. K., 1994, An Experimental Study of Critical Heat Flux in very High Heat Flux Subcooled Boiling, International Journal of Heat and Mass Transfer, 37 (Supplement 1), pp. 161-173.

Venkatesan, M., Das, Sarit K., and Balakrishnan, A. R., 2011, Effect of Diameter on Two-Phase Pressure Drop in Narrow Tubes, Experimental Thermal and Fluid Science, 35 (3), pp. 531-541.

Wallis, G. B., 1969, One-Dimensional Two-Phase Flow, McGraw-Hill Book Company, New York.

Walsh, E. J., Muzychka, Y. S., Walsh, P. A., Egan, V., and Punch, J., 2009, Pressure Drop in Two Phase Slug/Bubble Flows in Mini Scale Capillaries, International Journal of Multiphase Flows, 35 (10), pp. 879-884.

Wang X. -Q., and Mujumdar A. S., 2008a, A Review on Nanofluids - Part I: Theoretical and Numerical Investigations, Brazilian Journal of Chemical Engineering, 25 (4), pp. 613-630.

Wang X. -Q., and Mujumdar A. S., 2008b, A Review on Nanofluids - Part II: Experiments and Applications, Brazilian Journal of Chemical Engineering, 25 (4), pp. 631-648.

Wei, W., Ding, G., Hu, H., and Wang, K., 2007, Influence of Lubricant Oil on Heat Transfer Performance of Refrigerant Flow Boiling inside Small Diameter Tubes. Part I: Experimental Study, Experimental Thermal and Fluid Science, 32 (1), pp. 67–76.

Weisman, J., Duncan, D., Gibson, J., and Crawford, T., 1979, Effects of Fluid Properties and Pipe Diameter on Two-Phase Flow Patterns in Horizontal Lines, International Journal of Multiphase Flow 5 (6), pp. 437-462.

Whalley, P. B., 1987, Boiling, Condensation, and Gas-Liquid Flow, Clarendon Press, Oxford.

Whalley, P. B., 1996, Two-Phase Flow and Heat Transfer, Oxford University Press, UK.

White, F. M., 2005, Viscous Fluid Flow, 3rd edition, McGraw-Hill Book Co, USA.

Wilson, M. J., Newell, T. A., Chato, J. C., and Infante Ferreira, C. A., 2003, Refrigerant Charge, Pressure Drop and Condensation Heat Transfer in Flattened Tubes, International Journal of Refrigeration, 26 (4), pp. 442–451.

Yan, Y.-Y., and Lin, T.-F., 1998, Evaporation Heat Transfer and Pressure Drop of Refrigerant R134a in a Small Pipe, International Journal of Heat and Mass Transfer, 41 (24). pp. 4183–4193.

Yang, C., Wu, Y., Yuan, X., and Ma, C., 2000, Study on Bubble Dynamics for Pool Nucleate, International Journal of Heat and Mass Transfer, 43 (18), pp. 203-208.

Yarin, L. P., Mosyak, A., and Hetsroni, G., 2009, Fluid Flow, Heat Transfer and Boiling in Micro-Channels, Springer, Berlin.

Yun, J., Lei, Q., Zhang, S., Shen, S., and Yao, K., 2010, Slug Flow Characteristics of Gas–Miscible Liquids in a Rectangular Microchannel with Cross and T-Shaped Junctions, Chemical Engineering Science, 65 (18), pp. 5256–5263.

Zhang, W., Hibiki, T., and Mishima, K., 2010, Correlations of Two-Phase Frictional Pressure Drop and Void Fraction in Mini-Channel, International Journal of Heat and Mass Transfer, 53 (1-3), pp. 453-465.

Zhao, L., and Rezkallah, K. S., 1993, Gas–Liquid Flow Patterns at Microgravity Conditions, International Journal of Multiphase Flow, 19 (5), pp. 751–763.

Zivi, S. M., 1964, Estimation of Steady-State Void Fraction by Means of the Principle of Minimum Energy Production, ASME Journal of Heat Transfer, 86 (2), pp. 247-252.

Chapter 12

Heat Generation and Removal in Solid State Lasers

V. Ashoori, M. Shayganmanesh and S. Radmard

Additional information is available at the end of the chapter

http://dx.doi.org/10.5772/2623

1. Introduction

Based on the type of laser gain medium, lasers are mostly divided into four categories; gaseous, liquid (such as dye lasers), semiconductor, and solid state lasers. In recent past years, solid state lasers have been attracted considerable attentions in industry and scientific researches to achieve the high power laser devices with good beam quality. In solid state lasers the gain medium might be a crystal or a glass which is doped with rare earth or transition metal ions. These lasers can be made in the form of bulk [1, 2], fiber [3-7], disk [8, 9] and Microchip lasers [10,11].

Optical pumping is associated with the heat generation in solid state laser materials [12]. Moving of heat toward the surrounding medium which is mostly designed for the cooling management causes thermal gradient inside the medium [13]. This is the main reason of appearance of unwanted thermal effects on laser operation. Thermal lensing [14], thermal stress fracture limit [15], thermal birefringence and thus thermal bifocusing [16-19] are some examples of thermal effects.

Optimizing the laser operation in presence of thermal effects needs to have temperature distribution inside the gain medium. Solving the heat differential equation beside considering boundary conditions gives the temperature field.

Boundary conditions are directly related to the cooling methods which lead to convective or conductive heat transfer from gain to the surrounding medium.

In the case of bulky solid state laser systems (such as rod shape gain medium), water cooling is the most common method which is almost used in high power regime. Design of optimized cooling cavity to achieve the most effective heat transfer is the first step to scale up high power laser devices. Then determination of temperature distribution inside the laser gain medium is essential for evaluation of induced thermo-optic effects on laser operation.

This chapter is organized according to the requirements of reader with thermal considerations in solid state laser which are mainly dependent on several kinds of laser materials, pumping procedure and cooling system. We hope the subjects included in this chapter will be interesting for two guilds of scientific and engineering researchers. The first category relates to the laser scientists, who need enough information about the recent cooling methods, their benefits and disadvantages, thermal management and effects of utilizing a specific cooling method on laser operation. And the second one is the mechanical or opto-mechanical engineers who are responsible for designing and manufacturing of the cooling systems. In this chapter our efforts directed such a way to satisfy both the mentioned categories of researchers.

At First, the principle of heat generation process inside the laser gain medium due to the optical pumping are introduced. Then, heat differential equation in laser gain medium and relating boundary conditions are introduced in detail. Formulation of heat problem for a specific form of gain medium such as bulk, disk, fiber and Microchip lasers and details of solution are presented through individual subsections.

2. Principle of heat generation in solid state laser gain medium

2.1. Laser pumping

A laser device is composed of three essential components which are the "active medium", "pumping source" and the "optical resonator". In the case of solid state lasers, the active medium which is made of a definite glass or crystal, is placed inside the optical resonator and receives energy from another external optical source through the pump beam light. Then it can itself emit an amplified laser beam light delivering a completely modified energy and wavelength [20]. The act of energy transfer from the external source to the active medium is called the laser pumping.

In recent years, diode lasers [21-22] have attracted considerable attention between laser scientists as available, high power and beam quality pumping sources. In this chapter we just concentrate on this kind of pumping sources rather than traditional flash lamp pump sources [15].

The pumping process commonly performs in two methods, which are continuous wave (CW) and pulse pumping laser systems. Furthermore, diode pumped solid state lasers (DPSS lasers) can be divided into side pumped and end pumped configurations. Figure 1, Shows schematically typical solid state lasers, including gain medium, optical resonator and diode pumping in the case of end and side configurations. In the end-pumped geometry, the pump light mostly transfers from diode Laser (DL) to the laser material through either optical system or fiber optics which yields a desirable pump beam shape and size. Then it focuses to the gain medium longitudinally, collinear to the propagation of laser light. In the side-pumped geometry, the diode arrays locate along the laser material in a definite arrangement around it, such that the pump light is perpendicular to the propagation of laser light.

The pumping geometry and the resultant pump characteristics (such as beam shape and size along the gain medium) play an important role in heat generation and therefore thermal gradient inside the gain medium. The issue will review in details in the following sub-sections for each kind of laser gain mediums.

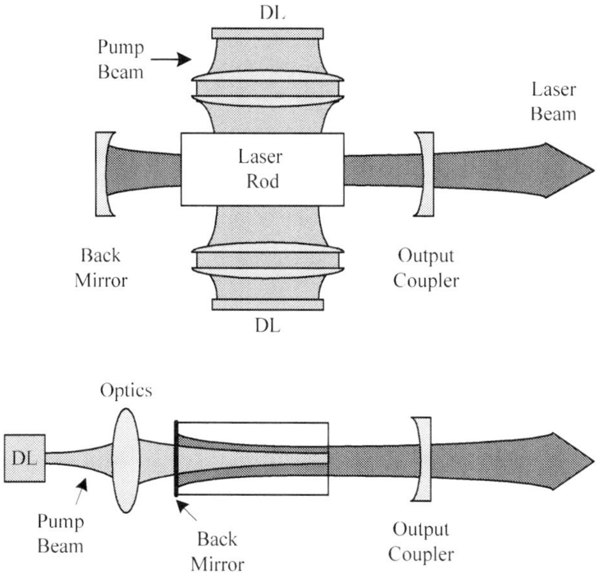

Figure 1. Simple drawing for two common pumping methods of solid state laser gain medium; a) side pumping b) end pumping. The blue color used for marking the pump light carrying energy from laser diode to the gain medium, and the red beam concerns to the laser resonator mode.

2.2. Heat generation

In solid state lasers, a fraction of the pump energy converts to heat which acts as the heat source inside the laser material [23, 24]. Spatial and time dependence of the heat source causes important effects on temperature distribution and warming rate of the gain medium, respectively. The spatial form is assumed to be the same shape as pumping light [23, 24] and time dependence relates to the pumping procedure, which may perform in CW or pulsed regime. Furthermore, depending on the gain medium configuration and cooling geometry, deposited heat may mostly flow through a preferable direction inside the gain medium and therefore causes thermal gradient. For instance, in traditional rod shape laser mediums with water cooling configuration and also fiber lasers, the main proportion of heat removal occurs through the radial direction which leads to the considerable radial thermal gradient inside the medium. Figure 2 shows a schematic setup of pumping procedure for several kinds of solid state lasers. The dominant directions of heat removal which are associated with the gain medium and cooling system geometry are illustrated.

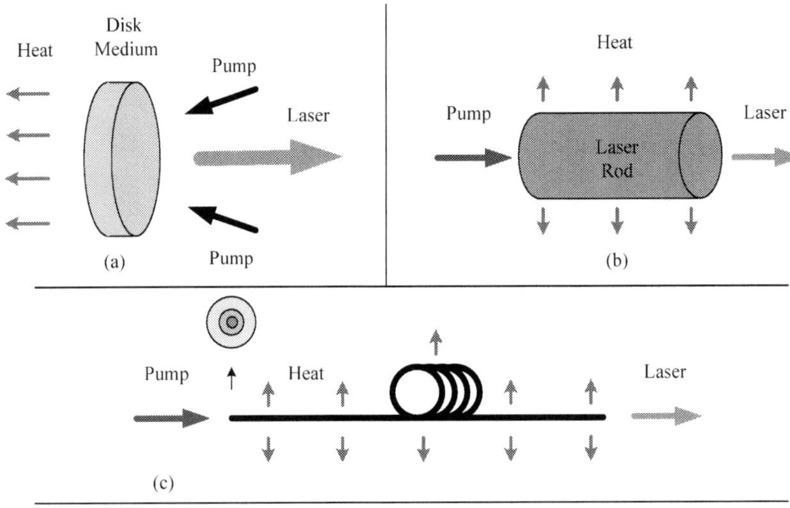

Figure 2. Schematic figure of preferable direction of heat transfer in three common types of solid state lasers; a) disk laser, b) rod, and c) fiber laser

Heat differential equation should be solved for evaluation of temperature and thermal gradient induced by optical pumping in solid state lasers. The general form of heat differential equation in cylindrical coordinate system is given by [25]

$$\frac{1}{r}\frac{\partial}{\partial r}\left(r\frac{\partial T(r,\varphi,z,t)}{\partial r}\right)+\frac{1}{r^2}\frac{\partial^2 T(r,\varphi,z,t)}{\partial \varphi^2}+\frac{\partial^2 T(r,\varphi,z,t)}{\partial z^2}+\frac{Q(r,\varphi,z,t)}{k_c}=\frac{\rho c}{k_c}\frac{\partial T(r,\varphi,z,t)}{\partial t} \tag{1}$$

Where, $Q(r,\varphi,z,t)$ denotes heat source density (W/m^3), k_c is thermal conductivity, ρ and c are the density (Kg/m^3) and specific heat ($J/Kg.^{o}C$) of laser gain medium, respectively. Eq.1 denotes to the transient heat differential equation and can specify the time dependence temperature in the case of pulsed pumped laser systems. As we mentioned before, $Q(r,\varphi,z,t)$ can be determined according to the pumping characteristics in several kinds of solid state lasers.

The overall thermal load in the laser gain medium due to the optical pumping can be obtained from

$$P_h=\int_v Q(r,z)dv=\xi P_o \tag{2}$$

In which, P_o ξ is the pump power and is the fractional thermal load [12]. In the case of diode pumping, the fractional thermal load, originates from two basic phenomenon, which show the main role in heat generation; quantum defect heating [15] and energy transfer upconversion (ETU) [26]. In most cases, the first is responsible for the heat generation and therefore has the main contribution. However it must be noted that, influence of the second phenomena cannot be ignored in some cases such as Er doped laser materials. The fractional thermal load in the gain medium is due to the quantum defect and related to the pumping and laser wavelength which are shown by λ_p and λ_L respectively.

$$\xi = 1 - \frac{\lambda_p}{\lambda_L} \tag{3}$$

2.3. Bulk solid state lasers

2.3.1. Temperature distribution

2.3.2. Side pumping

Pumping configuration performing by one module is illustrated in Figure3.a [27]. The pump beam emitted by diode laser is focused on the rod by the interfacing optics which consist of two lenses. End view of side-pumping geometry is depicted in Figure 3.b. Transversal directions of pump and signal beams are easily observable.

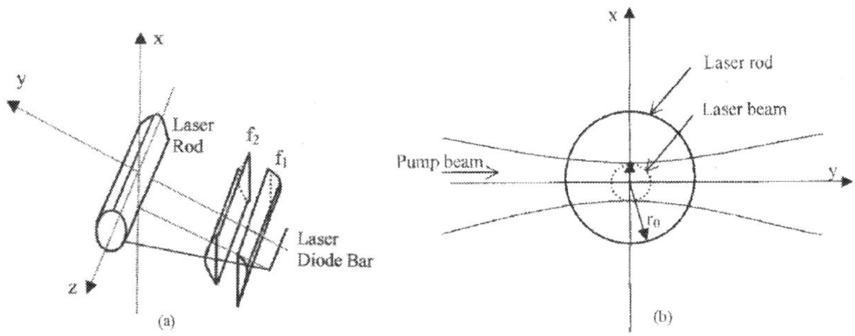

Figure 3. side pumping geometry; a) Pumping of a laser rod by one module, b) end view of side pumping geometry [27].

In the case of CW pumping of a laser rod, the steady state heat equation can be written as [27]

$$\nabla . h(r,z) = Q(r,z) \tag{4}$$

Where, h is heat flux and associated with the temperature in the rod by

$$h(r,z) = -k_c \nabla T(r,z), \tag{5}$$

And $Q(r,z)$ is determined according to spatial variation of pump intensity and is given by

$$Q(r,z) = I_0 \exp\left(\frac{-2r^2}{\omega_p^2}\right) \tag{6}$$

In which, I_0 is the heat irradiance on axis. Integration of Eq.4 over rod cross section yields

$$h(r)=\frac{\omega_p^2 I_0}{4}\frac{1-\exp\left(\frac{-2r^2}{\omega_p^2}\right)}{r} \tag{7}$$

Substituting Eq.7 into Eq. 5 and integrating to the rod radius r_0 gives the temperature difference inside the rod

$$\Delta T(r)=\frac{I_0\omega_p^2}{8k_c}\left[\ln\left(\frac{r_0^2}{r^2}\right)+E_1\left(\frac{2r_0^2}{\omega_p^2}\right)-E_1\left(\frac{2r^2}{\omega_p^2}\right)\right] \tag{8}$$

Where $\Delta T(r)=T(r)-T(r_0)$ and E_1 is the exponential integral function [27].

2.3.3. *End pumping*

One of the common pumping configurations which are used in diode-pumped solid state lasers is end-pumping or longitudinal pump scheme. The pumping beam is coaxial with the resonator beam in end-pumped lasers; it leads to highly efficient lasers with good beam quality.In this geometry, the pumping beam of diode laser(s) is delivered to the end of the active medium by optical focusing lenses or optical fibers. In lower power operations (less than a few watts), end-pumping yields more acceptable results [15]. Today's development of diode lasers and new techniques such as using micro-lenses in beam shaping of diode laser bars make end-pumped lasers very promising specially in commercial lasers [28]. Although end-pumping is a common configuration in solid state lasers which include many types of active medium geometries such as slab and microchip, but it is more commonly used in rod shape lasers. Many end-pumping investigations and results have reported about rod lasers and most of commercial end pump systems involve rod lasers [29-33]. Thus, the discussions in this section are focused on end-pumped rod lasers. Schematic diagram of an end-pumped system is shown in figure 4.

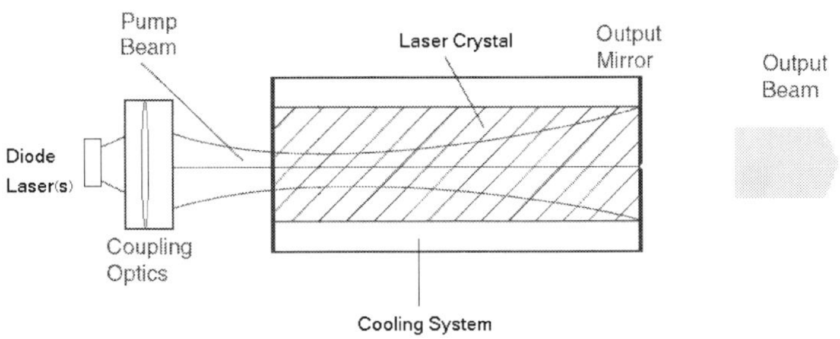

Figure 4. End pump systems major elements [15]

In order to establish high matching efficiency between resonator beam and pumping modes, in end pumped systems the pumping beam is focused in the active medium with the small beam waist. This issue leads to generation of an intense local heating and then, creation of refracting index gradient inside the laser crystal [15]. As a consequence, the laser rod acts as a thermal lens inside the resonator, which can destroy the beam quality and decrease the output efficiency. Additionally, in contrast to side pumped lasers, heat distribution within the laser material is inhomogeneous in high-power end-pumped lasers, leading to increased in stress and strain [33]. The restricting factors in end-pumped lasers in high power regime are thermal lensing and thermal fracture limit for the laser crystal. The aforementioned restrictions make thermal problems very important in end-pumped lasers especially in high-power systems.

From the thermal point of view, the flat top pumping profile is superior to Gaussian profile in high power end-pumped systems, due to the creation of lower thermal gradient inside the laser crystal leading to lower thermal distortions [28]. However, Gaussian pumping profiles is more investigated because of practical considerations in laser resonator design. One of the earliest thermal analysis in end pumped systems is presented in [15], which relates to the solution of steady state heat differential equation for Nd:YAG crystal with the assumption of Gaussian pumping profile. (Figure 5).

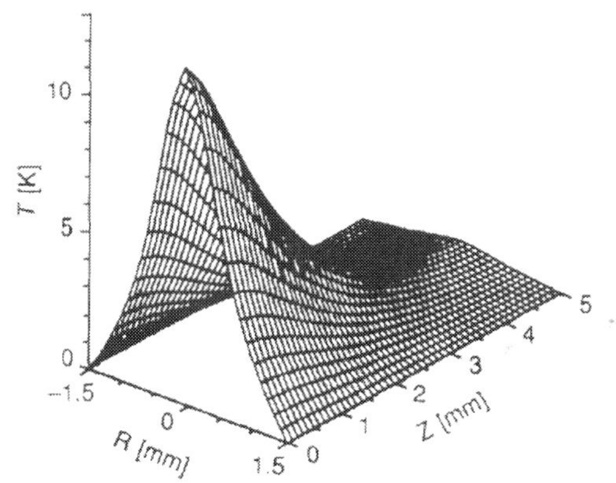

Figure 5. Temperature distribution in an end pumped Nd:YAG laser in the case of end pumping configuration. The pumping and laser beams propagate in z direction of cylindrical coordinate [15].

The temperature profile and thus associated thermal effects in bulky solid state lasers had been the subject of consideration in past years by various authors. In the case of CW pumping, a collection of excellent literatures discussing numerical and analytical thermal analysis can be found in [13, 23, 34-39]. Similar works concerning the transient heat analysis are available in [40].

A famous work which presents analytical expressions for temperature and describes the behavior of temperature inside the rod belongs to Innocenzi [13]. In this work, the laser rod is surrounded by a copper heat sink and exposed to the pump beam with the Gaussian intensity profile as

$$I = \frac{2P_h}{\pi\omega_p}\exp\left(\frac{-2r^2}{\omega_p^2}\right)\exp(-\alpha z) \tag{9}$$

Neglecting axial heat flux, the steady state heat differential equation can be solved analytically. The derived temperature difference is obtained as

$$\Delta T(r,z) = \frac{\alpha P_h \exp(-\alpha z)}{4\pi k_c}\left[\ln\left(\frac{r_0^2}{r^2}\right) + E_1\left(\frac{2r_0^2}{\omega_p^2}\right) - E_1\left(\frac{2r^2}{\omega_p^2}\right)\right] \tag{10}$$

As would be expected, the temperature decays exponentially along the laser rod and have the highest temperature on the pumping surface (z=0).

A common cooling method for end-pumped systems concerns to utilize of water jacket or copper tube surrounding the laser rod and keep the cylindrical surface at the definite temperature. Heat generated inside the gain medium flows to this surface through the heat conduction process in radial direction. Although this method is considered as a simple efficient technic providing considerable heat removal from gain medium, but the uncooled pumped surface of the rod which is in direct contact with air, performs very week heat transfer. This issue may cause undesirable effects, especially in high power regimes [34]. The thermal load on this surface not only increases thermal lensing effects but also restricts the maximum pumping power because of the fracture limit of the crystal, and damage threshold of optical coatings.

One of the effective strategies to reduce thermal effects is based on the cooling of pumping surface of laser rod. In this respect, three technics to achieve more efficient cooling process have been presented in [33], which are schematically shown in figure 6. In the first method (b), the cooling water is directly in contact with the pumping surface. In the second method (c), a cooling plate cooled by water is mounted in close contact with the rod pumping surface. The cold plate should have large Yang's module and high heat conductivity. The other method utilizes an un-doped cap on the pumping surface (d). The pumping power does not absorb in the undoped cap so there is not heat load in this region, but using this cap increases the effective cooling surface as well as the ratio of cooling surface to heat generation volume.

The influence of thermal effects on laser gain medium in the mentioned methods has been analyzed using FE method [33]. The maximum temperature decreased almost 30%, and 25% using an undoped cap and a sapphire cooling plate in pumping surface respectively. The maximum stress occurred in the configuration with water cooled pumped surface and reduced to half of uncooled system value using an undoped cap or cooling plate. According to recently developed ceramic laser materials using composite rods with undoped cap could be very promising as the best choice for high power end pumped systems. The undoped

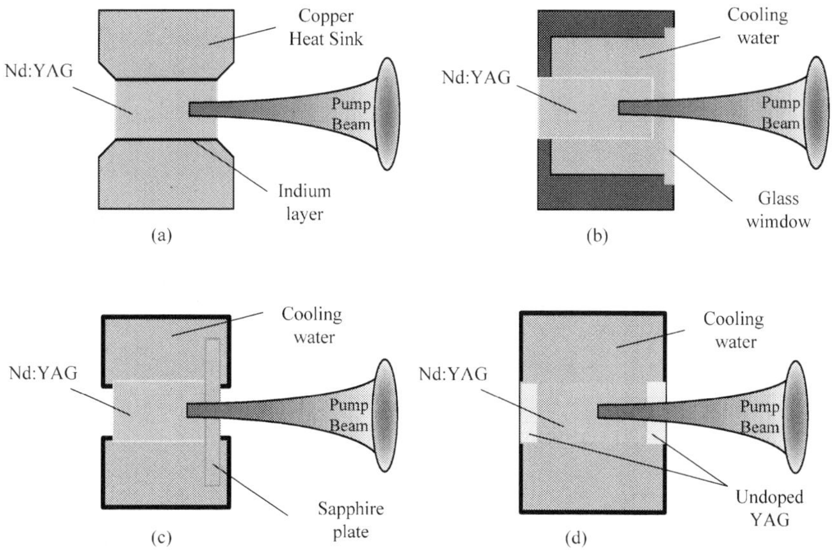

Figure 6. Different cooling configurations in end pumped systems. In these geometries the pumped surface (a) uncooled, (b) water cooled, (c) sapphire cold plate cooled, (d) undoped cap rod [33]. The figures are repainted in color version just for better realization.

end cap considerably lowers the thermal stress in the entrance facet of an end-pumped laser. This not only reduces the thermal lensing effects and thermal stresses but also lowers the maximum temperature of the laser rod and so removes some constraints imposed on the coatings. Prominent role of composite rod in reduction of thermal destructive effects on laser operation have been frequently examined and reflected in literatures [41-43]. Figure 7 illustrates the pump model of the dual-end- pumped geometry of composite $Nd:YVO_4$

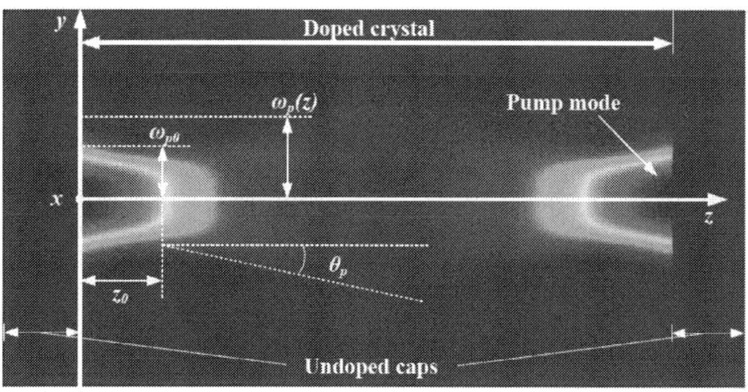

Figure 7. Pump modeling of the dual-end-pumped geometry [44].

laser[44]. The Nd:YVO4 as the laser gain medium is connected to two YVO4 caps at two ends and the pump energy lunches to it from both ends. As can be seen, every point inside the rod absorbs pump power and experiences heat generation. No absorption inside the caps takes place and therefore, they can show important role in axial heat transfer from end surfaces. Numerical calculations of temperature distribution in composite laser rod can be found in [45]

In the case of pulsed pumping laser rod, interesting numerical analysis has been done by Wang published in [46]. In this work the laser rod is surrounded by a cylindrical heat sink which leads to conductive heat transfer from rod surface to the ambient medium. Schematic figure of the rod and cooling system geometry is depicted in Figure 8-a.

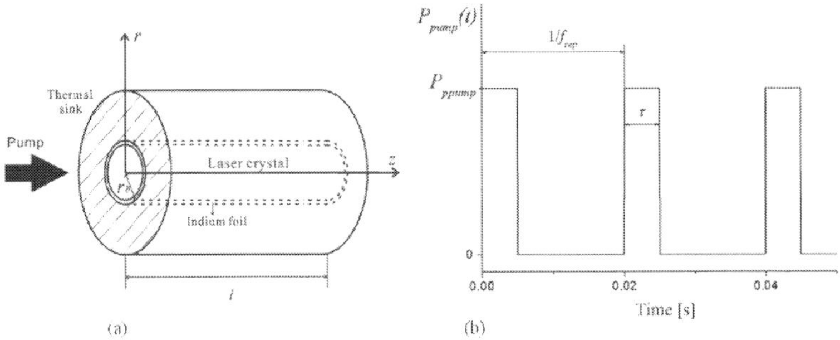

Figure 8. a) simple drawing of laser rod which is surrounded by a cylindrical heat sink, b) time variation of pump power [46]

The laser rod is assumed to be coupled by a fiber-optic to a laser diode; therefore the pump intensity profile with a good approximation has the top-hat shape. Thus the heat source density can be described by

$$Q(r,z,t)=\begin{cases}\dfrac{\xi P(t)}{\pi\omega_p^2}\alpha e^{-\alpha z}, & r\le\omega_p\\ 0, & r>\omega_p\end{cases} \tag{11}$$

Where P(t) is a periodic function of time describing pump power in a repetitively pumping regime given by

$$P(t)=\begin{cases}P_o, & 0\le t\le\tau\\ 0, & \tau\le r\le 1/f_{rep}\end{cases}$$
$$P(t+1/f_{rep})=P(t) \tag{12}$$

In which P_o is the pump peak power, f_{rep} is the repetition frequency, and τ is the pump duration time. Figure 8.b tries to simply specify the P(t). Detailed information of solving the transient heat differential equation can be found in [46].

Figure 9 shows the temperature distribution inside a Nd:YAG laser rod exposing to the pump beam with pulse duration of 300μs, and repetition frequency of 50 Hz. According to the results, the temperature at first increases by passing the time and then becomes nearly constant with small fluctuations which lead the temperature to the steady state condition.

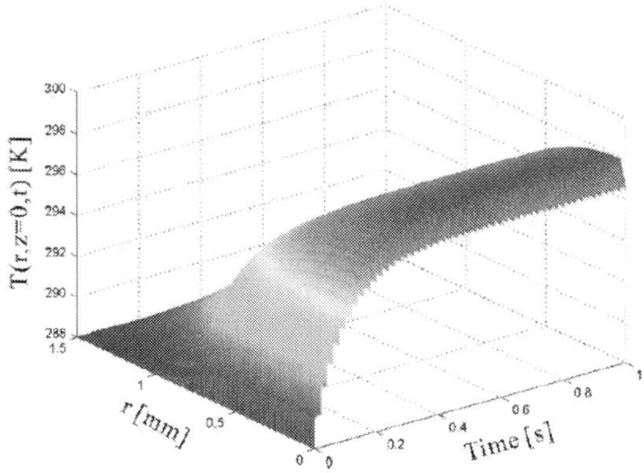

Figure 9. a) Time dependent temperature distribution in laser rod at position z = 0 at 50 Hz repetition rate [46].

2.4. Thin disk lasers

Thin disk lasers are one of the recent frontiers in solid state lasers. The most important features which make the thin disk laser distinguishable between solid state lasers are power scalability, good beam quality and minimal thermal lensing [47,48]. These features are related to the thermal characteristics of the thin disk laser. In disk lasers, active medium is cooled from the disk face. The surface to volume ratio of the disk is large due to the disk geometry, therefore cooling is very efficient and as a result thermal distortion of the active medium is very low. Considering axial heat flow in a thin disk laser there is no thermal lensing in a first-order approximation. In fact, however weak thermal lensing occurs because of two residual effects: first, pumped diameter is typically smaller than the diameter of the crystal and second, thermo-mechanical contribution to the thermal lensing from bending of the disk due to thermal expansion [49].

Thermal lensing is important issue in laser design and operation. This factor can be calculated theoretically using thermo-mechanical modeling softwares. In thin disk lasers, the disk is mounted with a cold plate on a heat sink (figure 10). At the same time the other side of disk is radiated by pumping laser, accordingly there is a temperature difference between two faces of the disk. This will cause a temperature distribution through the disk bulk. Generally the refractive index of materials is depended on temperature; accordingly the refractive index of disk will be a function of position. The other effect is the expansion of the disk due to the temperature distribution formed in it. Also mounting the disk on heat sink causes deformation and stress in the disk. The stress itself will affect the refractive index of the disk crystal. To complete analysis of the effects of the disk on the laser and pump beam, one must calculate cumulative effects of expansion and deformation of the disk, also thermo-optical and stress dependent variations of refractive index [50]. Total effect of the disk on laser beam phase can be calculated by [51]:

$$\Phi(r) = 2[\int_0^h [n_0 + \frac{\partial n}{\partial T}(T(r,z) - T_0) + \Delta n_s(r,z) - 1].[1 + \varepsilon_z(r,z)]dz - z_0(r)] \quad (13)$$

In which n_0 is refractive index of disk at reference temperature T_0, $\partial n/\partial T$ is the thermo-optical coefficient, Δn_s is the changes of refractive index due to stresses, ε_z is the strain in direction of thickness of the disk, z_0 is the displacement of back side (High Reflection coated) of the disk and h is the thickness of the disk. As relation (1) shows, optical behavior of the disk is strongly depended to the temperature distribution of disk. The temperature distribution is result of optical pumping and cooling of the disk.

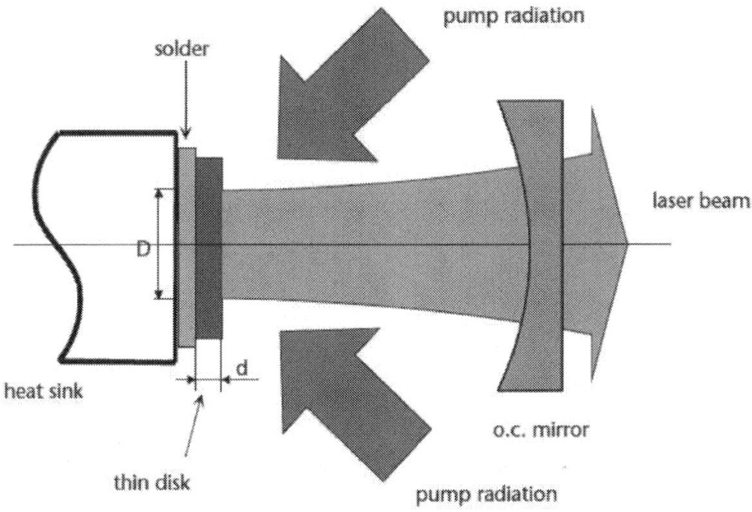

Figure 10. Schematic setup of a thin disk laser; end pump configuration [52]

2.4.1. *Pump and cooling configurations*

There are two conventional methods for pumping disk lasers; first is (quasi) end pumping and the second is edge pumping. Schematic diagram of end pumping is shown in figure 10. Also figure 11 shows a schematic setup of the edge pumped thin disk laser.

In both mentioned methods, the disk is cooled from the face. The disk can be cooled by jet impingement (figure 12) or cryogenic technique. The disk is mounted with a cold plate on the heat sink. In jet impingement a jet of cooling liquid is sprayed to the cold plate. Different liquids can be considered as coolant which most common is water.

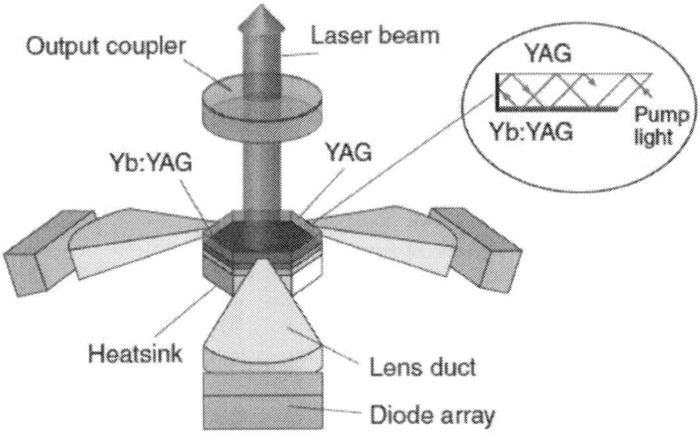

Figure 11. Schematic setup of edge pumped thin disk laser [53]

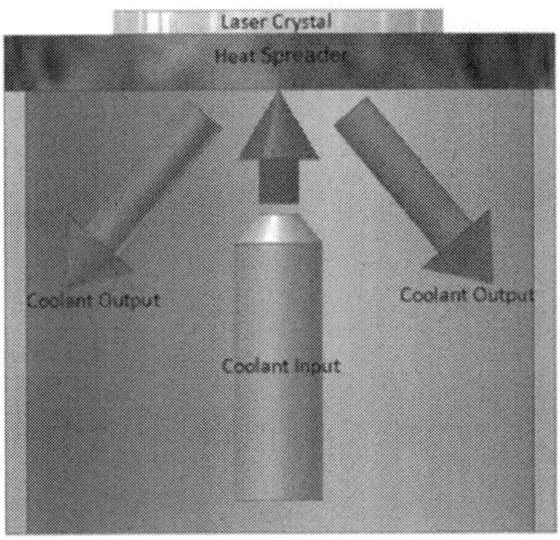

Figure 12. Schematic diagram of jet impingement cooling system for thin disk laser [54]

Laser cooling has been an important problem from the invention of the first practical laser in 1960. After invention of laser, cryogenic cooling of solid state lasers has been interested and first time proposed by Bowness [55] in 1963 and then by McMahon [56] in 1969. In mentioned references the conduction cooling is used and laser element was placed in contact with a material with very high thermal conductivity. That material was, in turn, in contact with a cryogen such as liquid nitrogen near 77 K, liquid Ne near 27 K, or He near 4 K.

At cryogenic temperatures the absorption and emission cross sections increases and the Yb:YAG absorption band near 941 nm narrows at 77 K, however it is still broad enough for pumping with practical diode lasers. At 77 K, Yb:YAG crystal behaves as a four-level active medium however in room temperature Yb:YAG is a quasi three-level material. Significant problems associated with quasi-three-level materials like Yb:YAG, such as the need to provide a significant pump density to reach transparency, a high pump threshold power, and the associated loss of efficiency, disappear at 77 K [57]. When the Yb:YAG is cooled from room to cryogenic temperatures, the lasing threshold decreases and slope efficiency increases. Figure 13 shows drop of lasing threshold from 155 W to near 10 W and increasing slope efficiency from 54% to a 63% for a typical thin disk laser [58]. The inset spectral peak in Figure 13 is the 80 K laser output and is centered near 1029.1 nm. At room temperatures this peak is near 1030.2 nm.

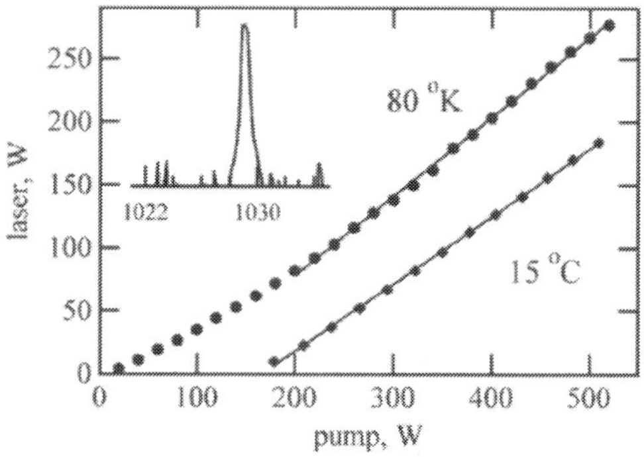

Figure 13. Lasing power versus pump power at 15°C (288 K) and 80 K. [58]

2.4.2. Temperature distribution

Specifications of disk laser beam are tightly related to the active medium geometry. The precise geometry of the active medium geometry is also strongly depended to the thermo-mechanical and opto-mechanical properties of the disk and the temperature distribution in the disk. Temperature distribution of the disk can be obtained by solving the heat conduction equation using proper boundary conditions.

The flow of heat generated by the pumping diode radiation through a laser gain media in general form is described by a non-homogeneous partial differential equation:

$$\frac{1}{k_c(T,dop)}\frac{\partial T(r,\theta,z)}{\partial t} - \nabla^2 T(r,\theta,z) = \frac{Q(r,\theta,z)}{k_c(T,dop)} \tag{14}$$

In which, K_c is heat conductivity that is assumed to be isotropic and is the heat source density in the laser crystal. In CW regime of output laser, the steady-state temperature distribution obeys the heat diffusion equation

$$\nabla^2 T(r,\theta,z) = -\frac{Q(r,\theta,z)}{k_c(T,dop)} \tag{15}$$

As it is seen the heat conductivity is depended on the temperature and doping concentration (dop) of the crystal. Temperature dependency of YAG crystal in room temperature is not significant and it can be considered as a constant [37] however this approximation would not be valid anymore at cryogenic temperature. Heat conductivity of Yb:YAG crystal which is conventional active medium of thin disk laser can be given by [59]:

$$k_c(T,dop) = (7.28 - 7.3 \times dop).\left(\frac{204}{T-94}\right)^{(0.48-0.46\times dop)} \qquad W/m^{-1}K^{-1} \tag{16}$$

Characterizing the behavior of the thermally induced lensing effect of the thin disk gain medium is not a trivial task. In order to fully analyze the dynamics of the heat flow and thus the induced $\partial n/\partial T$ stresses and strains on the gain medium one must solve the 3-D heat equation with appropriate boundary conditions. This can be accomplished in several ways. The most common is to employ a finite element analysis (FEA) method. Another method is to solve the 3-D heat equation using the Hankel transform. For more details the reader can refer to [60].

Initial estimation of disk thermal behavior can be carried out by calculation of maximum and average temperature of the disk. In thin disk lasers, the thickness of the disk is very low. When the pump spot size is very larger than the disk thickness, one dimensional heat conduction is a good approximation. If pump power of P_{pump} radiates to the disk in a pump spot with radius of r_p, the heat load per area can be given by [61]:

$$I_{heat} = \frac{P_{pump}\eta_{abs}\eta}{\pi r_p^2} \tag{17}$$

In which η_{abs} is absorption efficiency and η is heat generation coefficient in the disk. The heat generation coefficient in the disk is due to the quantum defect and related to the pumping and laser wavelength which are shown by λ_p and λ_l respectively.

$$\eta = 1 - \frac{\lambda_p}{\lambda_l} \tag{18}$$

A parabolic temperature profile will be formed along the axis inside the disk due to the loaded heat which is given by:

$$T(z) = T_0 + I_{heat}R_{th,disk}\left(\frac{z}{h} - \frac{1}{2}\frac{z^2}{h^2}\right) \tag{19}$$

in which $R_{th,disk} = h/k_c$ is the heat resistance of the disk material and T_0 is the temperature of the disk's cooled face. Also z is the distance along the disk axis in the thickness of the disk and h is the thickness of the disk. In particular, one can calculate maximum temperature from relation (19) which is given by

$$T_{\max} = T_0 + \frac{1}{2} I_{heat} R_{th,disk} \tag{20}$$

also the average temperature in the disk thickness can be given by

$$T_{av} = T_0 + \frac{1}{3} I_{heat} R_{th,disk} \tag{21}$$

In this way using relations (19) to (21) one can evaluate, temperature distribution and maximum and average temperature in the disk in one dimensional heat conduction approximation.

2.5. Fiber laser and fiber amplifiers

2.5.1. Introduction to fiber geometry and cooling methods

In recent years, design and manufacturing of high efficient fiber lasers which deliver excellent beam quality, has made them as the main adversary of other types of high power solid state lasers, such as bulk and thin-disk lasers. Achieving to multi-kilowatt output powers [7, 62] with diffraction limited laser beam could be considered as the unique record in laser technology. This progress can be attributed directly to the capability of more efficient cooling procedure in fiber lasers, which originates from inherent large surface to volume ratio. In fact, thermal load spreads over meters or tens of meters of fibers, which causes convenient and efficient cooling management and therefore avoids thermo-optic problems.

Fiber lasers are consists of fiber core which is mostly surrounded by two coaxial fiber cladding (double clad fiber lasers) and is pumped by diode bars or diode laser at one or both ends. The laser light can only propagate through the fiber core and doesn't have any role in heat generation inside the fiber.

There are two general methods for lunching pump light into the fiber laser which are called as "core pumping" and "cladding pumping". The conventional core pumping was initially used to achieve single mode output laser, in which the pump light was coupled into the small core. On the other hand, small core causes a serious restriction on pump power level [63]. Furthermore, the core size leads to highly localized pump intensity which usually induces thermal damage at the fiber ends. Therefore, cladding pumping has been developed as the proper solution which ensures lunching high pump power into the double clad fiber lasers. In this method the pump light couples into the inner cladding and propagates through it and gradually absorbs in doped core. In both cases, the pump light only absorbs within the core, where heat generation takes place. Figure 14 shows a simple diagram of cladding pumping of fiber laser geometry [63].

In most cases, cooling procedure in fiber lasers does not need any special cooling system and are called passive cooling, which easily can be performed by the air through the convectional process [62-64,65]. However, in modern fiber lasers an active cooling system is

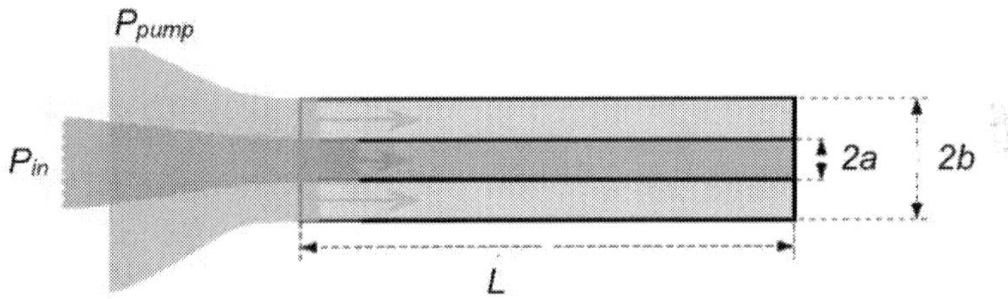

Figure 14. Cladding pumped fiber amplifier [63]

considered to scale up high power lasers, which ensures a forced heat dissipation process. Liquid cooled fiber lasers [66] is an example of new cooling methods in which, the whole or some part of the fiber is placed inside a liquid with a definite temperature. Therefore, heat removal occurs through the convection from the fiber periphery to the cooling liquid. This technique is usually applied to the long fiber lasers. Another technique which is often utilized for cooling of short fiber lasers, concerns to the thermoelectric cooling system (TEC). In this case, the fiber medium surrounds by a copper heat sink and therefore heat removal performs through the heat conduction process. The three common methods which imply to the passive and active cooling techniques are introduced in more details in follow.

Example of conductive boundary condition in which a short length fiber is surrounded by a temperature controlled copper heat sink can be found in [67]. Ignoring of axial heat flux as an approximation, heat differential equation can be solved numerically by means of Finite Element (FE) method. Figure 15, shows a drawing of a TEC-cooled fiber assembly.

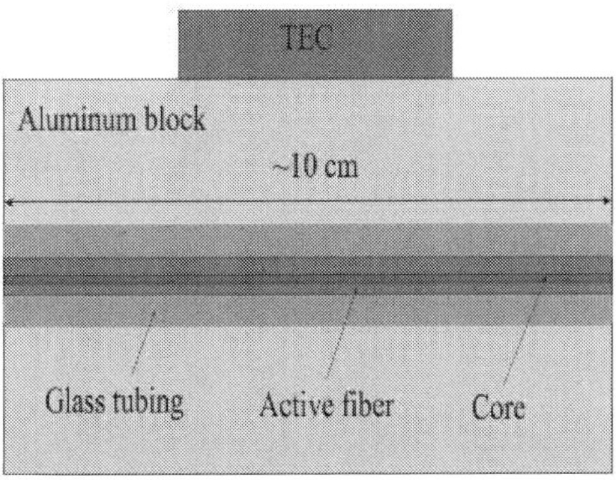

Figure 15. side view of a TEC-cooled short-length fiber laser [67].

Cooling of long fiber laser based on the conductive heat transfer is reported in [67]. The fiber is placed between aluminum plates with constant temperature caused by water cooling.

Practical models of high power fiber lasers with the unforced convective heat transfer from fiber to air are reported in [62-64, 65]. Figure16, illustrates the experimental set up for an Er:

ZBLAN double-clad fiber laser. Fiber is placed inside the resonator and is pumped from one end by a diode laser after passing the pump beam from the designed optics.

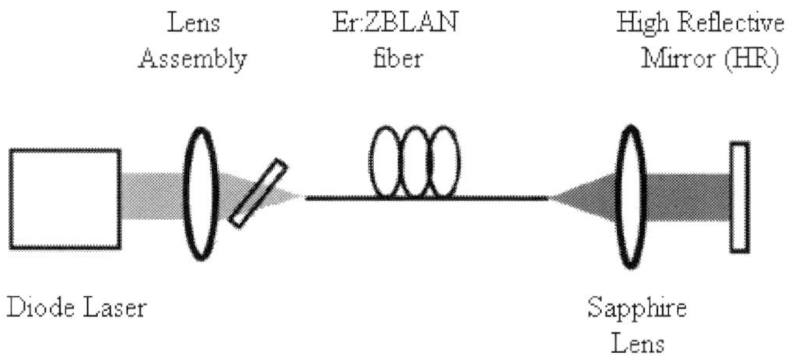

Figure 16. experimental set up for high power Er: ZBLAN fiber laser [64]. The fiber is pumped by a diode laser at one end.

Figure 17 shows another example relating to high power Yb doped Fiber laser (YDFL), which is pumped from both ends [62]. Convectional cooling process from fiber to air is freely established. Pump power is delivered from two diode stacks propagate from the both ends toward the fiber center and cause two individual heat sources inside the fiber.

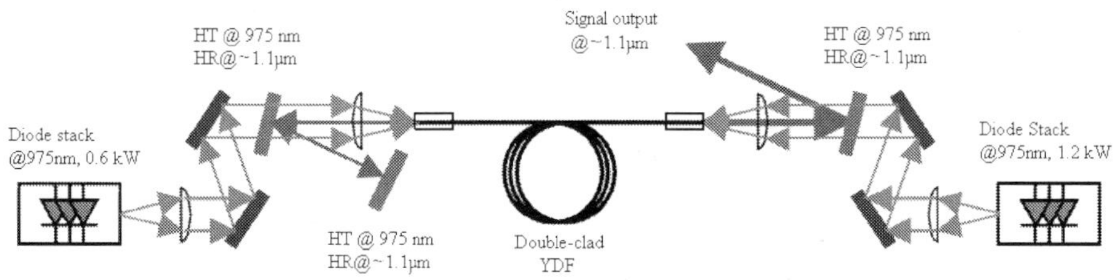

Figure 17. experimental arrangement of double clad YDFL pumping with two diode stacks [62]. Freely convective heat transfer to the surrounding air is considered.

Efficient fiber cooling leads to scale up high power lasers without thermal damage and avoiding destructive thermal effects on laser operation. A new technique for thermal management of fiber was examined in [66], which called direct liquid cooling. In this method the fiber was in direct contact with the fluorocarbon liquid. Furthermore, the both ends of fiber facet are in physical contact with the CaF_2 windows. This leads to conductive heat transfer from fiber facet to the window and considerable axial heat removal, which allows increasing pump power without thermal damage. This technique was already used in composite bulky solid state laser mediums [24, 41-43] and had found highly operative [68, 69]. Figure 18, shows a drawing of system assembly. The fiber is pumped by two fiber coupled diode lasers from both ends.

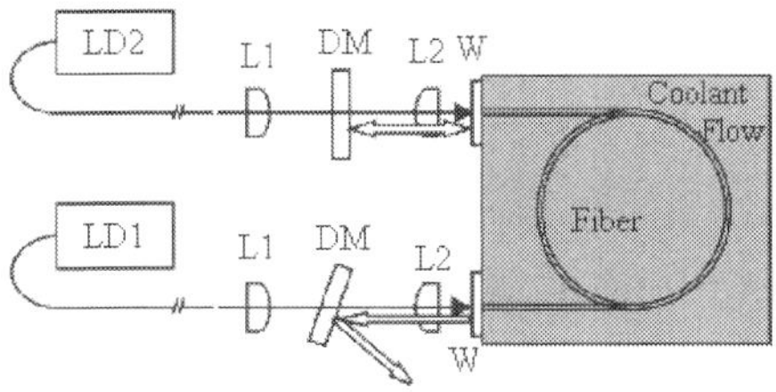

Figure 18. Liquid cooling of long fiber laser. Ld1 and Ld2 fiber coupled diode lasers, L1 and L2 aspheric lenses, W, CaF_2 windows, DM dichroic mirrors [66].

2.5.2. *Continuous Wave (CW) pumping conditions*

Using CW pump sources lead to generation of time independent heat source density in fiber core. Therefore, Eq.1 turns to steady state heat differential equation as

$$\frac{1}{r}\frac{\partial}{\partial r}\left(r\frac{\partial T(r,z)}{\partial r}\right)+\frac{\partial^2 T(r,z)}{\partial z^2}=-\frac{Q(r,z)}{k_c} \tag{22}$$

In which, the azimuthal part of the temperature is omitted due to the cylindrical symmetry of pumping spatial distribution. As we mentioned before, spatial form of heat source density obeys from the spatial form of pump intensity profile lunched to the fiber. "Top hat" and "Gaussian" are two common shapes for the pump beam profile, which are usually considered as the spatial form of heat source density in thermal analysis. Different considerations lead to different differential equations and therefore, needs different solutions. Analytical and Numerical solutions of Eq.1, to specify the temperature behavior inside fiber medium, are reported in various literatures with different approximations and methods.

As we mentioned before, different cooling arrangements lead to different boundary conditions which are conductive or convective heat transfer from fiber periphery to the surrounding medium.

In the case of fiber coupled fiber laser, pump intensity distribution with a good approximation has a top hat profile across the beam. Entering the pump beam inside the fiber core and propagating along the fiber length causes exponential decay in axial direction. Therefore, the heat source density Q(r,z), inside the fiber can be expressed by [70]

$$Q(r,z)=\begin{cases}\dfrac{1}{\pi a^2\,\mathrm{L}_{eff}}\,\xi P_0 \mathrm{e}^{-\alpha z} & ;\ \mathrm{r}\le a\\ 0 & ;\ \mathrm{a}\le \mathrm{r}<b\end{cases} \tag{23}$$

Where, P_o is the pump power, a is radius of the fiber core, b is the radius of outer cladding, $L_{eff}=\dfrac{(1-e^{-\alpha L})}{\alpha}$ is the effective fiber length, L is the geometrical fiber length and α is the effective pump absorption coefficient. Substitution of $Q(r,z)$ from Eq. 23 into Eq. 22, the heat differential equation for two regions can be written as

$$\frac{1}{r}\frac{\partial}{\partial r}\left(r\frac{\partial T(r,z)}{\partial r}\right)+\frac{\partial^2 T(r,z)}{\partial z^2}=-\frac{1}{\pi a^2 k_c L_{eff}}\xi P_0 e^{-\alpha z} \qquad ;0\leq r\leq a \qquad (24\text{-a})$$

$$\frac{1}{r}\frac{\partial}{\partial r}\left(r\frac{\partial T(r,z)}{\partial r}\right)+\frac{\partial^2 T(r,z)}{\partial z^2}=0 \qquad ;a\leq r<b \qquad (24\text{-b})$$

Equation 24-a corresponds to the fiber core, where exposes to the pump beam and therefore experiences the heat generation. Equation 24-b relates to fiber cladding that is just responsible for heat transmission to the surrounding medium. Solution of these differential equations give the steady state three dimensional temperature at any point in fiber core $T_1(r,z)$ and cladding $T_2(r,z)$.Figure 19, illustrates schematically double clad circular fiber geometry under pumping process [70].

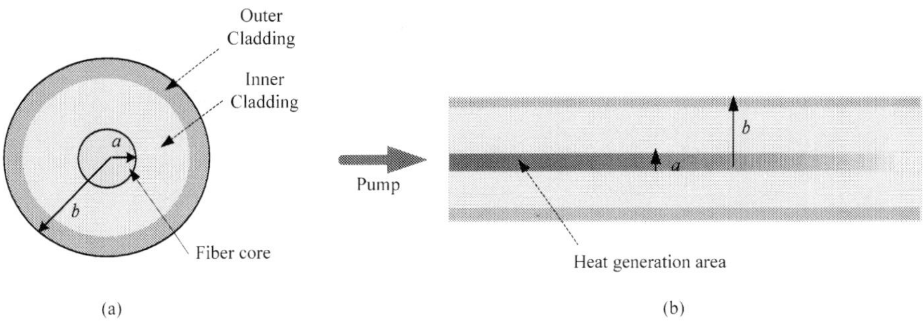

Figure 19. End (a) and side (b) view of circular fiber laser. Absorption of pump power inside the fiber core causes heat generation which reduces exponentially along the fiber axis [70].

Assuming the outer surface of fiber is in direct contact with a liquid or gas such as air, the convective boundary condition could be defined according to Newton's law as

$$k_c\left.\frac{\partial T_2(r,z)}{\partial r}\right|_{r=b}=h(T_c-T_2(b,z)) \qquad (25)$$

Where, h is heat transfer coefficient [71],T_c is the coolant temperature andT_2 (b,z) represents the temperature along the fiber on its cylindrical surface and the first derivative is taken with respect to the surface normal. This equation expresses that the radial thermal flux which arrives at the fiber periphery via the conduction method, removes by the coolant through the heat convection process. Further information about the analytical solution of Equation and some technical points on numerical approach can be found in [70]. The resultant temperature expressions are

$$T_1(r,z)=\frac{\xi P_0\alpha\exp(-\alpha z)}{\pi K_c a^2(1-\exp(-\alpha L))}(CJ_0(\alpha r)-\frac{1}{\alpha^2})+T_c \qquad ;r\le a \qquad (26\text{-a})$$

$$T_2(r,z)=\frac{\xi P_0\alpha\exp(-\alpha z)}{\pi K_c a^2(1-\exp(-\alpha L))}(A_1J_0(\alpha r)-A_2Y_0(\alpha r))+T_c \qquad ;r>a \qquad (26\text{-a})$$

Where C, A_1 and A_2 are constant coefficients which are derived from the boundary conditions as follows

$$C=A_1-\frac{Y_0(\alpha a)}{J_0(\alpha a)}A_2+\frac{1}{\alpha^2 J_0(\alpha a)} \qquad (27\text{-a})$$

$$A_1=\frac{\pi a^2}{4h}(h.Y_0(\alpha b)-K_c.\alpha.Y_1(\alpha b)) \qquad (27\text{-b})$$

$$A_2=\frac{\pi a^2}{4h}(hJ_0(\alpha b)-K_c\alpha J_1(\alpha b)) \qquad (27\text{-c})$$

The temperature is reduced exponentially along the fiber axis due to the exponential decay of heat deposition. Radial dependence in the core includes zero order Bessel function and inside the cladding, linear combination of the first and second kind of zero order Bessel functions are established. This is illustrated in Figure 20 which shows the calculated temperature distribution in the r-z plane of ZBLAN (ZrF_4-BaF_2-LaF_3-AlF_3-NaF) double clad fiber. More information about the fiber geometry and pumping characteristics is given in Table 1.

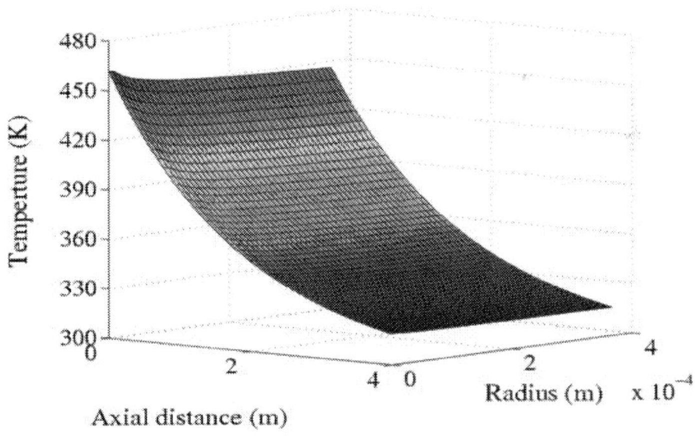

Figure 20. Temperature distribution in r-z plane for the double clad ZBLAN fiber laser [70].

Quantity	Magnitude
Core radius, a	15 μm
Clad Radius, b	370 μm
Fiber length, L	4 m
thermal conductivity, K_c	0.628 W/(m.K)
absorption coefficient, α	2.3dB/m ($0.54 m^{-1}$)
Pump Power, P_0	42.8 W
heat transfer coefficient, h	50 W $m^{-2}K^{-1}$
ambient air temperature, T_c	300 K

Table 1. amount of physical quantities are used in temperature calculations [70].

In the case of short length fiber laser, analytical approach of temperature expressions for an Er^{3+}/Yb^{3+} co-doped fiber laser can be found in [26]. Schematic drawing of fiber pumping is illustrated in Figure 21.

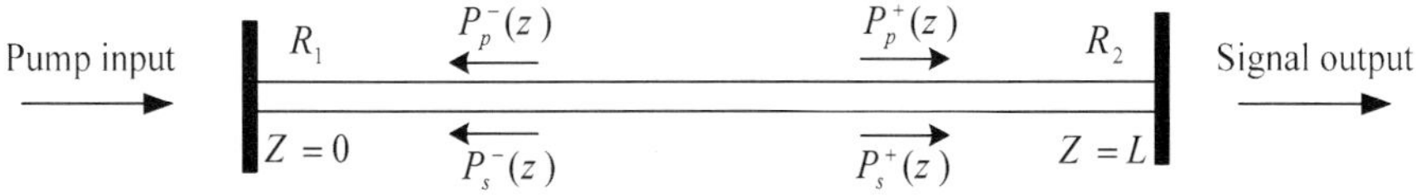

Figure 21. Schematic figure of short length fiber laser with propagation of traveling pump and signal light into positive and negative axial direction[26] .

In short fiber lasers, the pumping energy doesn't absorb completely by one passing of light along the fiber length and thus, the reflected part from the coupler acts as a pump beam delivers inside the fiber at the end face. Heat source function is composed of the proportion of the both the left and right traveling pump beams. The heat source density function with the assumption of Gaussian pump beam shape is given by [26]

$$Q(r,z)=\frac{2(\alpha\eta+\alpha_{ps})[P_p^+(z)+P_p^-(z)]+2\alpha_s[P_s^+(z)+P_s^-(z)]}{\pi\omega_p^2}\exp\left(-2r^2/\omega_p^2\right) \tag{28}$$

where $P_p^{\pm}(z)$ and $P_s^{\pm}(z)$ are the pump power and signal power in the positive and negative directions, α_{ps} and α_s are scattering loss coefficients at the pump wavelength and signal wavelength respectively and ω_p is the Gaussian radius of the pump light. Figure 22 illustrates an end view drawing of the modeled fiber. As it can be seen, the active fiber is surrounded by a ceramic ferrule which is placed inside a copper tube. Heat conduction from fiber cladding to ceramic ferrule occurs through the fiber cylindrical surface.

The steady state heat differential equations are

$$\frac{1}{r}\frac{\partial}{\partial r}[r\frac{\partial T(r,z)}{\partial r}]+\frac{\partial^2 T(r,z)}{\partial z^2}=-\frac{Q(r,z)}{k_c},0\le r\le a \tag{29}$$

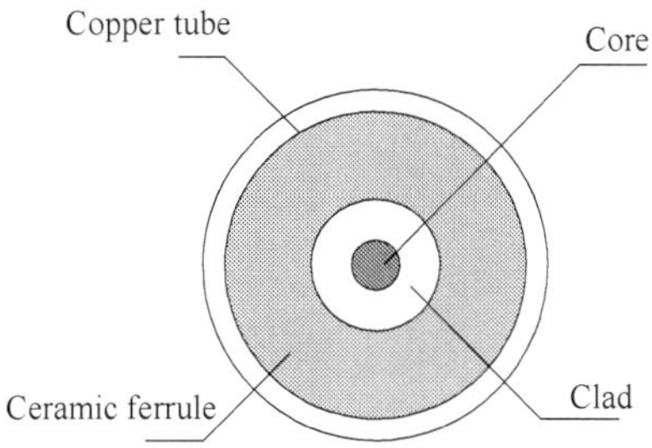

Figure 22. Geometry of modeled fiber laser

$$\frac{1}{r}\frac{\partial}{\partial r}[r\frac{\partial T(r,z)}{\partial r}]+\frac{\partial^2 T(r,z)}{\partial z^2}=0,\ a\le r\le b \tag{30}$$

and corresponding boundary conditions are

$$k_c\frac{\partial T(r,z)}{\partial r}=h[T_c-T(r,z)],r=b \tag{31}$$

$$T_1=T_2,\ \frac{\partial T_1}{\partial r}=\frac{\partial T_2}{\partial r},\ r=a \tag{32}$$

$$\frac{\partial T(r,z)}{\partial z}=0,\ z=0,z=L \tag{33}$$

Where, T_1 and T_2 are the temperatures in fiber core and cladding regions. Eq. 30 relates to the heat removal into the ambient air which is hold at the constant temperature of T_c, Eq. 31 ensures the same temperature value at the fiber cladding and core joint boundary (r = a) and Eq.35 expresses the negligible heat transfer from fiber ends into the air because of the small amount of air heat transfer coefficient..

Solution of equations (29)-(33) gives the temperature expressions in fiber core and cladding are given by

$$T_1(r,z)=T_c+C_0\ln a+D_0+\sum_{n=1}^{\infty}\sum_{m=0}^{\infty}A_{nm}\cos\left(\frac{m\pi}{L}z\right)J_0\left(\frac{\mu_n^0}{a}r\right)+\sum_{m=1}^{\infty}\left[C_m I_0\left(\frac{m\pi}{L}a\right)\cos\left(\frac{m\pi}{L}z\right)\right] \tag{34}$$

$$T_2(r,z)=T_c+C_0\ln r+D_0+\sum_{m=1}^{\infty}\left[C_m I_0\left(\frac{m\pi}{L}r\right)+D_m K_0\left(\frac{m\pi}{L}r\right)\right]\cos\left(\frac{m\pi}{L}z\right) \tag{35}$$

Where, μ_n^0 is nth zero point of the zero order Bessel function of the first kind, J_0 and J_1 are the zero and first order Bessel function of the first kind, I_0, I_1 and K_0, K_1 are the zero and first order modified Bessel function of the first and second kind, respectively [26]. C_m , D_m and A_{nm} are the constants which derived from boundary conditions, via solution of coupled equations and are given in [26]. The above equations are plotted in figure 23, illustrating the temperature distribution in fiber core and cladding in a phosphate fiber. As would be expected, the maximum temperature achieved at the center of the fiber facet and is 479.85K. Amount of some parameters used in calculations are summarized in Table 2.

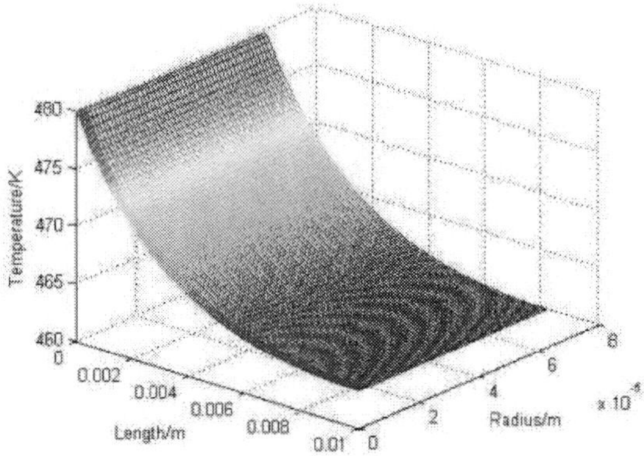

Figure 23. Temperature distribution in r-z plane of the short-length co-doped phosphate fiber laser [26].

Quantity	**Magnitude**
Core radius, a	2.7μm
Clad Radius, b	62.5 μm
Fiber length, L	1cm
thermal conductivity, K_c	0.55 W/(m.K)
Pump Power, P_0	100 W
heat transfer coefficient, h	10 W $m^{-2}K^{-1}$
ambient air temperature, T_c	300 K

Table 2. Amount of physical quantities are used in temperature calculations [26].

2.5.3. *Pulsed pumping conditions*

An example of pulsed pumped fiber laser which causes transient temperature field is presented in [72]. This work relates to short length fiber laser and introduces analytical temperature expressions for solution of transient heat differential equations which are

$$\frac{\partial^2 T(z,r,t)}{\partial r^2}+\frac{1}{r}\frac{\partial T(z,r,t)}{\partial r}]+\frac{\partial^2 T(z,r,t)}{\partial z^2}+\frac{Q(z,r,t)}{k_c}=\frac{\rho c}{k_c}\frac{\partial T(z,r,t)}{\partial t} \quad ,0 \le r \le a \tag{36}$$

$$\frac{\partial^2 T(z,r,t)}{\partial r^2}+\frac{1}{r}\frac{\partial T(z,r,t)}{\partial r}]+\frac{\partial^2 T(z,r,t)}{\partial z^2}=\frac{\rho c}{k_c}\frac{\partial T(z,r,t)}{\partial t} \quad ,a \le r \le b \tag{37}$$

In fact, Eq. 36 and Eq. 37 are is written for the fiber core and cladding regions. In this case, the heat source density function is defined as

$$Q(z,r,t) = \frac{2\eta\alpha P_{in}}{\pi\omega_p^2}\exp\left(\frac{-2r^2}{\omega_p^2} - \alpha z\right)g(t) \tag{38}$$

Where, P_{in} is the pulse energy and g(t) is temporal shape of pump pulse. The other parameters were previously introduced. The equations (38)-(40) are applicable here as the boundary conditions. Furthermore, the laser system is assumed to be in thermal equilibrium with the ambient air before pumping process which expresses by

$$T(z,r,t) = T_c \qquad ,t = 0 \tag{39}$$

Solution of equations (36) and (37) with the aid of boundary conditions through the integral transform method introduced by Özisik [73] gives the temperature expressions as

$$\begin{aligned}\theta &= T - T_0\\ &= \sum_{p=1}^{\infty}\frac{k_c}{\rho c l N_p}J_0(\beta_p r)\exp(-\frac{k_c}{\rho c}\beta_p^2 t)\int_0^t g_{0p}(\tau)\exp(\frac{k_c}{\rho c}\beta_p^2\tau)d\tau\\ &+\sum_{m=1}^{\infty}\sum_{p=1}^{\infty}\frac{2k_c}{\rho c l N_p}J_0(\beta_p r)\cos(\eta_m z)\exp[-\frac{k_c}{\rho c}(\beta_p^2+\eta_m^2)t]\int_0^t g_{mp}(\tau)\exp[\frac{k_c}{\rho c}(\beta_p^2+\eta_m^2)\tau]d\tau\end{aligned} \tag{40}$$

Where

$$g_{0p}(\tau) = \int_0^l\int_0^{r_1}\frac{Q(z,r,t)}{k_c}J_0(\beta_p r)rdrdz \tag{41}$$

$$g_{mp}(\tau) = \int_0^l\int_0^{r_1}\frac{Q(z,r,t)}{k_c}\cos(\frac{m\pi}{L}z)J_0(\beta_p r)rdrdz \tag{42}$$

$$\begin{aligned}&\int_0^t g_{0p}(\tau)\exp(\frac{k_c}{\rho c}\beta_p^2\tau)d\tau\\ &= \frac{2\eta P_{in}[1-\exp(-\beta l)]}{k_c\pi\omega_p^2}\int_0^{r_1}\exp(-2r^2/\omega_p^2)J_0(\beta_p r)rdr\int_0^t g(\tau)\exp(\frac{k_c}{\rho c}\beta_p^2\tau)d\tau\end{aligned} \tag{43}$$

$$\begin{aligned}&\int_0^t g_{mp}(\tau)\exp[\frac{k_c}{\rho c}(\beta_p^2+\eta_m^2)\tau]d\tau\\ &= \frac{2\eta\beta^2 l^2 P_{in}[1-\exp(1-\beta l)\cos(m\pi)]}{k_c\pi\omega_p^2(\beta^2 l^2+m^2\pi^2)}\int_0^{r_1}\exp(-2r^2/\omega_p^2)J_0(\beta_p r)rdr\times\\ &\int_0^t g(\tau)\exp[\frac{k_c}{\rho c}(\beta_p^2+\eta_m^2)\tau]d\tau\end{aligned} \tag{44}$$

$$\eta_m = m\pi / l \tag{45}$$

$$N_p = r_2^2 (k_c^2 \beta_p^2 + h^2) J_0^2 (\beta_p r_2) / 2k_c^2 \beta_p^2 \tag{46}$$

$$h J_0(\beta_p r_2) - k_c \beta_p J_1(\beta_p r_2) = 0 \tag{47}$$

Where J_0 and J_1 are the zero and first rank Bessel function of the first kind, respectively. Assuming square shape for the temporal dependence of pump pulses as bellow, temperature distribution was calculated.

$$g(t) = \begin{cases} 1 & (n-1)T_0 \le t \le (n-1)T + t_0 \\ 0 & (n-1)T_0 + t_0 \le t \le nT_0 \end{cases} \tag{48}$$

In which, n denotes number of pulses, t_0 is the pulse duration and T_0 is pulse period. Figure 24, shows the temperature distribution achieved from the analytical temperature expressions for a typical Er^{3+}/Yb^{3+}co-doped fiber laser. Characteristics of the fiber laser which was introduced in Table 2 were considered in calculations. The fiber is in encounter of pulses with the pulse pump of $P_{in} = 1W$.

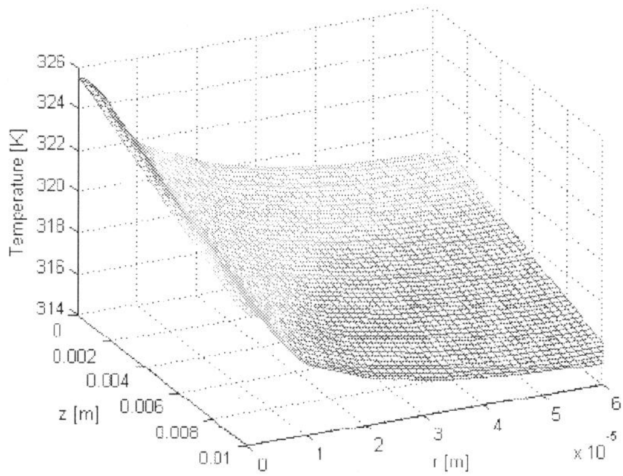

Figure 24. Temperature distribution in r-z plane of a pulsed pumped fiber laser at the time of 0.1s [72].

Entering of each pulse into the fiber core causes definite heat load and thus leads to increase of temperature. Figure 25-a, shows the temperature evolution at fiber end facet during the space time of pulse generation. The curves correspond to the time increment of 1ms.

If the pulse period is longer than the space time of heat removal, the temperature will return to the initial value before loading the next pulse. But if the pulse period is shorter than the time which is necessary for completely remove of generated heat, the heat will gradually deposit inside the fiber medium and leads to increase of the fiber temperature. Figure25-b illustrates reduction of fiber facet temperature after the first pulse. At the time of 15ms temperature profile has nearly flat shape but the temperature in all positions is 2°C over the initial temperature.

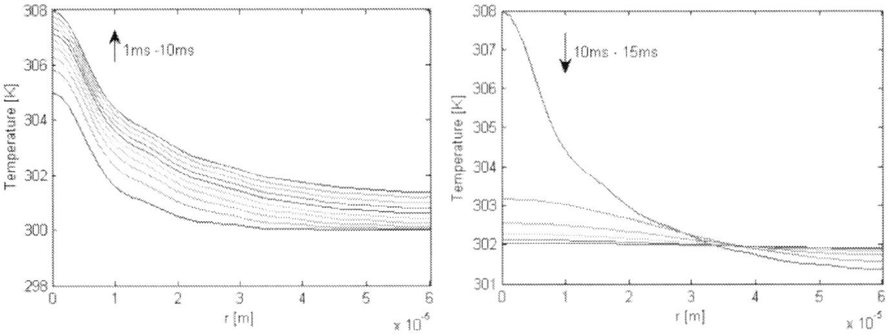

Figure 25. a) radial temperature distribution inside fiber laser at the pump facet during the first pulse pump time, b) reduction of temperature generated by the first pulse during 5ms [73].

Figure 26, illustrates the temperature evolution of the fiber facet at the center versus time. According to that, temperature rises gradually with increasing the time until reaching to almost constant temperature with the small and regular fluctuations. Each fluctuation corresponds to heat generation and then heat removal to outer space during the space time of pulse formation and pulse period time.

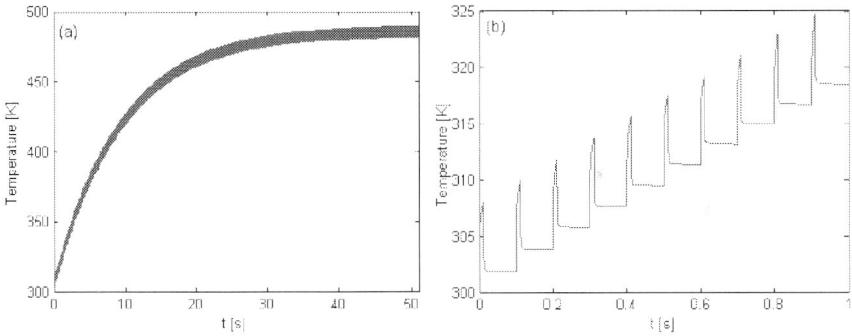

Figure 26. a) time evaluation of temperature at the end surface of fiber, b) enlarged drawings of the part of graph a which shows regular fluctuations in temperature profile [73].

2.6. Microchip lasers

In microchip lasers the coated surface of active medium is used as laser's mirror. Popular types of microchip lasers have less than 1 Watt output power. However, recently some reports are demonstrated microchip lasers with several hundred Watts [74-78]. Due to the importance of thermal problems in high power operation, we focus our investigation on this regime. Generally, composite active media are used in high power microchip lasers. The laser material includes a central doped region as the active medium surrounded by an un-doped rim as pumping waveguide.

The microchip laser has a thin active medium with geometry very similar to that of thin disk lasers. Considering the uniform absorbed power distribution, the thermal considerations assumed in thin disk lasers are still valid in microchip lasers. Most common microchip lasers in high power operations are composite microchips with an un-doped rim surrounding active region acting as pump wave guide. The main drawback in this geometry is the non-uniformity of absorbed power distribution which leads to non-uniform temperature distribution. Regarding boundary conditions and non-uniform heat deposition, the temperature distribution in the active medium should be investigated.

An example of thermal analysis in high power microchip laser is presented in [79]. Utilizing optimum geometry and dimensions for active medium, one can consider uniform absorbed power distribution in microchip lasers which leads to relatively uniform temperature distribution and can overcome the common disadvantage of microchip lasers. A common geometry of the Yb:YAG/YAG composite microchip lasers is illustrated in figure 27 [79].

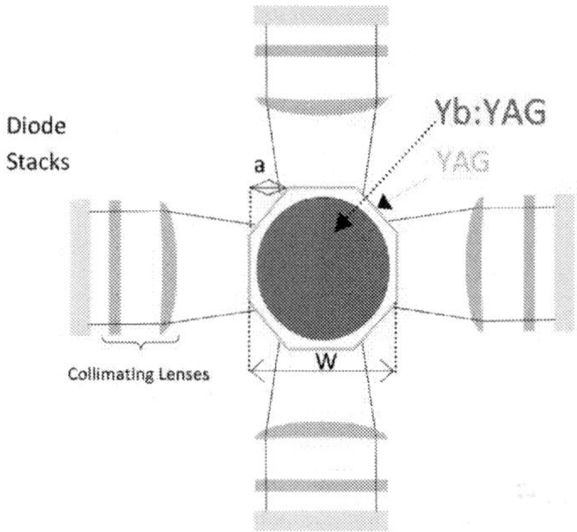

Figure 27. Schematic Diagram of a composite Yb:YAG/YAG side pumped microchip laser elements [79]

In this arrangement, a thin disk Yb:YAG core enclosed by an irregular symmetric eight sided un-doped YAG rim is bonded on a water-cooled Cu-W cold-plate.

Fraction of absorbed pump power in the gain medium produces heat in the core and raises the core temperature. The temperature distribution in the laser material is calculated using the, which is

$$\nabla.(k_c \nabla T) + Q = 0 \tag{49}$$

Thermal conductivity is strongly dependent on temperature and atomic doping concentration of Yb ions C_{Yb} as follows [80]

$$k(T,C_{Yb}) = (7.28 - 7.3C_{Yb})\left(\frac{204\ ^{\circ}K}{T - 96\ ^{\circ}K}\right)^{0.48-0.46C_{Yb}} \tag{50}$$

The absorbed power distribution in active medium is calculated using Mont-carlo ray tracing of pumping photons through the optical elements of the system. Heat generation Q is estimated based on the energy difference between the pump and laser photons, given by [81]

$$Q = (1 - \frac{\lambda_p}{\lambda_l})P_p \tag{51}$$

In which P_p is local absorbed power density. The heat transfer between the gain medium and air is week, which is described by

$$\left.\frac{dT}{dz}\right|_{z=t} = 0 \tag{52}$$

In which z axis coincides on microchip optical axis and t denotes the microchip thickness. Moreover, the temperature of the back face of the cold-plate and contact cooling water were considered as the same. The numerical values used in calculations are given in Table 3.

Quantity	**magnitude**
Pumping wavelength	942nm
Laser wavelength	1030nm
active medium thickness	0.3mm
cold plate thickness	1mm
thermal conductivity [9-82] cold plate	1.9 W/cm.K
cold plate heat transfer coefficient	12 W/cm2.K

Table 3. The microchip laser parameters used in numerical calculations.

Heat differential equation (49), was calculated in numerical Finite- Difference method for the described gain medium [79]. This code is used to determine the temperature distribution in the active medium. Figure 28 shows the maximum temperature variation versus a/w ratio for various core diameters with 9 at.% Yb3+ doping concentration. Moreover, the temperature distribution profiles of these cores are also shown in Figure 29.

According to figure 29, the laser gain mediums with the smaller core size experience higher temperatures at the same pump power, which is attributed to the higher thermal load density.

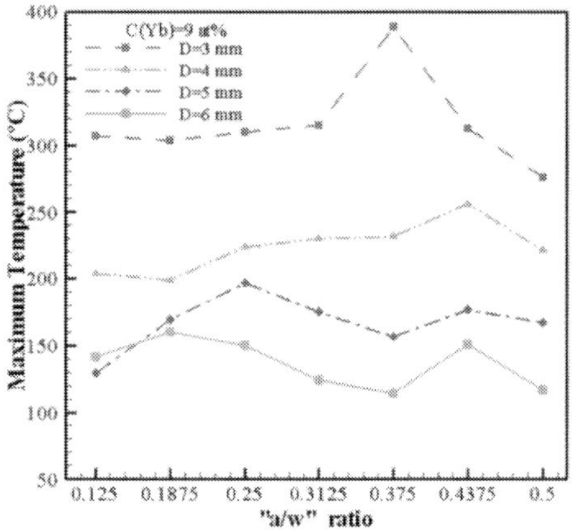

Figure 28. Maximum temperature of different microchip core sizes [79].

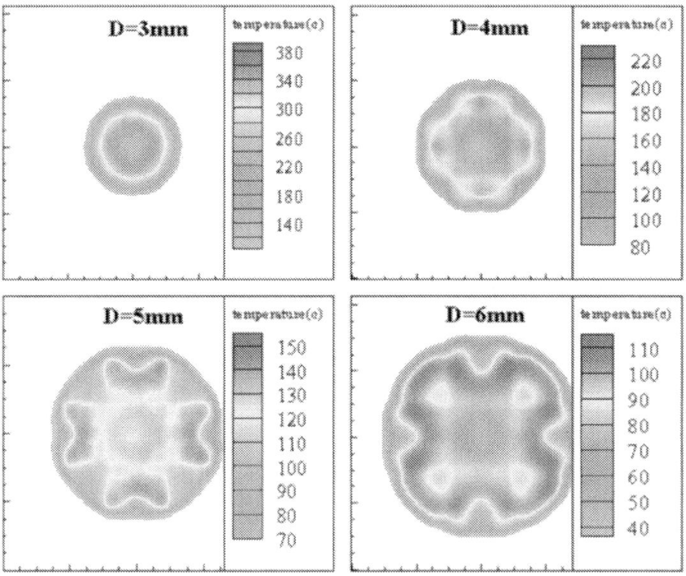

Figure 29. Temperature distributions of diffrent cores[79].

Summary

In this work, at first the basic principles of heat generation and removal in all kind of solid state lasers which are Bulk, Disk, Fiber and Microchip configurations are introduced. Then we tried to present a complete collection of solution for heat transfer equation in these types of gain medium regarding to the modern cooling systems, which are available in the literatures. Therefore, this work gives the reader a straight forward way to reach the latest progress in thermal problem in solid state lasers applicable for high power laser design.

Author details

V. Ashoori and M. Shayganmanesh

Department of physics, Iran University of science & Technology, Narmak, Tehran, Iran

S. Radmard

Iranian National Center for Laser Science and Technology (INLC), Tehran, Iran

3. References

[1] Guangyuan H, Jing G, Biao W, and Zhongxing J (2011) Generation of radially polarized beams based on thermal analysis of a working cavity. Opt. Express. 19: 18302-18309.

[2] Wan-Jun H, Bao-Quan Y, You-Lun J, and Yue-Zhu W (2006) Diode-pumped efficient Tm,Ho:GdVO4 laser with near-diffraction limited beam quality. Opt. Express. 14: 11653-11659.

[3] Zhu X, Peyghambarian N (2010) High-Power ZBLAN Glass Fiber Lasers. Review and Prospect Advances in OptoElectronics.1-23.

[4] Richardson D, Nilsson J, and Clarkson W (2010) High power fiber lasers: current status and future perspectives. J. Opt. Soc. Am. B.27: B63-B92.

[5] Chaitanya Kumar S ,Samanta G and Ebrahim-Zadeh M. (2009) High-power, single-frequency, continuous-wave second-harmonic-generation of ytterbium fiber laser in PPKTP and MgO:sPPLT. Opt. Express. 17: 13711-13726.

[6] Brilliant N. and Lagonik K (2001) Thermal effects in a dual-clad ytterbium fiber laser. Optics Letters. 26: 1669-1671.

[7] Jeong Y, Boyland A, Sahu J, Chung S, Nilsson J, and Payne D (2009) Multi-kilowatt Single-mode Ytterbium-doped Large-core Fiber Laser. J. Opt.Soc. Korea. 13: 416-422.

[8] Ahmed M, Haefner M, Vogel M, Pruss M, Voss A, Osten W and Graf T (2011) High-power radially polarized Yb:YAG thin-disk laser with high efficiency. Opt. Express. 19: 5093-5104.

[9] Beil K, Fredrich-Thornton S, Tellkamp F, Peters R, Kränkel C, Petermann K and Huber G (2010)Thermal and laser properties of Yb:LuAG for kW thin disk lasers. Opt. Express. 18: 20712-20722.

[10] Dascalu T, Taira T (2006) Highly efficient pumping configuration for microchip solid-state laser. Opt. Express. 14: 670-677.

[11] Bhandari R and Taira T (2011) Megawatt level UV output from [110] Cr4+:YAG passively Q-switched microchip laser. Opt. Express. 19: 22510-22514.

[12] Fan T.Y (1993) Heat generation in Nd: YAG and Yb: YAG. IEEE J. Quantum Electron. 29: 1457–1459.

[13] Innocenzi M E, Yura H T, Fincher C L and Fields R A (1990) Thermal modeling of continuous-wave end-pumped solid-state lasers. Appl. Phys. Lett. 56: 1831–3.

[14] Fan S, Zhang X, Wang Q, Li S, Ding S, Su F (2006) More precise determination of thermal lens focal length for end-pumped solid-state lasers. Opt. Com. 266: 620-626.

[15] Koechner Walter (2006) Solid-State Laser Engineering 6th edn (New York: Springer) chapter 7.

[16] Moshe I, Jackel S and Meir A (2003)Production of radially or azimuthally polarized beams in solid-state lasers and the elimination of thermally induced birefringence effects. Opt. Lett. 28: 807-809.

[17] Moshe I and Jackel S (2005)Influence of birefringence-induced bifocusing on optical beams. J. Opt. Soc. Am. B. 22: 1228-1235.

[18] Roth M. S, Wyss E. W, Glur H and Weber H. P (2005) Generation of radially polarized beams in a Nd:YAG laser with self-adaptive overcompensation of the thermal lens. Opt. Lett. 30: 1665- 1667.

[19] He G, Guo J, Wang B and Jiao Z (2011) Generation of radially polarized beams based on thermal analysis of a working cavity. Opt. Express. 19: 18302-18309.

[20] Thyagarajan K, Ghatak A (2010) Lasers, Fundamental and Applications: Springer. 467 p.

[21] Knight P. L, Miller. A () Diode Laser Arrays: Cambridge University Press. 464 P.

[22] Bachmann F, Loosen P, Poprawe R. (2007) High Power Diode Lasers, Technology and Applications: Springer. 552p.

[23] Pfistner C, Weber R, Weber H. P, Merazzi S, and Gruber R (1994) Thermal beam distortions in end-pumped Nd:YAG, Nd: GSGG, and Nd: YLF Rods. IEEE J. Quantum Electronics. 30: 1605-1614.

[24] Weber R, Neuenschwander B, Mac Donald M, Roos M.B, and Weber H. P (1998) Cooling Schemes for Longitudinally diode laser-pumped Nd:YAGRods. IEEE J. Quantum Electronics. 34: 1046-1052.

[25] CarlslawH.S, and Jaeger (1959) Conduction of Heat in solids: Oxford University Press. 189 p.

[26] Liu T, Yang Z. M and Xu S. H (2008) 3-Dimensional heat analysis in short-length Er3+/Yb3+ co-doped phosphate fiber laser with upconversion. Opt. Express. 17: 235-247.

[27] Xie W, Tam S. C,Y. L, Liu J, Yang H, Gu J, and Tan W (2000) Influence of the thermal effect on the TEM00 mode output power of a laser-diode side-pumped solid-state laser . Appl. Opt. 39: 5482-5487.

[28] Clarkson W A (2001) Thermal effects and Their Mitigation in End-Pumped Solid-State Lasers. J. Phys. D. Appl. Phys. 34: 2381-2395.

[29] Huang Y, Tsai H, Chang F (2007) Thermo-optics effects affecting the High Pump power End Punped Solid State Lasers: Modeling and Analysis Opt. Communications. 273: 515-525.

[30] Shi P, Chen W, Li L, Gan A (2007), Semianalytical Thermal Analysis of Thermal Focal Length on Nd:YAG Rods. Appl. Optics. 46: 6655-6661.

[31] Innocenzi M E, Yura H T, Fincher C L, and Fields R A (1990) Thermal Modelling of Continuous-Wave End-Pumped Solid-State Lasers. Appl. Phys. Lett. 56 (19): pp. 1831-1833.

[32] Wang S and etc. (2009) Diode End-Pumped Nd:YAG Laser at 946 nm With High Pulse Energy Limited by Thermal Lensing. Appl. Physics B. 95:721-730.

[33] Weber R and etc. (1998) Cooling Schemes for Longitudinally Diode Laser-Pumped Nd:YAG Rods. IEEE Journal of Quantum Electronics. . 34: 1046-1053.

[34] MacDonald M E, Graf Th., Balmer J. E., and Weber H. P. (1990) Reducing Thermal Lensing in Diode-Pumped Laser Rods. Appl. Phys. Lett. 56: 1831-1833.

[35] Koechner W. (1970) Absorbed pumped power, thermal profile and stresses in a cw pumped Nd:YAG crystal. Appl.Opt. 9: 1429-1434.

[36] Cousins A. K. (1992) Temperature and thermal stress scaling in finite-length end-pumped laser rods.IEEE Journal of Quantum Electronics. 28: 1057-1069.

[37] Brown D. C (1997) Ultahigh-average power diode-pumped Nd:YAG and Yb:YAG lasers. IEEE J. Quantum Electron. 33: 861-873.

[38] Sovizi M, Massudi R. (2007) Thermal distribution calculation in diode pumped Nd:YAG rod by boundary element method. Opt. Laser.Thec.39: 46-52.

[39] Bourdet G. L and Yu H. (2007) Longitudinal temperature distribution in an end-pumped solid-state amplifier medium: application to a high average power diode pumped Yb: YAG thin disk amplifier. Appl. Opt. 46: 6033-6041.

[40] Farrukh U. O, Buoncristiani A. M and Byvik C. E (1988) An analysis of the temperature distribution in finite solid-state laser rods. IEEE Journal of Quantum Electronics. 24: 2253-2263.

[41] Kracht D, Wilhelm R, Frede M, Dupré F. K, Ackermann L (2005)407 W End-pumped Multi-segmented Nd:YAG Laser. Opt. Express. 13: 10140-10144.

[42] Kracht D, Frede M, Wilhelm R, Fallnich C (2005)Comparison of crystalline and ceramic composite Nd:YAG for high power diode end-pumping. Opt. Express. 13: 6212-6216.

[43] Frede M, Wilhelm R, Brendel M, Fallnich C, Seifert F, Willke B and DanzmannK (2004) High power fundamental mode Nd:YAG laser with efficient birefringence compensation. Opt. Express. 12: 3581-3589.

[44] Yan X, Liu Q, Fu X, Chen H, Gong M and Wang D (2009) High repetition rate dual-rod acousto-optics Q-switched composite Nd:YVO4 laser. Opt. Express. 17: 21956-21968.

[45] Wilhelm R, Freiburg D, Frede M, Kracht D, and Fallnich C (2009) Design and comparison of composite rod crystals for power scaling of diode end-pumped Nd:YAG lasers. Opt. Express.17: 8229-8336.

[46] Wang S,Eichler H. J, Wang X, Kallmeyer F, Riesbeck J. Ge. T, Chen J (2009)Diode end pumped Nd:YAG laser at 946 nm with high pulse energy limited by thermal lensing. Appl. Phys. B. 95: 721-730.

[47] Giesen A, Hügel H, Voss A, et al. (1994) Scalable concept for diode-pumped high-power solid-state lasers. Appl. Phys. B. 58: 365-372.

[48] Giesen A and Speiser J (2007) Fifteen Years of Work on Thin-Disk Lasers: Results and Scaling Laws. IEEE J. Sel. Top. Quantum Electron. 13: 598-609

[49] Karszewski M, Erhard S, Rupp T and Giesen A (2000) Efficient high-power TEM00 mode operation of diode-pumped Yb:YAG thin disk lasers. OSA TOPS Advanced Solid State Lasers. 34: 70-77.

[50] Shayganmanesh M, Daemi M.H, Osgoui Zh, Radmard S and Kazemi S.Sh (2012) Measurement of thermal lensing effects in high power thin disk laser. Opt. Laser. Technol. 44(7): 2292-2296

[51] Speiser J and Giesen A (2008) Scaling of thin disk pulse amplifiers Proc. SPIE 6871 68710J.

[52] Giesen A (2005) Thin disk lasers power scalability and beam quality. Laser Technik Journal 2(2): 42-45.

[53] Liu Q, Fu X, Ma D, et al. (2007) Edge-pumped asymmetric Yb:YAG/YAG thin disk laser. Laser Phys. Lett. 4(10): 719–721.

[54] Saravani M, Jafarnia A.F.M, Azizi M (2012) Effect of heat spreader thickness and material on temperature distribution and stresses in thin disk laser crystals. Opt. Laser Technol. 44: 756–762.

[55] Bowness C (1967) Liquid Cooled Solid State Laser. U.S. Patent 3339150.

[56] McMahon D.H (1972) Cooling System for Laser Media. U.S. Patent 3676798.

[57] Brown D.C (2005) The Promise of Cryogenic Solid-State Lasers. IEEE J. Sel. Topics Quantum Electron. 11(3): 587-599.

[58] Vretenar N, Newell T.C, Carson T, etal. (2012) Cryogenic ceramic 277 watt Yb:YAG thin-disk laser. Opt. Eng. 51(1): 014201.

[59] Contage S.K, Larinove M, Giesen A and Hugel H (2000) A 1-kW CW thin disc laser. IEEE J. Sel. Topics Quantum Electron. 6(4): 650–657.

[60] J.A. Sulskis, Design and characterization of a diode pumped, solid-state, thin-disk Yb:YAG laser, M.Sc. thesis, University of Illinois (Chicago, 2005).

[61] H. Injeyan and G.D. Goodno (editors), High power laser handbook (McGraw-Hill, 2011).

[62] Jeong Y, Sahu J. K, Payne D. N, and Nilsson J (2004) Ytterbium-doped large-core fiber laser with1.36 kW continuous-wave output power. Opt. Express. 12: 6088-6092.

[63] Dawson J. W, Messerly M. J, Beach R. J, Shverdin M. Y, Stappaerts E. A, Sridharan A. K, Pax P. H, Heebner J. E, Siders C. W and Barty C. P. J (2008) Analysis of the scalability

of diffraction-limited fiber lasers and amplifiers to high average power. Opt. Express. 16: 13240- 13266.

[64] Zhu X and Jain R (2007)10-W-level diode-pumped compact 2.78μm ZBLAN fiber laser.Opt. Lett. 32: 26-28.

[65] Gorjan M, Marinček M, and Čopič M (2009) Pump absorption and temperature distribution in erbium-doped double-clad fluoride-glass fibers. Opt. Express. 17: 19814-19822.

[66] Tokita S, Murakami M, Shimizu S, Hashida M and Sakabe S (2009) Liquid-cooled 24 W mid-infrared Er:ZBLAN fiber laser. Opt. Lett. 34: 3062-3064.

[67] Li L, Qiu H. L. T,Temyanko V. L, Morrell M. M,Schülzgen A,Mafi A, Moloney J. V,Peyghambarian N (2005) 3-Dimensional thermal analysis and active cooling of short-length high-power fiber lasers. Opt. Express.13: 3420-3428.

[68] Frede M, Wilhelm R, Brendel M, Fallnich C (2004) High power fundamental mode Nd:YAG laser with efficient birefringence compensation. Opt. Express. 12: 3581-3589.

[69] Wilhelm R, Freiburg D, Frede M, Kracht D and Fallnich C (2009) Design and comparison of composite rod crystals for power scaling of diode end-pumped Nd:YAG lasers. Opt. Express. 17: 8229-8236.

[70] Ashoori V and Malakzadeh A (2011) Explicit exact three-dimensional analytical temperature distribution in passively and actively cooled high-power fibre lasers. J. Phys. D: Appl. Phys. 44:35103-35109.

[71] H. W. McAdams, Heat Transmission, 3th ed. (McGraw-Hill, 1954).

[72] Liu T, Yang Z. M and XuS. H(2009) Analytical investigation on transient thermal effects in pulse end-pumped short-length fiber laser. Opt. Express. 17: 12875-12890.

[73] M. N. Özisik, Heat Conduction (Wiley, New York, 1980).

[74] Dascalu T (2008) edged-pump high power microchip Yb:YAG laser. Rom. Rep. Phys. 60: 977–994.

[75] Tsunekane M, Dascalu T, and Taira T (2005) High-power operation of diode edge-pumped, microchip Yb:YAG laser composed with YAG ceramic pump wave-guide. OSA TOPS on Advanced Solid- State Photonics. 98: 603-607.

[76] Dascalu T and Dascalu C (2007) High-power lens-shape diode edge-pumped composite laser. Proc. SPIE. 6785, B67850

[77] Dascalu T and Taira T (2006) Highly efficient pumping configuration for microchip solid-state laser. Opt. Express. 14: 670-677.

[78] Mohammadzahery Z, Jandaghi M, S, Alipour, Dadras S, Kazemi Sh., Sabbaghzadeh J (2012) Theoretical study on thermal behavior of passively Q-switched microchip Nd:YAG laser. Opt. Laser Technol. 44: 1095-1100

[79] Radmard S, Hagparast A, Arabgari S and Mehrabani M (2012) High-Powerb:YAG/YAG Microchip Laser Using Octagonal-Shape Waveguide with Uniform Absorbed Power distribution. Opt. Laser Technol. http://dx.doi.org/10.1016/j.optlastec.2012.09.02

[80] Stewen C, Contag K, Giesen A (2000) A 1-kW CW thin disc laser. IEEE Journal of Selected Topics in Quantum Electronics. 6: 650-657.

[81] Fan T Y (1993) Heat Generation in Nd:YAG and Yb:YAG. IEEE Journal of Quantum Electronics. 29: 1457-1459.

Section 5

Heat Transfer Augmentation

Chapter 13

Single and Two-Phase Heat Transfer Enhancement Using Longitudinal Vortex Generator in Narrow Rectangular Channel

Yan-Ping Huang, Jun Huang, Jian Ma, Yan-Lin Wang,
Jun-Feng Wang and Qiu-Wang Wang

Additional information is available at the end of the chapter

http://dx.doi.org/10.5772/53713

1. Introduction

In the past several decades, heat transfer enhancement techniques have rapidly developed and have been widely employed in many industrial fields. Up to now, these techniques have entered a stage of the so-called third generation, in which the use of longitudinal vortex generator is one of representative methods. When fluid flows over a barrier, different kinds of vortices will come into being. Among these vortices, longitudinal vortex (LV) has a swirling axis parallel to the main flow direction and moves downstream swirly around this axis and exhibits strong three-dimensional characteristics. Such a barrier is technically defined as longitudinal vortex generator (LVG). From the heat transfer perspective, the three-dimensional swirling movement of LV is useful for heat transfer enhancement (Sohankar, A. & Davidson, L. 2001), and can be employed in many industrial fields such as high temperature vane cooling, convective heat transfer in narrow channels, fin tube heat transfer enhancement, and etc. Therefore, the heat transfer enhancement of LV has been paid close attention and a large number of relevant references can be found, especially in the gas heat transfer cases (Schubauer, G.B. & Spangenberg, 1960; Johnson T.R. & Joubert P.N., 1969).

However, very few researchers chose water as working medium in their studies on heat transfer enhancement by LV in narrow rectangular channels (Chen, Q.Y., et al., 2006; Islam, M.S., et al., 1998; Sohankar, A. & Davidson, L., 2001; Wang, Q.W., et al., 2007). In fact, water is the most important working fluid and widely used in power industry field. For example in some plate-fin water-water heat exchangers, water flows through a series of narrow rectangular channels to release or absorb heat from the other side. If LVG is chosen to increase the heat transfer capability and enhance superficial heat transfer efficiency for such type of heat exchanger, the single and two-phase heat transfer coefficient must be available during design process. In the case of the rod-type fuel assembly used in a pressurized-water nuclear reactor, the vanes on spacers play a role similar to LVG, which reconstructs the velocity and temperature fields and enhances the heat transfer. In this case, the improvement of Critical Heat Flux (CHF) of fuel assembly must also be taken into account (Crecy F., 1994).

For the above mentioned reasons and experiences, the single and two-phase flow and heat transfer and CHF in a narrow rectangular channel with LVG have been studied in this chapter.

2. CFD method and theoretical basis for LVG design

In this study, the LVG was composed of four pairs of rectangular ribs. To optimize the LVG's configuration, the dimensions of a single rib and the whole layout of all ribs were elaborately designed with the help of computational fluid dynamics (CFD) simulation. In the simulation, the computation domain was a narrow rectangular channel which had an aspect ratio of 13 (the channel width to narrow gap ratio), a narrow gap of 3.0mm, and a length to hydraulic diameter ratio of 80.0. And, four pairs of ribs were periodically installed on the unilateral wall in the narrow rectangular channel. In our previous work (Wang L. & Wang Q.W., 2005), each single rectangular rib (with a 50°attack angle, 1.2mm in height, 2.0mm in width, 10.0mm in length, 10.0mm in transverse distance, 100.0mm in axial distanc) had the same original parameters.

CFX5-Build was used as a geometry and mesh generation pre-processor module. Fig.1 shows the computation domain including adjacent two pairs of LVGs and the mesh that were constructed for the simulation using CFX-5 computer code. Mesh generation was based on multi-block structured hexahedral grid. Number of the independent elements in the simulated mesh are 232,632. The surface mesh was created using a Delaunay method and the volume mesh was created through advancing front and inflation method. The mesh elements on the wall surfaces were refined to improve the accuracy in these regions.

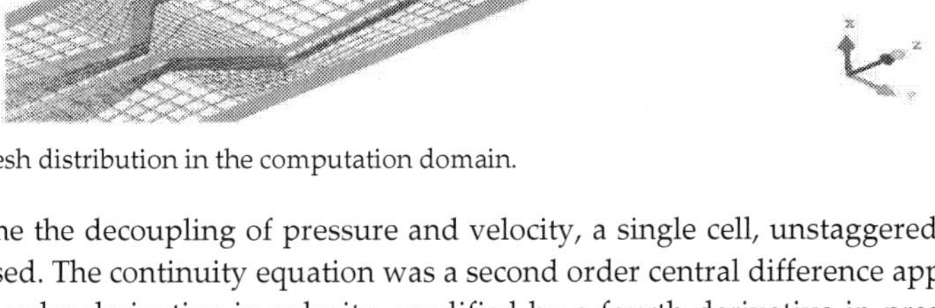

Figure 1. Mesh distribution in the computation domain.

To overcome the decoupling of pressure and velocity, a single cell, unstaggered, collocated grid was used. The continuity equation was a second order central difference approximation to the first order derivative in velocity, modified by a fourth derivative in pressure which acts to redistribute the influence of the pressure. This overcame the problem of checker board oscillations. Transient term took the second order backward Euler scheme, which is robust, implicit, conservative in time, and does not create a time step limitation. Following the standard finite element approach, shape functions were used to evaluate the derivatives for all the diffusion terms. Pressure gradient term was evaluated using the shape functions. Advection term took high Resolution Scheme which does not violate boundedness principles.

The boundary normal velocity was specified at the inlet. The Reynolds numbers based on the channel hydraulic diameter varied from 3200 to 31400. No slip boundary condition was applied to the wall surfaces. Only the wall, on which the ribs stands, was provided with constant heat flux while the other wall were heat insulated. The heat flux was initialized to the value of 100 kW/m^2. The average static pressure was specified at the outlet. De-ioned water was chosen as working fluid.

The Shear Stress Transport (SST) model was used for turbulence modeling. Scalable Wall-Functions were used for near wall treatment. In the calculations, the coupled solver was used to solve the governing equations. During the simulation process, only a single parameter varied while the other parameters kept constant. In each case, friction factor and Nusselt number were calculated as below and they were used for evaluating flow drag and convective heat transfer in the channel.

$$f = \frac{8\tau_{\mathrm{w}}}{\rho u^2} \tag{1}$$

$$Nu = \frac{q_{\mathrm{w}} D_{\mathrm{h}}}{\lambda(T_{\mathrm{w}} - T_{\mathrm{f}})} \tag{2}$$

The weighted factor for every parameter could be calculated as bellow:

$$\varphi_{Nu} = \left|\left(Nu_{\mathrm{av}} - Nu_{\mathrm{min}}\right) / \Delta x\right| \tag{3}$$

$$\varphi_f = \left|\left(f_{\mathrm{av}} - f_{\mathrm{min}}\right) / \Delta x\right| \tag{4}$$

where Δx represents every normalized parameter as $\beta/180°$, X/L, h/H, $b/(B/2)$, s/B and $a/(B/2)$.

The computation result at Reynolds number of 30000 and Prandtl number of 3.7 was shown in Fig.2, but it was difficult to directly determine which parameter has the most marked effect on Nusselt numbers and friction factors. Therefore, it should need a compromise between flow drag and convective heat transfer to optimize the LVG's configuration. The orthogonal method for multi-parameters optimization was proposed by Taguchi and was used for the current case. By choosing dimensionless *JF* number proposed by Yun (Yun J. Y., & Lee K. S., 2000) as target function, the general performance evaluation criteria of heat transfer enhancement with LVG could be expressed as:

$$JF = \left(j_{\mathrm{LVG}} / j_{\mathrm{S}}\right) / \left(f_{\mathrm{LVG}} / f_{\mathrm{S}}\right)^{1/3} \tag{5}$$

where $j = Nu / (Re \cdot Pr^{1/3})$.

By calculation and comparison based on Eq. (5), the final optimized LVG's configuration was obtained as shown in Fig.3. Each single rib had an uniform dimension of 14.0mm×2.2mm×1.8mm, each pair of ribs had an uniform attack angle of 44° and an uniform transverse distance of 4mm, and the total four pairs of ribs had an uniform axial distances of 100.0mm.

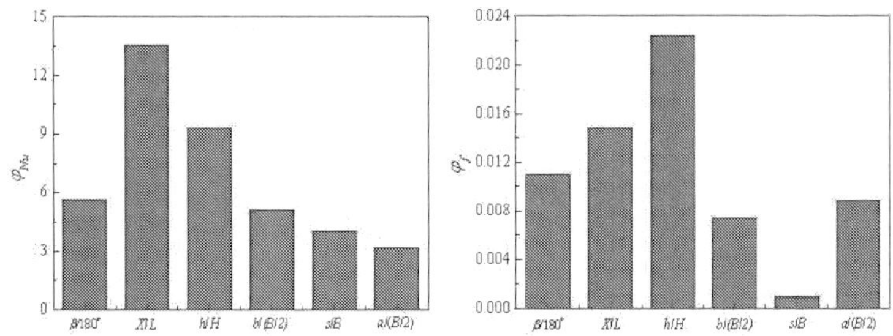

Figure 2. Weighted Nusselt number and friction factor of different parameters.

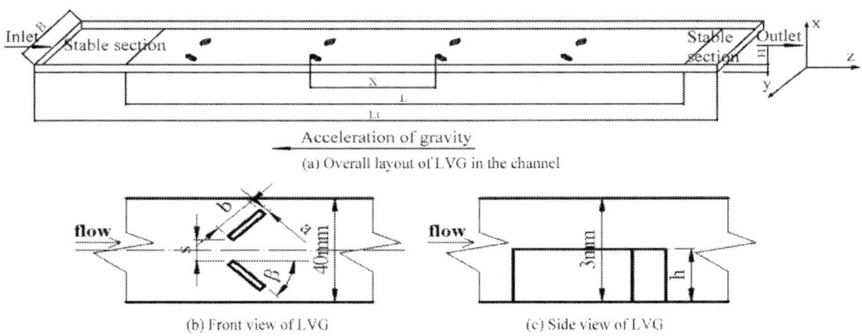

Figure 3. Schematic of the final LVG configuration.

3. Experimental system and method

The experiments were conducted at the flow and heat transfer experimental platform in CNNC Key Laboratory on Nuclear Reactor Thermal Hydraulics Technology, Nuclear Power Institute of China. The experimental apparatus mainly consisted of experimental loop, test section, instruments, power supplier, and data acquisition system. Each part was described in details as following.

1. Experimental loop

It was shown in Fig.4 that the experimental loop contains pumps, pre-heater, pressurizer, flow-meters, test section, mixer, heat exchanger and some valves. Among these parts, the piston pump supplied the loop with de-ioned water from the water tank, the pressurizer kept system pressure steady, the circulating pump drived de-ioned water flow in the loop, the mixer mixed hot water from the outlet of the test section with cold water from the cyclic pump, the flow-meters were used for measuring flow rates in the cyclic pump and test channel, and the heat exchanger cooled water from the outlet of the mixer.

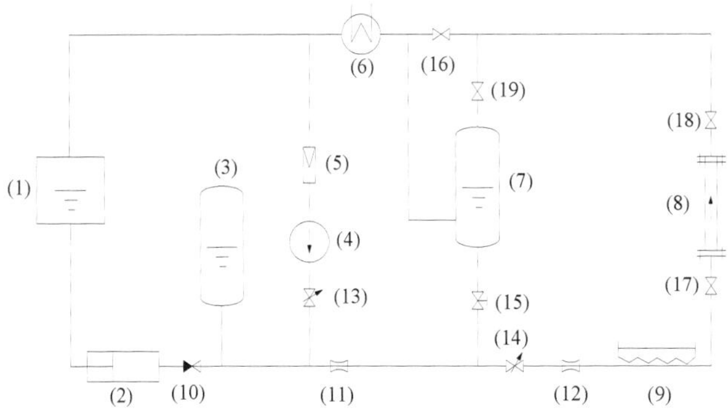

(1) water tank (2) piston pump (3) pressurizer (4) circulating pump (5) filter (6) heat exchanger (7) mixer (8) test section (9) preheater (10) check valve (11),(12) flow-meter (13),(14) flow regulator (15),(16),(17),(18) automatic control valve (19) manual valve

Figure 4. Flow chart of the experimental Loop.

2. Test section

The test section was designed for the two different kinds of experiments, viz., flow and heat transfer experiment and visualization experiment, and mainly consisted of two holders, quartz glass and heating plate. Both the holders and the heating plate were made of 0Cr18Ni10Ti stainless steel, and the other accessories material was chosen dependent on their functions. The total four pairs of rectangular ribs as LVG were machined on the heating plate surface according to the final LVG configuration shown in Fig.3. The whole structure of the test section was shown in Fig.5.

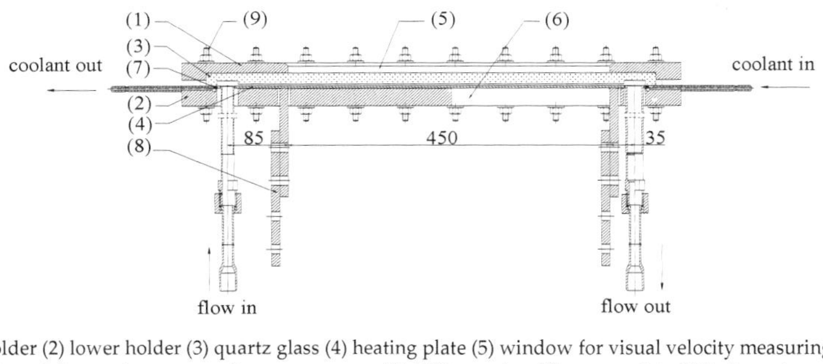

(1) upper holder (2) lower holder (3) quartz glass (4) heating plate (5) window for visual velocity measuring (6) empty space for infrared thermo-image recording temperature (7) "O" type ring (8) copper plate (9) bolts, nuts, and washers

Figure 5. Schematic of the test Section.

There was a narrow rectangular groove with the dimension of 600mm×40mm×3mm on a piece of quartz glass. The heating plate was 610mm×50mm×3mm in dimension. So the close combination between the narrow rectangular groove and the heating plate could form a narrow channel with the cross section of 40mm×3mm in dimension, and the effective heating length was 450mm in dimension as the two ends occupation of the copper plates for power inputs. Thus, the narrow channel had a hydraulic diameter of 5.58mm and a length to diameter ratio of 80.65. The waterproof of the narrow channel was guaranteed by a silicon latex "O" type ring. There was a coolant channel on the surface of the lower holder contacting the outer wall of the heating plate to keep the "O" type ring at an acceptable temperature and prevent from heat concentration on the edges and destruction of heating plate in case of high heat flux. The electrical insulation between the heating plate and the holders was achieved by a piece of isinglass paper with the thickness of 0.3mm. To facilitate recording the outer wall temperature by the use of infrared thermo-imager, a rectangular empty space of 235mm×40mm was machined on the tail half lower holder below the heating plate and located in the region of downstream two pairs of LVG. The cross sectional details of the test section with the empty space were shown in Fig.6.

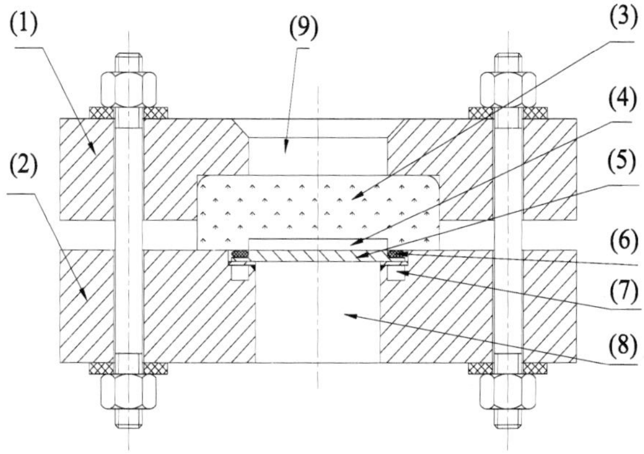

(1) upper holder (2) lower holder (3) quartz glass (4) narrow rectangular channel (5) heating plate (6) "O" type ring (7) cooling channel (8) empty space for thermal infrared imager recording temperature (9) window for visual velocity measuring

Figure 6. Schematic of the cross section of the test section with an empty space.

To accurately measure the outer wall temperatures, the total 45 orifices were drilled in 6 rows and in 15 columns respectively in the span-wise and stream-wise directions on the lower holder for the access of 45 thermocouples(24 were located in the region of the empty space) to the outer wall; each orifice was 3mm in diameter, and each span-wise distance was 9mm while each stream-wise distance is 30mm. To measure the pressure drop across the test section, a hole with 13mm in diameter was drilled near the first end copper plate on the lower holder for the access of the first pressure tap to the channel, there was no need to drill another hole downstream 424mm because the second pressure tap crossed the empty space and could be directly connected with the hole on the heating plate.

After the thermocouples being fixed, the two bakelite plates were used for holding the thermocouples, and many layers of asbestos piece as heat insulation stuff filled the empty space between the two bakelite plates. This measure could prevent from heat loss brought by the empty space. Before conducting the flow and heat transfer experiment, the test section was wrapped with a thick layer of asbestos piece for further heat insulation. It was noticed that the empty space on the lower holder need to be exposed to atmosphere so that the camera could easily receive the infrared thermo-images in the visualization experiments.

3. Instruments, power supplier, and data acquisition system

The pressures at the inlet and outlet of the test channel were measured using SMART3000 type pressure transducer. The pressure drop across the test channel was measured using SMART3000 type differential pressure transducers. The pressure transducer and the differential pressure transducers were respectively connected to the fittings attached to the hole with 13mm in diameter on the lower holder and to the other one downstream 424mm on the heating plate. The accuracy in the measurement of the inlet and outlet pressures was about ±3kPa. The accuracy in the measurement of the pressure drop across the test channel was about ±30Pa. The pressure at the inlet of coolant channel was measured using a manometer with an accuracy of ±10kPa.

The flow rates of the test fluid and coolant were respectively measured using a Venturi-tube type flow-meter which coupled with a SMART3000 type differential pressure transducer with an accuracy of ±30Pa. The Venturi-tubes were calibrated prior to their installation, and had a calibration of accuracy of ±2%.

The temperatures at the inlet and outlet of the test channel and the coolant channel and the outer wall temperature on the heating plate were measured using sheathed N-type thermocouples with 1mm in diameter. All thermocouples were calibrated prior to their installation. All thermocouples used in the experiments had a calibration accuracy of ±0.3℃.

The pre-heater had a power capacity of 100kW and was able to achieve the appropriate temperature at the inlet of the test channel. The heating plate was directly heated using a DC power supply with a power capacity of 200kW.

All the above mentioned parameters were recorded by a data acquisition system (AT-96) connected to a computer.

4. Single-phase flow and heat transfer

During the experiments, the parameters varied as shown in Table.1. Firstly, the single-phase flow and heat transfer with and without LVG in the channel were discussed to quantitatively learn heat transfer enhancement accompanied by flow drag increase by LV, then the visual velocity and temperature distributions in the two cases were shown to qualitatively explain heat transfer enhancement mechanism related to LV.

Case	With LVG	Without LVG
Thermal boundary	Uniform heat flux on the heating plate	
System pressure	0.25-0.72MPa	0.25-0.72MPa
Mass flow flux	47.5-592.6kg/(m^2 s)	49.1-669.7kg/(m^2 s)
Inlet temperature	22.4-88.0℃	23.1-76.4℃
Heat flux	0-201.5kW/m^2	0-191.4kW/m^2
Reynolds numbers	310-9909	311-9247
Prandtl numbers	4.12-5.27	4.44-5.79

Table 1. Experimental parameters ranges for single-phase flow and heat transfer.

4.1. Flow behavior with and without LVG

It was shown in Fig.7 that the variation of friction factor as a function of Reynolds number in the smooth channel. The laminar-to-turbulent transition occurred at Reynolds number being around 1900. The best fits for data respectively in laminar and turbulent regime as below:

$$f_0 = 176.78 / \mathrm{Re}, \quad \text{where,} \quad \mathrm{Re} < 1900 \tag{6}$$

$$f_0 = 0.3887 / \mathrm{Re}^{0.25}, \quad \text{where,} \quad \mathrm{Re} > 1900 \tag{7}$$

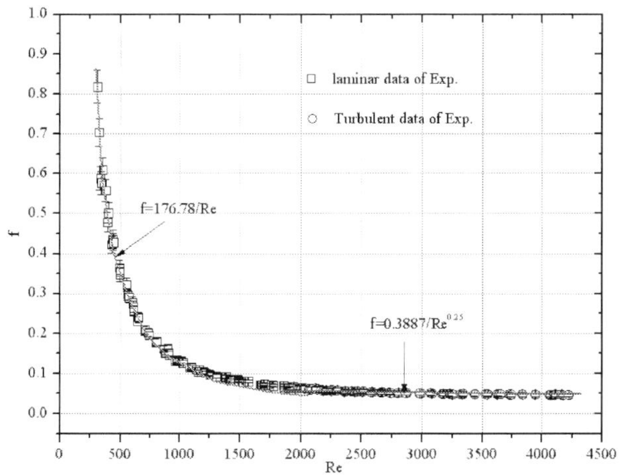

Figure 7. Variation of friction factors with Reynolds numbers in the smooth channel.

It was noticed that, because the channel flow was hydro-dynamically developing, the laminar friction factors were much higher than the corresponding analytical solutions proposed by Kays and Clark (Kays, W.M. & Clark. S.H., 1953), and the turbulent ones were a little higher than the values calculated by Blasius correlation.

In the channel with LVG, the laminar-to-turbulent transition occured at Reynolds number being around 1650, which was shown in Fig.8. The best fits for data respectively in laminar and turbulent regime as below:

$$f = 196.86 / \mathrm{Re}, \quad \text{where,} \quad \mathrm{Re} < 1650 \tag{8}$$

$$f = 0.7786 / \mathrm{Re}^{0.25}, \quad \text{where,} \quad \mathrm{Re} > 1650 \tag{9}$$

In comparison with the smooth channel case, the laminar-to-turbulent transition occurred earlier, and the flow drag became higher in the channel with LVG.

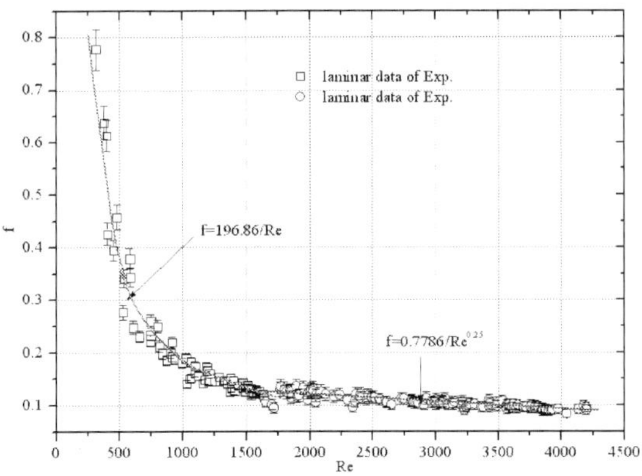

Figure 8. Variation of friction factors with Reynolds numbers in the channel with LVG.

4.2. Heat transfer with and without LVG

It was shown in Fig.9 that the variation of mean Nusselt number as a function of Reynolds number in the smooth channel. The best fits for data respectively in laminar and turbulent regime as below:

$$Nu_0 = 0.0117\,\mathrm{Re}\,\mathrm{Pr}^{1/3}, \quad \text{where,} \quad \mathrm{Re} < 1900 \tag{10}$$

$$Nu_0 = 0.023\,\mathrm{Re}^{0.9}\,\mathrm{Pr}^{0.4}, \quad \text{where,} \quad \mathrm{Re} > 1900 \tag{11}$$

It was also noticed that, because the channel heat transfer was thermally developing and the fluid property varied, the laminar Nusselt numbers weren't constant and were much higher than the corresponding analytical solutions proposed by Shah and London (Shah, R.K. & A.L., 1978), and the turbulent ones were a little higher than the values calculated by Dittus-Boelter correlation.

It was shown in Fig.10 that the variation of mean Nusselt number as a function of Reynolds number in the channel with LVG. The best fits for data respectively in laminar and turbulent regime as below:

$$Nu = 0.0235\,\mathrm{Re}\,\mathrm{Pr}^{1/3}, \quad \text{where,} \quad \mathrm{Re} < 1650 \tag{12}$$

$$Nu = 0.023\,\mathrm{Re}^{0.983}\,\mathrm{Pr}^{0.4}, \quad \text{where,} \quad \mathrm{Re} > 1650 \tag{13}$$

In comparison with the smooth channel case, the heat transfer was enhanced obviously in the channel with LVG.

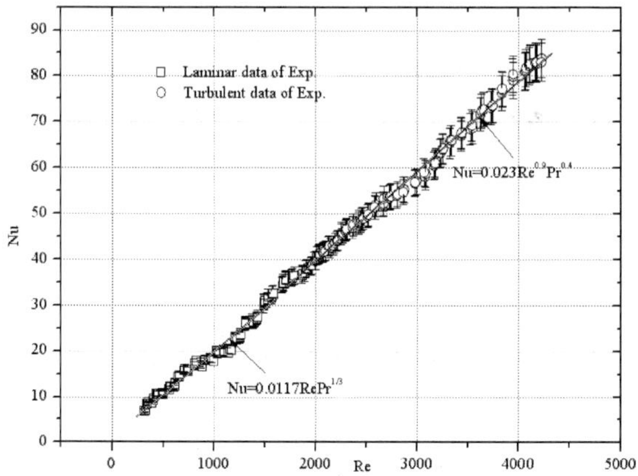

Figure 9. Variation of mean Nusselt numbers with Reynolds numbers in the smooth channel.

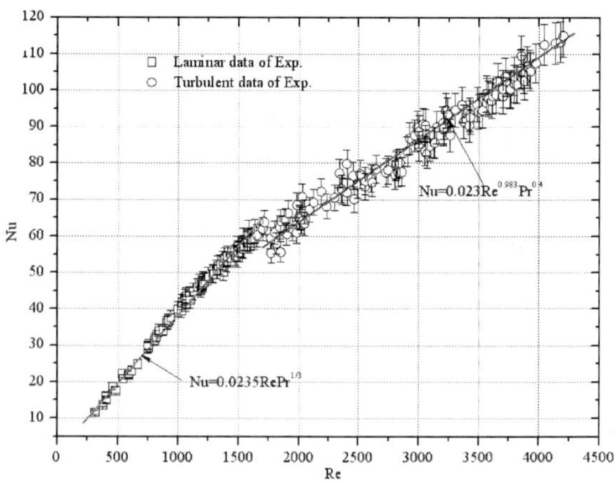

Figure 10. Variation of mean Nusselt numbers with Reynolds numbers in the channel with LVG.

4.3. Evaluation of heat transfer enhancement

Under the conditions of this project, the channel flow and heat transfer in the two cases (with LV and without LV) were thermally and hydro-dynamically developing owing to entry length effects. With the above discussed, the further interesting results could be introduced.

In the laminar regime with Reynolds numbers lower than 1650, the comparative differences of friction factor and mean Nusselt number between the two different cases were calculated as:

$$\frac{f - f_0}{f_0} \times 100\% = 11.4\% \tag{14}$$

$$\frac{Nu - Nu_0}{Nu_0} \times 100\% = 100.9\% \tag{15}$$

Also, the integral index for evaluating heat transfer enhancement by LV was given as:

$$\frac{Nu / Nu_0}{f / f_0} = 1.8 \tag{16}$$

The above three values indicated that LV could enhance laminar heat transfer up to 100.9% while flow drag increased 11.4% in the test channel, and heat transfer enhancement was 1.8 times against flow drag increase. In a word, the degree of heat transfer enhancement was superior to that of flow drag increase in laminar regime.

In the turbulent regime with Reynolds numbers higher than 1900, the calculated values were as below:

$$\frac{f - f_0}{f_0} \times 100\% = 100.3\% \tag{17}$$

$$\frac{Nu - Nu_0}{Nu_0} \times 100\% = (\mathrm{Re}^{0.083} - 1) \times 100\% > 87.1\% \tag{18}$$

$$\frac{Nu / Nu_0}{f / f_0} = \frac{\mathrm{Re}^{0.083}}{2.003} \tag{19}$$

Therefore, LV could enhance turbulent heat transfer over 87.1% while flow drag increased 100.3% in the test channel, and the heat transfer enhancement to the flow drag increase ratio was a power law function of Reynolds numbers, and the performance of heat transfer enhancement would become better with increase Reynolds numbers in turbulent regime.

4.4. Velocity distribution with and without LVG

To learn LV's behavior related to flow drag increase, a Phase Doppler Particle Analyzer (PDPA) was employed to measure velocity distribution in the two different channels. Each measuring plane was perpendicular to the channel axial direction; the measuring coordinates and data acquisition dots were shown in Fig.11.

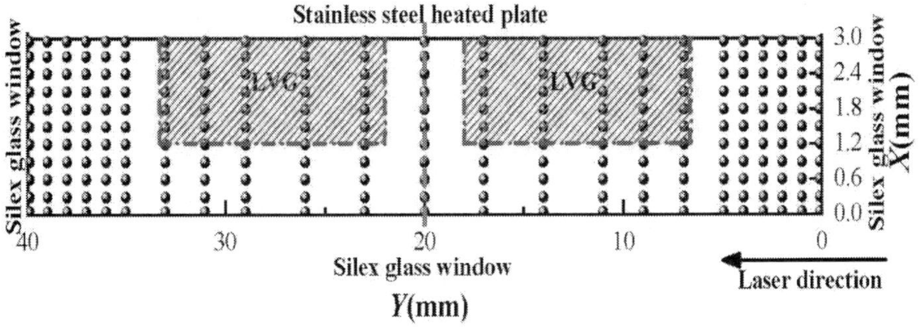

Figure 11. Schematic of measuring coordinates and data acquisition dots using PDPA

The first measuring plane was located at Z=65mm in the axial direction and was 15mm far from the front corners on the first pair of ribs. The data measured in the two different cases were shown in Fig.12. It was seen that LVG caused the normal velocity distribution distort (still symmetrical) at different locations. This phenomenon indicated that the LVG could produce LV and reconstruct velocity field in the channel. In Fig.12(a), a stagnation domain formed immediately behind the ribs, the bulk velocity decreased, and the boundary layer became thicker; when moving downstream, LV was stronger because of upstream superimposition. In Fig.12(b) and Fig.12(c), the wall effect became weaker, the bulk velocity increased, the boundary layer became thinner, and the bulk velocity in the channel with LVG was bigger than that in the smooth channel. During moving downstream, LV swirled from channel center to both sides in an involute path, thus exhibited strong three-dimensional characteristics. In comparison with the case of smooth channel, the form drag was the reason for flow drag increase in the channel with LVG.

4.5. Temperature distribution with and without LVG

To learn LV's behavior related to heat transfer enhancement, a thermal infrared imager was employed to measure outer wall temperature distribution in the two different channels. The measuring region was on the heating plate where the tail two pairs of ribs were against the empty space.

The comparison of measuring images from the two different cases was shown in Fig.13. In the smooth channel, the outer wall temperature distribution was homogeneous, which was coincidence with the thermal characteristics under constant heat flux wall boundary. In the channel with LVG, the symmetrical lower temperature region formed immediately behind the ribs, and periodically appeared and disappeared in the axial direction. This phenomenon indicated that LV could reconstruct temperature field and improve local heat transfer capability periodically in the channel.

The more accurate result was shown in Fig.14. It could be found that the mean outer wall temperature in the channel with LVG was lower than that in the smooth channel by 18.1%. Therefore, LV could obviously enhance heat transfer in the channel.

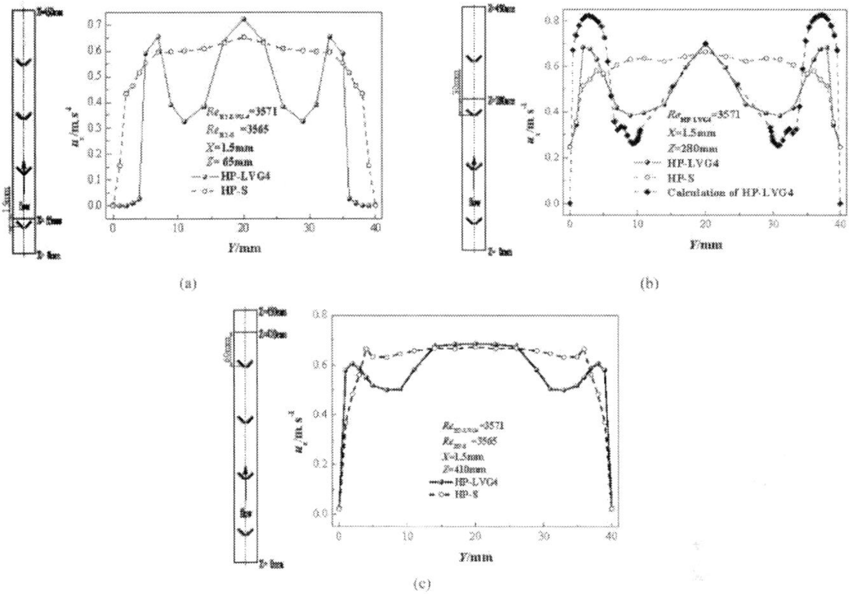

Figure 12. Comparison of velocity distribution in the two cases.(p_{in}=0.465Mpa, u_{in}=0.493m/s, and T_{in}=23.3◎)

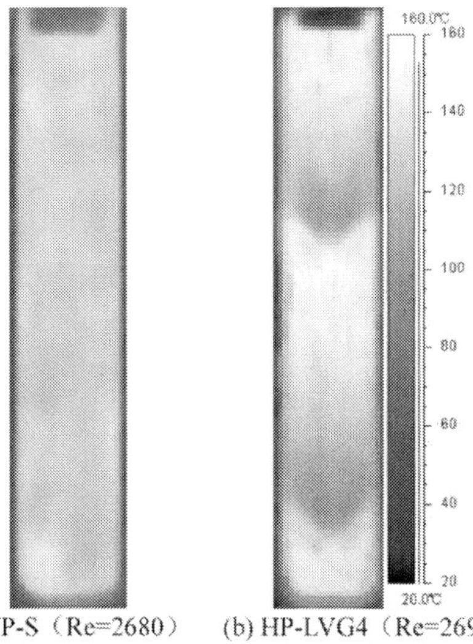

Figure 13. Outer wall temperature fields in the two cases in turbulent flow

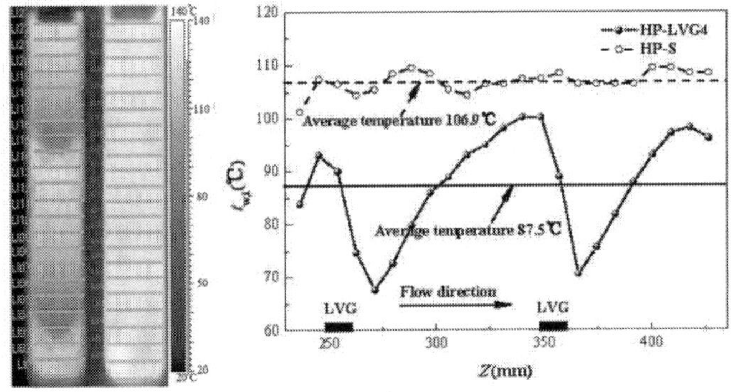

Figure 14. Quantitative comparison of wall temperature between the two cases.(p_{in}=0.461MPa, G=435.9kg/(m^2·s), $T_{l,m}$=34.1◎,and q=240.9kW/m^2)

5. Vapor-liquid two-phase flow and heat transfer

During the experiments, the parameters varied as shown in Table.2. Firstly, bubble behaviors with and without LVG in the channel were observed to learn the two-phase heat transfer enhancement accompanied by flow drag increase, then two-phase flow and heat transfer in the two cases was discussed.

Case	With LVG	Without LVG
Thermal boundary	Uniform heat flux on the heating plate	
System pressure	0.44-0.81MPa	0.45-0.82MPa
Mass flow flux	65.9-415.9kg/(m^2 s)	50.7-330.4kg/(m^2 s)
Inlet temperature	93.6-120.8℃	93.2-128.1℃
Heat flux	8.7-885.9kW/m^2	7.0-834.6kW/m^2
Reynolds numbers	1353-7364	1596-6758
Prandtl numbers	1.23-1.57	1.25-1.73

Table 2. Experimental parameters ranges for two-phase flow and heat transfer.

5.1. Bubble behaviors with and without LVG

In order to learn LV behavior related to two-phase flow drag increase, a high-speed camera was employed to record bubble behaviors in the two different cases, and the measuring region contained all the four pairs of ribs. It was shown in Fig.15 that the bubble evolve between z=45mm and z=75mm near the first pair of ribs, where Onset of Nucleate Boiling (ONB) began to appear. Obviously, the number of bubbles attached on the wall was small, and bubbles mainly distributed in both sides and the central region of the channel; this phenomenon could be ascribed to the bubble production being small and LV stir. The arrowed line denoted the dominant region by LV. In this region, bubbles couldn't congregate on the wall, but moved downstream with bulk fluid, thus the heat transfer capability between the cold and hot fluid was improved.

Figure 15. Bubbles distribution near the first pair of ribs in the channel with LVG.(pin=0.507MPa, G=413.6kg/(m² • s), ΔTsub=44.6℃,and q=296.7kW/m²)

It was illustrated respectively in Fig.16 and Fig.17 that the bubbles evolve with heat flux increase in the two different cases. In the smooth channel, a huge number of coalescent bubbles moved downstream without transverse stir. In the channel with LVG, the number of bubbles obviously decreased a lot, the bubble transverse stir was stronger, the bubble growth was inhibited, and bubbles weren't easy to coalesce, thus more heat on the heated surface was transferred. As being bounded in narrow gap, distorted bubbles almost occupied the whole space in the transverse direction; the big bubbles directly pushed superheated liquid layer depart from the wall surface; the small ones were baffled by these slow rising big bubbles and moved around; a single bubble couldn't exist; a large number of bubbles coalesced and improved vapor quality. All of this facilitated the two-phase mixture to speed up. Therefore, the bubble departure diameter became smaller while the bubble growth frequency became higher, and the bubble stir was strengthened, which was one of the important factors for boiling heat transfer enhancement.

5.2. Two-phase friction pressure drop with and without LVG

Two-phase friction pressure drop was an important hydrodynamic parameter to measure flow drag increase in channel flow. In this study, the two-phase friction pressure drop in the channels was most dependent on the exit quality in all related parameters. The best fits for two-phase friction pressure drops in the two different channels were listed as the following correlations:

$$\Delta p_{\mathrm{f,smooth}} = 3959.118\left(-3.664x^2 + 3.409x + 0.020\right)^{1.919} \tag{20}$$

$$\Delta p_{\mathrm{f,LVG}} = 1100.514\left(171.642x^2 - 7.266x + 0.168\right)^{0.277} \tag{21}$$

where x denoted the channel exit quality and varied between 0.179 and 0.6. The predicted results by the two equations deviated to the corresponding experimental data by ±30%, which was shown in Fig.18.

It was calculated using Eq.(20) and Eq.(21) that the two-phase friction pressure drop ratio between the two cases was 1.04-1.506, or the value in the channel with LVG .

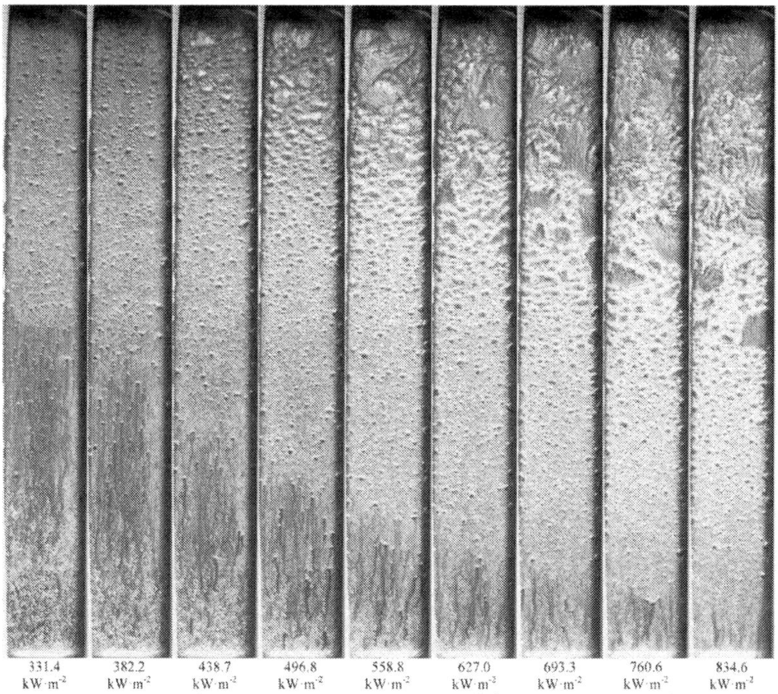

Figure 16. Bubbles evolution with heat flux increase in the smooth channel.(pin=0.620MPa, G=270.7kg/(m² · s), and △Tsub=57.9℃)

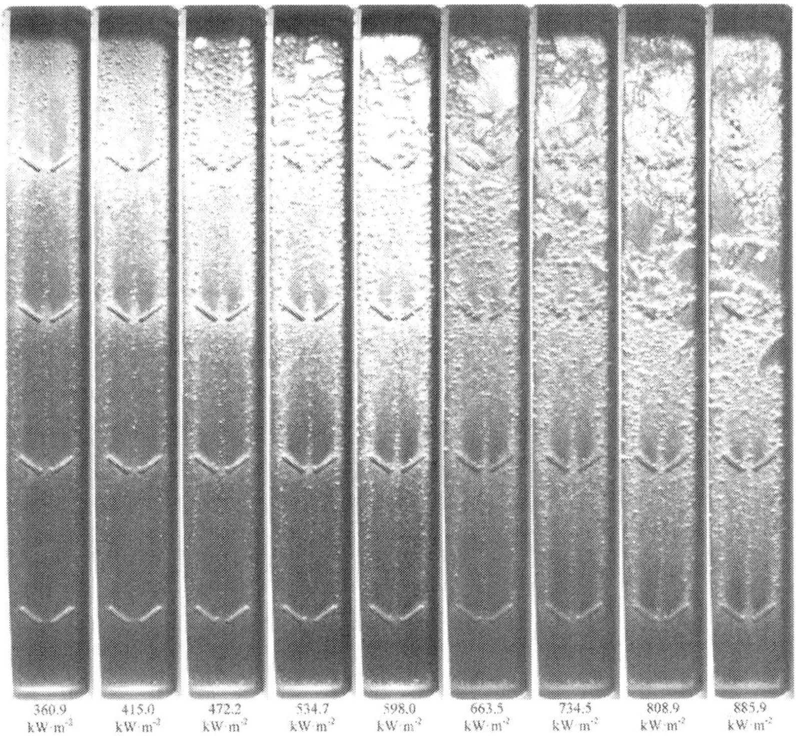

Figure 17. Bubbles evolution with heat flux increase in the channel with LVG.(pin=0.654MPa, G=278.9kg/(m2 · s), and ΔTsub=58.2℃)

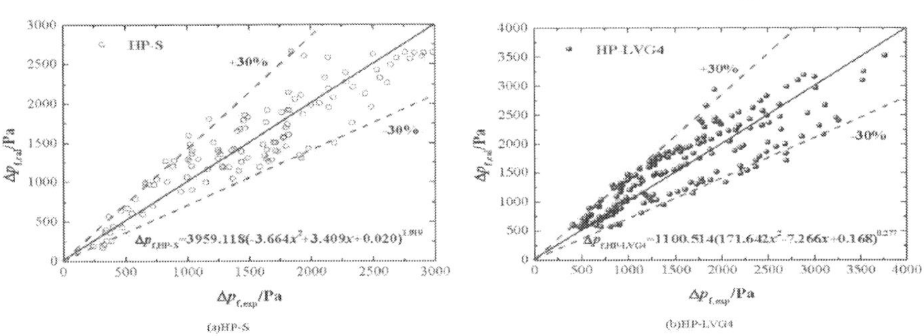

Figure 18. Friction pressure drop comparison between prediction and experiment in the two cases.

5.3. Two-phase boiling heat transfer with and without LVG

In this study, the two-phase boiling heat transfer behavior was discussed. The heat transfer coefficients were correlated with Reynolds number, Boiling number, thermal conductivity and hydraulic diameter by taking the form proposed by Gungor and Winterton(Gungor K.E., Winterton R.H.S., 1987), which was adaptable to narrow rectangular channel with a big aspect ratio. The best fits for heat transfer coefficient in the two different channels were listed as below:

$$h_{\mathrm{TP,without\ LVG}} = 1.40 Re \cdot Bo^{0.349} \lambda / D_{\mathrm{hy}} \tag{22}$$

$$h_{\mathrm{TP,with\ LVG}} = 3.05 Re \cdot Bo^{0.440} \lambda / D_{\mathrm{hy}} \tag{23}$$

where the Reynolds number varied between 1596 and 6758, and the Boiling number varied between 2.15×10^{-4} and 2.137×10^{-3}.

The prediction by the two equations deviated to the corresponding experimental data by ±30%, which was shown in Fig.19.

It was calculated using Eq.(22) and Eq.(23) that the two-phase boiling heat transfer coefficient ratio between the two cases was 1.011-1.258, or the value in the channel with LVG was 1.1%-25.8% higher than that in the smooth channel.

6. Critical Heat Flux

In this section, CHF with and without LVG in the test channel was intentionally studied. Firstly, the observation of bubble and liquid film behaviours was conducted when CHF occurred in the two different channels; then the quantitative CHF in the two cases was discussed especially in the parameter dependence and experimental correlations; finally the analytical models for CHF in the two cases were proposed and validated. The varied parameters during the experiment were listed in Table.3.

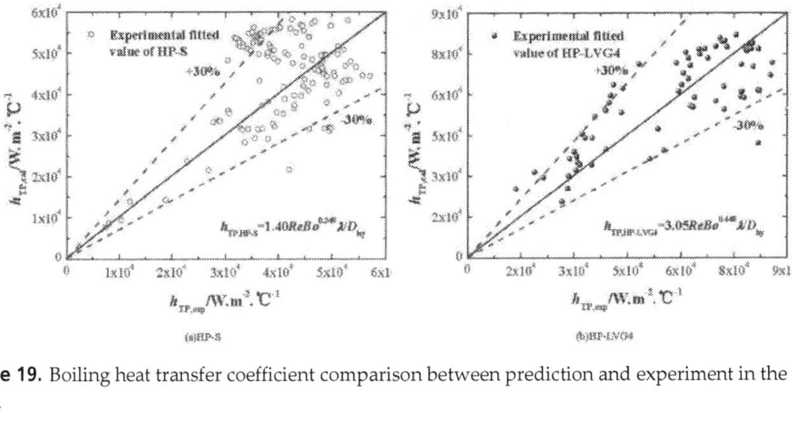

Figure 19. Boiling heat transfer coefficient comparison between prediction and experiment in the two cases.

Case	With LVG	Without LVG
Thermal boundary	Uniform heat flux on the heating plate	
System pressure	0.43-0.85MPa	0.44-0.82MPa
Mass flow flux	40.2-690.1kg/(m²s)	40.2-745.7kg/(m²s)
Subcooling	46.8-104.2℃	51.7-100.2℃
Quality	0.203-1.000	0.183-0.997
CHF	0.299-2.316MW/m²	0.294-2.263MW/m²

Table 3. Experimental parameters ranges for CHF experiments

6.1. Bubble and liquid film behaviours

In order to learn LV's behaviour during CHF occurrence, the high-speed camera was employed to record bubble behaviours in the two different cases, and the measuring region contained all the four pairs of ribs.

By comparison between Fig.20 and Fig.21, it was found that ONB in the channel with LVG appeared later than that in the smooth channel when CHF occurred. The main reason for this phenomenon was that the bubbles growth was inhibited by LV, and the number of bubbles decreased with the generation of LV, and the bubbles were very difficult to coalesce under LV dominance. Simultaneously, the disturbance of liquid film on the wall in the channel with LVG was much stronger than that in the smooth channel. As a result, CHF was greatly improved with the help of LV.

In more details, it was found in Fig.22 that the void fraction increased in the axial direction, and the transverse distribution of vapour-phase varied under LV dominance. The channel flow and heat transfer would greatly change with such heterogeneousness. First of all, the disturbance of bubbles, energy transfer and momentum transfer in the channel were all strengthened by LV. Secondly, the bubbles growth and coalescence were inhibited by LV, also the bubbles collision frequency was low. Thirdly, the enhanced bubbles disturbance facilitated to destroy thermal boundary layer. Finally, the bubbles would be brought from the side wall to the central region by LV. All these factors would help improve heat transfer and CHF in the channel.

On the one hand, it was shown in Fig.20 that CHF in the smooth channel occurred at the moment of 0.380s, prior to which, the wall surface had been covered with a layer of stable liquid film. When CHF occurred, the liquid film began to lose stability owing to the rapid pressure drop variation with heat flux, though the heating power had been cut down by 50%. On the other hand, Fig.21 showed that CHF in the channel with LVG occurred at the moment of 0.400s, prior to which, the liquid film had been disturbed by LV. This phenomenon was beneficial to accelerate the liquid film evaporation, thus more energy was transferred, and CHF was improved.

It was also seen in Fig.22 that liquid film evolved near the fourth pair of ribs during CHF occurrence, the vapour-phase occupied the channel central region while the liquid-phase concentrated on the side wall, and thus a typical annular flow pattern came into being. Under LV dominance, the liquid film distribution became heterogeneous on the wall surface, and the wall temperatures fluctuated periodically, while annular flow and slug flow alternately appeared in the channel. Such a reverse flow pattern transition visually indicated CHF enhancement in the channel with LVG.

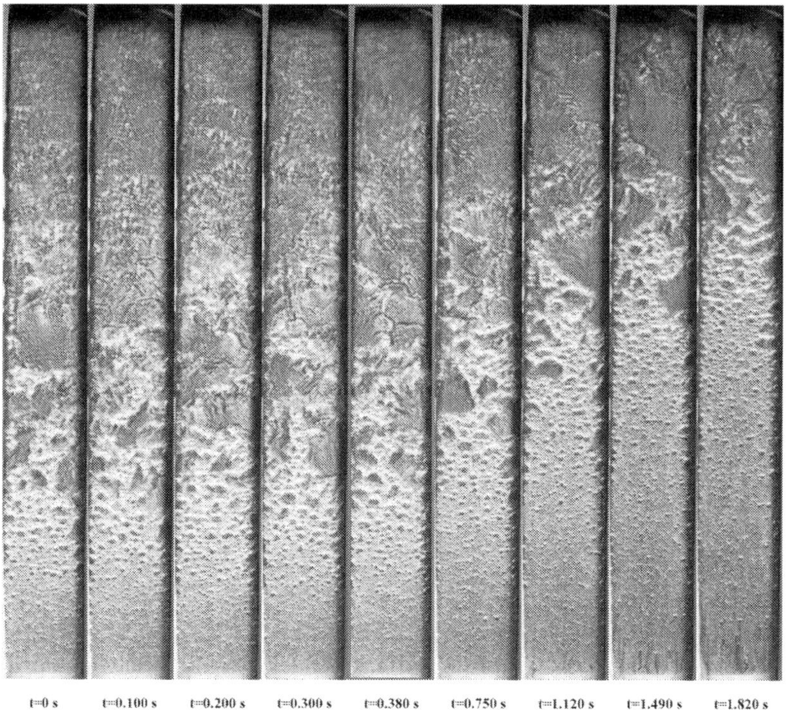

Figure 20. Bubble evolution during CHF occurrence in the smooth channel. (pin=0.713MPa, G=217.1kg/(m2 • s), ΔTsub=52.1℃,and qc=1.287MW/m2)

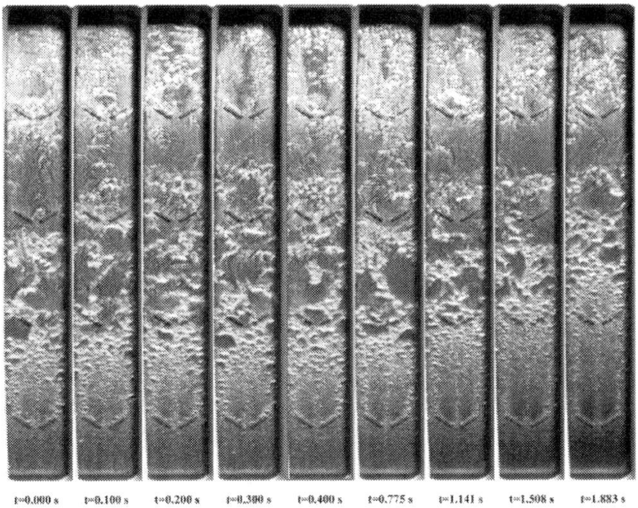

Figure 21. Bubble evolution during CHF occurrence in the channel with LVG. (pin=0.763MPa, G=198.9kg/(m2・s), ΔTsub=61.4℃, and qc=1.327MW/m2)

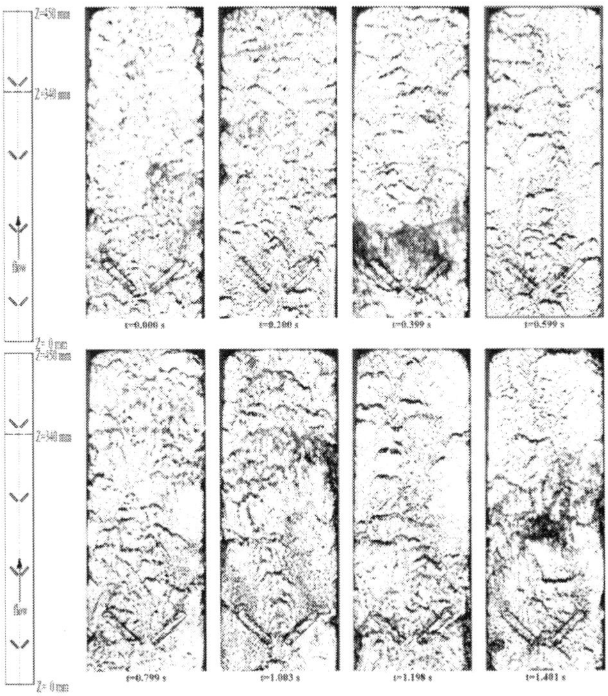

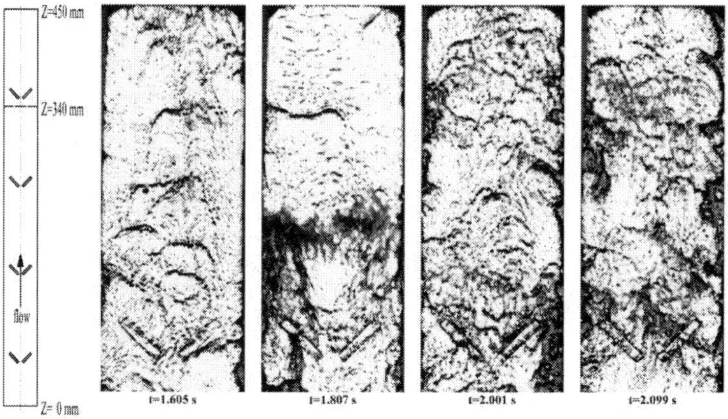

Figure 22. Liquid film evolution during CHF occurrence near the fourth pair of ribs in the channel with LVG. (pin=0.663MPa, G=188.2kg/(m2 • s), △Tsub=53.1℃, and qc=1.238MW/m2)

6.2. Parameter dependence and experimental correlations

By reduction on experimental data, it could be found in Fig.23 that the CHF value varied with different parameters in the two different channels. In these two cases, the CHF value monotonously decreased with exit quality increase, and monotonously increased with mass flow flux and pressure drop increase, but had no obvious variation with system pressure and inlet subcooling. It seemed that the difference between the two cases was difficult to find, therefore, the more detailed discussion needed to be conducted.

Taking the above mentioned parameter dependence into consideration, the best fits for CHF in the two cases were obtained as below:

$$q_{c,\mathrm{smooth}} = \left(0.0022G - 0.0767\right)^{0.5002}\left(0.0005\Delta t_{\mathrm{sub,in}} + 0.6695\right)^{1.6000} \times \left(-10.9828p_{\mathrm{in}}^2 + 15.9420p_{\mathrm{in}} - 1.9973\right) \quad (24)$$

$$q_{c,\mathrm{LVG}} = \left(0.0315G - 0.7638\right)^{0.5385}\left(0.0142\Delta t_{\mathrm{sub,in}} - 0.4344\right)^{0.2413} \times \left(-1.9409p_{\mathrm{in}}^2 + 2.9748p_{\mathrm{in}} - 0.5006\right) \quad (25)$$

The prediction by the two equations deviated to the corresponding experimental data by ±10%, which was shown in Fig.24.

It was calculated using Eq. (24) and Eq. (25) that the CHF value in the channel with LVG was 24.3% higher than that in the smooth channel within the present experimental parameter range.

Also, the best fits for pressure drop in the two cases were obtained as below:

$$\Delta p_{\text{c,smooth}} = \left(2.1854G - 46.0350\right)^{0.5842}\left(-1.0776\Delta t_{\text{sub,in}} + 165.4139\right)^{0.3839} \times\left(-216.8849p_{\text{in}}^2 + 278.0483p_{\text{in}} - 49.6720\right) \quad (26)$$

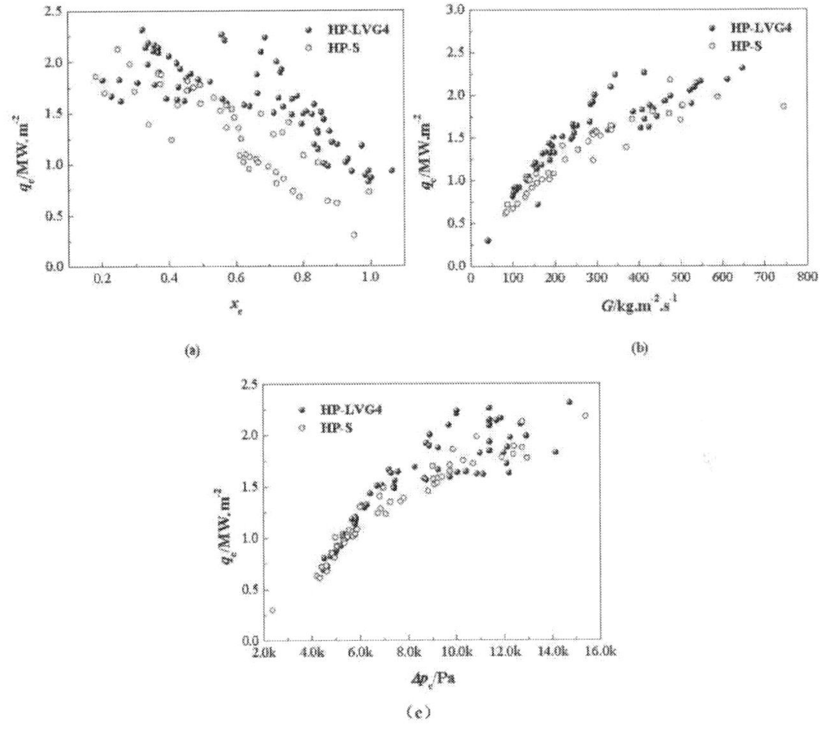

Figure 23. CHF variation with several thermal-hydraulic parameters in the two channels

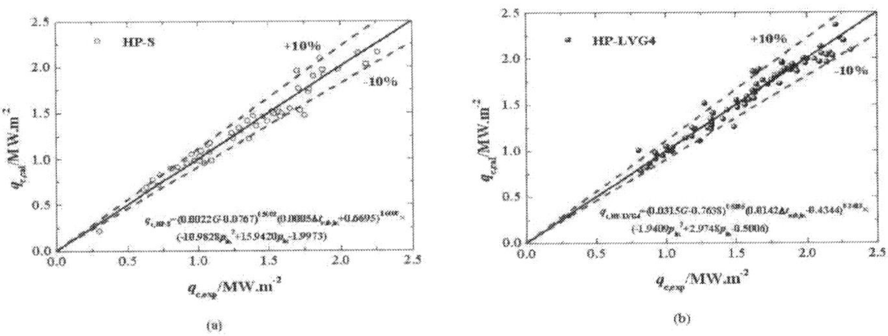

Figure 24. Comparison between the experimental data with the predicted values of CHF in the two channels.

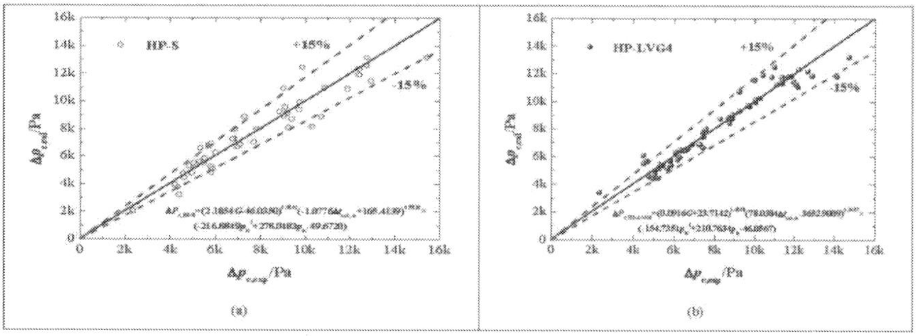

Figure 25. Comparison between the experimental data with the predicted value of pressure drops in the two channels.

$$\Delta p_{c,\text{HP-LVG4}} = \left(0.0916G + 23.7142\right)^{1.4808}\left(78.0384\Delta t_{\text{sub,in}} - 3652.9009\right)^{0.0057} \times\left(-154.7351p_{\text{in}}^{2} + 210.7634p_{\text{in}} - 46.0567\right) \tag{27}$$

The prediction by the two equations deviated to the corresponding experimental data by ±15%, which was shown in Fig.25.

It was calculated using Eq. (26) and Eq. (27) that the pressure drop value in the channel with LVG was 62.9% higher than that in the smooth channel within the present experimental parameter range.

Obviously, the CHF increase must be at the cost of pressure loss in the channel with LVG, which well followed the second thermodynamic law.

6.3. Analytical models for CHF

CHF is a vital parameter for Nuclear Reactor Design and operation. Up to now, a large number of experimental investigations have been conducted for different channel type, and a large amount of experimental data and correlations have been obtained. However, the applicability of these data based correlations is strictly limited by experimental parameters. Analytical model for CHF is encouraged due to its advantages on physical mechanism and applicability.

The annular flow liquid film dry-out mechanism for CHF (Joel, 1992) has been known very well. Fig.26 exhibited the CHF occurrence process related to this mechanism. Owing to the simultaneous effects from the droplet deposition, droplet entrainment, and liquid film evaporation, the liquid film on the wall surface become more and more thin and even disappear, CHF will occur at some point. On the basis of this mechanism, the multi-fluid model for CHF had been proposed, but the modelling was complicated as too many field equations need to be solved. In addition, several constitutive correlations were not accurate enough because the studies on two-phase interface transfer had been still premature. On the contrary, Celata and Zummo's analytical model incorporated the appropriate droplet deposition and entrainment, and was able to predict the CHF in annular flow with a comparative high accuracy, therefore the analytical model for CHF in this project will refer to this model.

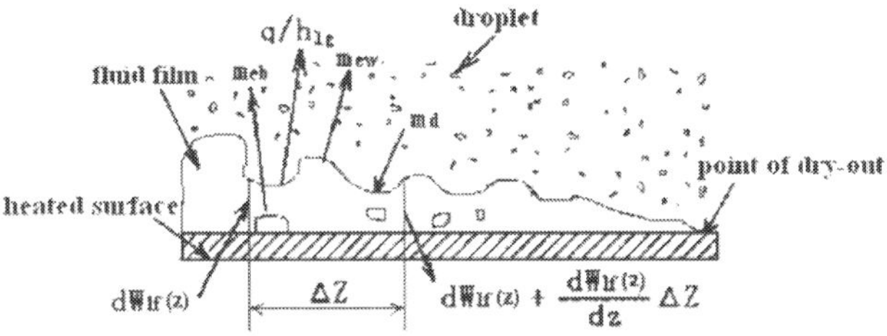

Figure 26. Schematic of liquid film dry-out mechanism of annular flow in narrow rectangular channel.(m_d, m_{ew}, m_{eb}, and q/h_{lg} were droplet deposition rate, vapour stream shear caused droplet entrainment rate, broken bubble-caused droplet entrainment rate and liquid film evaporation rate, respectively.)

Despite that the flow patterns and CHF analytical model for narrow rectangular channel had been difficult to find in open literature, several researchers (Jackey, et al., 1958; Kafengauz and Bocharov, 1959) suggested that the correlations for CHF in conventional pipe could be used for the case in narrow rectangular channel if the equivalent characteristics scale was same, which was shown in Fig.27.

6.3.1. *Onset of annular flow*

Mishima and Ishii's (1984) criterion was employed to predict the onset of annular flow in the present cases:

$$j_g \geq \left(\frac{\sigma g \Delta\rho}{\rho_g^2}\right)^{0.25} / \left[\mu_l \left(\rho_l \sigma \sqrt{\frac{\sigma}{g\Delta\rho}}\right)^{-0.5}\right]^{0.2} \tag{28}$$

Eq.(28) was tenable in the range as following: pressure between 0.1Mpa and 20.0 MPa, mass flow flux between 8 kg/(m²·s) and 15,000 kg/(m²·s), channel equivalent diameter between 0.001m and 0.025 m, channel length to equivalent diameters ratio between 1 and 400, inlet subcooling between 10◎ and 255◎, and exit quality between 0.1 and 1.

6.3.2. *Mass flow rate of liquid film in annular flow region*

If annular flow appeared in the channel, the onset location (z_{on}) could be obtained, and the mass flow rate of liquid film was calculated by Eq. (36) integral between the onset location and the channel outlet.

$$\frac{dW_{lf}}{dz} = \pi D_{hy}\left(m_d - m_{ew} - m_{eb} - \frac{q}{h_{lg}}\right) \tag{29}$$

The droplet deposition rate was calculated using Kaotaoka and Ishii's (1983) correlation.

$$m_{\rm d} = 0.22Re_{\rm l}^{0.74}E^{0.74}\left(\frac{\mu_{\rm g}}{\mu_{\rm l}}\right)^{0.26}\frac{\mu_{\rm l}}{D_{\rm hy}} \tag{30}$$

where $Re_{\rm l}$ was the liquid-phase Reynolds number; E was the droplet entrainment fraction in the channel cross section, and was calculated as below:

$$E = 1 - \frac{j_{\rm lf}}{j_{\rm l}} \tag{31}$$

where the superficial liquid film velocity was calculated as below:

$$j_{\rm lf} = \frac{4W_{\rm lf}}{\rho_{\rm l}\cdot\pi\cdot D_{\rm hy}^2} \tag{32}$$

Owing to the mass flow rate $W_{\rm lf}$ was an unknown quantity in Eq. (32), an iteration method was used to calculate E.

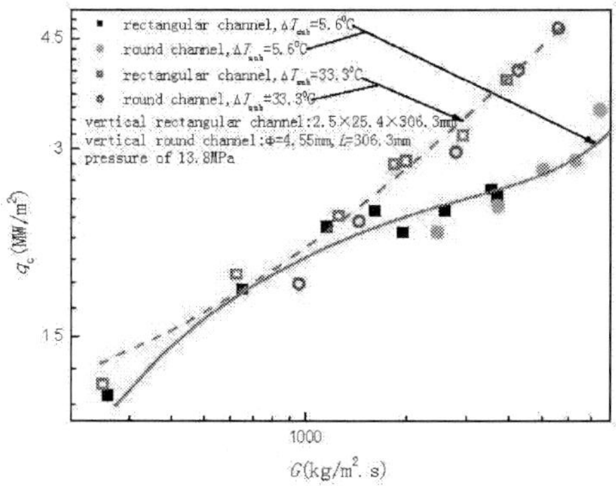

Figure 27. Critical heat flux trend of different type of channels with identical equivalent diameter

6.3.3. *Droplet entrainment rate*

(1) Broken bubble caused droplet entrainment

In this case, Ueda et al. (1981) correlation was used to calculate the droplet entrainment rate.

$$m_{\rm eb} = 477\left[\frac{q^2\delta}{h_{\rm lg}^2\sigma\rho_{\rm g}}\right]^{0.75}\frac{q}{h_{\rm lg}} \tag{33}$$

Where liquid film thickness δ was calculated by iteration from Eq. (34) to Eq. (37).

$$U_{\rm lf,m}^{+} = \frac{\int U^{+}dY^{+}}{\delta} \tag{34}$$

$$U_{\mathrm{lf,m}} = \frac{G_{\mathrm{lf}}}{A_{\mathrm{lf}}\rho_{\mathrm{l}}} \tag{35}$$

$$U_{\mathrm{lf,m}}^{+} = \frac{U_{\mathrm{lf,m}}}{\sqrt{\dfrac{\tau_{\mathrm{w}}}{\rho_{\mathrm{l}}}}} \tag{36}$$

$$A_{\mathrm{lf}} = \frac{D_{\mathrm{hy}}^{2}}{D_{\mathrm{hy}}^{2} - \left(D_{\mathrm{hy}} - 2\delta\right)^{2}} \tag{37}$$

(2) Vapour stream shear caused droplet entrainment

In this case, Kaotaoka's (1983) correlation was used to calculate the droplet entrainment rate.

$$m_{\mathrm{ew}} = 0.22 Re_{\mathrm{l}}^{0.74} E_{\infty}^{0.74} \left(\frac{\mu_{\mathrm{g}}}{\mu_{\mathrm{l}}}\right)^{0.26} \frac{\mu_{\mathrm{l}}}{D_{\mathrm{hy}}} \tag{38}$$

where E_{∞} was equilibrium droplet fraction and was calculated as below:.

$$E_{\infty} = \tan\mathrm{h}\left(7.25 \times 10^{-7} We^{1.25} Re^{0.25}\right) \tag{39}$$

where We is the Weber number.

The iteration process for CHF calculation was depicted as: Eq. (28) is used to estimate whether annular flow appear or not. If the supposed CHF was lower than the actual value, annular flow will not appear; contrarily, this flow pattern will appear. After the onset location (z_{on}) of annular flow being obtained, the parameters m_d, m_{ew}, m_{eb}, and W_{lf} were calculated at each time step from z_{on} to the channel outlet; if the supposed heat flux was lower than CHF, the mass flow rate of liquid film was over zero at the outlet; contrarily, this mass flow rate became zero before the outlet; the iteration finished until the supposed heat flux was equal to CHF.

6.3.4. *The assessment of the model*

Based on the analytical model, a computer code had been developed in FORTRAN language. The total fifty-seven groups of experimental CHF data in the smooth channel were used to validate the model. The validated result was shown in Fig.28. It was seen that 93% of prediction data fell within ±30% of discrepancy band. In general, the model had a good accuracy, and it could be used to predict CHF in the smooth channel.

The prior study (Wang L. & Wang Q.W., 2005) showed that the attack angle (β), longitudinal distance (X) and height (h) were the main parameters related to enhance heat transfer in the channel, so these three parameters were incorporated in the model.

Following the ribbed wall function methods proposed by Hanjalic and Launder (Hanjalic K. & Launder B.E., 1972), Donne and Meyer's wall (Donne M.D. & Meyer L., 1977), friction velocity in the channel with LVG could be modified as:

$$U'_{\mathrm{F}} = U_{\mathrm{F}}\left[\frac{1}{\kappa}\ln\left(\delta / h\right) + R\right] \tag{40}$$

where U'_{F} was modified friction velocity, and κ Karman constant, and R had an expression as:

$$R = 30\left(44^{\circ} / \beta\right)^{0.57}\left(X / \left(81h\right)\right)^{n}$$

where, $\mathrm{n} = -0.53\left(\beta / 90^{\circ}\right)^{-13}$ when $X / h < 81$; $n = 0.53\left(\beta / 90^{\circ}\right)^{-13}$ when $X / h \geq 81$.

Dimensionless temperature was modified as:

$$\left(t_1 - t_{\mathrm{w}}\right)\rho_1 c_{\mathrm{p}} U'_{\mathrm{F}}\left(h / X\right)^{0.5} / \left(15 q_{\mathrm{w}}\right) = Pr_{\mathrm{t}}\ln\left(\delta / h\right) / \kappa + \bar{G} \tag{41}$$

where $Pr_{\mathrm{t}} = 0.9$, and $\bar{G}$ had an expression as:

$$\bar{G} = 8\left(h^{+} / 105\right)^{C1}\left(\beta / 44^{\circ}\right)^{C2}$$

where $h^{+} = h \cdot U'_{\mathrm{F}} / \nu$ ($C1 = 0$ when $h^{+} < 105$; $C1 = -0.03$ when $h^{+} \geq 105$; $C2 = 0.5$ when $\beta < 45^{\circ}$; $C2 = -0.45$ when $\beta \geq 45^{\circ}$.

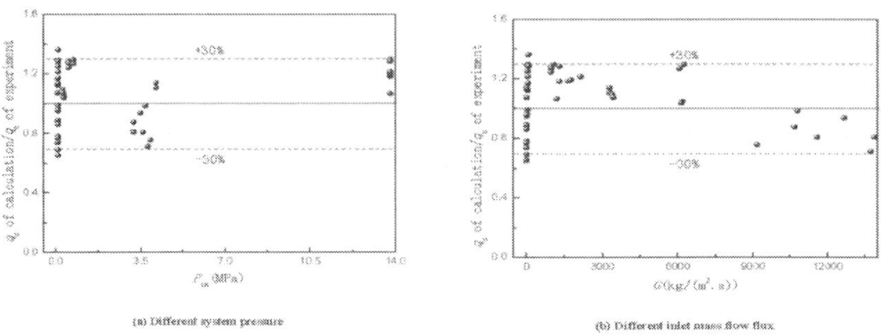

Figure 28. Comparison between calculation and experiment of CHF in the smooth channel.

The modified model was used to predict CHF in the channel with LVG, and the results were shown in Fig.29. It was seen that 93.7% of prediction data fell within ±30% of discrepancy band, the mean deviation was 3.6%, and the mean square root deviation of was 15.7%. In general, the modified model had a good accuracy, and it could be used to predict CHF in the channel with LVG.

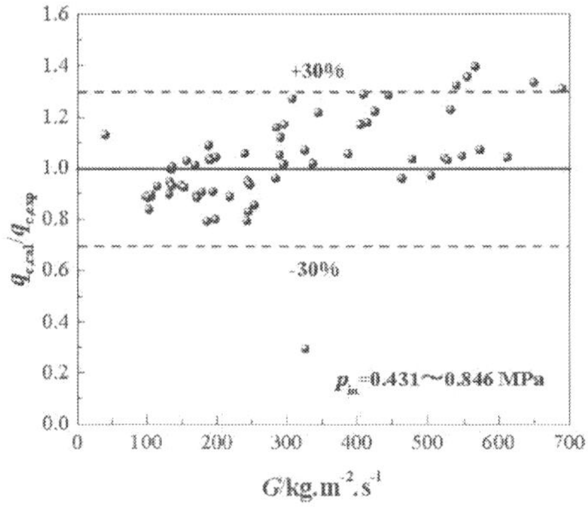

Figure 29. Comparison between calculation and experiment of CHF in the channel with LVG

7. Conclusions

In this chapter, aiming at thermal-hydraulical behavior in the smooth narrow rectangular channel and the LVGs machinednarrow rectangular channel, the systematical researches on the single-phase and two-phase Flow and heat transfer characteristics and their CHF behavior were carried out. The CFD method was used to optimize the configuration of the LVG machined narrow rectangular channel and to learn the Vortex behavior in LVGs channel and its effect on the flow and heat transfer characteristics with low pressure water being as working fluid in this project. In the single-phase flow state, the heat transfer capability in the LVG machined channel obviously increased with the friction pressure drop increase, but the effect of heat transfer enhancement is prominent. In the two-phase flow state, the boiling heat transfer coefficient increased also with the two-phase pressure drop increase. In the case of CHF, the CHF value in LVG machined channel was apparently improved with a quick two-phase pressure drop increase. But in the industrial application, the two-phase friction pressure drop in a equipment was not the majority of the whole system resistance, thus this heat transfer enhancement technology would have a bright future for industrial application. In this chapter, from the visualization experiments, the heat transfer enhancement mechanism of LV in the narrow rectangular channel was observed and analyzed and an analytical model for CHF in the narrow rectangular channel was represented and validated against the experimental data, the results could be used for similar researches.

Nomenclatures

A	area/m^2
a	single rib width/m
B	channel width/m
Bo	boiling number
b	single rib length/m
D	diameter/m

E	droplet entrainment fraction
f	friction factor
G	mass flow flux/$kg{\cdot}m^{-2}{\cdot}s^{-1}$
g	gravity acceleration/$m{\cdot}s^{-2}$
H	channel height/m
h	Enthalpy/$J{\cdot}kg^{-1}$; boiling heat transfer coefficient/$W{\cdot}m^{-2}{\cdot}$◎$^{-1}$; single rib height /m
j	dimensionless coefficient of heat transfer or superficial velocity/$m{\cdot}s^{-1}$
L	length/m
m	droplet entrainment flow flux/$kg{\cdot}m^{-2}{\cdot}s^{-1}$
Nu	Nusselt number
Pr	Prandtl number
p	pressure/MPa
q	heat flux/$W{\cdot}m^{-2}$
Re	Reynolds number
s	transverse distance in LVG configuration/m
T	temperature/◎
U , u	velocity/$m{\cdot}s^{-1}$
W	mass flow rate/$kg{\cdot}s^{-1}$
X	longitudinal distance in LVG configuration/m; Cartesian coordinate
Greek letters	
β	attack angle/°
δ	thickness/m
ϕ	effect factor
κ	Karman constant
λ	thermal conductivity/$W{\cdot}m^{-1}{\cdot}$◎$^{-1}$
μ	dynamic viscosity/$kg{\cdot}m^{-1}{\cdot}s^{-1}$
ν	kinematic viscosity/$m^2{\cdot}s^{-1}$
ρ	density/$kg{\cdot}m^{-3}$
σ	surface tension/$N{\cdot}m^{-1}$
τ	shear stress/$N{\cdot}m^{-2}$
Superscripts	
+	dimensionless symbol
Subscripts	
an	annular flow
av	average
c	critical
cal	calculation
d	droplet deposition
eb	broken bubble caused
ew	Vapor stream shear caused
exp	experiment
F	friction
g	gas phase

hy	hydraulic equivalent
in	inlet
l	liquid phase
lf	liquid film
lg	evaporation
m	mean
min	minimum
sub	subcooled
TP	two-phase
w	wall
∞	bulk flow

Author details

Yan-Ping Huang*, Jun Huang, Jian MA, Yan-Lin Wang and Jun-Feng Wang
CNNC Key Laboratory on Nuclear Reactor Thermal Hydraulics Technology, Chengdu, China

Qiu-Wang WANG
State Key Laboratory of Multiphase Flow in Power Engineering Xi'an Jiaotong University, Xi'an, China

Acknowledgement

This project is supported by the National Natural Science Fund (NO.50576089) in 2005, the Fund of Key Laboratory of Bubble Physics and Natural Circulation (NO.9140C7101030602) in 2006 and the National Natural Science Fund (NO.51176176) in 2012.

8. References

Chen, Q.Y., Zeng, M., Wang L., et al. (2006). Effect Longitudinal Vortex Generators on Convective Heat Transfer in Narrow Rectangular Channel, *J. Xi'an Jiaotong University*, vol. 40(9), pp. 1010-1013 (in Chinese).

Crecy F. (1994). The effect of grid assembly mixing vanes on critical heat flux values and azimuthal location in fuel assemblies, *Nuclear Engineering and Design*, Vol. 149, pp. 233-249.

Donne M.D. & Meyer L. (1977). Turbulent convective heat transfer from rough surfaces with two-dimensional rectangular ribs. *Int. J. Heat and Mass Transfer*, 20: 583-620.

Fedorov A.G. & Viskanta R. (2000). Three-Dimensional Conjugate Heat Transfer in the Microchannel Heat Sink for Electronic Packaging, *Int. J. Heat and Mass Transfer*, Vol. 43(3), pp. 399-415.

Gungor KE, Winterton RHS. Simplified General Correlation for Saturated Flow Boiling and Comparisons of Correlations with Datas[J]. Chem Eng Res Des, 1987, 365: 148-165.

* Corresponding Author

Hanjalic K. & Launder B.E. (1972). Fully developed asymmetric flow in a plane channel. *J. Fluid Mech.* 51, Part 2: 301-335.

Islam, M.S., Hino, R., Haga, K., et al. (1998). Experimental Study on Heat Transfer Augmentation for High Heat Flux Removal, *J. NUCLEAR SCIENCE and TECHNOLOGY*, vol. 35(9), pp. 671-678.

Jackey, H.S., Roarty, J.D. and Zerbe, J. E.(1958). Investigation of Burnout Heat Flux in Rectangular Channels at 2000 Psia, Trans. ASME, vol. 80, 391.

Johnson T.R. & Joubert P.N. (1969). The Influence of Vortex Generators on Drag and Heat Transfer from a Circular Cylinder Normal to an Airstream, *Heat Transfer*, Vol. 91, pp. 91-99.

Joel, W. (1992). The current status of theoretically based approaches to the prediction of the critical heat flux in flow boiling, Nuclear Technology, vol. 99, pp. 1-21.

Kays, W.M. & Clark. S.H. (1953). TR no.17, Department of Mechanical Engineering, Stanford University, Stanford ,California, August 15.

Kafengauz, N. L. and Bocharov, I. D. (1959). Effect of Heat Transfer for Water by Gap Width of Narrow Rectangular Channel, Teplocncrgetika, vol. 3, pp. 76-78.

Kaotaoka, I., Ishii, M., Mishima, K. (1983). Genertation and size distribution of droplet in gas-liquid annular two-phase flow, ASME J Fluids Engineering, vol. 105, pp. 230-238.

Ma J. (2008). Investigation on heat transfer enhancement of longitudinal vortices in narrow rectangular channel with single-phase water, Nuclear Power Institute of China (in Chinese).

Mishima, K., Ishii, M. (1984). Flow Regime Transition Criteria for Upward Two-phase Flow in Vertical Tubes, Int. J. Heat and Mass Transfer, vol. 27(5), pp. 723-737.

Metha R.D. & Bradshaw P. (1988). Longitudinal vortices imbedded in turbulent boundary layers, Part 2. Vortex Pair with 'Common Flow' Upwards, *J. Fluid. Mech.*, Vol. 188, pp. 529-546.

Schubauer, G.B. & Spangenberg. (1960). WG. Forced Mixing in Boundary Layers, *Fluid Mech*, Vol. 8, pp. 10-31.

Shah, R.K. & A.L. (1978). London: Laminar Flow Forced Convention in Ducts, *Advances in Heat Transfer*, Academic Press, New York.

Sohankar, A. & Davidson, L. (2001). Effect of inclined vortex generators on heat transfer enhancement in a three-dimensional channel, *Numerical Heat Transfer, Part A: Applications*, vol. 39, no. 5, pp. 433-448.

Veda, T., Inoue, M., Nagatome, S. (1981). Critical heat flux and droplet entrainment rate in boiling of falling liquid films, Int. J. Heat and Mass Transfer, vol. 24(7), pp. 1257-1266.

Wang, L., Chen, Q.Y., Zhou, Y.G. (2005). Heat Transfer Enhancement in Rectangular Narrow Channel with Periodically Mounted Longitudinal Vortex Generators on One Sidewall, *Nuclear Power Engineering*, vol. 26(4), pp. 344-347 (in Chinese).

Wang, Q.W., Chen, Q.Y., Wang, L., et al. (2007). Experimental study of heat transfer enhancement in narrow rectangular channel with longitudinal vortex generators, *Nuclear Engineering and Design*, vol. 237, pp. 686–693.

Yun JY, Lee KS. (2000). Influence of design parameters on the heat transfer and flow friction characteristics of the heat exchanger with slit fins. *International Journal of Heat and Mass Transfer*, 43: 2529-2539.

Chapter 14

Application of Nanofluids in Heat Transfer

P. Sivashanmugam

Additional information is available at the end of the chapter

http://dx.doi.org/10.5772/52496

1. Introduction

A wide variety of industrial processes involve the transfer of heat energy. Throughout any industrial facility, heat must be added, removed, or moved from one process stream to another and it has become a major task for industrial necessity. These processes provide a source for energy recovery and process fluid heating/cooling.

The enhancement of heating or cooling in an industrial process may create a saving in energy, reduce process time, raise thermal rating and lengthen the working life of equipment. Some processes are even affected qualitatively by the action of enhanced heat transfer. The development of high performance thermal systems for heat transfer enhancement has become popular nowadays. A number of work has been performed to gain an understanding of the heat transfer performance for their practical application to heat transfer enhancement. Thus the advent of high heat flow processes has created significant demand for new technologies to enhance heat transfer

There are several methods to improve the heat transfer efficiency. Some methods are utilization of extended surfaces, application of vibration to the heat transfer surfaces, and usage of micro channels. Heat transfer efficiency can also be improved by increasing the thermal conductivity of the working fluid. Commonly used heat transfer fluids such as water, ethylene glycol, and engine oil have relatively low thermal conductivities, when compared to the thermal conductivity of solids. High thermal conductivity of solids can be used to increase the thermal conductivity of a fluid by adding small solid particles to that fluid. The feasibility of the usage of such suspensions of solid particles with sizes on the order of 2 millimeters or micrometers was previously investigated by several researchers and the following significant drawbacks were observed (Das and Choi, 2006).

1. The particles settle rapidly, forming a layer on the surface and reducing the heat transfer capacity of the fluid.
2. If the circulation rate of the fluid is increased, sedimentation is reduced, but the erosion of the heat transfer devices, pipelines, etc., increases rapidly.

3. The large size of the particles tends to clog the flow channels, particularly if the cooling channels are narrow.
4. The pressure drop in the fluid increases considerably.
5. Finally, conductivity enhancement based on particle concentration is achieved (i.e., the greater the particle volume fraction is, the greater the enhancement—and greater the problems, as indicated above).

Thus, the route of suspending particles in liquid was a well known but rejected option for heat transfer applications.

However, the emergence of modern materials technology provided the opportunity to produce nanometer-sized particles which are quite different from the parent material in mechanical, thermal, electrical, and optical properties.

1.1. Emergence of nanofluids

The situation changed when Choi and Eastman in Argonne National Laboratory revisited this field with their nanoscale metallic particle and carbon nanotube suspensions (Choi and Eastman (1995); Eastman et al. (1996)). Choi and Eastman have tried to suspend various metal and metal oxides nanoparticles in several different fluids (Choi (1998); Choi et al. (2001); Chon et al. (2005); Chon et al. (2006); Eastman et al. (2001); Eastman et al. (1999); Eastman et al. (2004)) and their results are promising, however, many things remain elusive about these suspensions of nano-structured materials, which have been termed "nanofluids" by Choi and Eastman.

Nanofluid is a new kind of heat transfer medium, containing nanoparticles (1–100 nm) which are uniformly and stably distributed in a base fluid. These distributed nanoparticles, generally a metal or metal oxide greatly enhance the thermal conductivity of the nanofluid, increases conduction and convection coefficients, allowing for more heat transfer

Nanofluids have been considered for applications as advanced heat transfer fluids for almost two decades. However, due to the wide variety and the complexity of the nanofluid systems, no agreement has been achieved on the magnitude of potential benefits of using nanofluids for heat transfer applications. Compared to conventional solid–liquid suspensions for heat transfer intensifications, nanofluids having properly dispersed nanoparticles possess the following advantages:

- High specific surface area and therefore more heat transfer surface between particles and fluids.
- High dispersion stability with predominant Brownian motion of particles.
- Reduced pumping power as compared to pure liquid to achieve equivalent heat transfer intensification.
- Reduced particle clogging as compared to conventional slurries, thus promoting system miniaturization.
- Adjustable properties, including thermal conductivity and surface wettability, by varying particle concentrations to suit different applications.

The first test with nanofluids gave more encouraging features than they were thought to possess. The four unique features observed are listed below (Das and Choi, 2006).

- *Abnormal enhancement of thermal conductivity.* The most important feature observed in nanofluids was an abnormal rise in thermal conductivity, far beyond expectations and much higher than any theory could predict.
- *Stability.* Nanofluids have been reported to be stable over months using a stabilizing agent.
- *Small concentration and Newtonian behavior.* Large enhancement of conductivity was achieved with a very small concentration of particles that completely maintained the Newtonian behavior of the fluid. The rise in viscosity was nominal; hence, pressure drop was increased only marginally.
- *Particles size dependence.* Unlike the situation with microslurries, the enhancement of conductivity was found to depend not only on particle concentration but also on particle size. In general, with decreasing particle size, an increase in enhancement was observed.

The above potentials provided the thrust necessary to begin research in nanofluids, with the expectation that these fluids will play an important role in developing the next generation of cooling technology. The result can be a highly conducting and stable nanofluid with exciting newer applications in the future.

2. Thermo physical properties of nanofluids

Thermo physical properties of the nanofluids are quite essential to predict their heat transfer behavior. It is extremely important in the control for the industrial and energy saving perspectives. There is great industrial interest in nanofluids. Nanoparticles have great potential to improve the thermal transport properties compared to conventional particles fluids suspension, millimetre and micrometer sized particles. In the last decade, nanofluids have gained significant attention due to its enhanced thermal properties.

Experimental studies show that thermal conductivity of nanofluids depends on many factors such as particle volume fraction, particle material, particle size, particle shape, base fluid material, and temperature. Amount and types of additives and the acidity of the nanofluid were also shown to be effective in the thermal conductivity enhancement.

The transport properties of nanofluid: dynamic thermal conductivity and viscosity are not only dependent on volume fraction of nanoparticle, also highly dependent on other parameters such as particle shape, size, mixture combinations and slip mechanisms, surfactant, etc. Studies showed that the thermal conductivity as well as viscosity both increases by use of nanofluid compared to base fluid. So far, various theoretical and experimental studies have been conducted and various correlations have been proposed for thermal conductivity and dynamic viscosity of nanofluids. However, no general correlations have been established due to lack of common understanding on mechanism of nanofluid.

2.1. Thermal conductivity

A wide range of experimental and theoretical studies were conducted in the literature to model thermal conductivity of nanofluids. The existing results were generally based on the definition of the effective thermal conductivity of a two-component mixture. The Maxwell (1881) model was one the first models proposed for solid–liquid mixture with relatively large particles. It was based on the solution of heat conduction equation through a stationary random suspension of spheres. The effective thermal conductivity (Eq.1) is given by

$$k_{eff} = \frac{k_p + 2k_{bf} + 2\varphi\left(k_p - k_{bf}\right)}{k_p + 2k_{bf} - \varphi\left(k_p - k_{bf}\right)} k_{bf} \tag{1}$$

Where k_p is the thermal conductivity of the particles, k_{eff} is the effective thermal conductivity of nanofluid, k_{bf} is the base fluid thermal conductivity, and ϕ is the volume fraction of the suspended particles.

The general trend in the experimental data is that the thermal conductivity of nanofluids increases with decreasing particle size. This trend is theoretically supported by two mechanisms of thermal conductivity enhancement; Brownian motion of nanoparticles and liquid layering around nanoparticles (Ozerinc et al, 2010). However, there is also a significant amount of contradictory data in the literature that indicate decreasing thermal conductivity with decreasing particle size.

Published results illustrated neither agreement about the mechanisms for heat transfer enhancement nor a unified possible explanation regarding the large discrepancies in the results even for the same base fluid and nanoparticles size. There are various models available for the measurement of effective thermal conductivity of nanofluids (Wang and Mujumdar, 2007) but there exists large deviations among them. Currently, there are no theoretical results available in the literature that predicts accurately the thermal conductivity of nanofluids.

2.2. Viscosity

Compared with the experimental studies on thermal conductivity of nanofluids, there are limited rheological studies reported in the literature for viscosity. Different models of viscosity have been used by researchers to model the effective viscosity of nanofluid as a function of volume fraction. Einstein (1956) determined the effective viscosity of a suspension of spherical solids as a function of volume fraction (volume concentration lower than 5%) using the phenomenological hydrodynamic equations (Eq.2). This equation was expressed by

$$\mu_{eff} = \left(1 + 2.5\varphi\right)\mu_{bf} \tag{2}$$

Where μ_{eff} is the effective viscosity of nanofluid, μ_{bf} is the base fluid viscosity, and ϕ is the volume fraction of the suspended particles.

Later, Brinkman (1952) presented a viscosity correlation (Eq.3) that extended Einstein's equation to suspensions with moderate particle volume fraction, typically less than 4%.

$$\mu_{eff} = \mu_{bf} \frac{1}{(1-\phi)^{2.5}} \tag{3}$$

The effect of Brownian motion on the effective viscosity in a suspension of rigid spherical particles was studied by Batchelor (1977). For isotropic structure of suspension, the effective viscosity was given by(Eq.4):

$$\mu_{eff} = \left(1 + 2.5\varphi + 6.2\phi^2\right)\mu_{bf} \tag{4}$$

2.3. Specific heat and density

Using classical formulas derived for a two-phase mixture, the specific heat capacity (Pak and Cho, 1998) and density (Xuan and Roetzel, 2000) of the nanofluid as a function of the particle volume concentration and individual properties can be computed using following equations(Eqs 5,and 6) respectively:

$$\rho_{eff} = (1-\varphi)\rho_{bf} + \phi\rho_p \tag{5}$$

$$\left(\rho C_p\right)_{eff} = (1-\varphi)\left(\rho C_p\right)_{bf} + \varphi\left(\rho C_p\right)_p \tag{6}$$

3. Applications of nanofluids

The novel and advanced concepts of nanofluids offer fascinating heat transfer characteristics compared to conventional heat transfer fluids. There are considerable researches on the superior heat transfer properties of nanofluids especially on thermal conductivity and convective heat transfer. Applications of nanofluids in industries such as heat exchanging devices appear promising with these characteristics. Kostic reported that nanofluids can be used in following specific areas:

- Heat-transfer nanofluids.
- Tribological nanofluids.
- Surfactant and coating nanofluids.
- Chemical nanofluids.
- Process/extraction nanofluids.
- Environmental (pollution cleaning) nanofluids.
- Bio- and pharmaceutical-nanofluids.
- Medical nanofluids (drug delivery and functional tissue–cell interaction).

Nanofluids can be used to cool automobile engines and welding equipment and to cool high heat-flux devices such as high power microwave tubes and high-power laser diode arrays. A nanofluid coolant could flow through tiny passages in MEMS to improve its efficiency. The measurement of nanofluids critical heat flux (CHF) in a forced convection loop is useful for nuclear applications. Nanofluids can effectively be used for a wide variety of industries, ranging from transportation to energy production and in electronics systems like microprocessors, Micro-Electro-Mechanical Systems (MEMS) and in the field of biotechnology. Recently, the number of industrial application potential of nanofluids technology and their focus for specific industrial applications is increasing. This chapter deals the some of the important application of nanofluids in the field of heat transfer.

4. Heat transfer applications

The increases in effective thermal conductivity are important in improving the heat transfer behavior of fluids. A number of other variables also play key roles. For example, the heat transfer coefficient for forced convection in tubes depends on many physical quantities related to the fluid or the geometry of the system through which the fluid is flowing. These quantities include intrinsic properties of the fluid such as its thermal conductivity, specific heat, density, and viscosity, along with extrinsic system parameters such as tube diameter and length and average fluid velocity. Therefore, it is essential to measure the heat transfer performance of nanofluids directly under flow conditions. Researchers have shown that nanofluids have not only better heat conductivity but also greater convective heat transfer capability than that of base fluids. The following section provides the wide usage and effective utilization of nanofluids in heat exchangers as heat transfer fluids.

4.1. Tubular (circular pipe) heat exchangers

Pak and Cho (1998) investigated experimentally the turbulent friction and heat transfer behaviors of dispersed fluids (i.e., ultrafine metallic oxide particles suspended in water) in a circular pipe. Two different metallic oxide particles, γ-alumina (Al_2O_3) and titanium dioxide (TiO_2) with mean diameters of 13 and 27 nm, respectively, were used as suspended particles. In their flow loop, the hydrodynamic entry section and the heat transfer section was made using a seamless, stainless steel tube, of which the inside diameter and the total length were 1.066 crn and 480 crn, respectively. The hydrodynamic entry section was long enough (i.e., $x/D = 157$) to accomplish fully developed flow at the entrance of the heat transfer test section. They observed that the Nusselt number for the dispersed fluids increased with increasing volume concentration as well as Reynolds number. But at constant average velocity, the convective heat transfer coefficient of the dispersed fluid was 12% smaller than that of pure water.

They proposed a new correlation (Eq.7) for the Nusselt number under their experimental ranges of volume concentration (0-3%), the Reynolds number (10^4 - 10^5), and the Prandtl number (6.54 - 12.33) for the dispersed fluids γ-Al_2O_3 and TiO_2 particles as given below

$$Nu = 0.021\,\mathrm{Re}^{0.8}\,\mathrm{Pr}^{0.5} \qquad (7)$$

Xuan and Li (2003) built an experimental rig to study the flow and convective heat transfer feature of the nanofluid flowing in a tube. Their test section was a straight brass tube of the inner diameter of 10 mm and the length of 800 mm. Eight thermocouples were mounted at different places of the heat transfer test section to measure the wall temperatures and other two thermocouples were respectively located at the entrance and exit of the test section to read the bulk temperatures of the nanofluid. They investigated convective heat transfer feature and flow performance of Cu-water nanofluids for the turbulent flow. The suspended nanoparticles remarkably enhance heat transfer process and the nanofluid has larger heat transfer coefficient than that of the original base liquid under the same Reynolds number. They found that at fixed velocities, the heat transfer coefficient of nanofluids containing 2.0 vol% Cu nanoparticles was improved by as much as 40% compared to that of water. The Dittus–Boelter correlation failed to predict the improved experimental heat transfer behavior of nanofluids. The heat transfer feature of a nanofluid increases with the volume fraction of nanoparticles.

They have proposed the following correlation (Eq.8) to correlate the experimental data for the nanofluid. The Nusselt number Nu for the turbulent flow of nanofluids inside a tube are obtained as follows

$$Nu_{nf} = 0.0059(1.0 + 7.6286\,\varphi^{0.6886}\,Pe_d^{0.001})\,\mathrm{Re}_{nf}^{0.9238}\,\mathrm{Pr}_{nf}^{0.4} \qquad (8)$$

They found good coincidence between the results calculated from this correlation and the experimental ones.

The Peclet number Pe describes the effect of thermal dispersion caused by micro convective and micro diffusion of the suspended nanoparticles. The case $c_2 = 0$ refers to zero thermal dispersion, which namely corresponds to the case of the pure base fluid. The particle Peclet number Pe_d, Re_{nf} and Pr_{nf} in (Eq.9) are defined as

$$\text{Particle Peclet number } Pe_d = \frac{u_m d_p}{\alpha_{nf}} \quad \text{(i)}$$

$$\text{The Reynolds number of the nanofluid } \mathrm{Re}_{nf} = \frac{u_m D}{\upsilon_{nf}} \quad \text{(ii)}$$

$$\text{The Prantdl number of the nanofluid } \mathrm{Pr}_{nf} = \frac{\upsilon_{nf}}{\alpha_{nf}} \quad \text{(iii)} \qquad (9)$$

$$\alpha_{nf} = \frac{k_{nf}}{(\rho c_p)_{nf}} = \frac{k_{nf}}{(1-\varphi)\left(\rho c_p\right)_f + \varphi\left(\rho c_p\right)_d} \quad \text{(iv)}$$

The thermal diffusivity of the nanofluid in Eq.8 is defined as Eq 8.iv

They defined the friction factor (Eq.10) as

$$\lambda_{nf} = \frac{\Delta P_{nf} D}{L} \frac{2g}{u_m^2} \tag{10}$$

It should be noted that, correlations developed by Pak and Cho (1998) and Xuan and Li (2003) were of a form similar to that of well known Dittus - Boelter formula. In both the works, the nanofluid was treated as a single phase fluid for the calculation of nanofluid Nusselt number

Wen and Ding (2004) were first to study the laminar entry flow of nanofluids in circular tubes. A straight copper tube with 970 mm length, 4.5 mm inner diameter, and 6.4 mm outer diameter was used as the test section. The whole test section was heated by a silicon rubber flexible heater. Their results showed a substantial increase in the heat transfer coefficient of water-based nanofluids containing γ-Al_2O_3 nanoparticles in the entrance region and a longer entry length is needed for the nanofluids than that for water. They concluded that the enhancement of the convective heat transfer could not be solely attributed to the enhancement of the effective thermal conductivity. Particle migration is proposed to be a reason for the enhancement, which results a non-uniform distribution of thermal conductivity and viscosity field and reduces the thermal boundary layer thickness.

Yang et al., (2005) measured the convective heat transfer coefficients of several nanofluids under laminar flow in a horizontal tube heat exchanger. A small circular tube of inner diameter 0.457 cm, outside diameter of 0.635 cm and length 45.7 cm was used as test section. The whole system was heavily insulated to reduce heat loss. Pipes were wrapped with insulation material, and plastic fittings were attached at both ends of the test area to thermally isolate the connection. The average diameter of the disk-shaped graphite nanoparticles used in this research was about 1 to 2μm, with a thickness of around 20 to 40 nm.

They applied the correlations for the convective heat transfer of the single-phase fluid to predict heat transfer coefficient of a nanofluid system, if the volume fraction of particles is

very low. They used the following correlation (Eq.11) to identify the impact of Reynolds number on the heat transfer coefficient

$$Nu.\mathrm{Pr}^{-1/3}\left(\frac{L}{D}\right)^{1/3}\left(\frac{\mu_b}{\mu_w}\right)^{-0.14} = 1.86\,\mathrm{Re}^{1/3} \tag{11}$$

Their results indicated that the increase in the heat transfer coefficient of the nanofluids is much less than that predicted from a conventional correlation. Near-wall particle depletion in laminar shear flow is one possible reason for the phenomenon. However, there is a doubt whether this work falls in the category of nanofluids at all because the particle diameter is too large for the particles to be called nanoparticles.

Maiga et al., (2005) presented the numerical study of fully developed turbulent flow of Al_2O_3 - water nanofluid in circular tube at uniform heat flux of 50 W/cm^2. The classical k-ε model was used for turbulence modeling and their study clearly showed that the inclusion of nanoparticles into the base fluids has produced a considerable augmentation of the heat transfer coefficient that clearly increases with an increase of the particle concentration. However, the presence of such particles has also induced drastic effects on the wall shear stress that increases appreciably with the particle loading. Among the mixtures studied, the ethylene glycol γ-Al_2O_3nanofluid appears to offer a better heat transfer enhancement than water– γ-Al_2O_3. The following correlations(Eqs 12 and 13) have been proposed for computing the averaged Nusselt number for the nanofluids considered for both the thermal boundary conditions, valid for $Re \leq 1000$, $6 \leq Pr \leq 7.53$ and $\phi \leq 10\%$

$$Nu_{nf} = 0.086\,\mathrm{Re}_{nf}^{0.55}\,\mathrm{Pr}_{nf}^{0.5} \quad \text{for constant wall flux} \tag{12}$$

$$Nu_{nf} = 0.28\,\mathrm{Re}_{nf}^{0.35}\,\mathrm{Pr}_{nf}^{0.36} \quad \text{for constant wall temperature} \tag{13}$$

Maiga et al., (2006) studied the hydrodynamic and thermal behavior of turbulent flow in a tube using Al_2O_3 nanoparticle suspension at various concentrations under the constant heat flux boundary condition. Assuming single-phase model, governing equations were solved by a numerical method of control volume. The following correlation (Eq.14) was proposed to calculate the heat transfer coefficient in terms of the Reynolds and the Prandtl numbers, valid for $10^4 \leq Re \leq 5x10^5$, $6.6 \leq Pr \leq 13.9$ and $0 \leq \phi \leq 10\%$.

$$Nu_{nf} = 0.085\,\mathrm{Re}_{nf}^{0.71}\,\mathrm{Pr}_{nf}^{0.35} \tag{14}$$

Ding et al., (2006) were first to study the laminar entry flow of water-based nanofluids containing multiwalled carbon nanotubes (CNT nanofluids). The experimental system for measuring the convective heat transfer coefficient was similar to the one reported by Wen and Ding (2004). Significant enhancement in the convective heat transfer was observed in relation to pure water as the working fluid. The enhancement depends on the flow condition, CNT concentration and the pH level, and the effect of pH is observed to be small. They stated that the enhancement in convective heat transfer is a function of the axial distance from the inlet of the test section. This enhancement increases first, reaches a maximum, and then decreases with increasing axial distance. For nanofluids containing only 0.5 wt% CNTs, the maximum enhancement in the convection heat transfer coefficient reaches over 350% at Re = 800. Such a high level of enhancement could not be attributed purely to enhanced thermal conductivity. They proposed possible mechanisms such as particle rearrangement, reduction of thermal boundary layer thickness due to the presence of nanotubes, and the very high aspect ratio of CNTs. They also concluded that, the observed large enhancement of the convective heat transfer could not be attributed purely to the enhancement of thermal conduction under the static conditions. Particle re-arrangement, shear induced thermal conduction enhancement, reduction of thermal boundary layer thickness due to the presence of nanoparticles, as well as the very high aspect ratio of CNTs are proposed to be possible mechanisms.

Heriz et al., (2006) investigated laminar flow convective heat transfer through circular tube with constant wall temperature boundary condition for nanofluids containing CuO and

Al_2O_3 oxide nanoparticles in water as base fluid. The experimental apparatus consisting of a test chamber constructed of 1 m annular tube with 6 mm diameter inner copper tube and with 0.5 mm thickness and 32 mm diameter outer stainless steel tube. Nanofluid flows inside the inner tube while saturated steam enters annular section, which creates constant wall temperature boundary condition. The fluid after passing through the test section enters heat exchanger in which water was used as cooling fluid. The experimental results emphasized that the single phase correlation with nanofluids properties (Homogeneous Model) was not able to predict heat transfer coefficient enhancement of nanofluids. The comparison between experimental results obtained for CuO/ water and Al_2O_3 / water nanofluids indicated that heat transfer coefficient ratios for nanofluid to homogeneous model in low concentration were close to each other but by increasing the volume fraction, higher heat transfer enhancement for Al_2O_3/water was observed. They concluded that heat transfer enhancement by nanofluid depends on several factors including increment of thermal conductivity, nanoparticles chaotic movements, fluctuations and interactions.

The flow and heat transfer behavior of aqueous TiO_2 nanofluids flowing through a straight vertical pipe was carried out by He et al., (2007) under both the laminar and turbulent flow conditions. Their experimental system consisted of a flow loop, a heating unit, a cooling unit, and a measuring and control unit. The test section was a vertically oriented straight copper tube with 1834 mm length, 3.97 mm inner diameter, and 6.35 mm outer diameter. The tube was heated by two flexible silicon rubber heaters. There was a thick thermal isolating layer surrounding the heaters to obtain a constant heat flux condition along the test section. Two pressure transducers were installed at the inlet and outlet of the test section to measure the pressure drop. They investigated the effects of nanoparticles concentrations, particle size, and the flow Reynolds number. They reported that, addition of nanoparticles into the base liquid enhanced the thermal conduction and the enhancement increased with increasing particle concentration and decreasing particle size. Their results also showed that the convective heat transfer coefficient increases with nanoparticle concentration in both the laminar and turbulent flow regimes and the effect of particle concentration seems to be more considerable in the turbulent flow regimes for the given flow Reynolds number and particle size. Pressure drop of nanofluids was very close to that of the base liquid for given flow Reynolds number. Predictions of the pressure drop with the conventional theory for the base liquid agree well with the measurements at relatively low Reynolds numbers. Deviation occurs at high Reynolds numbers possibly due to the entrance effect.

Kulkarni et al., (2008) investigated heat transfer and fluid dynamic performance of nanofluids comprised of silicon dioxide (SiO_2) nanoparticles suspended in a 60:40 (% by weight) ethylene glycol and water (EG/water) mixture. The heat transfer test section was a straight copper tube with outside diameter of 4.76 mm, inside diameter of 3.14 mm, and a length of 1 m. The wall temperature was measured by means of six thermocouples mounted on the tube surface along the length. The inlet and outlet temperatures of the nanofluid were measured using two thermowells at the inlet and outlet of the test section. Two plastic fittings at inlet and outlet section of the copper tube provide a thermal barrier to axial heat conduction. The test section was heated electrically by four strip heaters to attain the constant heat flux boundary condition. The test section was insulated by 10 cm of fiber glass to minimize the heat loss from the heat transfer test system to ambient air. A four-pass shell and tube counter flow heat exchanger cools the nanofluids to keep the inlet fluid temperature constant using shop water. The effect of particle diameter (20 nm, 50 nm, 100 nm) on the viscosity of the fluid was investigated. They performed experiments to investigate the convective heat transfer enhancement of nanofluids in the turbulent regime by using the viscosity values measured. They observed increase in heat transfer coefficient due to nanofluids for various volume concentrations and loss in pressure was observed with increasing nanoparticle volume concentration.

Hwang et al., (2009) investigated flow and convective heat transfer characteristics of water-based Al_2O_3 nanofluids flowing through a circular tube of 1.812 mm inner diameter with the constant heat flux in fully developed laminar regime. Water-based Al_2O_3 nanofluids with various volume fractions ranging from 0.01% to 0.3% are manufactured by the two-step method. They also measured physical properties of water-based Al_2O_3 nanofluids such as the viscosity, the density, the thermal conductivity and the heat capacity. They presented that the nanoparticles suspended in water enhance the convective heat transfer coefficient in the thermally fully developed regime, despite low volume fraction between 0.01 and 0.3 vol%. Especially, the heat transfer coefficient of water-based Al_2O_3 nanofluids was increased by 8% at 0.3 vol% under the fixed Reynolds number compared with that of pure water and the enhancement of the heat transfer coefficient is larger than that of the effective thermal conductivity at the same volume concentration. Based on their experimental results, it was shown that the Darcy friction factor of water-based Al_2O_3 nanofluids experimentally measured has a good agreement with theoretical results from the friction factor correlation for the single-phase flow ($f = 64/Re_D$).

Sharma et al., (2009) conducted experiments to evaluate heat transfer coefficient and friction factor for flow in a tube and with twisted tape inserts in the transition range of flow with Al_2O_3 nanofluid. Hydro dynamically and thermally developed heat transfer test section is having 1.5 m long with an L/D ratio of 160. The tube was heated uniformly for a length of 1.5 m by wrapping with two nichrome heaters of 1 kW electrical rating. Their twisted tapes were made from 1 mm thick and 0.018 m width aluminum strip. The two ends of the strip are held on a lathe and subjected to 180° twist by turning the chuck manually and obtained twist ratios of 5, 10 and 15. The results showed considerable enhancement of convective heat transfer with Al_2O_3 nanofluids compared to flow with water. They found that the effect of inclusion of twisted tape in the flow path gives higher heat transfer rates compared to flow in a plain tube. They also observed the equation of Gleninski(1976) applicable in transitional flow range for single-phase fluids exhibited considerable deviation when compared with values obtained with nanofluid. The heat transfer coefficient of nanofluid flowing in a tube with 0.1% volume concentration was 23.7% higher when compared with water at number of 9000.

Heat transfer coefficient and pressure drop with nanofluid were experimentally determined with tapes of different twist ratios and found to deviate with values obtained from equations(Eqs 15 and 16) developed for single-phase flow. The data of Al_2O_3 nanofluid for flow in plain tube and with twisted tape insert is fit to a regression equation with average deviation of 4.0% and standard deviation of 5.0%.

$$Nu = 3.138 \times 10^{-3} \left(\mathrm{Re}\right)\left(\mathrm{Pr}\right)^{0.6} \left(1.0 + H/D\right)^{0.03} \left(1 + \varphi\right)^{1.22} \quad (15)$$

0 < H/D < 15, 3500 < Re <8500, 4.5 < Pr <5.5 and 35 < T_b < 40.

The data of friction factor for flow of fluids a plain tube and with tape insert is also subjected to regression with the assumption that nanofluid behaves as single-phase fluid in the low volume concentration given by

$$f = 172\,\mathrm{Re}^{-0.96} (1.0 + \varphi)^{2.15} \left(1.0 + H/D\right)^{2.15} \quad (16)$$

Valid for water ($\phi = 0$) and nanofluid of $\phi < 0.1$ volume concentration

Yu et al., (2009) measured the heat transfer rates in the turbulent flow of a potential commercially viable nanofluid consisting of a 3.7% volume of 170-nm silicon carbide particles suspended in water. Their test facility was a closed-loop system with major components consisting of a pump with variable speed drive, pre heater, horizontal tube test section, heat exchanger (cooler), and flow meter. The test section itself was a stainless steel circular tube with dimensions of 2.27-mm inside diameter, 4.76-mm outside diameter, and 0.58-m heated length. Heat transfer coefficient increase of 50–60% above the base fluid water was obtained when compared on the basis of constant Reynolds number. This enhancement is 14–32% higher than predicted by a standard single-phase turbulent heat transfer correlation pointing to heat transfer mechanisms that involve particle interactions. The data were well predicted by a correlation modified for Prandtl number dependence although experiments in the present study did not support the postulated mechanisms of Brownian diffusion and thermophoresis. The pumping power penalty of the SiC/water nanofluid was shown to be less than that of an Al_2O_3/water nanofluid of comparable particle concentration. The two nanofluids were compared using a figure of the merit(Eq.17) consisting of the ratio of heat transfer enhancement to pumping power increase.

$$\mathit{Figure\ of\ merit} = \frac{h_{nanofluid} / h_{base\ fluid}}{Power_{nanofluid} / Power_{base\ fluid}} \quad (17)$$

The merit parameter was 0.8 for the SiC/water nanofluid compared to 0.6 for the Al_2O_3/water nanofluid, which is favorable to the SiC/water nanofluid for applications that are pumping power sensitive.

Torii and Yang (2009) studied the convective heat transfer behavior of aqueous suspensions of nanodiamond particles flowing through a horizontal tube heated under a constant heat flux condition. Their experimental system consisting of a flow loop, a power supply unit, a cooling device, and a flow measuring and control unit. The flow loop includes a pump, a digital flow meter, a reservoir, a collection tank, and a test section. A straight seamless stainless tube with 1000 mm length, 4.0 mm inner diameter, and 4.3 mm outer diameter was used as the test section. The whole test section was heated with the aid of the Joule heating method through an electrode linked to a dc power supply. They reported that (i) significant enhancement of heat transfer performance due to suspension of nanodiamond particles in the circular tube flow was observed in comparison with pure water as the working fluid, (ii) the enhancement was intensified with an increase in the Reynolds number and the nanodiamond concentration, and (iii) substantial amplification of heat transfer performance is not attributed purely to the enhancement of thermal conductivity due to suspension of nanodiamond particles.

Effect of particle size on the convective heat transfer in nanofluid by Anoop et al., (2009) in the developing region of pipe flow with constant heat flux showed that the enhancement in heat transfer coefficient was around 25% whereas for the 150 nm particle based nanofluids it was found to be around 11%. The heated test section was made of copper tube of 1200 mm length and 4.75 ± 0.05 mm inner diameter and the thickness of the tube was around 1.25 mm. Electrically insulated nickel chrome wire was uniformly wound along the length giving a maximum power of 200 W. They found that, with increase in particle concentration and flow rate the average heat transfer coefficient value was increased. They also observed that at the developing region the heat transfer coefficient is more than that at nearly developed region. It was further observed that the nanofluid with 45 nm particles shows higher heat transfer coefficient than that with 150 nm particles. For instance at x/D = 147, for 45 nm particle based nanofluid (4 wt%) with Re = 1550, the enhancement in heat transfer coefficient was around 25% whereas for the 150 nm particle based nanofluids it was found to be around 11%. After conducting sufficient number of experiments, they proposed the following correlation (Eq.18)

$$Nu_x = 4.36 + \left[6.219\times10^{-3}\, x_+^{1.1522}\left(1+\varphi^{0.1533}\right).\exp^{\left(-2.5228x_+\right)}\right]\left[1+0.57825\left(\frac{d_p}{d_{ref}}\right)^{-0.2183}\right] \tag{18}$$

Where, d_{ref} = 100 nm and x_+ is the dimensionless distance.

Rea et al., (2009) investigated laminar convective heat transfer and viscous pressure loss for alumina–water and zirconia–water nanofluids in a flow loop. The vertical heated test section was a stainless steel tube with an inner diameter (ID) of 4.5 mm, outer diameter (OD) of 6.4 mm, and length of 1.01 m. The test section had eight sheathed and electrically insulated T-type thermocouples soldered onto the outer wall of the tubing along axial locations of 5, 16, 30, 44, 58, 89, 100 cm from inlet of the heated section. Two similar T-type thermocouples were inserted into the flow channel before and after the test section to measure the bulk fluid temperatures. The heat transfer coefficients in the entrance region and in the fully developed region were found to increase by 17% and 27%, respectively, for alumina–water nanofluid at 6 vol % with respect to pure water. The zirconia–water nanofluid heat transfer coefficient increases by approximately 2% in the entrance region and 3% in the fully developed region at 1.32 vol %. The measured pressure loss for the nanofluids was in general much higher than for pure water and in good agreement with the traditional model predictions for laminar flow

Garg et al., (2009) used a straight copper tube of 914.4 mm length, 1.55 mm inner diameter and 3.175 mm outer diameter. The whole section was heated by an AWG 30 nichrome 80 wire wound on the tube. Both ends of the copper tube were connected to well-insulated plastic tubing to insulate the heat transfer section and fluid from axial heat conduction, and to avoid heat losses. The experiments were run under constant heat flux conditions using a current of 0.2 A. The test section was insulated to prevent loss of heat to the surroundings. Four surface-mount thermocouples were mounted on the test section at axial positions of 19 cm, 39.5 cm, 59 cm and 79 cm from the inlet of the section to measure wall temperatures. Additionally, two thermocouples were mounted on individual, unheated, and thermally insulated, short copper tubes located before and after the heat transfer section to measure the fluid bulk temperature at the inlet and outlet of the heat transfer section. De-ionized (DI) water, Gum Arabic (GA) and multi-walled carbon nanotubes (MWCNT) were used to produce aqueous suspensions. The nanotubes procured had a specified average outside diameter of 10–20 nm, length of 0.5–40 lm and purity of 95%. They observed a maximum percentage enhancement of 32% in heat transfer coefficient at Re - 600 ± 100. This percentage enhancement in heat transfer coefficient was found to continuously increase with axial distance. The percentage enhancement in heat transfer coefficient was found to continuously increase with axial distance. The reason behind the phenomenon is explained by the contribution from a considerable increase in thermal conductivity with an increase in bulk temperature with axial distance.

Lai et al., (2009) experimentally investigated the convection heat transfer performance of 20-nm, γ Al_2O_3 water-based nanofluids in a single 1.02-mm inner diameter, and constant heat flux stainless steel tube for laminar flow in both the developing and fully developed regions. Overall experimental results showed that the heat transfer coefficient increases with volume flow rate and nanoparticle volume fraction. In the developing region, the heat transfer coefficient enhancement decreased with increasing axial distance from the test section entrance. These results also showed that the higher the volume fraction, the longer is the thermal entrance length.

Chandrasekar et al., (2010) carried out experimental investigations on convective heat transfer and pressure drop characteristics of Al_2O_3/water nanofluid in the fully developed laminar region of pipe flow with constant heat flux with and without wire coil inserts. Their test loop consisting of a pump, calming section, heated test section, cooling section, a collecting station and a reservoir. Calming section of straight copper tube 800 mm long, 4.85 mm inner diameter, and 6.3 mm outer diameter was used to eliminate the entrance effect and to ensure fully developed laminar flow in the test section. A straight copper tube with 1200 mm length, 4.85 mm inner diameter, and 6.3 mm outer diameter was used as the test section. The test section was first wound with sun mica to isolate it electrically. Then, ceramic beads coated electrical SWG Nichrome heating wire giving a maximum power of 300W was wounded over it. Over the electrical winding, thick insulation consisting of layers of ceramic fiber, asbestos rope, glass wool and another layer of asbestos rope at the outer surface was provided to prevent the radial heat loss. The test section was isolated thermally from its upstream and downstream sections by plastic bushings to minimize the heat loss resulting from axial heat conduction. Two types of wire coil inserts were used which were fabricated using 0.5 mm stainless steel wire having a coil diameter of 4.5 mm and coil pitch ratio (defined as the ratio of pitch of the coil to diameter of tube) of 2 and 3. Dilute 0.1% Al_2O_3/water nanofluid increased the Nusselt number by 12.24% at Re = 2275 compared to that of distilled water. Further enhancements in Nusselt numbers was observed when Al_2O_3/water nanofluid is used with wire coil inserts. Nusselt numbers were increased by 15.91% and 21.53% when Al_2O_3/water nanofluid was used with their two types of wire coil inserts respectively at Re = 2275 compared to those of distilled water.

The Nusselt number and friction factor experimental results have been correlated by the following equations (Eqs 19 and 20).

$$Nu = 0.279\left(\mathrm{Re}\,\mathrm{Pr}\right)_{nf}^{0.558}\left(\frac{p}{d}\right)^{-0.447}\left(1+\varphi\right)^{134.65} \tag{19}$$

$$f = 530.8\,\mathrm{Re}_{nf}^{-0.909}\left(\frac{p}{d}\right)^{-1.388}\left(1+\varphi\right)^{-512.26} \tag{20}$$

The regression equation coefficients were assessed with the help of classical the least square method and the correlation is valid for laminar flow with Re < 2300, dilute Al_2O_3/water nanofluid with volume concentration $\phi = 0.1\%$ and wire coil inserts with $2 \leq p/d \leq 3$. They also found that, when compared to the pressure drop with distilled water, there was no significant increase in pressure drop for the nanofluid.

Amrollahi et al., (2010) measured the convective heat transfer coefficients of water-based functionalized multi walled nano tubes (FMWNT) nanofluid under both laminar and turbulent regimes flowing through a uniformly heated horizontal tube in entrance region. The straight copper tube with 11.42 mm inner diameter and 1 m length was used as the test section. The tube surface is electrically heated by an AC power supply to generate constant heat 800W and was insulated thermally by about 150 mm thick blanket to minimize the heat loss from the tube to the ambient. Five thermocouples were soldered on at different places along the test section to measure the wall temperature of the copper tube and the mean temperature of the fluids at the inlet, and two thermocouples were inserted at the inlet and outlet of the test section. They compared effective parameters to measure the convective heat transfer coefficients for functionalized MWNT suspensions such as Re, mass fraction and temperature altogether in entrance region for the first time. Their experimental results indicated that the convective heat transfer coefficient of these nanofluids increases by up to 33–40% at a concentration of 0.25 wt. % compared with that of pure water in laminar and turbulent flows respectively. Their results also showed that, increasing the nanoparticles concentration does not show much effect on heat transfer enhancement in turbulent regime in the range of concentrations studied. Also the ratio of heat transfer coefficient decreased with increasing Reynolds number. It was observed that the wall temperature of the test tube decreased considerably when the nanofluid flowed in the tube. Furthermore, this coefficient of nanofluids at the entrance of the test tube increases with Reynolds number, contrary to the fully developed laminar region that is constant.

Xie et al., (2010) reported on investigation of the convective heat transfer enhancement of nanofluids as coolants in laminar flows inside a circular copper tube with constant wall temperature. Nanofluids containing Al_2O_3, ZnO, TiO_2, and MgO nanoparticles were prepared with a mixture of 55 vol. % distilled water and 45 vol. % ethylene glycol as base fluid. It was found that the heat transfer behaviors of the nanofluids were highly depended on the volume fraction, average size, species of the suspended nanoparticles and the flow conditions. MgO, Al_2O_3, and ZnO nanofluids exhibited superior enhancements of heat transfer coefficient with the highest enhancement up to 252% at a Reynolds number of 1000 for MgO nanofluid. They also demonstrated that these oxide nanofluids might be promising alternatives for conventional coolants.

Fotukian and Esfahany (2010a) experimentally investigated the CuO/water nanofluid convective heat transfer in turbulent regime inside a tube. The test section was constructed of 1 m annular tube with inner copper tube of 5 mm inner diameter and 0.5 mm thickness and 32mm diameter outer stainless steel tube. Nanofluid flowed inside the inner tube while saturated steam entered annular section. They used dilute nanofluids with nanoparticles volume fractions less than 0.3%. They got excellent agreement between the measured heat transfer coefficients of pure water and the Dittus–Boelter predictions. They observed that heat transfer coefficients for nanofluids were greater than that of water and increasing the nanoparticle concentration showed a very weak effect on heat transfer coefficient. In such low concentrations of nanofluid investigated, the augmentation of heat transfer coefficient could not be attributed to the increase of thermal conductivity. The heat transfer coefficient increased about 25% compared to pure water. They concluded that, increasing nanoparticles concentration does not show much effect on heat transfer enhancement in turbulent regime in their studied range of concentrations. Also, the ratio of convective heat transfer coefficient of nanofluid to that of pure water decreased with increasing Reynolds number. It was also reported that the wall temperature of the test tube decreased considerably when the nanofluid flowed in the tube.

Fotukian and Esfahany (2010b) investigated turbulent convective heat transfer and pressure drop of γ Al_2O_3 /water nanofluid inside a circular tube, the same as described previously. The volume fraction of nanoparticles in base fluid was less than 0.2%. Their results indicated that addition of small amounts of nanoparticles to the base fluid augmented heat transfer remarkably. Increasing the volume fraction of nanoparticles in the range studied did not show much effect on heat transfer enhancement. Their experimental measurements showed that pressure drop for the dilute nanofluid was much greater than that of the base fluid.

Experimental investigations on convective heat transfer and pressure drop characteristics of three different concentration of CuO/water nanofluid was carried out by Suresh et al., (2010) in the fully developed turbulent region of pipe flow with constant heat flux. Experiments were done with a dimpled tube having dimensions of 4.85 mm diameter and 800 mm length. They reported that i) the relative viscosity of nanofluids increase with an increase in concentration of nanoparticles. ii) The thermal conductivity of nanofluid increases nonlinearly with the volume concentration of nanoparticles. iii) The convective heat transfer coefficient increases with increasing Reynolds number and increasing volume concentration in plain tube, and increases further with a dimpled tube. The Nusselt number and friction factor experimental results of nanofluids with dimpled tubes have been correlated by the following expressions (Eqs 21 and 22) using the least squares regression analysis

$$Nu = 0.00105 \mathrm{Re}^{0.984} \mathrm{Pr}^{0.4} (1+\varphi)^{-80.78} \left(1+\frac{p}{d}\right)^{2.089} \tag{21}$$

$$f = 0.1648 \mathrm{Re}^{0.97} (1+\varphi)^{107.89} \left(1+\frac{p}{d}\right)^{-4.463} \tag{22}$$

Pathipakka and Sivashanmugam (2010) numerically estimated the heat transfer behavior of nanofluids in a uniformly heated circular tube fitted with helical inserts in laminar flow. They used Al_2O_3 nanoparticles in water of 0.5%, 1.0% and 1.5% concentrations and helical twist inserts of twist ratios (ratio of length of one twist to diameter of the twist) 2.93, 3.91 and 4.89 for the simulation. Assuming the nanofluid behave as a single phase fluid, they investigated three dimensional steady state heat transfer behavior using Fluent 6.3.26. They concluded that the heat transfer increases with Reynolds number and decrease in twist ratio with maximum for the twist ratio 2.93. The increase in Nusselt number was 5%_31% for helical inserts of different twist ratio and nanofluids of different volume concentrations. The heat transfer enhancement was 31% for helical tape insert of twist ratio 2.93 and Al_2O_3volume concentration of 1.5% corresponding to the Reynolds number of 2039.

Suresh et al., (2011) presented a comparison of thermal performance of helical screw tape inserts in laminar flow of Al_2O_3/water and CuO/water nanofluids through a straight circular duct with constant heat flux boundary condition. Their experimental set up consists of a test section, calming section, pump, cooling unit, and a fluid reservoir. Both the calming section and test sections were made of straight copper tube with the dimension 1000 mm long, 10 mm ID and 12 mm OD. The calming section was used to eliminate the entrance effect. The test section tube was wounded with ceramic beads coated electrical SWG Nichrome heating wire. Over the electrical winding a thick insulation is provided using glass wool to minimize heat loss. They used three types of helical screw tape inserts with various twist ratio (1.78, 2.44, and 3) was made by winding uniformly a copper strip of 3.5 mm width over a 2.5 mm copper rod. The twist ratio 'Y', defined as the ratio of length of one twist (pitch, P) to diameter of the twist.

They used their experimental results of heat transfer to derive the following correlations(Eqs 23 and 24) of Nusselt number using least square method of regression analysis. The correlations are valid for laminar flow (Re < 2300) of 0.1% volume concentration of Al_2O_3/water and CuO/water nanofluids and for helical screw tape inserts of twist ratio ranging from 1.78 to 3.

$$\text{For } Al_2O_3 \text{ / water nanofluid; } Nu = 0.5419(\text{Re}\,\text{Pr})^{0.53}\left(\frac{P}{D}\right)^{0.594} \tag{23}$$

$$\text{For CuO / water nanofluid; } Nu = 0.5657(\text{Re}\,\text{Pr})^{0.5337}\left(\frac{P}{D}\right)^{0.6062} \tag{24}$$

Their results showed thermal performance factor of helical screw tape inserts using CuO/water nanofluid is found to be higher when compared with the corresponding value using Al_2O_3/water.

The experimental results on convective heat transfer of non-Newtonian nanofluids flowing through a horizontal uniformly heated tube under turbulent flow conditions by Hojjat et al., (2011a) states that convective heat transfer coefficient and Nusselt number of nanofluids are remarkably higher than those of the base fluid. Their experimental setup consists of a flow loop comprised of three sections: cooling unit, measuring and control units. The test section consists of a straight stainless steel (type 316) tube, 2.11-m long, 10-mm inner diameter, and 14-mm outer diameter. The test section was electrically heated by an adjustable DC power supply in order to impose a constant wall heat flux boundary condition. Ten K-type thermocouples were mounted on the tube outside wall to measure the wall temperature at different axial locations. The locations of the thermocouples were placed at the following axial positions from the test section inlet: 100, 150, 200, 350, 550, 800, 1100, 1400, 1700, and 2000 mm. The test section was thermally insulated from the upstream and downstream sections by thick Teflon bushings in order to reduce the heat loss along the axial direction. Two K-type thermocouples were also inserted in the calming chamber and the mixing chamber to measure the inlet and outlet bulk temperatures of the nanofluid, respectively. The whole test section including the calming and mixing chambers were heavily insulated. Three different types of nanofluids were prepared by first dispersing γ-Al_2O_3, TiO_2 and CuO nanoparticles in deionized water. The solution were subjected to ultrasonic vibration to obtain uniform suspensions, and then appropriate amounts of concentrated Carboxy Methyl Cellulose (CMC) solution were added to the suspension and mixed thoroughly with a mechanical mixer to achieve homogeneous nanofluids with the desired concentration. Average sizes of γ-Al_2O_3, TiO_2 and CuO nanoparticles were 25, 10, and 30-50 nm, respectively. Their results showed that Convective heat transfer coefficient and Nusselt number of nanofluids are remarkably higher than those of the base fluid. These enhancements of nanofluids were directly proportional to the particle concentration and Peclet number. Since the enhancement of heat transfer coefficient of nanofluids was much higher than that attributed to the improvement of the thermal conductivity, it was expected that the enhancement of heat transfer coefficient of nanofluids was affected by some other factors. Based on the experimental results, they proposed the following empirical correlation (Eq.25) to predict the heat transfer coefficients of non-Newtonian nanofluids.

$$Nu = 7.135 \times 10^4\, \text{Re}^{0.9545}\, \text{Pr}^{0.913}\left(1 + \varphi^{0.1358}\right) \tag{25}$$

$2800 < \text{Re} < 8400$; $40 < \text{Pr} < 73$.

Hojjat et al., (2011b) experimentally investigated the forced convection heat transfer of non-Newtonian nanofluids in a circular tube with constant wall temperature under turbulent flow conditions. Three types of nanofluids were prepared by dispersing homogeneously γ-Al_2O_3, TiO_2 and CuO nanoparticles into the base fluid. An aqueous solution of carboxymethyl cellulose (CMC) was used as the base fluid. Nanofluids as well as the base fluid show shear-thinning (pseudoplastic) rheological behavior. The test section consists of two 2-m long concentric tubes. The internal diameter of inner tube was 10 mm and a thickness 2 mm. The internal diameter of outer tube was 48 mm. Both tubes were made of stainless steel (type 316). The nanofluid flows through the inner tube whereas hot water was circulated through the annular section at high flow rates in order to create constant wall temperature boundary condition. Results indicated that the convective heat transfer coefficient of nanofluids is higher than that of the base fluid. The enhancement of the convective heat transfer coefficient increases with an increase in the Peclet number and the nanoparticle concentration. The increase in the convective heat transfer coefficient of nanofluids was greater than the increase that would be observed considering strictly the increase in the effective thermal conductivity of nanofluids. Experimental data were compared to heat transfer coefficients predicted using available correlations for purely viscous non-Newtonian fluids. Results showed poor agreement between experimental and predicted values. Hence they proposed a new correlation(Eq.26) to successfully predict Nusselt numbers of non-Newtonian nanofluids as a function of Reynolds and Prandtl numbers.

$$Nu = 0.00115\,\mathrm{Re}^{1.050}\,\mathrm{Pr}^{0.693}\,(1+\varphi^{0.388}) \tag{26}$$

Mahrood et al., (2011) experimentally investigated free convection heat transfer of non Newtonian nanofluids under constant heat flux condition. Two different kinds of non-Newtonian nanofluids were prepared by dispersion of Al_2O_3 and TiO_2 nanoparticles in a 0.5 wt. % aqueous solution of carboxy methyl cellulose (CMC). Experimental investigation of natural convection heat transfer behavior of non- Newtonian nanofluids in a vertical cylinder was attempted. Test section was a vertical cylindrical enclosure made up of PTFE (Poly Tetra Fluoro Ethylene). Fluid in the test section was heated from below by a heating system which consists of an aluminum circular plate and an electrical heater. In order to achieve a constant wall heat flux, the heater was placed between the aluminum plate and a thick PTFE circular plate. The PTFE plate also acts as insulation. Their results showed that the heat transfer performance of nanofluids is significantly enhanced at low particle concentrations. Increasing nanoparticle concentration has a contrary effect on the heat transfer of nanofluids, so at concentrations greater than 1 vol. % of nanoparticles the heat transfer coefficient of nanofluids is less than that of the base fluid. Indeed it seems that for both nanofluids there exists an optimum nanoparticle concentration that heat transfer coefficient passes through a maximum. The optimum concentrations of Al_2O_3 and TiO_2 nanofluids are about 0.2 and 0.1 vol. %, respectively. It is also observed that the heat transfer enhancement of TiO_2 nanofluids is higher than that of the Al_2O_3 nanofluids. The effect of enclosure aspect ratio was also investigated and the heat transfer coefficient of nanofluids as well as the base fluid increases by increasing the aspect ratio as expected.

Corcione et al., (2012) theoretically studied the heat transfer of nanoparticle suspensions in turbulent pipe flow. Both constant pumping power and constant heat transfer rate have been investigated for different values of the Reynolds number of the base fluid in the range between 2300 and $5x10^6$, the diameter of the suspended nanoparticles in the range between 25 nm and 100 nm, the length-to-diameter ratio of the pipe in the range between 50 and 1000, the nanofluid bulk temperature in the range between 303 K and 343 K, as well as for three different nanoparticle materials (i.e., CuO, Al_2O_3, and TiO_2) and two different base liquids (i.e., water and ethylene glycol). The significant findings of their study was the existence of an optimal particle loading for either maximum heat transfer at constant driving power or minimum cost of operation at constant heat transfer rate. In particular, for any assigned combination of solid and liquid phases, they found that the optimal concentration of suspended nanoparticles increases as the nanofluid bulk temperature is increased, the Reynolds number of the base fluid is increased, and the length-to-diameter ratio of the pipe is decreased, while it is practically independent of the nanoparticle diameter.

4.2. Double pipe heat exchanger

Chun et al., (2008) experimentally reported the convective heat transfer of nanofluids made of transformer oil and three kinds of alumina nanoparticles in laminar flow through a double pipe heat exchanger system. The experimental system consisted of two double-pipe heat exchangers for heating and cooling of nanofluid and was made of a non-corrosive stainless steel. Their experimental data showed that the addition of nanoparticles in the fluid increases the average heat transfer coefficient of the system in laminar flow. By non-linear regression of experimental data, the correlation (Eq.27) for heat transfer coefficient was decided as follows

$$h_i = \frac{k}{D} \times 1.7\,\mathrm{Re}^{0.4} \tag{27}$$

The surface properties of nanoparticles, particle loading, and particle shape were key factors for enhancing the heat transfer properties of nanofluids. They stated that these increases of heat transfer coefficients may be caused by the high concentration of nanoparticles in the wall side by the particle migration.

Duangthongsuk and Wongwises (2009) experimentally studied the heat transfer coefficient and friction factor of a nanofluid consisting of water and 0.2 vol. % TiO_2 flowing in a horizontal double-tube counter flow heat exchanger under turbulent flow conditions. Their test section was a 1.5 m long counter flow horizontal double-tube heat exchanger with nanofluid flowing inside the tube while hot water flows in the annular. The inner tube is made from smooth copper tubing with a 9.53 mm outer diameter and an 8.13 mm inner diameter, while the outer tube is made from PVC tubing and has a 33.9 mm outer diameter and a 27.8 mm inner diameter. The test section was thermally isolated from its upstream and downstream section by plastic tubes in order to reduce the heat loss along the axial direction.

They investigated the effects of the flow Reynolds number and the temperature of the nanofluid and the temperature and flow rate of the heating fluid on the heat transfer coefficient and flow characteristics. Their results showed that the convective heat transfer coefficient of nanofluid is slightly higher than that of the base liquid by about 6 -11%. The heat transfer coefficient of the nanofluid increased with an increase in the mass flow rate of the hot water and nanofluid, and increased with a decrease in the nanofluid temperature, and the temperature of the heating fluid had no significant effect on the heat transfer coefficient of the nanofluid. They also concluded that Gnielinski correlation for predicting the heat transfer coefficient of pure fluid is not applicable to a nanofluid. But, the Pak and Cho correlation (Eq. (7)) for predicting the heat transfer coefficient of a nanofluid agreed better with their experimental results than the Xuan and Li correlation (Eq. (8)).

Duangthongsuk and Wongwises (2010) experimentally studied the heat transfer coefficient and friction factor of the TiO_2-water nanofluids flowing in a horizontal double tube counter-flow heat exchanger under turbulent flow conditions. Their test fluid was TiO_2 nanoparticles with diameters of 21 nm dispersed in water with volume concentrations of 0.2 - 2 vol. %. The heat transfer coefficient of nanofluids was approximately 26% greater than that of pure water and the results also showed that the heat transfer coefficient of the nanofluids at a volume concentration of 2.0 vol.% was approximately 14% lower than that of base fluids for given conditions.

Their results showed that the Pak and Cho correlation (Eq. (7)) can predict the heat transfer coefficient of nanofluids and gives results that corresponded well only with the experimental results for the volume concentration of 0.2%. However, for the volume concentrations of 0.6% and 1.0%, the Pak and Cho equation fails to predict the heat transfer performance of the nanofluids. For the pressure drop, their results showed that the pressure drop of nanofluids was slightly higher than the base fluid and increases with increasing the volume concentrations.

New heat transfer and friction factor correlations(Eqs 28 and 29) for predicting the Nusselt number and friction factor of TiO2-water nanofluids were proposed in the form of

$$Nu = 0.074 \mathrm{Re}^{0.707} \mathrm{Pr}^{0.385} \varphi^{0.074} \tag{28}$$

$$f = 0.961 \varphi^{0.052} \mathrm{Re}^{-0.375} \tag{29}$$

The majority of the data falls within ±10% of the proposed equation. These equations are valid in the range of Reynolds number between 3000 and 18,000 and particle volume concentrations in the range of 0 and 1.0 vol. % for Nusselt number and 0 and 2.0 vol. % for friction factor.

Asirvatham et al., (2011) investigated the convective heat transfer of nanofluids using silver – water nanofluids under laminar, transition and turbulent flow regimes in a horizontal 4.3 mm inner-diameter tube-in-tube counter-current heat transfer test section. The volume concentration of the nanoparticles were varied from 0.3% to 0.9% in steps of 0.3% and the effects of thermo-physical properties, inlet temperature, volume concentration, and mass flow rate on heat transfer coefficient were investigated. Experiments showed that the suspended nanoparticles remarkably increased the convective heat transfer coefficient, by as much as 28.7% and 69.3% for 0.3% and 0.9% of silver content, respectively. Based on the experimental results a correlation (Eq.30) was developed to predict the Nusselt number of the silver–water nanofluid, with ±10% agreement between experiments and prediction.

$$Nu_{nf} = 0.023 \mathrm{Re}^{0.8} \mathrm{Pr}^{0.3} + (0.617\varphi - 0.135) \mathrm{Re}^{(0.445\varphi - 0.37)} \mathrm{Pr}^{(1.081\varphi - 1.305)} \quad (30)$$

4.3. Plate heat exchanger

Zamzamian et al., (2011) used nanofluids of aluminum oxide and copper oxide in ethylene glycol base fluid. They investigated the effect of forced convective heat transfer coefficient in turbulent flow, using a double pipe and plate heat exchangers. The inner pipe of the double pipe heat exchanger was made of copper, 12 mm in diameter and 1 mm in thickness, with a heat exchange length of 70 cm. The shell was made of green pipes, 50.8 mm in diameter. The flow inside the double pipe heat exchanger was arranged in opposite directions. The plate heat exchanger was a small, particularly manufactured model of common home radiators, 40 cm in height and 60 cm in length, exchanging heat freely with the ambience through four fins. The forced convective heat transfer coefficient of the nanofluids using theoretical correlations also calculated in order to compare the results with the experimental data. The effects of particle concentration and operating temperature on the forced convective heat transfer coefficient of the nanofluids were evaluated. The findings indicated considerable enhancement in convective heat transfer coefficient of the nanofluids as compared to the base fluid, ranging from 2% to 50%. Moreover, the results indicated that with increasing nanoparticles concentration and nanofluid temperature, the convective heat transfer coefficient of nanofluid increases.

4.4. Shell and tube heat exchanger

Farajollahi et al., (2010) measured the heat transfer characteristics of γ Al_2O_3 /water and TiO_2/water nanofluids in a shell and tube heat exchanger under turbulent flow condition. Water was allowed to flow inside the shell with 55.6 mm inside diameter and the nanofluid was passed through the 16 tubes with 6.1 mm outside diameter, 1 mm thickness, and 815 mm length. The tube pitch is 8 mm and the baffle cut and baffle spacing are 25% and 50.8 mm, respectively. The heat exchanger and pipe lines are thermally insulated to reduce heat loss to the surrounding. The effects of Peclet number, volume concentration of suspended nanoparticles, and particle type on heat transfer characteristics were investigated.

The observed the overall heat transfer coefficient of nanofluids increases significantly with Peclet number. For both nanofluids the overall heat transfer coefficient at a constant Peclet number increases with nanoparticle concentration compared to the base fluid. The experimental results for the Nusselt number of γ Al_2O_3/water and TiO_2/water nanofluids were compared with the prediction of Xuan and Li correlation (Eq. (6)). Results show that at

0.5 vol. % of γ Al_2O_3 nanoparticles and at 0.3 vol. % of TiO_2 nanoparticles a good agreement exists between the experimental results and the predicted values by Eq. (6) especially at higher Peclet numbers. They observed that the correlation is almost valid for the prediction of Nusselt number at low volume concentrations.

They reported that, adding of nanoparticles to the base fluid causes the significant enhancement of heat transfer characteristics. They experimentally obtained two different optimum nanoparticle concentrations for both the nanofluids. the heat transfer behavior of two nanofluids were compared and the results indicated that at a certain Peclet number, heat transfer characteristics of TiO_2/water nanofluid at its optimum nanoparticle concentration are greater than those of γ Al_2O_3 /water nanofluid while γ Al_2O_3 /water nanofluid possesses better heat transfer behavior at higher nanoparticle concentrations.

The emergence of several challenging issues such as climate change, fuel price hike and fuel security have become hot topics around the world. Therefore, introducing highly efficient devices and heat recovery systems are necessary to overcome these challenges. It is reported that a high portion of industrial energy is wasted as flue gas from heating plants, boilers, etc. Leong et al., (2012) focused on the application of nanofluids as working fluids in shell and tube heat recovery exchangers in a biomass heating plant. Heat exchanger specification, nanofluid properties and mathematical formulations were taken from the literature to analyze thermal and energy performance of the heat recovery system. It was observed that the convective and overall heat transfer coefficient increased with the application of nanofluids compared to ethylene glycol or water based fluids. In addition, 7.8% of the heat transfer enhancement could be achieved with the addition of 1% copper nanoparticles in ethylene glycol based fluid at a mass flow rate of 26.3 and 116.0 kg/s for flue gas and coolant, respectively.

4.5. Multi channel heat exchanger (MCHE)

Jwo et al., (2010) employed Al_2O_3 /water nanofluid to electronic chip cooling system to evaluate the practicability of its actual performance. Their experimental variables included nanofluids of different weight concentrations (0, 0.5, and 1.0 wt. %) and the inlet water temperature at different flow values. To determine if the addition of nanoparticles has any effects on overall heat transfer performance, they conducted a comparative experiment with water first. The control variables of their study were the mass flow rate, inlet water temperature, and heating power. Having completed the control experiment with water, nanofluids of different concentrations were used to carry out the same experiment. Using the same control variables, the ratio of the overall heat transfer performance of nanofluid to the overall heat transfer performance of water was calculated, and then acquired the overall heat transfer coefficient ratios under different conditions. Based on the collected temperature data for different mass flow rates, electric input powers, and nanofluid concentrations, the overall heat transfer coefficient ratio (r_U) of the MCHE (Eq.31) can be written as follows:

$$r_U = \frac{U_{nanofluid}}{U_{water}} = \frac{\left(T_{wall} - T_m\right)_{water}}{\left(T_{wall} - T_m\right)_{nanofluid}} \tag{31}$$

Where, $T_m = (T_{liq.in} + T_{liq.out})/2$ is the averaged temperature of liquid traversing the MCHE.

Results showed that the overall heat transfer coefficient ratio was higher at higher nanoparticle concentrations. In other words, the overall heat transfer coefficient ratio was higher when the probability of collision between nanoparticles and the wall of the heat exchanger were increased under higher concentration, confirming that nanofluids have considerable potential for use in electronic chip cooling systems. These results confirmed that nanofluid offers higher overall heat transfer performance than water, and a higher concentration of nanoparticles provides even greater enhancement of the overall heat transfer coefficient ratio.

4.6. Radial flow and electronic cooling devices

Gherasim et al., (2009) presented an experimental investigation of heat transfer enhancement capabilities of coolants with suspended nanoparticles (Al_2O_3 dispersed in water) inside a radial flow cooling device. Steady, laminar radial flow of a nanofluid between a heated disk and a flat plate with axial coolant injection has been considered. An experimental test rig was built consisting of the space between the two coaxial disks with central axial injection through the lower, high-temperature resistant PVC disk and the upper disk was machined from aluminum stock piece. They investigated the influence of disk spacing on local Nusselt number and proved that the local Nusselt number increases with a decrease in gap spacing. This behavior is obviously due to the increase of convection effects. They also analyzed the influence of particle volume fraction and Reynolds number on mean Nusselt number and found that the local Nusselt number increases with particle volume fraction. Their results showed that heat transfer enhancements are possible in radial flow cooling systems with the use of nanofluids. In general, it was noticed that the Nusselt number increases with particle volume fraction and Reynolds number and decreases with an increase in disk spacing.

Nguyen et al., (2007) investigated the heat transfer enhancement and behavior of Al_2O_3 nanoparticle - water mixture, for use in a closed cooling system that was destined for microprocessors or other heated electronic components. Their experimental liquid cooling system was a simple closed fluidic circuit which is mainly composed of a 5 l open reservoir and a magnetically driven pump that ensures a forced recirculation of liquid. An electrically heated block (aluminum body) was considered which simulates heat generated by a microprocessor. On top of this heated block, water-block (copper body) was installed. A thin film of high thermal conductivity grease was applied to minimize the thermal contact resistance at the interface junction between the heated block and the water-block. The assembly of heated block and water block has been thermally very well insulated with respect to the surrounding environment by means of fiberglass. Their data showed clearly that the inclusion of nanoparticles into distilled water produced a considerable enhancement of the cooling convective heat transfer coefficient. For a particular particle volume concentration of 6.8%, the heat transfer coefficient was found to increase as much as 40% compared to that of the base fluid. They observed that an increase of particle volume concentration has produced a clear decrease of the heated block temperature. Their experimental results also shown that a nanofluid with 36 nm particle size provides higher convective heat transfer coefficients than the ones given by nanofluid with 47 nm particles.

Gherasim et al., (2011) carried out a numerical investigation for heat transfer enhancement capabilities of coolants with suspended nanoparticles (Al_2O_3 dispersed in water) inside a confined impinging jet cooling device. They considered a steady, laminar radial flow of a nanofluid in an axis-symmetric configuration with axial coolant injection. A single phase fluid approach was adopted to numerically investigate the behavior of nanofluids. Good agreement was found between numerical results and available experimental data. Results indicated that heat transfer enhancement is possible in this application using nanofluids. In general, it was noticed that the mean Nusselt number increases with particle volume fraction and Reynolds number and decreases with an increase in disk spacing.

4.7. Double tube helical heat exchangers

G. Huminic and A. Huminic (2011) numerically studied heat transfer characteristics of double-tube helical heat exchangers using nanofluids under laminar flow conditions. CuO and TiO_2 nanoparticles with diameters of 24 nm dispersed in water with volume concentrations of 0.5–3 vol. % were used as the working fluid. The effect of particle concentration level and the Dean number on the heat transfer characteristics of nanofluids and water are determined. The mass flow rate of the nanofluid from the inner tube was kept and the mass flow rate of the water from the annulus was set at either half, full, or double the value. They showed the variations of the nanofluids and water temperatures, heat transfer rates and heat transfer coefficients along inner and outer tubes.

The effect of the nanoparticle concentration level on the heat transfer enhancement was calculated for different nanofluids and mass flow rate of the water. The results of the CFD analysis were used to estimate of the heat transfer coefficients and of the Dean number.

The numerical heat transfer coefficients of the nanofluid and water and Dean number were computed from the following equations (Eqs 32 and 33)

$$h_{nf} = \frac{q_{ave}}{(T_{nf,in} - T_{nf,out})}, \quad De_{nf} = \mathrm{Re}_{nf}\left(\frac{d_i}{2R_o}\right)^{0.5} \tag{32}$$

$$h_w = \frac{q_{ave}}{\left(T_{w,out} - T_{w,in}\right)}, \quad De_{nf} = \mathrm{Re}_{nf}\left(\frac{D_i - d_o}{2R_o}\right)^{0.5} \tag{33}$$

Where the average heat transfer rate is defined as $q_{ave} = \frac{q_{nf} + q_w}{2}$

Their results showed that for 2% CuO nanoparticles in water with the same mass flow rate in inner tube and annulus, the heat transfer rate of the nanofluid was approximately 14%

greater than that of pure water. They also showed that the convective heat transfer coefficients of the nanofluids and water increased with increasing of the mass flow rate and with the Dean number.

5. Conclusion

A detailed description of the state-of-the-art nanofluids research for heat transfer application in several types of heat exchangers is presented in this chapter. It is important to note that preparation of nanofluids is an important step in experiments on nanofluids. Having successfully engineering the nanofluids, the estimation of thermo physical properties of nanofluids captures the attention. Great quanta of attempts have been made to exactly predict them but large amount of variations were found. Research works on convective heat transfer using nanofluids is found to increase exponentially in the last decade. Almost all the works showed that the inclusion of nanoparticles into the base fluids has produced a considerable augmentation of the heat transfer coefficient that clearly increases with an increase of the particle concentration. It was reported by many of the researchers that the increase in the effective thermal conductivity and huge chaotic movement of nanoparticles with increasing particle concentration is mainly responsible for heat transfer enhancement. However, there exists aplenty of controversy and inconsistency among the reported results. The outcome of all heat transfer works using nanofluids showed that our current understanding on nanofluids is still quite limited. There are a number of challenges facing the nanofluids community ranging from formulation, practical application to mechanism understanding. Engineering suitable nanofluids with controlled particle size and morphology for heat transfer applications is still a big challenge. Besides thermal conductivity effect, future research should consider other properties, especially viscosity and wettability, and examine systematically their influence on flow and heat transfer. An in-depth understanding of the interactions between particles, stabilizers, the suspending liquid and the heating surface will be important for applications.

Author details

P. Sivashanmugam
Department of Chemical Engineering, National Institute of Technology, Tiruchirappalli, India

6. References

[1] S.K Das, S.U.S. Choi, H.E. Patel, Heat Transfer in Nanofluids-A Review, Heat Transf. Eng. 27 (10) (2006) 3-19.

[2] S.U.S. Choi, Enhancing thermal conductivity of fluids with nanoparticles, in: The Proc. 1995 ASME Int. Mech. Eng. Congr. Expo, San Francisco, USA, ASME, FED 231/MD 66, 1995, pp. 99-105.

[3] J.A. Eastman, S. U. S. Choi, S. Li, , L. J. Thompson, , and S. Lee, "Enhanced thermal conductivity through the development of nanofluids. Fall Meeting of the Materials Research Society (MRS), Boston, USA, 1996.

[4] S.U.S. Choi, Nanofluid technology: current status and future research. Korea-U.S. Technical Conference on Strategic Technologies, Vienna, VA, 1998.

[5] S.U.S. Choi, Z.G. Zhang, W. Yu, F.E. Lockwood, E.A. Grulke, Anomalous thermal conductivity enhancement in nanotube suspensions. Appl. Phys. Lett., 79(14) (2001), 2252-2254.

[6] C.H. Chon, K.D. Kihm, S.P. Lee, S.U.S. Choi, "Empirical correlation finding the role of temperature and particle size for nanofluid (Al_2O_3) thermal conductivity enhancement." Appl. Phys. Lett., 87(15), (2005), 153107-1531.

[7] C.H. Chon, S.W. Paik, J.B. Tipton, K.D. Kihm, Evaporation and Dryout of Nanofluid Droplets on a Microheater Array. J. Heat Transfer, 128(8), (2006) 735.

[8] J.A. Eastman., S. U. S. Choi, S. Li, W. Yu , L. J. Thompson, Anomalously increased effective thermal conductivities of ethylene glycol-based nanofluids containing copper nanoparticles." Appl. Phys. Lett., 78(6) (2001), 718-720.

[9] J.C. Maxwell, A Treatise on electricity and magnetism, second ed., Clarendon Press, Oxford, UK, 1881.

[10] S.Özerinç, S.Kakaç, A.G. Yazıcıoğlu, Enhanced Thermal Conductivity of Nanofluids: A State-of-the-Art Review, Microfluid. Nanofluid, **8**(2), (2010), 145-170.

[11] X.Q. Wang, A.S. Mujumdar, Heat transfer characteristics of nanofluids: a review, Int. J. Therm. Sci. 46 (2007) 1-19.

[12] Einstein, Investigation on the theory of brownian motion, Dover, New York, 1956.

[13] H.C. Brinkman, The viscosity of concentrated suspensions and solutions, J Chem Phys, 20 (1952), pp. 571–581.

[14] G. Batchelor, The effect of Brownian motion on the bulk stress in a suspension of spherical particles, J. Fluid Mech. 83 (1977) 97–117.

[15] Kostic, www.kostic.niu.edu/DRnanofluids; 2009 [14.11.2009].

[16] Y. Xuan, W. Roetzel, (2000). Conceptions for heat transfer correlation of nanofluids, International Journal of Heat and Mass Transfer, 43 (2000), 3701-3707

[17] B.C. Pak, Y.I. Cho, Hydrodynamic and heat transfer study of dispersed fluids with submicron metallic oxide particles, Exp. Heat Transf. 11 (1998) 151-170.

[18] Y. Xuan, Q. Li, Investigation on convective heat transfer and flow features of nanofluids, J. Heat Transf. 125 (2003) 151-155.

[19] D. Wen, Y. Ding, Experimental investigation into convective heat transfer of nanofluids at the entrance region under laminar flow conditions, Int. J. Heat Mass Transf 47 (2004) 5181-5188.

[20] Y. Yang, Z.G. Zhang, E.A. Grulke, W.B. Anderson, G. Wu, Heat transfer properties of nanoparticle-in-fluid dispersions (nanofluids) in laminar flow, Int. J. Heat Mass Transf. 48 (2005) 1107-1116.

[21] S.E.B. Maiga, S.J. Palm, C.T. Nguyen, G. Roy, N. Galanis, Heat transfer enhancement by using nanofluids in forced convection flows. Int. J. Heat Fluid Flow, 26 (2005) 530 - 546.

[22] S.E.B. Maiga, C.T. Nguyen, N. Galanis, G. Roy, T. Maré, M. Coqueux, Heat transfer enhancement in turbulent tube flow using Al_2O_3 nanoparticle suspension. Int. J. Numerical Methods Heat Fluid Flow, 16 (2006) 275 - 292.

[23] Y. Ding, H. Alias, D. Wen, R.A. Williams, Heat transfer of aqueous suspensions of carbon nanotubes (CNT nanofluids), Int. J. Heat Mass Transf. 49 (2006) 240 - 250.

[24] S.Z. Heris, S.G. Etemad, M.N. Esfahany, Experimental investigation of oxide nanofluids laminar flow convective heat transfer, Int. Commun. Heat Mass Transf. 33 (2006) 529-535.

[25] Y. He, Y. Jin, H. Chen, Y. Ding, D. Cang, H. Lu, Heat transfer and flow behavior of aqueous suspensions of TiO_2 nanoparticles (nanofluids) flowing upward through a vertical pipe, Int. J. Heat Mass Transf. 50 (2007) 2272-2281.

[26] D.P. Kulkarni, P.K. Namburu, H.E. Bargar, D.K. Das, Convective heat transfer and fluid dynamic characteristics of SiO_2-ethylene glycol/water nanofluid, Heat Transf. Eng. 29 (12) (2008) 1027-1035.

[27] K.S. Hwang, S.P. Jang, S. U. S. Choi, Flow and convective heat transfer characteristics of water-based Al_2O_3 nanofluids in fully developed laminar flow regime, Int. J. Heat Mass Transf. 52 (2009) 193-199.

[28] A. Sharma, S. Chakraborty, Semi-analytical solution of the extended Graetz problem for combined electro osmotically and pressure-driven microchannel flows with step-change in wall temperature, Int. J. Heat Mass Transf. 51 (2008) 4875-4885.

[29] W. Yu, D.M. France, D.S. Smith, D. Singh, E.V. Timofeeva, J.L. Routbort, Heat transfer to a silicon carbide/water nanofluid, Int. J. Heat Mass Transf 52 (2009) 3606-3612.

[30] S.Torii, W.J. Yang, Heat transfer augmentation of aqueous suspensions of nanodiamonds in turbulent pipe flow, J. Heat Transf. 131 (2009) 043203-1 - 043203-5.

[31] K.B. Anoop, T. Sundararajan, S.K. Das, Effect of particle size on the convective heat transfer in nanofluid in the developing region, Int. J. Heat Mass Transf. 52 (2009) 2189-2195.

[32] U. Rea, T. McKrell, L.W. Hu, J. Buongiorno, Laminar convective heat transfer and viscous pressure loss of alumina-water and zirconia-water nanofluids, Int. J. Heat Mass Transf. 52 (2009) 2042-2048.

[33] P. Garg, J. L. Alvarado, C. Marsh, T.A. Carlson, D.A. Kessler, K. Annamalai, An experimental study on the effect of ultrasonication on viscosity and heat transfer performance of multi-wall carbon nanotube-based aqueous nanofluids, Int. J. Heat Mass Transf. 52 (2009) 5090-5101.

[34] W.Y. Lai, S. Vinod, P.E. Phelan, P. Ravi, Convective heat transfer for water based alumina nanofluids in a single 1.02-mm tube, J. Heat Transf. 131(2009) 112401-1 - 112401-9.

[35] M. Chandrasekar, S. Suresh, A. Chandra Bose, Experimental studies on heat transfer and friction factor characteristics of Al_2O_3/water nanofluid in a circular pipe under laminar flow with wire coil inserts, Exp. Therm. Fluid Sci. 34 (2010) 122-130.

[36] A. Amrollahi, A.M. Rashidi, R. Lotfi, M.E. Meibodi, K. Kashefi, Convection heat transfer of functionalized MWNT in aqueous fluids in laminar and turbulent flow at the entrance region, Int. Commun. Heat Mass Transf. 37 (2010) 717- 723.

[37] H. Xie, Y. Li, W. Yu, Intriguingly high convective heat transfer enhancement of nanofluid coolants in laminar flows, Phys Lett. A 374 (2010) 2566-2568.

[38] S.M. Fotukian, M. Nasr Esfahany, Experimental study of turbulent convective heat transfer and pressure drop of dilute CuO/water nanofluid inside a circular tube, Int. Commun. Heat Mass Transf. 37 (2010a) 214–219.

[39] S.M. Fotukian, M. Nasr Esfahany, Experimental investigation of turbulent convective heat transfer of dilute γ -Al_2O_3/water nanofluid inside a circular tube, Int. J. Heat Fluid Flow 31 (2010b) 606-612.

[40] S. Suresh, M. Chandrasekar, S. Chandra sekhar, Experimental Studies on Heat Transfer and Friction Factor Characteristics of CuO/Water Nanofluid under Turbulent Flow in a Helically Dimpled Tube, Exp. Therm. Fluid Sci. 35 (2010) 542-549.

[41] G. Pathipakka, P. Sivashanmugam, Heat transfer behaviour of nanofluids in a uniformly heated circular tube fitted with helical inserts in laminar flow, Superlattices Microstruct. 47 (2010) 349 -360.

[42] S. Suresh, K.P. Venkitaraj, P. Selvakumar, Comparative study on thermal performance of helical screw tape inserts in laminar flow using Al_2O_3/water and CuO/water nanofluids, Superlattices Microstruct. 49 (2011) 608–622.

[43] M. Hojjat, S. Gh. Etemad, R. Bagheri, J. Thibault, Convective heat transfer of non-Newtonian nanofluids through a uniformly heated circular tube, Int. J. Therm. Sci. 50 (2011a) 525-531.

[44] M. Hojjat, S. Gh. Etemad, R. Bagheri, J. Thibault, Turbulent forced convection heat transfer of non-Newtonian nanofluids, Exp. Therm Fluid Sci. 35 (2011b) 1351–1356.

[45] M.R.K. Mahrood, S.G. Etemad, R. Bagheri, Free convection heat transfer of non Newtonian nanofluids under constant heat flux condition, Int. Commun. Heat Mass Transf. 38 (2011) 1449–1454.

[46] M. Corcione, M. Cianfrini, A. Quintino, Heat transfer of nanofluids in turbulent pipe flow, Int. J. Therm. Sci. 56 (2012) 58-69.

[47] B.H. Chun, H.U. Kang, S.H. Kim, Effect of alumina nanoparticles in the fluid on heat transfer in double-pipe heat exchanger system, Korean J. Chem. Eng. 25 (5) (2008) 966-971.

[48] W. Duangthongsuk, S. Wongwises, Heat transfer enhancement and pressure drop characteristics of TiO2–water nanofluid in a double-tube counter flow heat exchanger, Int. J. Heat Mass Transf. 52 (2009) 2059-2067.

[49] W Duangthongsuk, S Wongwises, An experimental study on the heat transfer performance and pressure drop of TiO2-water nanofluids flowing under a turbulent flow regime, Int. J.Heat Mass Transf. 53 (2010) 334-344.

[50] L.G. Asirvatham, B. Raja, D.M. Lal, S. Wongwises, Convective heat transfer of nanofluids with correlations, Particuology 9 (2011) 626– 631.

[51] A. Zamzamian, S.N. Oskouie, A. Doosthoseini, A. Joneidi, M. Pazouki, Experimental investigation of forced convective heat transfer coefficient in nanofluids of Al2O3/EG and CuO/EG in a double pipe and plate heat exchangers under turbulent flow, Exp. Therm Fluid Sci. 35 (2011) 495–502.

[52] B. Farajollahi, S.G. Etemad, M. Hojjat, Heat transfer of nanofluids in a shell and tube heat exchanger, Int. J. Heat Mass Transf. 53 (2010) 12-17.

[53] K.Y. Leong, R. Saidur, T.M.I. Mahlia, Y.H. Yau, Modeling of shell and tube heat recovery exchanger operated with nanofluid based coolants, Int. J. Heat Mass Transf. 55 (2012) 808–816

[54] C.S. Jwo, L.Y. Jeng, T.P. Teng, C.C. Chen, Performance of overall heat transfer in multi-channel heat exchanger by alumina nanofluid, J. of Alloy. Compd. 504 (2010) s385-s388.

[55] Gherasim, G. Roy, C.T. Nguyen, D. Vo-Ngoc, Experimental investigation of nanofluids in confined laminar radial flows, Int. J. Therm. Sci. 48 (2009) 1486-1493.

[56] C.T. Nguyen, G. Roy, C. Gauthier, N. Galanis, Heat transfer enhancement using Al_2O_3-water nanofluid for an electronic liquid cooling system, Appl. Therm. Eng. 27 (2007) 1501-1506.

[57] Gherasim, G. Roy, C.T. Nguyen, D. Vo-Ngoc, Heat transfer enhancement and pumping power in confined radial flows using nanoparticle suspensions (nanofluids), Int. J. Therm. Sci. 50 (2011) 369 - 377

[58] G. Huminic, A. Huminic, Heat transfer characteristics in double tube helical heat exchangers using nanofluids, International Journal of Heat and Mass Transfer 54 (2011) 4280–4287.

[59] V. Gnielinski, New equations for heat and mass transfer in turbulent pipe and channel flow, International Chemical Engineering 16 (1976) 359–368.

Chapter 15

Heat Transfer Enhancement of Impinging Jet by Notched – Orifice Nozzle

Toshihiko Shakouchi and Mizuki Kito

Additional information is available at the end of the chapter

http://dx.doi.org/10.5772/52029

1. Introduction

Jets are one of the most interesting and widely applied phenomena because they exhibit free or wall-bounded shear turbulent flows and large vortex structures that are essential in fluid dynamics and engineering. Jets are used for various industrial applications such as jet and rocket propulsion, cleaning, cutting, heating, cooling, atomizing, and burning (Shakouchi, 2004). Numerous studies have focused on gaining an understanding of free jets because this is essential for improving their heat transfer performance. For example, Livingood & Hrycak (1973), Hrycak (1981), Downs & James (1987), and Viskanta (1993) presented literature surveys of the impingement heat transfer of jets. Martin (1977) examined heat and mass transfer for single round nozzle, single slot nozzle, arrays of round and slot nozzles and provided extended reviews of heat transfer data. Yokobori et al. (1978) showed the role of large-scale eddy structures in the heat transfer enhancement of impinging slot jets. They also demonstrated that the present of a small disturbance in the upstream free shear layer enhanced the heat transfer in the stagnation region. Kataoka et al. (1987) showed the effects of surface renewal by large-scale eddies on the heat transfer of impinging jets. They explained that a secondary peak heat transfer occurs because of intermittent motions which entrain ambient fluid and break up the thermal boundary layer on the wall. Antonia & Zhao (2001) examined two circular jets—a pipe jet with a fully developed turbulent flow profile and a contraction jet with a laminar top-hat velocity profile at the nozzle exit—and discussed the similarities and differences between the two. Quinn (2006) demonstrated the effects of nozzle configurations on the mixing characteristics from the mean velocity and pressure distributions of an orifice plate and contoured nozzle jets. Mi et al. (2001) investigated the mixing characteristics and physical mechanisms of jets issuing from an orifice plate using flow visualization images and temperature measurements in comparison with a contoured nozzle and a pipe and found that the mixing rate of the orifice plate jet was higher than that of the others.

Impinging jets are applied in a wide variety of rapid cooling and heating processes because of their high efficiency and ability to provide high heat transfer rates. Many experiments and numerical simulations have been carried out to reveal the mechanism of impinging jets and to improve their performance. The effects of nozzle configurations, jet velocity, and exit conditions have been well recognized. Viskanta (1993) presented detailed reviews of the impingement heat transfer of single round and slot jets, a row of round and slot jets, inline and staggered arrays of round jets, square arrays of slot jets, a single row of round jets with cross flow, inline and staggered arrays of round jets with cross flow of spent air, an obliquely impinging single round jet, a slot jet with cross flow (jet inclined against cross flow), and a single round or slot jet impinging on concave and convex surfaces. He also discussed the impingement heat transfer of single flame jets and flame jet arrays.

Orifice jets are well known for their interesting characteristics—such as the vena contracta effect—and for having high mixing performance and good heat transfer characteristics; however, only a few studies have focused on these issues. Lee, J. & Lee, SJ. (2000) used three different orifice nozzles—square-, standard-, and sharp-edged—and demonstrated that the edge configuration of orifice nozzles affects the heat transfer characteristics of jet impingement. Quinn (2006) also demonstrated the effects of nozzle configurations on the mixing characteristics by measuring the mean velocity and pressure distributions for orifice and contoured-nozzle jets. Zhou et al. (2006) investigated the flow structure and the heat transfer characteristics of an orifice impinging jet with various mesh screens installed inside the nozzle. They also demonstrated the influence of the Reynolds number, nozzle-plate distance, and screen location on the heat transfer enhancement.

However, these studies have not provided an adequately detailed understanding of orifice jets. Therefore, this chapter provides an introduction to orifice jets and notched-orifice jets. This chapter has six sections. Sections 3 and 4 discuss orifice jets. Specifically, it describes the general characteristics of an orifice jet in comparison with a pipe and a quadrant jet, including the effects of the nozzle contraction area ratio on the free and heat transfer characteristics. Section 5 discusses the proposed notched-orifice jet. Specifically, it discusses the characteristics of a notched-orifice nozzle that is conically tapered on the inside to reduce the flow resistance and simultaneously increase the turbulent intensity.

2. Orifice jets

Before presenting a detailed analysis of orifice jets, we discuss their general characteristics in this section.

Figure 1 shows a pipe nozzle having an orifice nozzle with a 10-mm exit diameter and a pipe with a 29.75-mm inner diameter and a quadrant nozzle that provides a uniform velocity profile at the nozzle exit, which can be regarded as the standard profile for fundamental numerical and experimental studies. The quadrant nozzle contracted from a diameter of 29.75 mm to the exit diameter of 10.0 mm, which smoothly connected to a 3.0-mm-long straight contour [see Fig. 1 (c)]. Every nozzle had an exit diameter d_o = 10.0 mm, thickness of 1.0 mm, and length of 500 mm to ensure a fully developed turbulent pipe flow at the nozzle exit. Note that all the jets issued from a pipe having an exit velocity profile matching that of a fully developed turbulent pipe flow even in the case of a pipe with an inner diameter d_i = 29.75 mm, for which the pipe length ratio L/d_i = 16.8.

2.1. Nozzle exit characteristics

A constant temperature hot-wire anemometer was used to measure the mean and fluctuating velocity for an air jet at Reynolds number $Re = 1.5 \times 10^4$, which is based on the nozzle exit diameter and jet exit velocity.

The mean exit velocity profiles at $Re = 1.5 \times 10^4$, which should be sufficiently higher to describe an incompressible flow, are shown in the left-hand side of Fig. 2, and the fluctuating velocity for each jet is shown on the right-hand side because it was axisymmetric. The x- and r-axes starting at the origin of the jet were taken in the streamwise and spanwise directions, respectively. In the cross-stream direction, the origin $r = 0$ was taken at the centerline of the jet.

In the case of the pipe jet, the mean velocity profile of the pipe jet matched a fully developed turbulent pipe flow because the 500-mm-long pipe is sufficiently long for the boundary layer to grow completely. The fluctuating velocity reflecting the steep velocity gradient near the center of the jet is higher than that of the orifice jet that is discharged toward the center of the jet and accelerated.

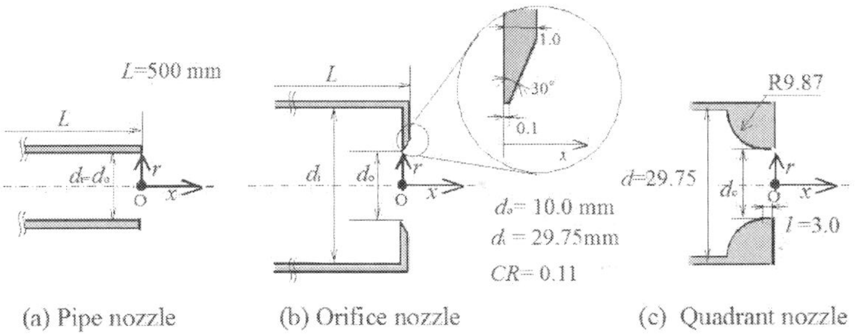

Figure 1. Nozzle configuration

The exit velocity of a quadrant nozzle also has a thin shear layer. However, the uniform velocity area over the nozzle exit is wider than that of the orifice jet because there is no vena contracta effect. It is therefore not surprising that the quadrant jet velocity u_c / u_m (u_m : mean velocity at the nozzle exit) is smaller than the others. The orifice jet velocity at the center is $u_c / u_m \cong 1.3$, and the quadrant jet velocity is approximately 15% smaller than the orifice jet velocity at the center. The maximum fluctuating velocities are observed at the largest velocity gradients.

On the other hand, the orifice jet has a considerable effect on the exit velocity profile. Jets issuing from an orifice nozzle are characterized by their large velocity gradient at the jet edge and the vena contracta effect, which decreases the jet diameter immediately after discharging and increases the centerline jet velocity, as shown in Fig. 3. The orifice jet has a saddle-shaped profile because of the vena contracta effect. The maximum fluctuating velocity of the orifice jet occurring at the jet edge is attributed to the exit velocity vector inside the nozzle toward the center because of the contraction of the orifice nozzle. Therefore, an orifice jet can potentially improve the mixing and spreading characteristics and the entrainment of ambient fluid.

However, it is true that the flow resistance of an orifice nozzle at the nozzle exit increases compared to that in a pipe and a quadrant nozzle. The flow resistance can be caused by the annular vortex inside the nozzle edge, as shown in Fig. 3. In order to avoid the increase in the flow resistance at the nozzle exit, a contour nozzle or a quadrant nozzle is commonly used. While the annular vortex causes a high flow resistance, it also increases the centerline velocity. The increase in the centerline velocity is one of major factors that enhance the heat transfer ratio on the impingement plate. There is no increase in the centerline velocity when the contour nozzle or quadrant nozzle is used because the jet produces a uniform flat velocity profile at the nozzle exit. In contrast, the orifice jet having a saddle-backed velocity profile because of the vena contracta effect appears to have a higher centerline velocity and enhances the heat transfer ratio. The detailed flow and heat transfer characteristics are presented in the following sections.

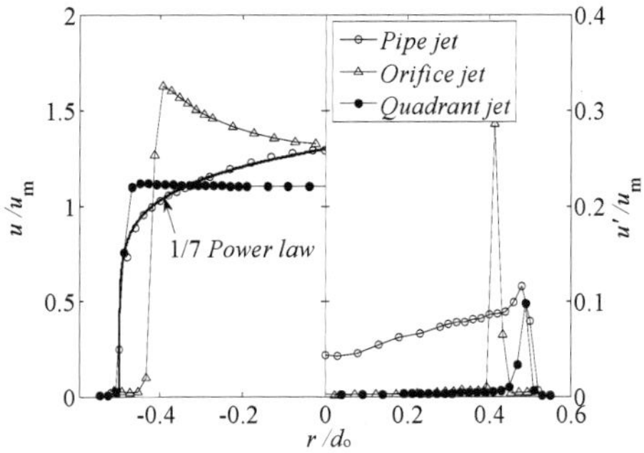

Figure 2. Mean and fluctuating velocities at the nozzle exit (x/d_o=0.2)

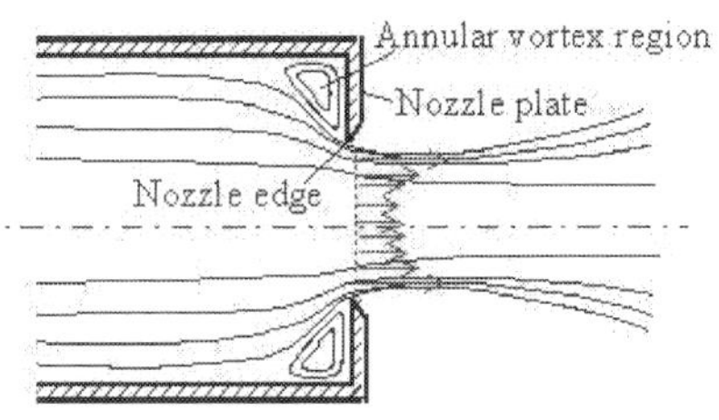

Figure 3. Orifice jet

2.2. Free flow jet characteristics

Figure 4 (a) and (b) show the normalized centerline velocity u_c/u_m and turbulent intensity u'_c/u_m for a pipe, an orifice, and a quadrant nozzle at $Re = 1.5 \times 10^4$. The orifice jet shows a different profile compared to the pipe and quadrant jets. The centerline velocity increases from the nozzle exit and reaches the maximum at $x/d_o \cong 2.0$ because of the vena contracta effect, whereas the velocity remains almost constant from the nozzle exit to $x/d_o \cong 4.0$ and 5.0 for the jet of the pipe nozzle and the quadrant nozzle, respectively. The velocity u_c/u_m decays according to the equation (1) in the fully developed region, which agrees with the equation for an axisymmetric circular jet.

$$u_c / u_m \propto \left(x / d_o\right)^{-1.0} \tag{1}$$

The turbulent intensity initially remained relatively small, following which it suddenly increased for all jets. The sudden increase occurs in the transition region in the case of the pipe jet. The quadrant jet shows the smallest turbulent intensity.

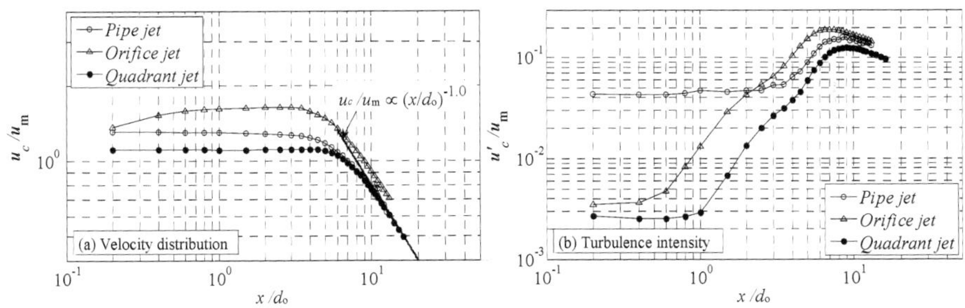

Figure 4. Mean and fluctuating velocity distributions along the jet centerline

2.3. Heat transfer characteristics

The local Nusselt number at $Re = 1.5 \times 10^4$ was calculated from the temperature distribution measured using thermocouples on the impinging plate and the jet temperature. The jet temperature was controlled carefully to within ±0.3°C of the atmospheric temperature. Thus, the entrainment effect claimed by Martin (1977) is neglected. To estimate the conduction heat flux, the heat loss due to radiation and natural convection were considered. The heat loss due to natural convection was estimated by measuring the temperature inside and on the surface of the impinging plate. The average Nusselt number is defined as the equation (2).

$$\overline{Nu} = \left(\frac{1}{\pi r^2}\right)\int_0^r Nu(2\pi r)dr \tag{2}$$

where r is an arbitrary location.

Figure 5 shows the local Nusselt number for $H/d_o = 2.0$ at $Re = 1.5 \times 10^4$. The thermocouples were mounted on the plate along the $\pm r$ axis to assure the center of impinging jet. The temperature distributions were axisymmetric. The only the half region of r is shown in Fig. 5. The Nusselt number changes considerably depending on the nozzle configurations, especially in the stagnation region. The orifice jet enhances the local Nusselt number significantly. For instance, it increases by 17% and 33% at the stagnation point as compared to those of the pipe and the quadrant jets, respectively. The reduction in the local Nusselt number for the quadrant jet is attributed to the exit velocity being measured to be ~15% smaller than that of the other jets.

Figure 6 shows the operation power W at $Re = 1.5 \times 10^4$. The operation power W was calculated by the equation (3).

$$W = Q(DP + ru_{pm}^{\ 2} / 2) \tag{3}$$

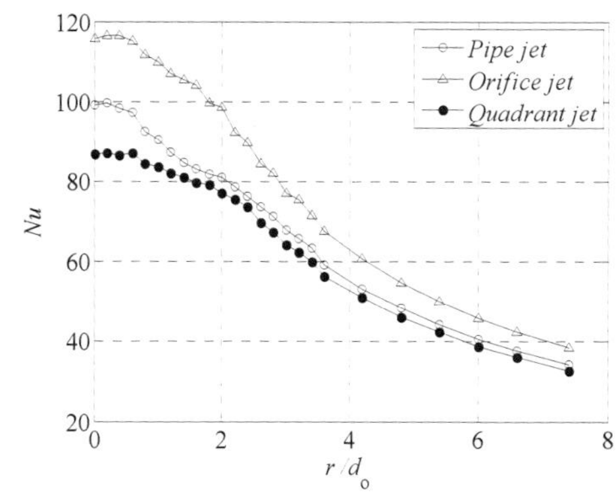

Figure 5. Local Nusselt number for $H/d_o = 2.0$ at $Re = 1.5 \times 10^4$

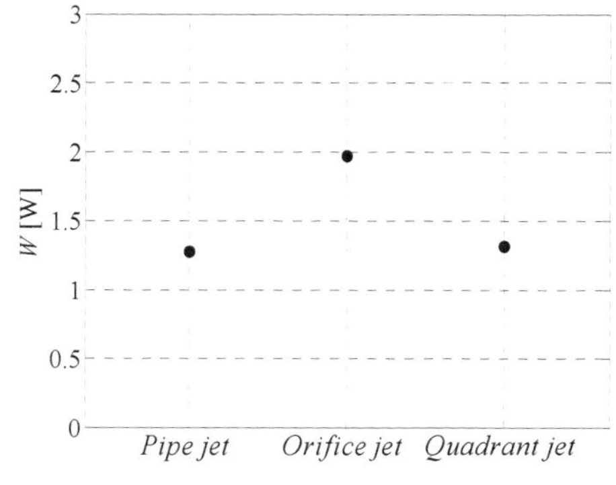

Figure 6. Operation power for the different nozzle types at $Re=1.5\times10^4$

where Q is the flow rate and ΔP, the pressure difference from atmospheric pressure. The static pressure was measured at $20d_o$ upstream from the nozzle exit using a 0.8-mm-diameter pressure tap mounted on the pipe nozzle wall at a flow rate of Q.

The required operation power of the orifice jet is larger than that of the pipe and the quadrant jets. As demonstrated above, the orifice jet provides better heat transfer performance because of the increase in its centerline velocity due to the vena contracta effect; however, it requires greater operation power because of its large flow resistance.

3. Flow control of orifice jet by changing nozzle contraction area ratio

Despite the evidence indicating that orifice jets improve the mixing and heat transfer characteristics, we were unable to find a systematic study of these jets that also focused on the nozzle contraction area ratio. Therefore, in this section, we consider the characteristics of orifice jets and the effects of nozzle configurations on their flow and heat characteristics by taking into account the nozzle contraction area ratio.

The nozzle geometry and the coordinate system are shown in Fig. 7 to reveal the effects of the nozzle contraction area ratio.

The contraction area ratio CR is defined as the equation (4).

$$CR = \left(\frac{A_o}{A_i}\right) = \left(\frac{d_o}{d_i}\right)^2, \tag{4}$$

where A_o is the cross-sectional area of the nozzle exit; A_i, the cross-sectional area of the pipe; d_o, the exit diameter of the nozzle; and d_i, the inner diameter of the pipe. The contraction area ratio was varied from CR = 0.11 to 1.00 (pipe nozzle). The various CRs were obtained by changing the pipe diameter under a constant nozzle exit diameter of d_o = 10.0 mm.

3.1. Nozzle exit characteristics

The mean exit velocity profiles at $Re = 1.5 \times 10^4$ are plotted in the left-hand side of Fig. 8 and the fluctuating velocity for each CR is plotted on the right-hand side because it was axisymmetric. CR was found to have a considerable effect on the exit velocity profile. The mean velocity profile for CR = 1.00 (pipe jet) agrees with the fully developed turbulent pipe flow, as shown in Fig. 2. For CR = 0.69, the profile was nearly top-hat-shaped, and a stronger effect of contraction was observed as CR decreased. The orifice jets had saddle-shaped profiles due to the vena contracta effect and the smaller CR and they had a thinner shear layer. The orifice jet velocity at the centre was $u_c/u_m \cong 1.3$ for all the orifices. The smaller CR caused a smaller orifice jet width, reflecting the maximum velocity value. The maximum fluctuating velocities are observed at the largest velocity gradients. The value of the turbulent intensity at the jet centre decreased from 4.3% to 0.2% with decreasing CR.

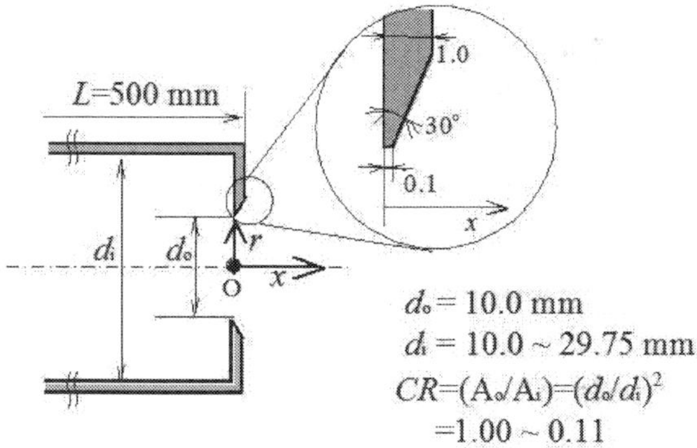

Figure 7. Nozzle configuration

3.2. Free flow jet characteristics

The centerline velocity u_c/u_m and turbulent intensity u'_c/u_m for various CR values at $Re = 1.5 \times 10^4$ are shown in Figs. 9(a) and (b), respectively. The centerline velocity increased from the nozzle exit and reached the maximum at $x/d_o = 2.0$ for all the orifice nozzles because of the vena contracta effect, whereas the velocity remained almost constant from the nozzle exit up to approximately $x/d_o = 4.0$ for the jet with $CR = 1.00$ (pipe nozzle).

The velocity growth from the exit increased with decreasing CR. The obtained turbulent intensity was apparently smaller at the exit and increased earlier as CR decreased. For example, for $CR = 0.11$, a sudden increase in turbulent intensity occurred at $x/d_o \cong 0.4$ and the maximum was at $x/d_o \cong 7.0$.

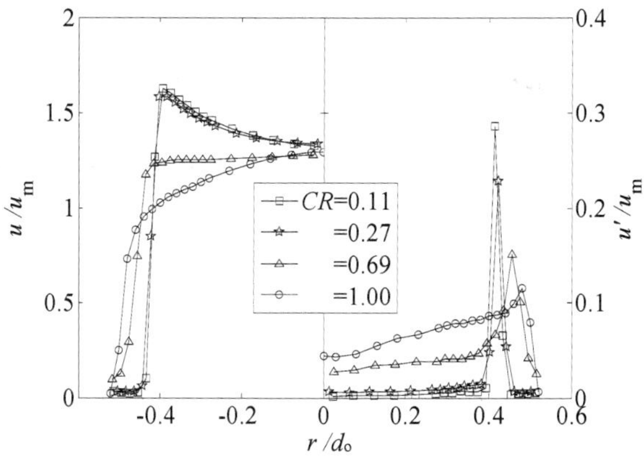

Figure 8. Mean and fluctuating velocities at the nozzle exit (x/d_o=0.2)

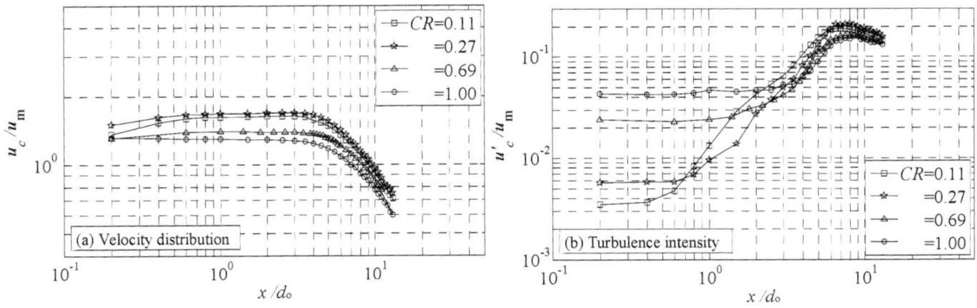

Figure 9. Mean and fluctuating velocity distributions along the jet centerline

The centerline maximum velocities for different values of *CR* at x/d_o = 2.0 are shown in Fig. 10. The centerline maximum velocity u_c/u_m can be expressed as follows:

$$u_c / u_m = 1.9CR^3 - 3.55CR^2 + 1.38CR + 1.53 \tag{5}$$

which has a maximum value of 1.7 at *CR* = 0.27. Velocity vectors towards the centerline produced by strong contraction at the orifice nozzle exit contributed to the increase in the centerline velocity, which had a maximum at *CR* = 0.27.

In order to obtain the dominant frequency of jet fluctuation where the jet reached its maximum velocity at x/d_o = 2.0, the power spectra were calculated from the velocity fluctuations at the jet center, as plotted in Fig. 11. The power spectra were normalized by integrating the function Φ_u^* as follows:

$\int \Phi_u^*(f^*)df^* = 1$, as proposed by Mi et al. (2001) for a comparison. The peak frequencies obtained for orifice jets at *CR* = 0.11, 0.27, 0.69, and 1.00 (pipe nozzle) and a quadrant nozzle were f^* = 0.71, 0.48, 0.55, 0.39, and 0.39, respectively, although the peak was not clear for *CR* = 1.00 (pipe nozzle). The values for the quadrant nozzle and *CR* = 0.11 are consistent with those reported by Mi et al. (2001). They presented values of $f^* \approx 0.40$ for a contoured nozzle and $f^* \approx 0.70$ for an orifice. Clearly, the peak frequency decreases with increasing *CR*. In addition, the magnitude of Φ_u^* at the peak also decreases and the peak becomes obscure, reflecting the large coherent vortex disintegrations. These results suggest that the same coherent vortex structures seen in Fig. 12 at low *Re* might exist at high *Re* as well.

The vortex structures of the submerged orifice water jet were visualized by the tracer method and the effects of *CR* on them were examined at low Reynolds number. Water jets seeded homogenously with Uranine (fluorescein) issued from the nozzle into a sufficiently large water tank. The images of the vortex structures were recorded using a digital video camera by illuminating the flow in the middle plane.

The results are shown in Fig. 12 for different *CR* values at $Re = 1.0 \times 10^3$. All of the flows were initially laminar and eventually became turbulent following the transition process.

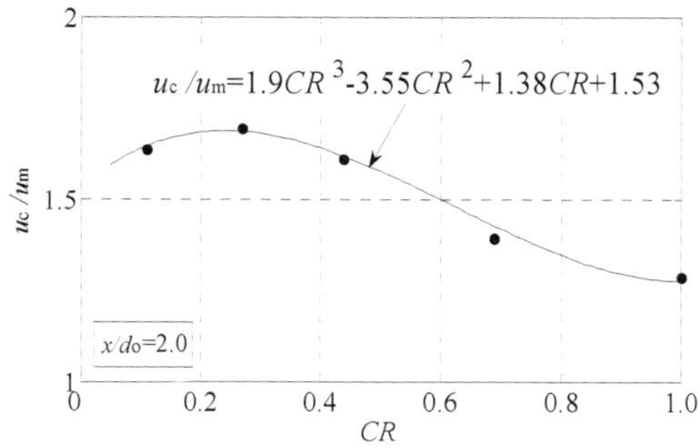

Figure 10. Centerline maximum velocity (x/do =2.0)

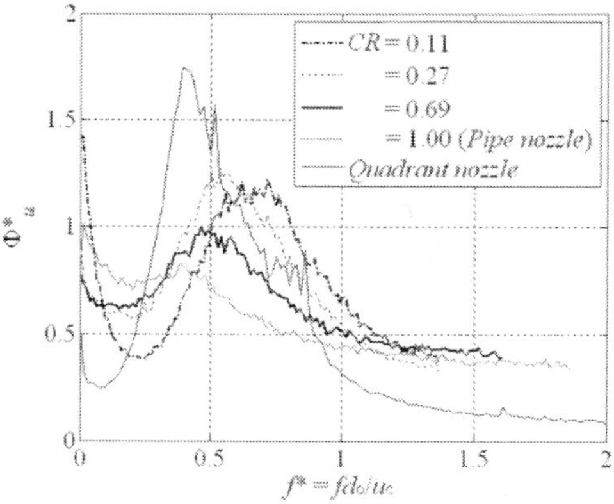

Figure 11. Power spectra at x/d_o =2.0.

Considerably large coherent vortexes were observed in Fig. 12 (a)–(b) and (d). Detailed observations of videos of the orifices jets revealed the production process of the large vortices. At first, many vortex rings were produced at the nozzle exit because of instability at the edge. A rapid decay in velocity in the downstream region prevented the formation of vortex rings, which yielded a large coherent vortex structure. *CR* caused significant changes in the formation of vortices. These changes must be related to the development of shear layer instability at the exit. Coherent vortex rings are observed in Fig. 12 (a) and (b), although those in Fig. 12 (b) are not as clear as those in the other cases. The coherent vortex rings produced for CR = 0.69 vanished in the downstream region. However, those produced for CR = 0.11 became incoherent in the downstream region, which promoted mixing with the ambient fluid. The mean velocity for CR = 0.11 increased and decreased more quickly than that for CR = 0.69, as explained in the next section. This may be the reason for the difference in the vortex formation. In contrast, there is no clear vortex in the pipe flow shown in Fig. 12 (c), suggesting the stability of the thick shear layer. It should be mentioned that even at $Re = 3.0 \times 10^3$, the flow was stable except in the most downstream region.

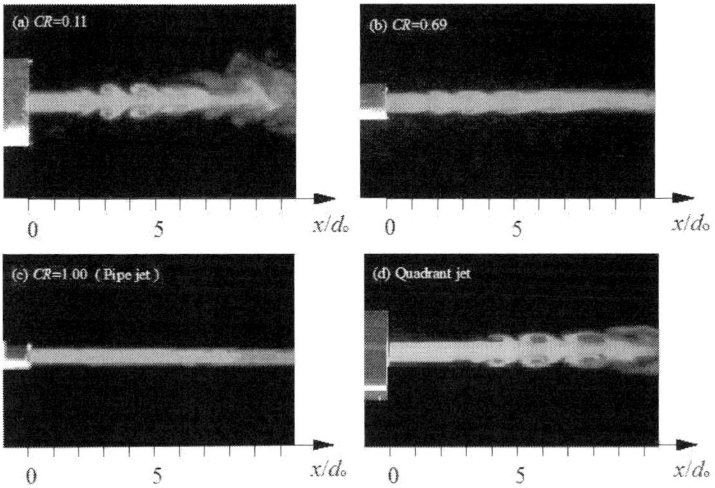

Figure 12. Flow visualization at Re=1.0×10^3

Figure 12 (d) also shows clear coherent vortexes produced by a quadrant nozzle because of instability at the edge of the nozzle exit, reflecting the thin shear layer shown in Fig. 8.

The large vortexes were stretched and appeared in the downstream region compared with the jet from the orifice nozzle that also had CR = 0.11, as shown in Fig. 12 (a). It is noteworthy that the jet from the quadrant nozzle did not grow as wide as those from the orifice nozzle. The flow for CR = 0.11 apparently spread wider and faster than the others at $Re = 3.0 \times 10^3$ as well. Thus, the images indicate that the smaller CR enhances the mixing rate more and the velocity difference in the jet core region promotes the earlier merging of the coherent vortex.

3.3. Heat transfer characteristics

Figure 13 shows the local Nusselt number at H/d_o = 2.0 at $Re = 1.5 \times 10^4$. The thermocouples were mounted on the plate along the r-axis on both sides in order to assure the center of impinging jet. The temperature distributed axisymmetrically after careful adjustment of the jet center. The distributions on only one side are shown in Fig. 13. The magnitude of the Nusselt number changed significantly with CR, especially at the stagnation point. The smaller CR enhanced the local Nusselt number significantly at $Re = 1.5 \times 10^4$. The obtained heat transfer rates of orifice jets for CR = 0.11 and 0.27 were similar. In the case of CR = 0.11 and 0.27, the heat transfer rates were increased by 17% and 33% at the stagnation point compared to those of the pipe and the quadrant nozzle, respectively.

The effects of CR on the heat transfer characteristics of the impingement plate for a constant nozzle exit mean velocity u_m are shown in Fig. 13. The flow loss of the nozzle, however, increases with decreasing CR in practice. Therefore, the operation power W for constant u_m

increases for an orifice jet. In contrast, the centerline velocity u_c of the orifice jet increases as a result of the contraction effect (Fig. 10). If W is not larger and the increase in u_c is significant, we expect that the heat transfer characteristics of the orifice impinging jet will improve more than those of the pipe jet.

Here, we discuss the heat transfer characteristics on the impinging plate under the same operation power.

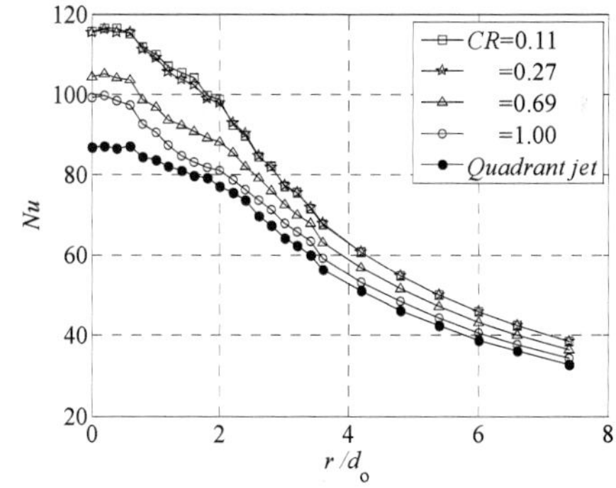

Figure 13. Local Nusselt number (H/d_o=2.0) at $Re = 1.5 \times 10^4$

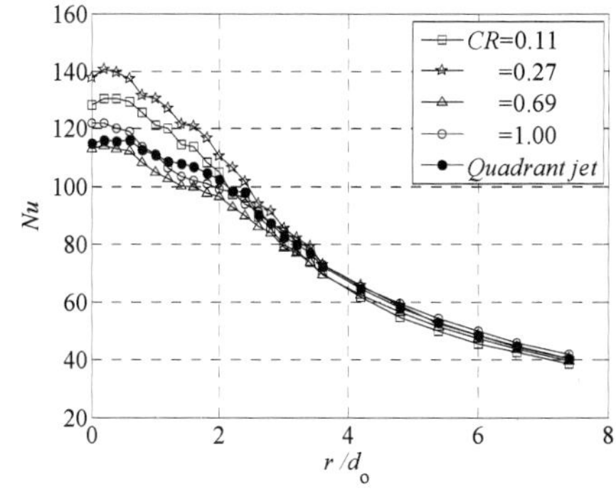

Figure 14. Local Nusselt number (H/d_o=2.0) under the same operation power

Figure 14 shows plots of the local Nusselt number distribution on the impingement plate for CR values ranging from 0.11 to 1.00 at $H/d_o = 2.0$ under the same operation power as that discussed in the previous section for $CR = 0.11$ ($u_m = 22.9$ m/s).

Because of the least flow loss from the nozzle, the nozzle exit mean velocity of the pipe jet was $u_m = 30.4$ m/s, which is larger than those of the orifice jets, under the same operation power as that required for $CR = 0.11$. However, because the orifice nozzle generates a contraction flow, the centerline velocity u_c increases from the nozzle exit and becomes maximum at $H/d_o = 2.0$. The maximum velocity ratio to u_m, $u/u_m = 1.7$, occurs for $CR = 0.27$.

An improved local Nusselt number is observed up to $r/d_o \cong 4$ for $CR = 0.27$ compared to that for the pipe. In addition, it should be noted that the local Nusselt number distribution for $CR = 0.11$ is smaller than that for the pipe. This is attributed to the large flow loss caused by the orifice nozzle for $CR = 0.11$.

4. Notched-orifice jet

We have seen the effects of the contraction area ratio $(d_o/d_i)^2$ on the flow characteristics of the orifice-free jets and the heat transfer enhancement in orifice impinging jets due to the increase in the centerline velocity that is caused by the vena contracta effect. However, it is found that the flow resistance of the orifice nozzle increases more than that of a nozzle without contraction. Reducing the flow resistance is essential for high performance. To reduce the flow resistance and improve the heat transfer characteristics using the vena contracta effect, we proposed a notched-orifice nozzle and demonstrated its flow and heat transfer characteristics.

In the previous section, the effects of the contraction area ratio of the orifice nozzle and operation power were discussed. These results suggest that the flow and heat transfer performances are governed by nozzle characteristics such as the jet impinging velocity, flow resistance, and turbulence intensity. Locating notches on the edge of the orifice nozzle exit will reduce the flow resistance and increase the turbulence intensity. Therefore, the heat transfer performance using the notched-orifice nozzle is expected to improve. The proposed nozzle geometry and coordinate system are shown in Fig. 15. Figure 15 (a) shows an orifice nozzle with a 10-mm exit diameter and a pipe with 19.23-mm inner diameter [$CR = (d_o/d_i)^2 = 0.27$]. Figure 15 (b) and (c) show V-shaped notched-orifice nozzles, namely, four- and eight-notched orifice nozzles, respectively. Four- or eight-notches having 60° openings and 1.56-mm depth are located on the edge of the orifice nozzle exit.

The nozzle exit diameter is generally selected as a reference length to represent the dynamical characteristics of jet flows. However, it is difficult to represent the universal dynamical characteristics for notched-orifice nozzles because there is little similarity in the flow patterns and characteristics even though the equivalent diameter based on the wet length and cross-sectional area is used. The investigations were performed considering the operation power calculated based on the flow resistance of each orifice jet under the same operation power as that of the orifice jet operated at $Re = 1.5 \times 10^4$.

4.1. Nozzle exit characteristics

Figure 16 shows the mean and fluctuating velocity distributions in the A-A′ and B-B′ directions at the nozzle exit $x/d_o = 0.2$ for the orifice, four-notched orifice, and eight-notched orifice jets. The mean exit velocity u/u_m is plotted in the left-hand side of Fig. 16 and the fluctuating velocity u'/u_m is plotted in the right-hand side because it is axisymmetric. A saddle-backed profile is observed for the orifice jet because of the vena contracta effect. The maximum fluctuating velocity of the orifice jet occurs at the jet edge attributed to the exit velocity vector inside the nozzle toward the center because of the contraction of the orifice nozzle. A similar profile can be seen for the notched-orifice jets, especially in the B-B′ direction. The jet widths of the orifice jets increase because of the notches in the A-A′ plane, whereas the vena contracta effects appear in the B-B′ plane. The differences between the profiles in the A-A′ and B-B′ planes become less significant when the number of notches increases. This might be attributable to the interference of vortices produced by the notches. The fluctuating velocity of the notched-orifice jets increases in the A-A′ plane.

Figure 17 shows the flow rate Q at the nozzle exit under the same operation power. The flow rate Q of the notched-orifice jet increases while the flow resistance decreases more than that in the case of the orifice jet without notches because of the increase in the cross-sectional area at the nozzle exit. The flow rates Q of four- and eight-notched orifice jets increased by 9.7% and 10.9%, respectively, compared with that of the orifice jet.

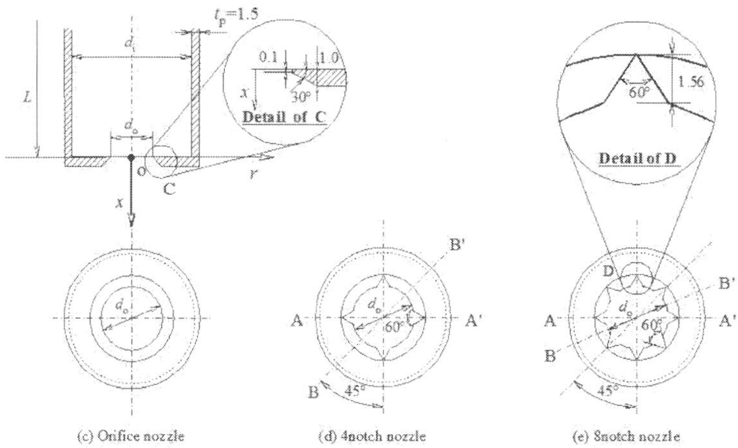

Figure 15. Nozzle configuration

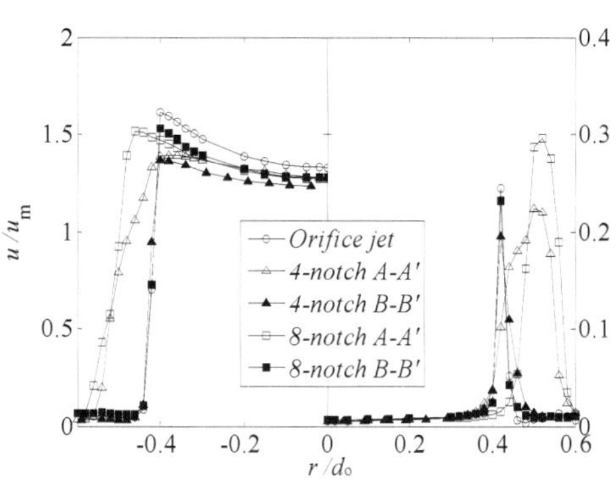

Figure 16. Mean and fluctuating velocities at the nozzle exit (x/d_o=0.2)

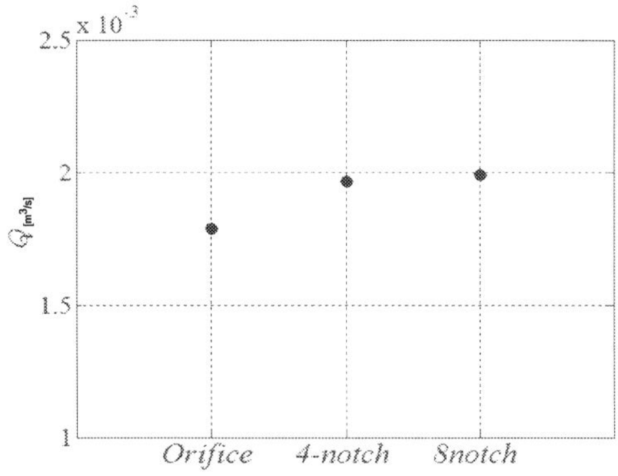

Figure 17. Flow rate at the nozzle exit

4.2. Free flow jet characteristics

The centerline velocity u_c and fluctuating velocity u_c' under the same operation power are plotted in Fig. 18. The centerline velocity increases from the nozzle exit for the orifice nozzles because of the vena contracta effect and remains nearly constant in a region called the potential core and then reaches the fully developed stage. Similar trends of development are also observed for the notched-orifice jets. In the case of a four-notched jet, the thick mixing layer reaches the center of the jet because of the significant difference between the velocity profiles in the A-A′ and B-B′ planes, as shown in Fig. 16, and causes the potential core length to be shorter than that of the other jets. The turbulence intensity along the jet centerline increases by ~47.2% at x/d_o = 3.5 compared with the other jets. In the case of the pipe jet, the turbulence intensity near the nozzle exit is larger than that of the other jets because the pipe jet had a fully developed turbulent velocity profile; however, the growth of the turbulence intensity is gradual and the turbulence intensity decreases at $x/d_o \geq 3$.

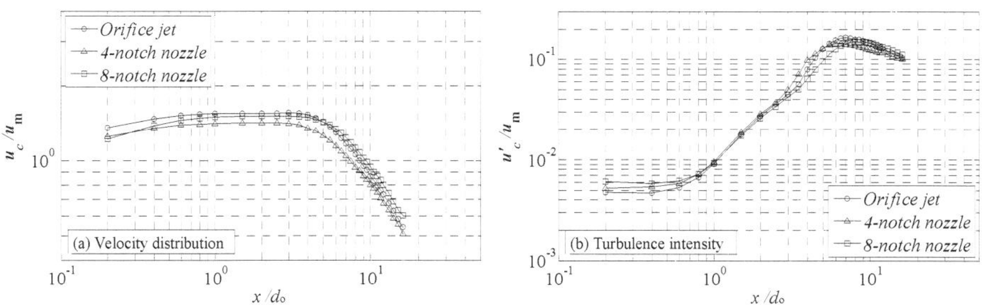

Figure 18. Mean and fluctuating velocity distributions along the jet centerline

4.2.1. Velocity profile and turbulence intensity distribution

Figure 19 (a) and (b) show the cross-sectional mean velocity u/u_m and turbulence intensity u'/u_m for the four-notched jet in the A-A′ and B-B′ directions under the same operation power, respectively. Because the profiles were axisymmetric, only half of the jet is shown in the figure. As mentioned, the nozzle configuration strongly influences the velocity profile at the nozzle exit and the influence can be seen in the downstream region up to $x/d_o = 7$. The jet width in the A-A′ plane is larger than that in the B-B′ plane at the nozzle exit $x/d_o = 0.2$, whereas it decreases at $x/d_o = 2$, which indicates an axis-switching phenomenon. The different development of the mixing layer in the process of jet spreading is observed between the A-A′ and B-B′ planes.

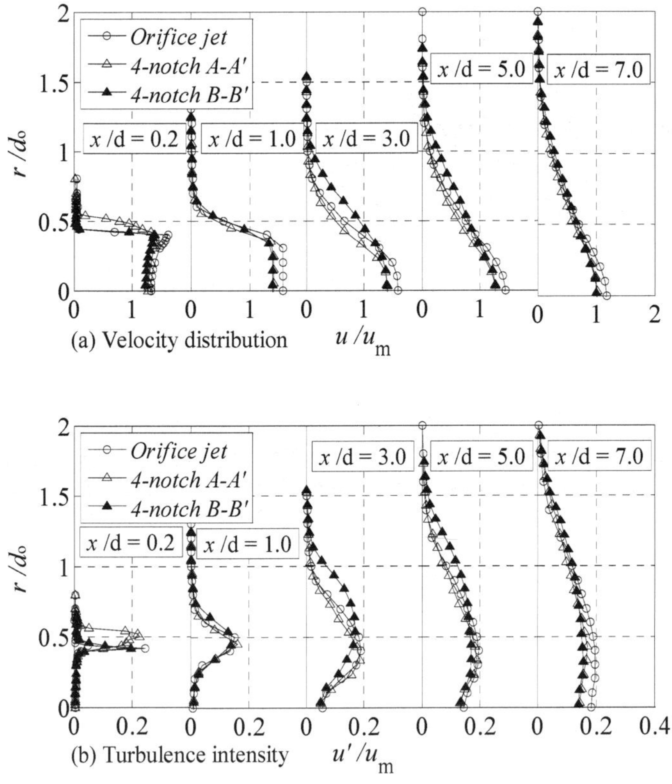

Figure 19. Centerline mean and fluctuating velocities for 4-notched orifice jet

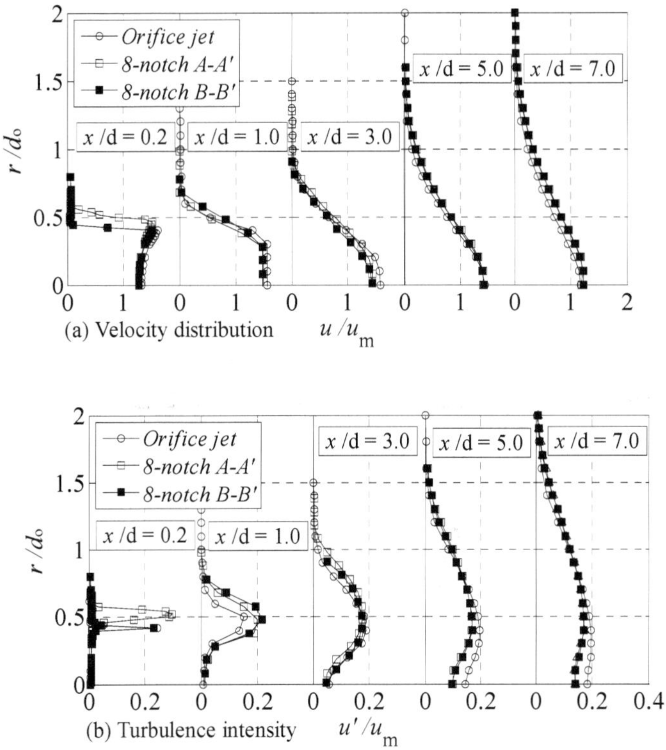

Figure 20. Centerline mean and fluctuating velocities for 8-notched orifice jet

Jets issuing from the eight-notched nozzle, which has a small spacing between the notches spread rapidly because of their instability in the mixing layer, as shown in Fig. 20 (a). The velocity profile in the A-A′ plane similarly coincides with that in the B-B′ plane at $x/d_o = 7$. The velocity profiles in the A-A′ and B-B′ planes at $x/d_o = 2$ are similar and larger than that of the orifice jet at $0.3 < r/d_o < 1.0$. Figure 20 (b) shows that the eight-notched jet flow in the A-A′ plane has high turbulence intensity at the nozzle exit, whereas the orifice jet has high turbulence intensity in the downstream region. This is attributed to the remarkable vena contracta effect and the large velocity gradient at the edge of the jet at $x/d_o = 2$ in the orifice jet flow shown in Fig. 20 (a), except at $0.5 < r/d_o < 1.0$, where the turbulence intensities of the eight-notched orifice jet both in the A-A′ and B-B′ planes are higher than that of the orifice jet.

The high jet velocity and turbulence intensity of the eight-notched jet at $x/d_o = 2$ may improve the heat transfer characteristics more than those of the orifice jets do.

4.2.2. *Flow spreading, switching phenomena*

Figure 21 shows the jet width b/d_o of the orifice and notched jets under the same operation power. The jet width b was determined from the spanwise location r where the velocity was u/u_{cm} = 0.1. The jet width for both notched jets is larger than that for the orifice jet not only because of the larger nozzle exit width with notches but also because of the higher mixing performance with ambient fluid due to an axis-switching phenomenon, which can be seen at $0.2 < r/d_o < 2.0$.

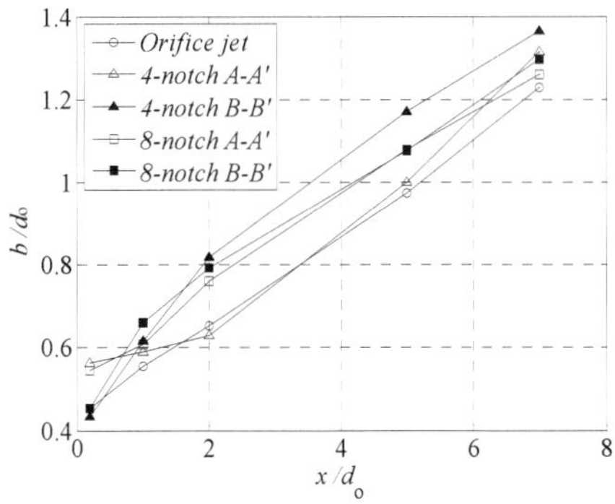

Figure 21. Jet width of the orifice and notched jets under the same operation power

Figures 22 and 23 show equivelocity profiles of the half jet width for four- and eight-notched orifice jets, respectively, that clearly show the axis-switching phenomenon. The notches are located on the axes in Fig. 15.

The pointed parts of the equivelocity profile at the nozzle exit are flattened at $x = d_o$ and the other parts protrude, as indicated by arrows in Fig. 22. The axis-switching phenomenon may be caused by longitudinal vortex flows produced at the edge of the notches. The jet flow entrains ambient flow, which causes the jet to accelerate and decelerate, distorting the vortex ring.

The axis-switching phenomenon, in which a major and minor axis switch to become the minor and major axis, respectively, in the downstream region, is well known in jet flows issued from elliptical or rectangular nozzles with various aspect ratios and cross-shaped nozzles. However, it is not well known that axis switching occurs in notched-orifice jets. We have demonstrated that an axis-switching phenomenon occurs in notched-orifice jets, although the notches are relatively small. Flows issuing from the orifice and notched-orifice nozzles at $Re = 1.0 \times 10^3$ were visualized to demonstrate the effects of the notches on the flow characteristics. Visualized flow patterns of the orifice and notched-orifice free jets using a laser-sheet technique are shown in Figs. 24 and 25, respectively. Figure 24 shows jets at a certain period in the same manner as in Fig. 12. There is no clear vortex in the pipe flow shown in Fig. 24 (a), suggesting the stability of the thick shear layer. In contrast, considerably large coherent vortexes were observed in Fig. 24 (b)–(f), although they were not clear in Fig. 24 (e) and (f) because of the earlier disintegration of the eight-notched orifice jet. In the case of the standard orifice jet, ring-shaped vortex formations in the shear layer of the jet can be clearly seen in Fig. 24 (b). The discharged ring vortices decayed in the velocity downstream, merging into a large vortex and then disintegrating. Figure 24 (c) and (e) show flow patterns of four- and eight-notched orifice jets, respectively, which were photographed in the middle of the A-A′ plane shown in Fig. 15. The flow patterns in the middle of the B-B′ plane are shown in Fig. 24 (d) and (f). The vortex shedding frequency of four- and eight-notched orifice jets was higher and lower, respectively, than that of the orifice jet. The jets directly issuing from notches such as those shown in Fig. 24 (c) and (e) spread earlier and wider than those in the B-B′ plane because of the high turbulent intensity, although those in the case of the eight-notched orifice jets are not clear because the clearance between the notches is not sufficient to indicate the difference between both planes. The comparison between flows in the A-A′ and B-B′ planes shows a slight difference between the jet widths at the nozzle exit; therefore, the jet width in the B-B′ plane narrower in the A-A′ plane represents high turbulence intensity at the nozzle exit, whereas the orifice jet represents high turbulence intensity in the downstream region. This is attributed to the remarkable vena contracta effect and large velocity gradient at the edge of the jet at $x/d_o = 2$ in the orifice jet flow shown in Fig. 20 (a), except at $0.5 < r/d_o < 1.0$, where the turbulence intensities of the eight-notched orifice jet in both the A-A′ and B-B′ planes are higher than that of the orifice jet. The high jet velocity and turbulence intensity of the eight-notched jet at $x/d_o = 2$ may improve the heat transfer characteristics more than do the orifice jets.

Figure 25(a) is photographs of the impinging pipe jet (CR=1.00). The four small outer holes in the images did not affect the flow since they were screw holes to fix the plate. The jet emitted from the pipe nozzle impinged on the plate with insignificant fluctuations and radically spread producing a thin film on the plate. No large vortex structures or fluctuations could be seen. Ring-shaped flow patterns were, however, observed at $r/d_o \cong 7$, indicating a transition from laminar to turbulent flow. Figure 25 (b) shows the standard orifice impinging jet flow pattern. In the case of the orifice impinging jet, the flow fluctuates at the nozzle edge and spreads radially to produce ring-shaped flow patterns that intermittently impinge on the plate, indicating a transition from laminar to turbulent flow. The interval between ring-shaped patterns corresponds to the vortex-shedding frequency. It is noteworthy that longitudinal vortex structures were observed around the ring-shaped flow patterns. Figure 25 (c) and (d) show flow patterns of four- and eight-notched orifice jets, respectively. The A-A′ and B-B′ planes defined in Fig. 15 are shown in Fig. 25 (c) and (d), respectively. It is clearly observed that the presence of notches strongly affects the flow patterns. The fluctuations at the nozzle edge produce large vortex structures in the downstream region that intermittently impinge on the plate, as observed in the orifice impinging jet. However, the large vortex structures of the notched-orifice jets are distorted by small notches. Diamond-shaped and eight-pointed-star patterns are seen in Fig. 25 (c) and (d), respectively, corresponding to the number of small notches located at the orifice nozzle exit. The fluctuations of the large vortex structure shown in Fig. 25 (c) are higher than those of the structure shown in Fig. 25 (d), indicating that the four-notched orifice nozzle has a high turbulence intensity. This result is consistent with that shown in Fig. 18.

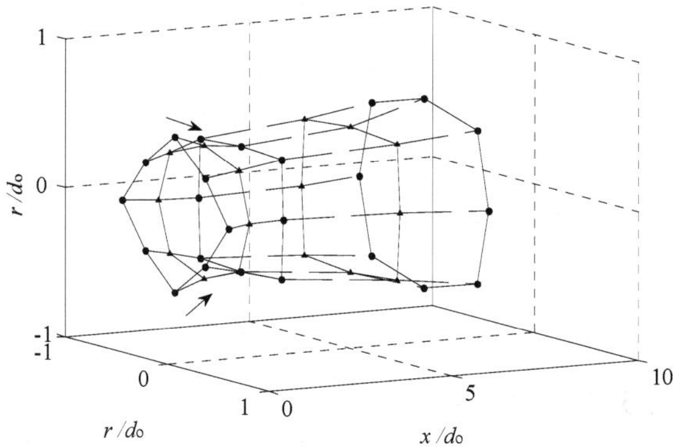

Figure 22. Equi-velocity profile of half width for 4-notched orifice jet

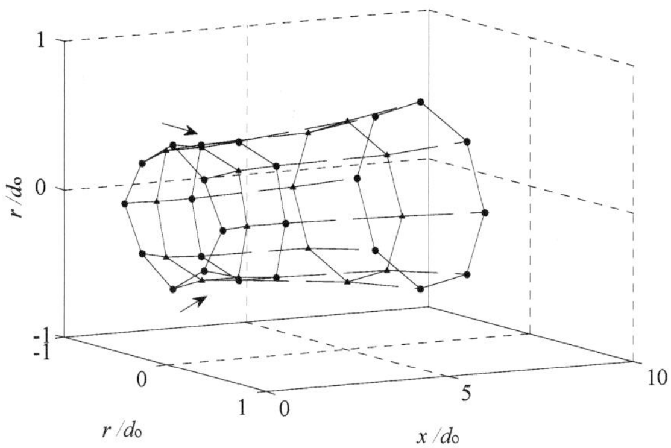

Figure 23. Equi-velocity profile of half width for 8-notched orifice jet

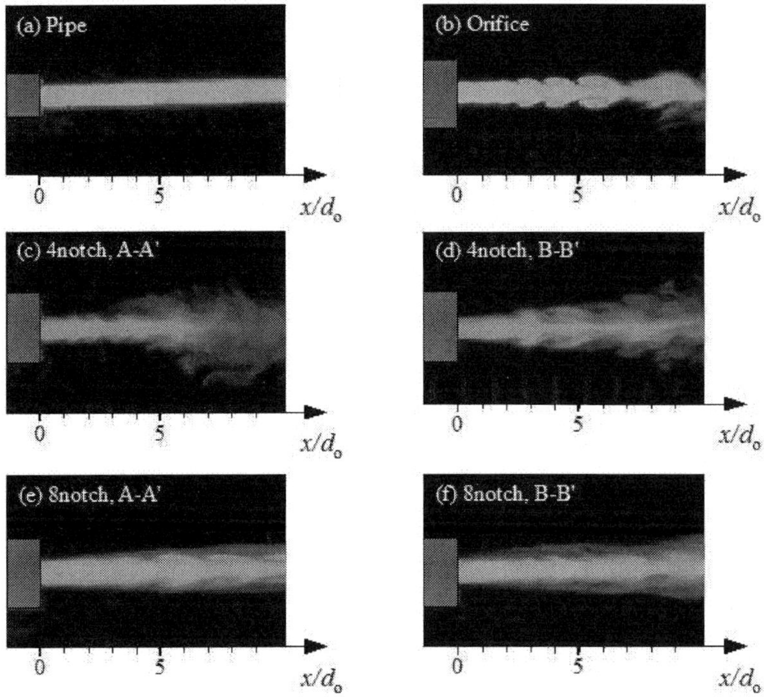

Figure 24. Flow visualization at Re=1.0×10^3

4.3. Heat transfer characteristics

Figures 26 shows the local Nusselt number distribution on the impingement plate at H/d_o = 2.0 under the same operation power. Because of the axisymmetry of the profile, only half the radial distribution from the stagnation point is plotted. The local Nusselt number reaches its maximum near the stagnation point and decreases in the outward direction. The local Nusselt number distribution of the notched-orifice jet is larger than that of the orifice jet in the downstream region because of the larger turbulence intensity caused by the notches.

4.4. Heat transfer enhancement of notched-orifice jet

We discussed the flow and heat transfer characteristics of notched-orifice jets in the previous section and demonstrated that placing notches on the edge of the orifice nozzle can reduce the flow resistance and increase the turbulent intensity. Therefore, the heat transfer performance improved with the use of a notched-orifice nozzle.

In this section, we propose a notched-orifice nozzle that is conically tapered on the inside and investigate the effects of the notches and tapering angle α on the flow and heat transfer characteristics of the proposed orifice jet.

A schematic diagram of the nozzle geometry is shown in Fig. 27. An orifice nozzle with a 10.0-mm exit diameter and a pipe with a 29.75-mm inner diameter [$CR = (d_o/d_i)^2 = 0.27$, see Fig. 27 (a)] was used unless otherwise stated.

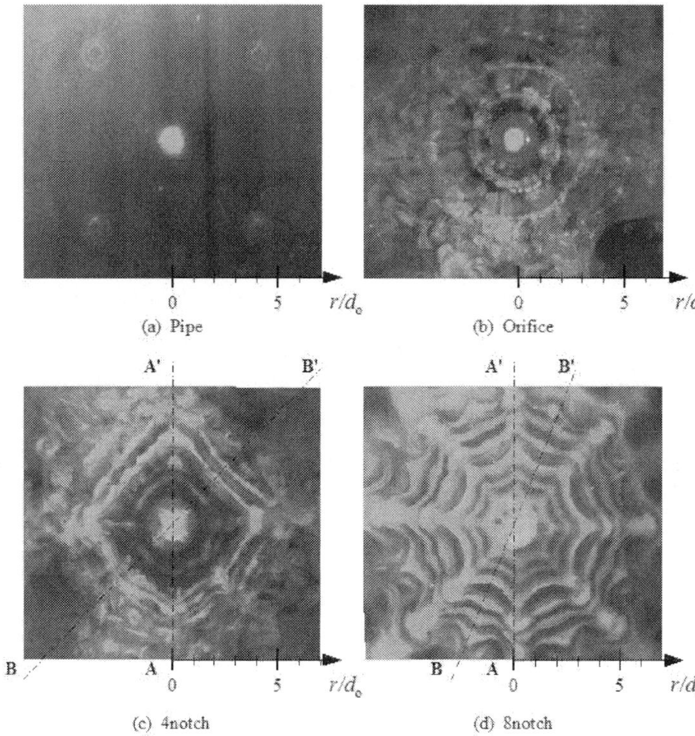

Figure 25. Flow visualization at $Re=1.0\times10^3$

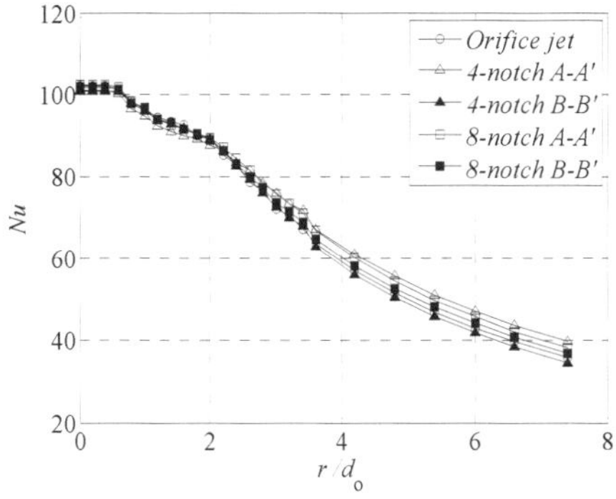

Figure 26. Local Nusselt number ($H/d_o=2.0$)

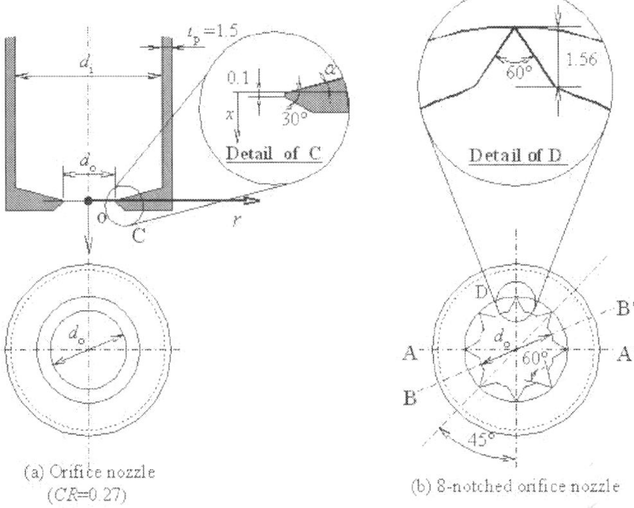

Figure 27. Nozzle configuration

Figure 27 (b) shows an orifice nozzle with eight notches having 60° openings and 1.56-mm depths. The inside of the eight-notched orifice nozzle is conically tapered at angles of α = 0°, 5°, 10°, and 15°. A pipe nozzle with a 10-mm inner diameter was also used for comparison. Every nozzle was 500-mm long and 1.5-mm thick.

Figure 28 shows the flow rate improvement rate, $(Q - Q_0)/Q_0$, for orifice jets that are conically tapered on the inside as compared to the orifice jet without notches at the nozzle exit under the same operating power. The flow rate Q of the notched-orifice jet increases while the flow resistance decreases compared to that of the orifice jet without notches because of the increase in the cross-sectional area at the nozzle exit.

The flow rate of notched-orifice jets Q increases linearly with the tapering angle α. Q of eight-notched orifice jets with α = 0°, 5°, 10°, and 15° increased by 10.9%, 11.3%, 13.1%, and 13.5% respectively, compared with that of the orifice jet.

Figure 29 shows the mean and fluctuating velocity distributions in the A-A′ direction at the nozzle exit x/d_o = 0.2 for notched-orifice jets. The mean exit velocity u/u_m is plotted on the left-hand side of Fig. 29 and the fluctuating velocity u'/u_m is plotted on the right-hand side because they were, respectively, axisymmetric. The results in the B-B′ direction at the nozzle exit are plotted in Fig. 30 in the same manner.

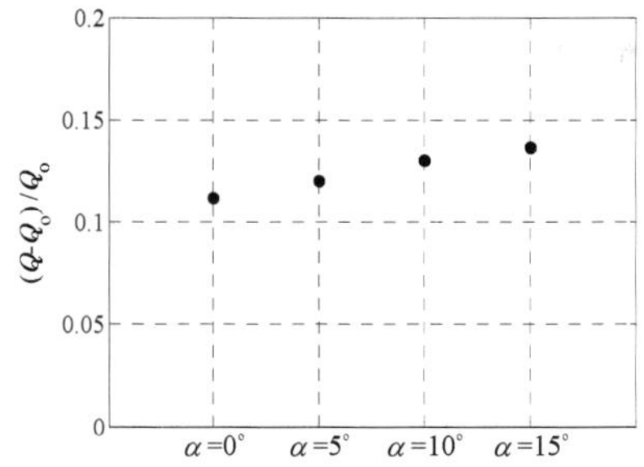

Figure 28. Increment rate of flow rate of notched-orifice nozzle to orifice nozzle

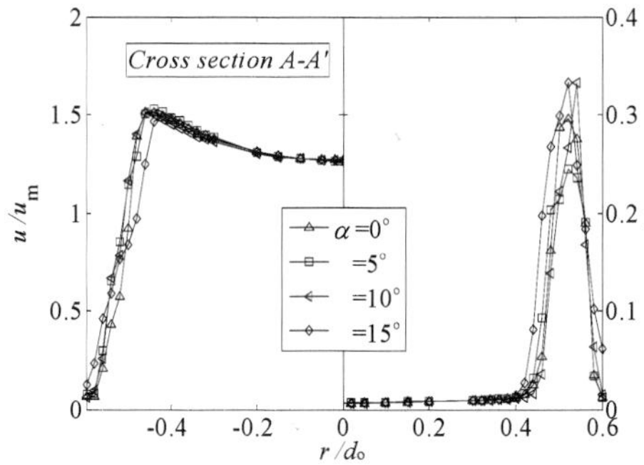

Figure 29. Mean and fluctuating velocities at the nozzle exit (x/d_o=0.2) in the A-A' plane

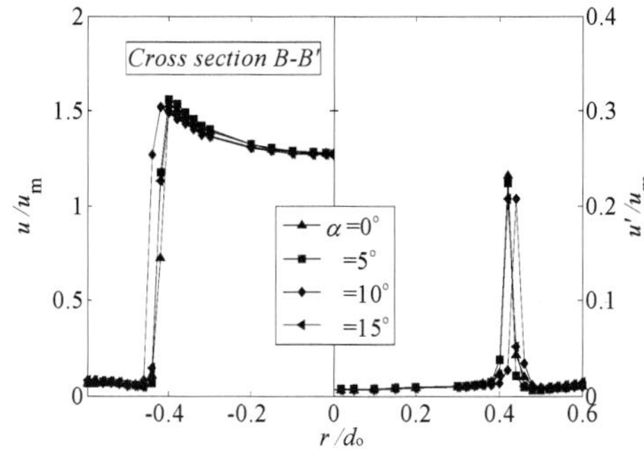

Figure 30. Mean and fluctuating velocities at the nozzle exit (x/d_o=0.2) in the B-B' plane

A similar profile can be seen for the notched-orifice jets, especially in Fig. 30. The jet widths of the notched-orifice jets increase because of the notches in the A-A′ plane, whereas the vena contracta effects appear in the B-B′ plane. The maximum velocity occurs at r/d_o = 0.45, which is smaller than the nozzle edge due to the vortex generated by the notches. The fluctuating velocity of the notched-orifice jets increases in the A-A′ plane.

The fluctuating velocity reflecting the steep velocity gradient near the center of the jet is higher than that of the notched-orifice jets that are discharged toward the center of the jet and accelerated. The turbulence intensity near the nozzle exit is larger than that of the other jets because the pipe jet has a fully developed turbulent velocity profile; however, the growth of the turbulence intensity is gradual and the turbulence intensity decreases at $x/d_o \geq 3$.

The heat transfer characteristics of the orifice and the notched-orifice impinging jets with tapering under the same operating power are shown in Fig. 31. The enhanced *Nu* number of the notched-orifice jets with tapering is observed in the downstream region due to the high turbulence intensity caused by the notches. However, in the case of the orifice jet, the heat transfer coefficients decrease more than in the case of the others in the downstream region because of the size of the nozzle exit.

Figure 32 shows the increasing ratio of the mean *Nu* number Nu_m in relation to that of the orifice jet for notched-orifice jets with tapering angle α under the same operating power. Nu_m was obtained by averaging the mean *Nu* number in the A-A′ and B-B′ directions considering the measured temperature as the reference temperature in the range of $\alpha = 2\pi/16$ [rad]. The mean *Nu* number for the notched-orifice jet has almost the same value at $r/d_o \approx 2$ as that of the orifice jet. The ratios of Nu_m of the notched-orifice jets with tapering in the downstream region $r/d_o > 3.3$ were larger than that of the orifice jet. The heat transfer for a notched-orifice jet with $\alpha = 10°$ increases by a maximum of ~4.5% in the downstream region. In addition, the increasing ratio of Nu_m to that of the pipe jet under the same operating power was also calculated. The ratio of Nu_m to that of the pipe jet was attained at $\alpha = 0°$ and reached a maximum of 6.7% and the heat transfer characteristics for $\alpha = 0°, 5°, 10°$, and $15°$ were enhanced by ~3.4%, 3.7%, 4.6%, and 4.5%, respectively, in the downstream region at $r/d_o = 7.4$.

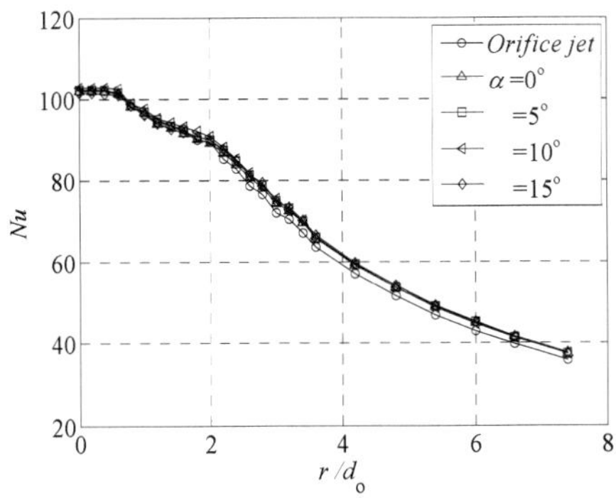

Figure 31. Local Nusselt number (H/d_o=2.0)

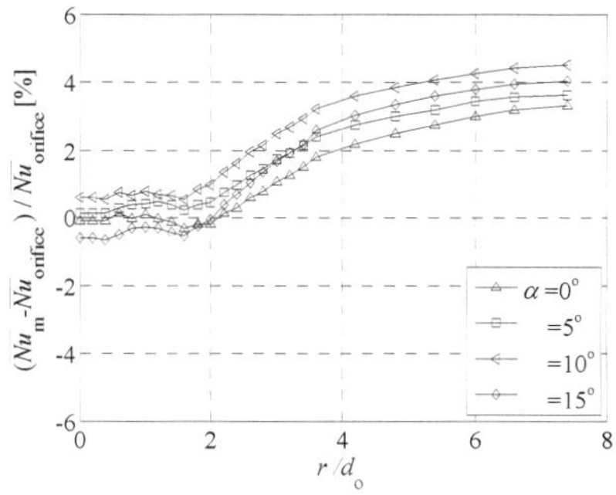

Figure 32. Increasing ratio of average Nu_m to the orifice jet (CR=0.27)

5. Conclusion

In this chapter, the reviews of the impingement heat transfer of various types of jets were introduced in brief, and then, the flow and impingement heat transfer of orifice jets were examined in detail. In particular, the effects of the nozzle contraction area ratio CR on the flow and heat transfer characteristics of the jet issuing from an orifice nozzle were examined experimentally and visually. The following conclusions were obtained.

1. CR was found to strongly depend on the large vortex structures of a submerged orifice water jet in visualized images. A more coherent vortex was produced with a decrease in CR. Smaller CR strongly promoted mixing with the ambient fluid and widened the jet spreading.
2. The dimensionless centerline maximum velocity u_c/u_m can be expressed as

$$u_c / u_m = 1.9CR^3 - 3.55CR^2 + 1.38CR + 1.53 ,$$

which has a maximum value of 1.7 at CR = 0.27.

Velocity vectors toward the centerline at the orifice nozzle exit that were produced by the contraction contributed to the increase in the centerline velocity, which was maximum for CR = 0.27.

Next, in order to simultaneously reduce the flow resistance, increase the turbulent intensity, and obtain the vena contracta effect, we proposed a notched-orifice nozzle. The effects of notches at the orifice nozzle exit edge on the flow and heat transfer characteristics were investigated. Moreover, we proposed notched-orifice nozzles with tapering to reduce the resistance further. The proposed tapering angle has also been taken into consideration, resulting in the following conclusions.

i. Small notches greatly affect the flow structure of jets and reduce the nozzle resistance or operating power as well as increase turbulence at the nozzle exit. The local Nusselt number distribution of the notched-orifice jet is larger than that of the orifice jet in the downstream region because of the larger turbulence intensity caused by the notches.

The turbulence intensity of the four-notched orifice jet increases along the centerline by ~47.2% at $x/d_o = 3.5$ compared with the other jets under the same operating power.

ii. Axis-switching phenomena occur in the notched-orifice jets, although the notches are relatively small.

iii. The heat transfer characteristics of the eight-notched orifice jet with a tapering angle of $\alpha = 10°$ were a maximum of 6.7% greater than those of the pipe jet at $r/d_o = 2.2$ and were greater in the downstream region compared to the orifice jet.

Nomenclature

A_i: Cross –sectional area of pipe [m^2]

A_o: Nozzle exit area [m^2]

b: Jet width where the velocity ratio $u/u_c = 0.1$ [m]

$b_{1/2}$: Half jet width [m]

CR: Contraction area ratio ($=A_o/A_i$)

d_i: Diameter of pipe [m]

d_o: Diameter of nozzle exit [m]

f: Frequency [Hz]

f^*: Normalized frequency ($= f\, d_o / u_c$)

H: Nozzle – plate distance [m]

h: Heat transfer coefficient [W/(m^2·K)]

Nu: Local Nusselt number ($= h\, d_o / \lambda_a$)

$\overline{Nu}$: Mean Nusselt number

$\overline{Nu_m}$: $\overline{Nu}$ obtained approximately by summing and averaging Nu calculated from the temperature distribution in the A-A′ and B-B′ directions considering the measured temperature as the reference temperature in the range of $\theta = 2\pi/16$ [rad.]

Q: Flow rate [m^3/s]

r: Spanwise distance from the center of nozzle [m]

p_a: Atmospheric pressure [Pa]

Re: Reynolds number ($= u_m\, d_o / \nu$)

u: Mean velocity [m/s]

u_c: Mean velocity along the centerline [m/s]

u_m: Mean velocity at the nozzle exit [m/s]

u_{max}: Maximum velocity [m/s]

u_{pm}: Mean velocity in the pipe [m/s]

u': Fluctuating velocity [m/s]

W: Operation power based on the pressure loss at the nozzle exit [W]

x: Streamline distance from the nozzle exit [m]

α: Taper angle of the nozzle plate

Δp: Pressure loss at the nozzle [Pa]

ρ: Air density [kg/m^3]

υ: Coefficient of kinematic viscosity [m^2/s]

λ_a: Air thermal conductivity [W/m·K]

ν: Kinetic viscosity [Pa·s]

Φ_y': Normalized power spectra density function

Author details

Toshihiko Shakouchi and Mizuki Kito

Graduate School of Engineering, Mie University/Suzuka National College of Technology, Japan

6. References

Antonia, R. A. & Zhao, Q. (2001). Effect of Initial Conditions on a Circular Jet, *Experiments in Fluids,* Vol.31, No.3, (September 2001), pp. 319-323, ISSN 0723-4864

Downs, S.J. & James, E.H. (1987). Jet Impingement Heat Transfer – A Literature Survey, *ASME National Heat Transfer Conference,* No.87-H-35, ASME, New York

Hrycak, P. (1981). Heat Transfer from Impinging Jets: A Literature Review, *AWAL*-TR-81-3054

Kataoka,K., Suguro,M., Degawa,H., Maruo,K. & Mihata,I. (1987). Effect of Surface Renewal Due to Large-Scale Eddies on Jet Impingement Heat Transfer, *International Journal of Heat and Mass Transfer,* Vol.30, pp.559-567, ISSN 0017-9310

Lee, J. & Lee, SJ. (2000). The effect of nozzle configuration on stagnation region heat transfer enhancement of axisymmetric jet impingement, *International Jouranal of Heat and Mass Transfer,* Vol.43, No. 18, (September 2000), pp. 3497-3509, ISSN 0017-9310

Livingood, J.N.B. & Hrycak, P. (1973). Impingement Heat Transfer from Turbulent Air Stream Jets to Flat Plates – A Literature Survey, *NASA* TM X-2778

Martin, H. (1977). *Heat and mass transfer between impinging gas jets and solid surface,* In: *Advances in Heat Transfer,* Hartnett, J.P. & Irvine, T.F. (Eds), 1-60, Vol.13, ISBN 0-12-020013-9, Academic press, New York

Mi, J., Nathan, G.J. & Nobes, D.S. (2001). Mixing Characteristics of Axisymmetric Free Jets from Contoured Nozzle, an Orifice Plate and a Pipe, *Jouranal of Fluids Enginerring-Transactions of the American Society of Mechanical Engineers,* Vol.123, No.4, (December 2001), pp. 878-883, ISSN 0098-2202

Quinn, W R. (2006). Upstream nozzle shaping effects on near field flow in round turbulent free jets. *European Journal of Mechanics B/ Fluids,* Vol.25, No.3, (May 2006), pp. 279-301, ISSN 0997-7546

Shakouchi, T. (2004). *Jet Flow Engineering –Fundamentals and Application–,* Morikita, ISBN 978-4-627-67201-7, Tokyo, Japan

Viskanta, R. (1993). Heat Transfer to Impinging Isothermal Gas and Flame Jets, *Experimental Thermanl and Fluid Science,* Vol.6, No. 2, (February 1993), pp. 111-134, ISSN 0894-1777

Yokobori, S., Kasagi, N., Hirata, M & Nishiwaki, N. (1978). Role of Large-Scale Eddy Structure on Enhancement of Heat Transfer in Stagnation Region of Two-Dimensional, Submerged, Impinging Jet, *Proceedings of 6th International Heat Transfer Conference,* pp.305-310, Toronto, Canada, August 1978

Zhou,D.W, Lee,S.J., Ma,C.F. & Bergles, A.E. (2006). Optimization of mesh screen for enhancing jet impingement heat transfer, *Heat and Mass Transfer,* Vol.42, No. 6, (February 2006), pp. 501-510, ISSN 0947-7411

Chapter 16

Conjugate Heat Transfer in Ribbed Cylindrical Channels

Armando Gallegos-Muñoz, Nicolás C. Uzárraga-Rodríguez and Francisco Elizalde-Blancas

Additional information is available at the end of the chapter

http://dx.doi.org/10.5772/49942

1. Introduction

In the last years, many technological advances have emerged in the turbo machinery industry, mainly in the area of power generation using gas turbines [1]. The main target in this science field consists on designing and building more efficient machines with a higher life-time by means of applied research. However, in order to achieve this, it is necessary that the gas turbine operates at high compression pressure ratios as well as high turbine inlet temperatures (TIT), but these operating conditions generate thermal consequences or degradations in the gas turbine components that are exposed to the high temperatures, like blades and vanes of the first stage. For this reason, it is necessary to have an internal cooling system in gas turbines to avoid the reduction of the useful life of their hot components, since the useful life of turbine blades is reduced to half with every 10 – 15 ºC rise in metal temperature [2]. Nowadays basic methods exist, which improve the gas turbines operating conditions, having as a result improvements of the external cooling [3], where the use of micro-jets with smaller diameters enhanced the overall heat transfer coefficient, or internal cooling where square ribbed channels are employed to study the thermal behaviour of the flow inside the channel [4], turbulence promoters with different geometries to study the temperature distribution in the gas turbine blades [5]. Also, it is possible using serpentine passages inside the turbine blade to improve internal convective cooling [6], ribs in the internal surface of the cooling passages where the rib-to-rib pitch and angle of attack that yield a maximum heat transfer and maximum thermal performance are determined [7] or ribs as turbulence promoters to increase the rate of heat transfer [8]. To increase the heat transfer with minimal friction in compact heat exchangers, the internal surfaces are ribbed with protuberances that have convex and concave forms [9]. To study the heat transfer characteristics of laminar flow in parallel-plate dimpled channels are used [10] or square-channel fitted with baffles [11].

However, a common way to increase the internal cooling efficiency in gas turbines is to add ribs, this method offers a better mixed fluid near to the hot internal surface of the channel thus increasing the thermal heat transfer. The present study shows a numerical analysis of the first stage blades in a gas turbine with internal cooling system (model MS7001E) applying the conjugated heat transfer. This method considers the direct coupling of fluid flow and solid body using the same mesh distribution and numerical principles for both domains. This coupling is achieved by using boundary conditions called double-side wall.

2. Mathematical formulation

The numerical analysis of a gas turbine at first stage blade with internal cooling system considers the solution of the conjugate heat transfer in steady state between the hot combustion gases flowing around the blade and the coolant flowing inside the cooling channels of gas turbine blades. The following assumptions are made to model the conjugate heat transfer problem:

a. Newtonian fluid,
b. Compressible and turbulent flow
c. Rotational frame with relative velocity formulation
d. Fluid is considered as an ideal gas.

2.1. Governing equations

The applied governing equations are the 3D Reynolds-averaged Navier-Stokes equations [12], which were solved by commercial Computational Fluid Dynamics software [13]. The governing equations solved by the model are:

Continuity equation

$$\frac{\partial}{\partial x_i}\left(\overline{\rho u_i} + \overline{\rho' u_i'}\right) = 0 \tag{1}$$

Momentum equation

$$\frac{\partial}{\partial x_j}\left(\overline{\rho u_i u_j}\right) = \rho \vec{g}_i - \frac{\partial p}{\partial x_i} + \frac{\partial\left(\tau_{ij}\right)}{\partial x_j} + F_i \tag{2}$$

where F_i is the source term which includes contributions due to the body force. Assuming constant rotational velocity with relative velocity formulation, the source terms due to rotation are:

$$a^{co} = 2\vec{\omega} \times \overline{u_i} \text{ and } a^{ce} = \vec{\omega} \times \vec{\omega} \times \vec{r} \tag{3}$$

The term τ_{ij} is the stress tensor, which is expressed as:

$$\tau_{ij} = \mu\left(\frac{\partial \overline{u_i}}{\partial x_j} + \frac{\partial \overline{u_j}}{\partial x_i} - \frac{2}{3}\delta_{ij}\frac{\partial \overline{u_k}}{\partial x_k}\right) - \rho\overline{u_i' u_j'} \tag{4}$$

Energy equations

The energy equation for the fluid domain is given by

$$\frac{\partial}{\partial x_i}\left(\rho C_P u_i T\right) = \frac{\partial}{\partial x_i}\left[C_P \frac{\mu}{Pr}\frac{\partial T}{\partial x_i} + \rho C_P \overline{u_i' T'}\right] + \mu\phi \tag{5}$$

where $\mu\phi$ is the viscous heating dissipation and $\rho C_P \overline{u_i' T'}$ represents the turbulent heat flux. The energy conservation equation for the solid is given by the conductive term in the energy equation. In the solid continuum, only the heat flux due to conduction is included inside itself. This is described by the heat diffusion equation:

$$\frac{\partial}{\partial x_i}\left(\lambda_{solid}\frac{\partial T_{solid}}{\partial x_i}\right) = 0 \tag{6}$$

where λ_{solid} is the thermal conductivity. Equations (5) and (6) are solved simultaneously by a conjugate heat transfer analysis. This permits to yield a fully coupled conduction-convection heat transfer prediction.

Turbulence model

Since the behaviour of the flow in the gas turbine is very chaotic, it is necessary to incorporate a turbulence model in the numerical analysis to determine the Reynolds stresses. The Standard k-ε model was used, which relates the Reynolds stresses to the mean velocity by the Boussinesq hypothesis [13]:

$$-\rho\overline{u_i' u_j'} = \mu_t\left(\frac{\partial \overline{u_i}}{\partial x_j} + \frac{\partial \overline{u_j}}{\partial x_i}\right) - \frac{2}{3}\left(\rho k + \mu_t \frac{\partial \overline{u_k}}{\partial x_k}\right)\delta_{ij} \tag{7}$$

The eddy turbulent viscosity, μ_t, is calculated by the combination of the turbulent kinetic energy (k) and the dissipation ratio (ε), as shown by equation (8).

$$\mu_t = \rho \cdot C_\mu \cdot \frac{k^2}{\varepsilon} \tag{8}$$

where C_μ is a constant. This model offers a reasonable accuracy for a wide range of turbulent flows in practical engineering problems, in which the turbulent kinetic energy, k, and its dissipation ratio, ε, are obtained from the following transport equations:

$$\frac{\partial}{\partial x_i}\left(\rho k u_i\right) = \frac{\partial}{\partial x_j}\left[\left(\mu + \frac{\mu_t}{\sigma_k}\right)\frac{\partial k}{\partial x_j}\right] + G_k + G_b - \rho\varepsilon - Y_m \tag{9}$$

$$\frac{\partial}{\partial x_i}(\rho \varepsilon u_i) = \frac{\partial}{\partial x_j}\left[\left(\mu + \frac{\mu_t}{\sigma_\varepsilon}\right)\frac{\partial \varepsilon}{\partial x_j}\right] + C_{1\varepsilon}\frac{\varepsilon}{k}(G_k + C_{3\varepsilon}G_b) - C_{2\varepsilon}\rho\frac{\varepsilon^2}{k} \tag{10}$$

In these equations, G_k represents the generation of turbulent kinetic energy due to the mean velocity gradients, G_b is the generation of turbulent kinetic energy due to buoyancy, and the quantities σ_k and σ_ε are the turbulent Prandtl numbers for k and ε, respectively. The empirical constants appearing in the above equations take the following values: $C_\mu = 0.09$, $C_{1\varepsilon} = 1.44$, $C_{2\varepsilon} = 1.92$, $\sigma_k = 1.0$ and $\sigma_\varepsilon = 1.3$ [13].

Equation of state

The density variation in both fluids, the hot combustion gases and cooling air, is assumed according to the gas ideal law:

$$\rho = \frac{P_{op} + P}{RT} \tag{11}$$

where R is the gas constant. This equation of state provides the linkage between the energy equation on one side, and the mass conservation and moment equations on the other. This linkage emerges from the density variations.

2.2. Computational model and grid

The computational model and mesh were generated in the pre-processor GAMBIT [14]. In order to avoid many simplifications in the computational model, such as the use of boundary conditions at surfaces and outlets of the cooling channels, the computational model was generated using the total blade geometry, which includes the plenum, gap in the tip, gaps between the seal and the plenum and the ribbed cooling channels. Figure 1 shows the rotor blade geometry at the first stage of the gas turbine MS7001E.

In this Figure the internal cooling system can be seen. This system has 13 cylindrical channels inside the blade. The ribs were placed on the inner surface of the cooling channels in order to increase the heat transfer from the solid body to the cooling air. Also, in Figure 1 can be seen that the ribs were only placed in the middle blade zone inside the cooling channels; because in this zone the largest temperature gradient is present [15], causing failure such as cracks in the blade structure [16].

Three different geometries of rib configurations are studied, which are square, triangular and semi-circular forms. The ribs are placed perpendicularly to the air flow. Figure 2 shows a sketch of the different rib parameters used in the square cross-section ribs. These parameters are used for the other two geometries (triangular and semi-circular). Moreover, these configurations are applied for both arrangements (full and half-ribs). Figure 3 shows the form and parameters of the ribs cross-section.

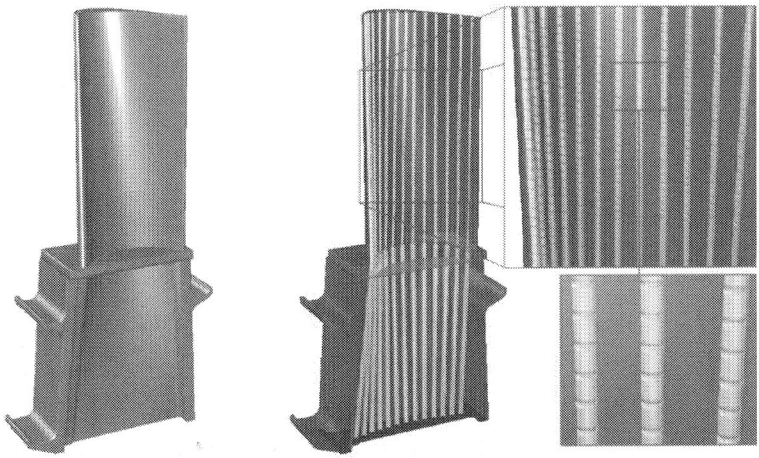

Figure 1. Blade geometry with plenum and its internal features (ribbed cooling channels)

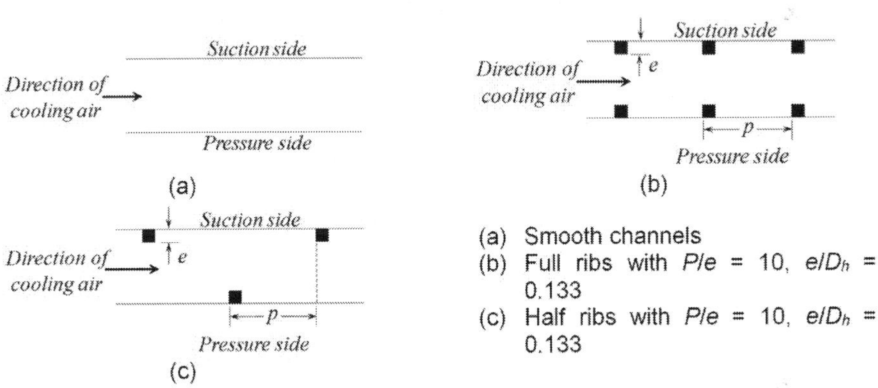

Figure 2. Sketch of different arrays to study the effect of the ribs

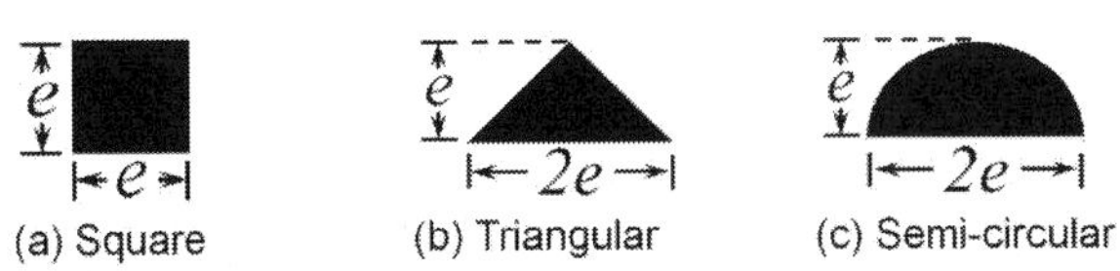

Figure 3. Cross-section of the ribs

Due to complexity of the geometry, different computational grid sizes are required. These grids are not uniform in all directions and were structured by mixing different types of cells (hexahedral and prismatic elements). For the grid used in the blade computational model with smooth cooling channels, the mesh density is high in the near-wall region of the blade body. The first wall-adjacent cells height in the vicinity of regions corresponding to the leading and trailing edge as well as in the suction and pressure side of the blade is 0.0012 mm, while in the region corresponding to the internal cooling channels, the wall-adjacent cell height is 0.0035 mm. This is developed in order to get a better solution into the boundary layer, obtaining y^+ values closer to unity along the surfaces (ranging from 1 to 5.5). An analysis to evaluate the grid independence of the numerical solution was developed. The computational grid has a total of 3809734 mixed cells, 80% of this total corresponding to fluid (33% air and 47% hot gases) and 20% to solid. Figure 4 shows the grid used in this case. It can be observed, that the hot combustion gases domain presents a high quantity of hexahedral elements.

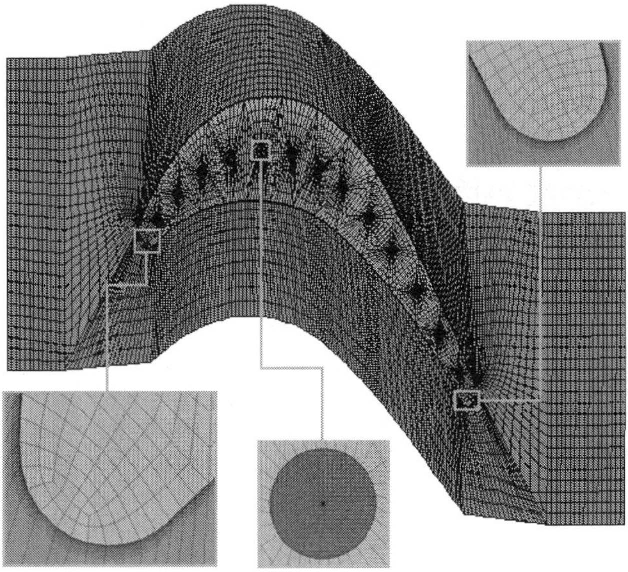

Figure 4. Computational grid for hot combustion gas (red), blade (gray) and cooling air (blue) domains

For the blade models with ribbed cooling channels, the same grid distribution was used at the region of hot combustion gases and the external blade surface. The height of the first wall-adjacent cells in the vicinity of regions corresponding to ribbed cooling channels was of 0.002 mm, obtaining y^+ values less than 5 for all the analyzed rib configurations. The computational grid used ranges from 6.5 to 7.5 million of mixed cells. Figure 5 shows the grids used in the internal cooling channels with full-ribs having an aspect ratio of P/e = 10.

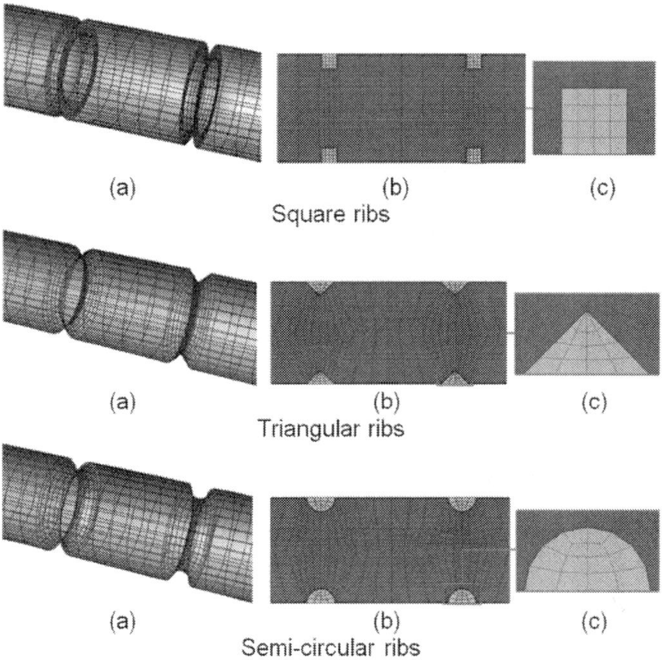

Figure 5. Computational grid for cooling channels domain with full-ribs with $P/e = 10$. (a) channels and surfaces, (b) cooling air and (c) ribs

For the models with half-ribs, it was used the same grid distribution showed in the Figure 5, having a small variation in the ribs domains, defining one half of the domain as fluid and the other half as solid. Figure 6 shows the grid used in the internal cooling channels with half-ribs having an aspect ratio of $P/e = 10$.

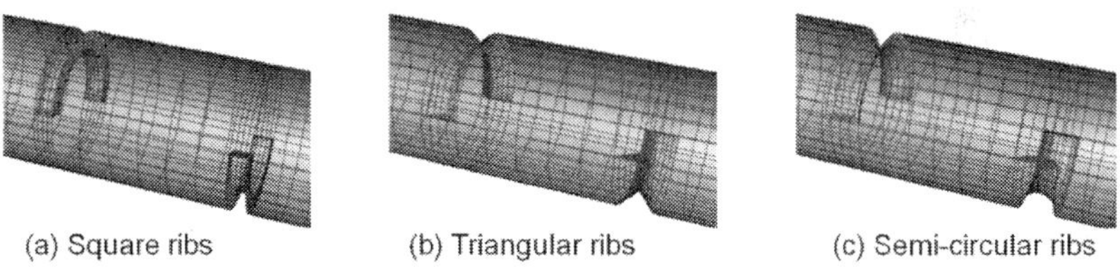

Figure 6. Computational grid for cooling channels domain with half-ribs with $P/e = 10$

2.3. Boundary conditions and thermal properties

The boundary conditions used for the computational model with smooth cooling channels are defined according to approximate values of the gas turbine operating conditions in steady state. Figure 7 shows the boundary conditions used in the computational model.

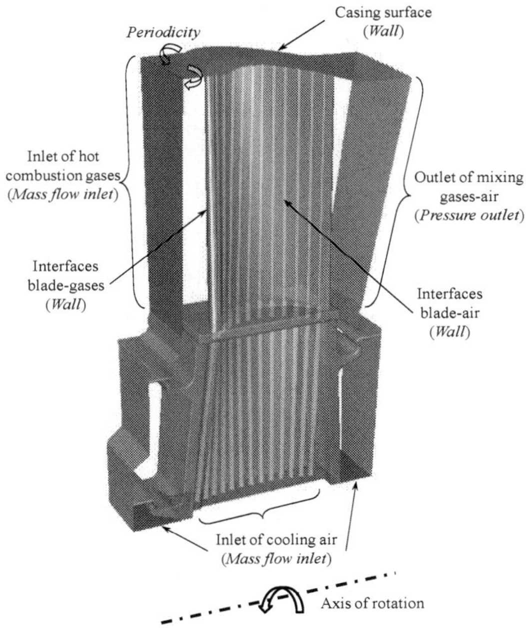

Figure 7. Boundary conditions in the model of a gas turbine blade

In the hot combustion gases inlet, the operational conditions are: mass flow of 2.47 kg/s, static pressure of 508700 Pa and total temperature profile is a function of the radial coordinate. This total temperature profile is described by the next equation [17]:

$$T(r) = -662195\,r^4 + 2\times10^6\,r^3 - 3\times10^6\,r^2 + 2\times10^6\,r - 435431 \tag{12}$$

This total temperature profile is imposed in order to match the oxidation mark of hot combustion gases over the airfoil of the blade, which has been operating until the end of its useful life [18], and because this profile is similar to the one obtained in the radial edge direction of the exit of the transition piece. Figure 8 shows the oxidation marks on the blade.

The units for the independent variables of Equation 12 are: (m) for the rotational radius and (K) for the total temperature. The turbulence quantities at the inlet of the model are defined using a turbulence intensity of 5% and 0.006 m for turbulent length scale.

A mass flow of 0.0048 kg/s of air, a static pressure of 897300 Pa, and a total temperature of 853.15 K were specified for each inlet zones of the thirteen cooling channels. Also, turbulence parameters are defined using a turbulence intensity of 5% and a hydraulic diameter of 0.004 m for these boundaries. At the left and right sides of the plenum inlets of cooling air were adjusted with a mass flow of 0.152 kg/s and 0.025 kg/s, respectively. As well, a turbulence intensity of 5% and turbulent length scale of 0.005 m was set. In the remaining inlet section of the cooling plenum, boundary conditions were adjusted to the same conditions used at the inlets of the cooling channels for the parameters of pressure and temperature. At the exit of the gas-air mixture a static pressure of 473170 Pa, and a backflow total temperature of 1226 K were specified.

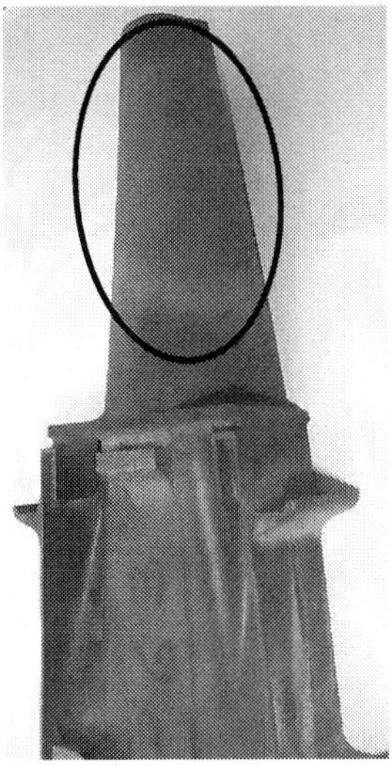

Figure 8. Oxidation (corrosion) marks on the blade [18].

For the solid surfaces the conditions were imposed as no-slip condition, while for the thermal condition were imposed as coupled. With these considerations it is possible to solve simultaneously the solid-fluid interfaces. At the interfaces, the temperature and heat flux could be continuous. These conditions are developed by the use of the boundary conditions denominated as two-side wall, which can be expressed as:

$$T_{fluid} = T_{solid} \tag{13}$$

$$\lambda_{fluid} \left.\frac{\partial T}{\partial n}\right|_{fluid} = \lambda_{solid} \left.\frac{\partial T}{\partial n}\right|_{solid} \tag{14}$$

Rotational periodic boundary conditions are defined for the suction and pressure side of the computational model and a nominal angular velocity vector were prescribed.

The flow and heat transfer analysis were performed under the assumption that the fluid behaviour is compressible and viscous. For the case of the air properties, these are temperature dependant, while the thermo-physical properties of the solid domain were assumed to behave like Inconel 738LC alloy. The thermo-physical properties of the fluid and solid domains are showed in Table 1 [19] and Table 2 [18], respectively.

T [K]	Cp [kJ/kg·K]	$\lambda \cdot 10^3$ [W/m·K]	$\mu \cdot 10^6$ [N·s/m^2]
200	1.007	18.1	13.25
300	1.007	26.3	18.46
400	1.014	33.8	23.01
500	1.030	40.7	27.01
600	1.051	46.9	30.58
700	1.075	52.4	33.88
800	1.099	57.3	36.98
900	1.121	62.0	39.81
1000	1.141	66.7	42.44
1100	1.159	71.5	44.90
1200	1.175	76.3	47.30
1300	1.189	82	49.6
1400	1.207	91	53.0

Table 1. Fluid properties

T [K]	Cp [J/kg·K]	λ [W/m·K]	ρ [kg/m^3]
294.260	420.100	11.83	8110
366.480	462.110	11.83	8110
477.590	504.120	11.83	8110
588.700	525.120	13.70	8110
699.810	546.130	15.58	8110
810.920	567.130	17.74	8110
922.030	588.140	19.76	8110
1033.150	630.150	21.50	8110
1144.260	672.160	23.37	8110
1255.370	714.170	25.39	8110
1366.480	714.170	27.27	8110

Table 2. Properties of Inconel 738LC alloy

2.4. Numerical method

Fluid flow and turbulent heat transfer analysis of the first stage gas turbine blade (MS7001E) with different ribs configurations placed in the internal cooling channels were realized using commercial Computational Fluid Dynamic software (FLUENT®). This code allows to solve the Reynolds averaged Navier-Stokes and the transport equations of the turbulent quantities for the compressible viscous flow. This CFD code solves the equations using the finite volume technique [20] to discretize the governing equations inside the computational domains. The Standard k- ε model [21] was used for all numerical simulations. This model is a semi-empirical linear eddy viscosity model based on the transport equations for the turbulent kinetic energy (k) and dissipation rate (ε). The model transport equation for k is derived from the exact equation, while the model transport equation for ε is obtained using physical reasoning and little resemblance to its mathematically exact counterpart. The SIMPLE algorithm was used to link the velocity field and pressure distribution inside the computational model. This algorithm uses a relation between the velocity and pressure in order to satisfy the mass conservation, getting a velocity field. The SIMPLE algorithm along with the implicit time treatment of the flow variables allow to obtain a steady solution or use rather time steps for unsteady flow computations [13].

The governing equations were solved simultaneously by the approach of the pressure-based solver. The pressure-based approach is recommended in the literature [13] to be used for flows with moderate compressibility, since it offers a better convergence. Due to the fact that the governing equations are non-linear and coupled, several iterations were needed to reach a converged solution. The Gauss-Seidel linear algorithm was used to solve the set of algebraic equations obtained by the discretization in FLUENT®. The convergence was reached when the residuals of the velocity components in the Reynolds averaged Navier-Stokes equations, continuity and turbulent quantities were smaller than 10^{-5}, while for the energy conservation equation the residuals were smaller than 10^{-6}.

Six computational equipments were used to solve the model. Each computer has a 3 GHz processor and 2 GB in RAM. These equipments were connected in a scheme of parallel processing. Figure 9 shows a sketch of the parallel processing equipment using a basic LAN topology.

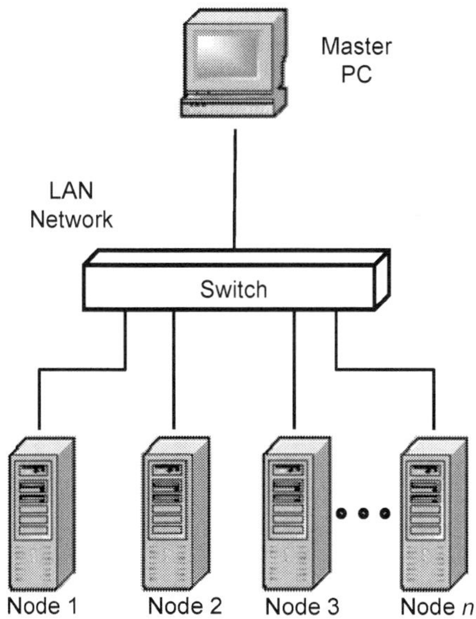

Figure 9. Sketch of parallel processing.

3. Results and discussions

In the first part, a comparison between the results obtained experimentally by Kwak [22] and numerically for the external flow is presented. Also, a comparison between results obtained from semi-empirical correlations derived from the *law of the wall* [23] and the numerical results of the internal flow are presented. In the second part the effects on internal flow through the internal cooling channels are presented. Finally, the temperature distribution inside the blade body and the surface temperature distribution in the blade body with and without ribs are showed.

3.1. Comparison with experimental and semi-empirical correlation data

In order to validate the external flow around the blade a qualitative comparison between the pressure distribution obtained numerically and experimentally [22], has been performed. In [22] the pressure distribution on the gas turbine blade of GE-E3 was measured. Figure 10 shows the comparison between the numerical and experimental pressure distribution at the middle of the blade. Several turbulence models were used. The turbulence models used in the comparison were Standard k-ε, RNG k-ε and SST k-ω models, which are known as two-equation eddy viscosity models (EVMs). The total inlet pressure and local static pressure around the blade (p_0/p_s) are plotted as a function of x/C_x, where x is the axial length measured from the leading edge (LE) of the blade, considered as the characteristic length.

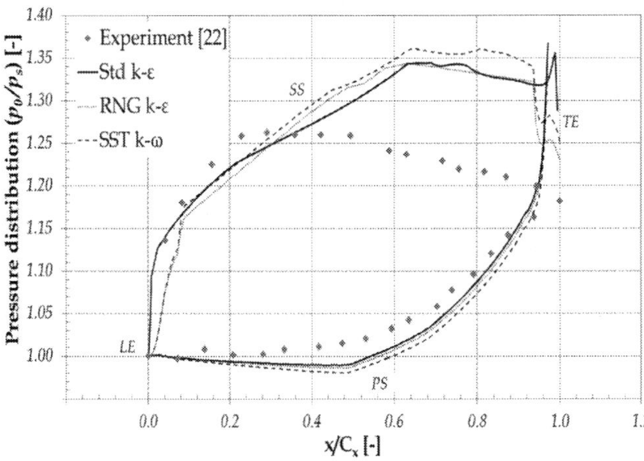

Figure 10. Pressure distribution around the blade

In Figure 10 can be seen that the pressure distribution, p_0/p_s, corresponding to numerical solution on the pressure side (PS) of the blade, agrees with the experimental data. While some differences for the pressure distribution, p_0/p_s, on the suction side (SS) of the blade, are observed.

For comparison purposes of the internal flow in the cooling channels, the pressure drop given by the section with square full-ribs having an aspect ratio of $P/e = 10$ for the central cooling channel was determined. This pressure drop was calculated by the friction factor for ribbed tubes, using semi-empirical correlations derived from the law of wall. Nikuradse [23] developed a friction factor correlation to be used in geometries with sand-grain roughness. His results were excellent for a wide range of roughness sizes. This correlation is expressed by Equation (15).

$$R\left(e^{+}\right)=\sqrt{2/f}+2.5\,ln\left(2e/D_h\right)+3.75 \tag{15}$$

The term $e^{+} = eu^{*}/v$ is the roughness Reynolds number and D_h is the hydraulic diameter which can be expressed as:

$$D_h=\frac{4A}{P_w} \tag{16}$$

where A is the cross section area and P_w is the wetted perimeter of the cooling channel. Webb [24] used Equation (15) to calculate frictional data for turbulent flows in ribbed tubes with circular cross-section, obtaining excellent results. They found that the roughness function could be correlated as Equation (17).

$$R\left(e^{+}\right)=0.95\left(P/e\right)^{0.53} \tag{17}$$

This equation is valid in the range $e^{+} > 35$. By solving Equations (15) and (17) the friction factor can be found from the geometrical parameters of the internal structure of the ribbed channels. Equation (18) shows the result obtained.

$$f=\frac{2}{\left[R-\left(2.5\,ln\left(2e/D_h\right)+3.75\right)\right]^2} \tag{18}$$

Equation (18) is valid for channels with ribs placed 90º to the flow direction and an aspect ratio of $P/e = 10$. Table 3 shows the pressure drop calculated by the Equation (18) and the numerical results obtained in this work.

	Correlation Eq. (18)	Numeric
Δp (kPa)	3893.90	3917.20

Table 3. Pressure drop comparison between analytical and numerical results

The pressure drop calculated numerically presents a good approximation, having an absolute difference of 3.25%.

3.2. Effects of the internal flow through internal cooling channels

In order to determine the effects generated by the ribs, a line along the central cooling channel was created. This centerline is dimensionless with a range from 0 to 1, where y is the dimensionless distance of the axial length of the internal cooling channel, measured from the base of the blade until the outlet of the cooling channel. This centerline was used to obtain data of the flow parameters, such as temperature, velocity, Mach number and pressure loss.

Temperature distribution

Figure 11 shows the temperature distributions of the coolant flow along the centerline of the central cooling channels. Figures 11a and 11c show the results for different types of full-ribs, and Figures 11b, and 11d show the results for the half-ribs studies, having a ratio of $P/e = 10$ and $P/e = 20$, respectively. The higher and smaller temperatures are presented for the configurations of full-ribs in the ribbed section (Figs. 11a and 11c.), reaching temperatures from 937 K to 741 K, respectively. While the half-ribs configurations offer a smaller difference of temperature in this section. This range is between temperature values of 927 K and 791 K. As can be seen in Figure 11, the temperature presents a moderate increase at the smooth inlet section. In the ribbed section, the temperature presents a periodic variation, due to the acceleration and deceleration of the flow inside the cooling channels caused by the presence of the ribs. At the end of the ribbed section, the temperature strongly decreases. This effect is similar to compressible flow with heat transfer (Rayleigh curve) [25], where the temperature decreases to reach a Mach number larger than one. In the smooth outlet section, the temperature increases at the beginning of the section, due to the decrement in the Mach number. The temperature suffers a decrement while the flow gets closer to the outlet channel.

The temperature contours along the central cooling channel at a longitudinal plane are shown in Figure 12. Figures 12a, 12c and 12e show the results for different types of full-ribs, and Figures 12b, 12d and 12f show results for the half-ribs, both models have a ratio of $P/e = 10$. The fluid temperature increases while it goes along the channel for all cases (arrow indicates the direction of flow). The surface temperature of the channels is higher in the cases with half-ribs. Thus, the flow is heated at surface near the ribs. For the cases with full-ribs, the surface temperature is lower, showing a more uniform temperature distribution near to the wall.

The triangular ribs configuration presents the lowest temperatures, because this configuration offers the best cooling design inside the blade body. These effects are similar to the configurations with ribs whose $P/e = 20$.

Velocity and mach number distribution

Figure 13 shows the comparison between velocity magnitude and Mach number distributions obtained for the cases of blades with smooth and ribbed channels.

In the ribbed channel corresponding to the configuration of square full-ribs with an aspect ratio of $P/e = 10$, the highest Mach number and velocity are obtained, whose values are 1.45 and 823 m/s, respectively. The smooth channel presents a continuous acceleration of the flow through the channels. In the case of the ribbed channel, the flow is moderately accelerated in the smooth inlet section. The flow becomes unstable in the ribbed section; experiencing acceleration and deceleration continuously. At the end of this section, in the last rib, the flow is strongly accelerated as can be seen in Figure 13. High velocities decrease at the beginning of smooth outlet section, and then accelerate again towards the outlet of the smooth channel. This effect is due to the rotational force.

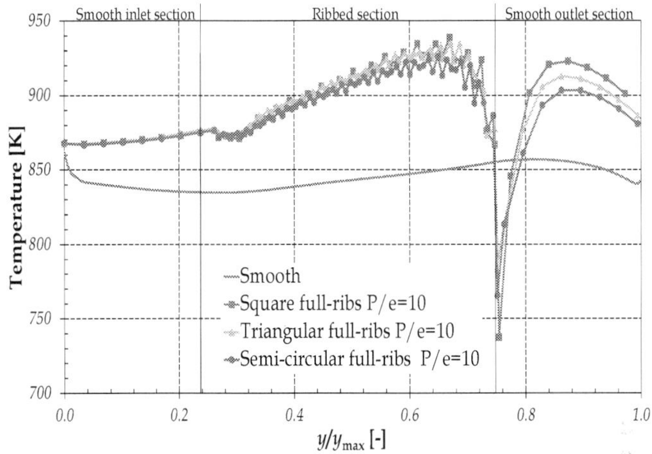

(a)

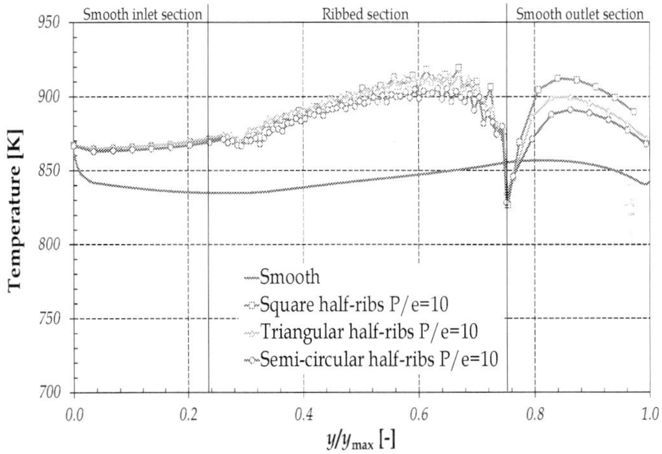

(b)

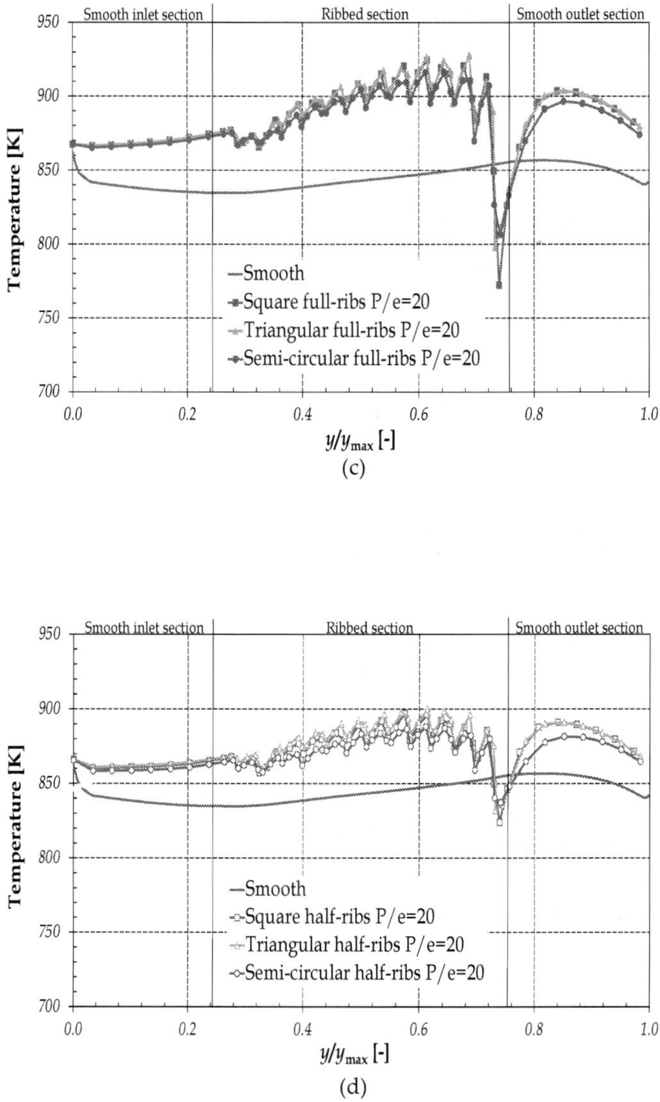

Figure 11. Temperature distribution at the centerline inside of the central cooling channel

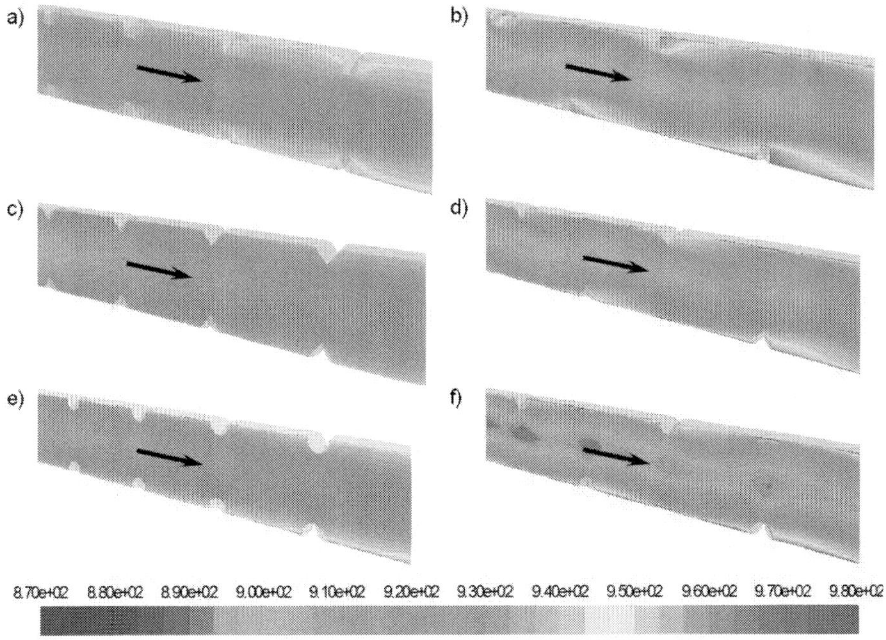

Figure 12. Temperature Contours [K] along the central cooling channel for the different types of ribs with $P/e = 10$

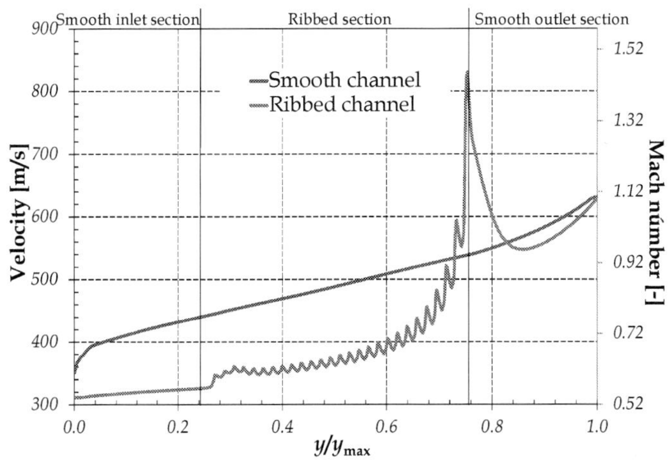

Figure 13. Velocity magnitude [m/s] and Mach number distributions along the centreline inside of the central cooling channels.

In order to have a better description of the effects caused by the acceleration and deceleration of the flow mentioned above, Figure 14 shows the contours of the axial velocity through the central cooling channel at a longitudinal plane. Figures 14a, 14c and 14e show the results for the different types of full-ribs. Figures 14b, 14d and 14f show the results for the half-ribs, both models have a ratio of $P/e = 10$. As it can be observed, the flow is strongly accelerated every time that it passes between the rib tips, which provoke that the fluid increases its velocity due to area reduction. Then, the area increases again to decelerate the fluid flow. These fluctuations of acceleration are presented periodically in the cooling channel, generating variations in the flow parameters. This effect is more noticeable in the cases of half-ribs, having the higher bulk velocity in the channel at each rib. Also, the higher velocity is extended downstream of the ribs tip, where the flow follows a wavy path in the bulk section of the channel. Also, it is observed that in the half-rib cases, the flow is forced towards the surface of the opposite rib.

On the other hand, the square and semi-circular ribs produce recirculation zones as well as stagnation points over the upstream and downstream rib surfaces, respectively. The triangular ribs only produce recirculation zones in the downstream surfaces. These effects are similar to configurations with a ratio of $P/e = 20$.

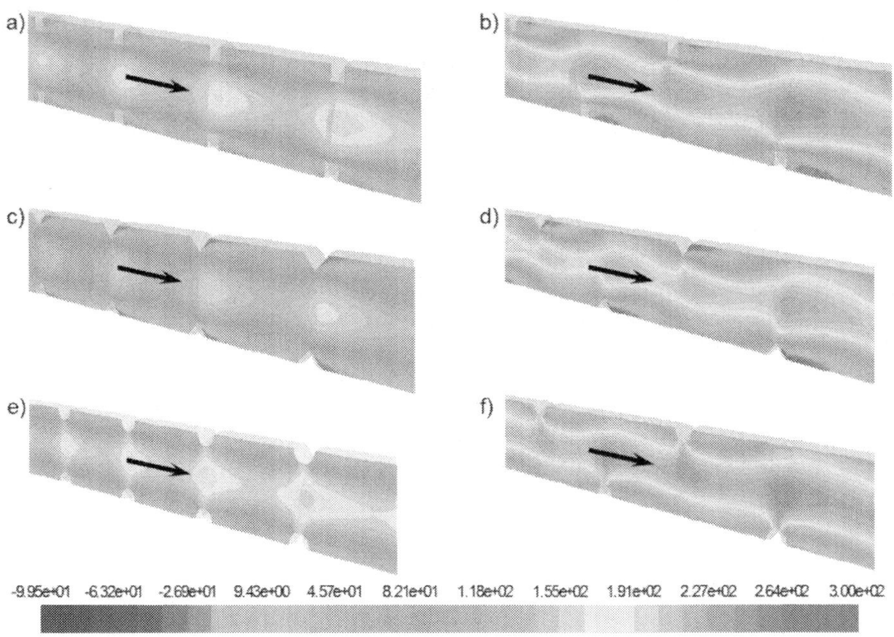

Figure 14. Contours of axial velocity [m/s] along to central cooling channel for the different types of ribs with $P/e = 10$

Pressure loss

The local static pressure is presented in terms of the normalized pressure difference, calculated by the equation (19)

$$\Delta p = \frac{p - p_{exit}}{0.5 \rho u^2} \qquad (19)$$

where p is the local static pressure, p_{exit} is the pressure at the outlet of the cooling channel, where the cooling air is mixed with the hot combustion gases and the average velocity u is calculated by the channels mass flow rate. Figures 15 and 16 show the local normalized static pressure distribution for the different rib configurations with an aspect ratio, P/e, of 10 and 20, respectively.

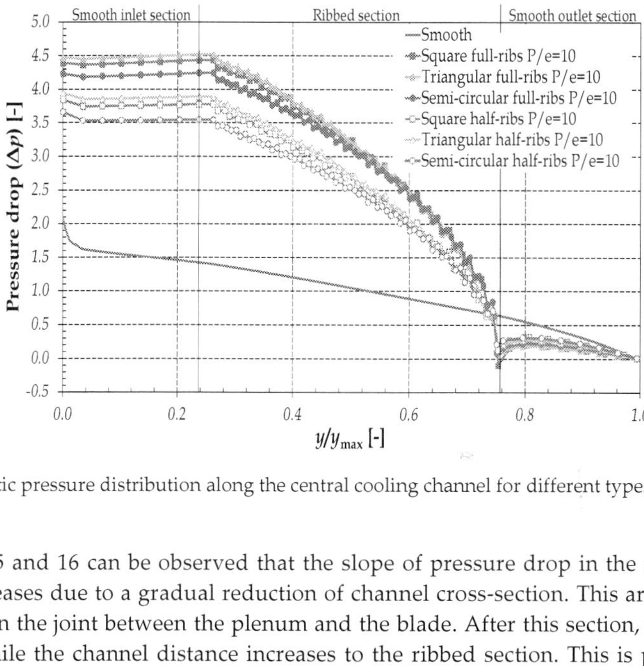

Figure 15. Static pressure distribution along the central cooling channel for different types of ribs with $P/e = 10$

In Figures 15 and 16 can be observed that the slope of pressure drop in the smooth inlet section decreases due to a gradual reduction of channel cross-section. This area reduction is localized in the joint between the plenum and the blade. After this section, the pressure increases while the channel distance increases to the ribbed section. This is produced by the stagnation point when the flow shocks with the first ribs. In the ribbed section, the slope of the pressure drop becomes unstable, presenting periodical increments and decrements due to the cross-section variation, producing accelerations and decelerations of the flow. At the smooth outlet section, the slope of the pressure drop is relatively higher than that in the smooth inlet section. This is due to the increase of the flow velocity at this zone due to the rotational force, ejecting the flow inside the hot gases stream in the tip of the blade.

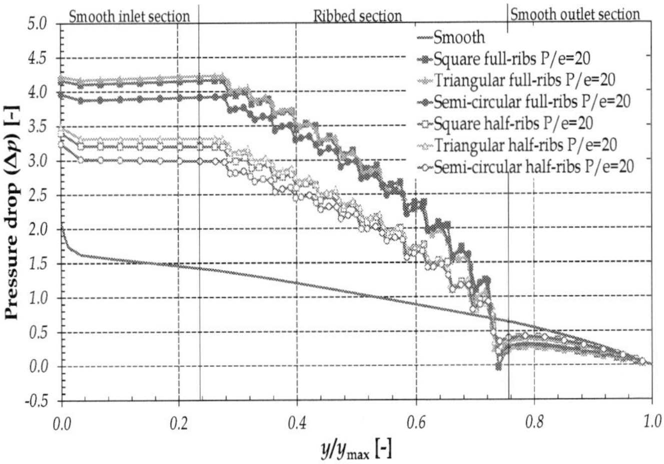

Figure 16. Static pressure distribution along the central cooling channel for different types of ribs with $P/e = 20$

3.3. Temperature distribution inside the blade body

Figure 17 shows the temperature contours at a transversal plane of the blade body right in the middle of the blade for the cases with smooth and full-rib channels with a P/e = 10. Comparing the results obtained, it is possible to observe an improvement in the blade internal cooling, allowing a decrement in the internal surface temperature of the cooling channels. Thus, the heat removed by the coolant is increased.

As can be seen in Figure 17, the maximum temperature decreases, approximately about 10 to 20 degrees and is reached close to the internal surfaces for the cases of blade with ribbed channels (Figures 17b, 17c and 17d), noticing that the cooling zone covers the major part of the internal cooling channels, propagating to the leading and trailing edge. In the cases of the blade with smooth channels, it is only present a smaller cooling zone at the three central cooling channels (Figure 17a).

Models with square and triangular cross-section full-ribs show a similar temperature distribution and major heat dissipation compared with the semi-circular full-ribs.

Mazur [16] performed an analysis of a gas turbine bucket failure made of Inconel 738LC super alloy. This bucket operated for 24,000 hours. Mazur et al. [16] found that the maximum stresses are present in the blade cooling channels, producing cracks. Figure 18 shows that kind of cracks. These start in the coating of the cooling channels, propagating and following intergranular trajectories, reaching a depth up to 0.4 mm.

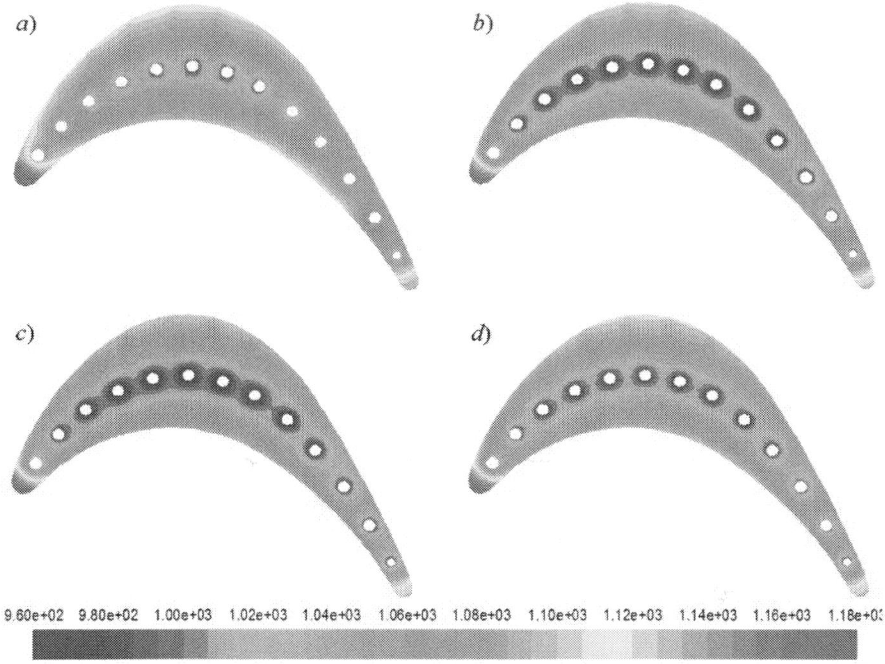

Figure 17. Temperature contours [K] in the metal, a) smooth, b) square full-ribs, c) triangular full-ribs and d) semi-circular full-ribs, cooling channels

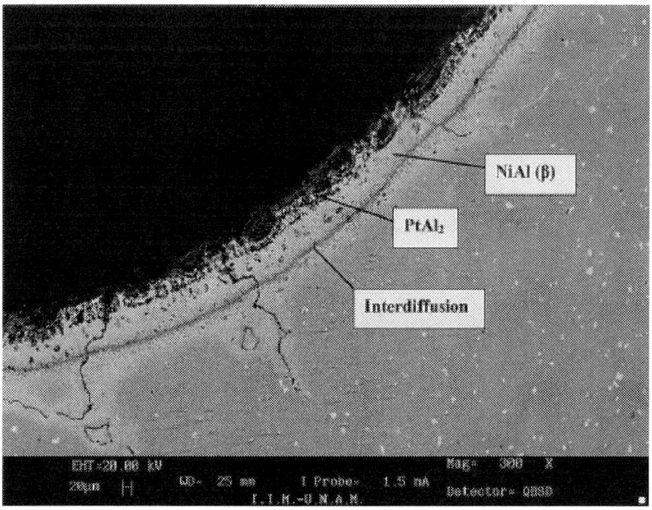

Figure 18. Cracks in the central cooling channels [16]

In this way, the effect generated by increasing the internal cooling zone produces an increment in the useful life of the blade, since the useful life of gas turbine blades is reduced to half with every 10-15 °C rise in metal temperature [2]. On the other hand, the use of ribs increases the heat transfer, generating an increase in thermal gradients at internal surface of cooling channels. In the leading edge another interesting effect is presented. There is a minor penetration of the blade body temperature (Figures 17b to 17d) caused by the use of the ribs. However, it cannot be adequate due to the fact that the thermal gradients at the leading edge are increasing. Due to these thermal effects, these zones must be taken into account to be protected by means of the film cooling method.

Figure 19 shows the blade profile right in the middle along the blade height. In this section, a perpendicular line to the blade chord is created to obtain the temperature distribution inside the solid body as well as in the cooling air through the central cooling channel. The distance is a dimensionless parameter, taking values between 0 and 1, starting in the suction side and ending in the pressure side, respectively.

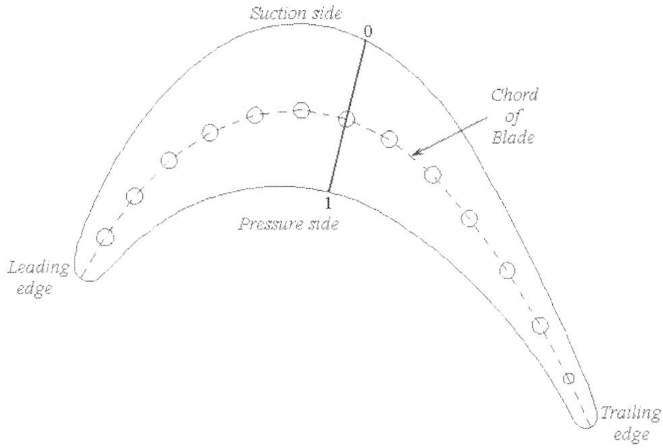

Figure 19. Perpendicular line to the blade chord where data is obtained

In Figures 20a and 20b the temperature distributions for the cases with full and half-ribs are presented, respectively. Both models have an aspect ratio between pitch and height of the ribs (*P*/*e*) of 10. Figure 20a shows that the cases of ribbed channels present a significant decrement in the temperatures inside the solid body in comparison with smooth channels. The triangular full-ribs model presents a higher temperature decrement, reaching a temperature decrement up to 20 K closer to the channel surface and 10 K in the pressure and suction sides. Figure 20b shows that square and semi-circular half-ribs do not offer a considerable decrement in the temperature inside the blade body, since the temperature distributions are very similar to the case with smooth cooling channels. The case with triangular half-ribs presents a smaller temperature decrement than the case with triangular full-ribs (Figure 20a). The decrement of temperature achieved by this configuration is between 10 K near to the channel surface and 4 K in the pressure and suction sides. In the fluid section, the temperature is bigger in all the ribbed cases than in the smooth case, obtaining improved heat dissipation to the cooling air.

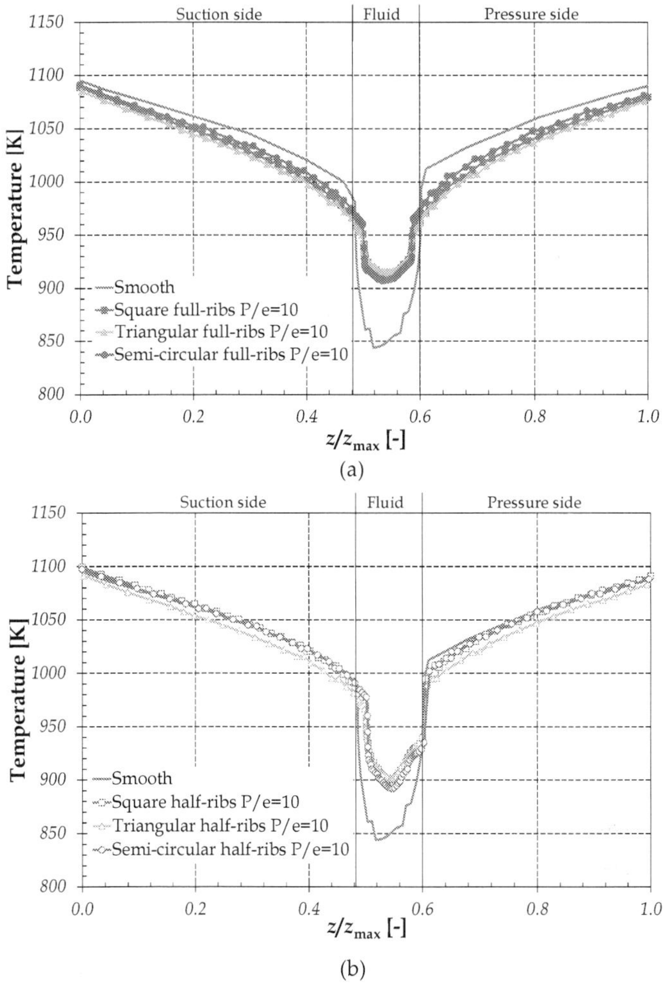

Figure 20. Temperature distributions in the blade with ribbed cooling channels with $P/e = 10$

Figures 21a and 21b show the temperature distributions for the cases of cooling channels with full and half-ribs with an aspect ratio of $P/e = 20$. Figure 21a shows that the temperature distributions in the solid body are similar for the three ribbed cases, having the lowest temperature in the square ribs model. With these configurations a higher penetration of the cooling blade using any rib geometry is achieved. These configurations present a similar behaviour, in comparison with the results presented in Figure 20a. These cases present a temperature reduction up to 22 K in regions close to the channel surface and up to 10 K in the pressure and suction sides. In Figure 21b can be observed a uniform behaviour of the temperature distribution for the three types of ribs. This behaviour is basically the same for all the cases. However, the blade cooling is affected due to the temperature distributions obtained for all the cases with different tendency to be similar for the smooth case, having a smaller improvement on the blade temperature when compared with the temperature profiles shown in Figure 20b.

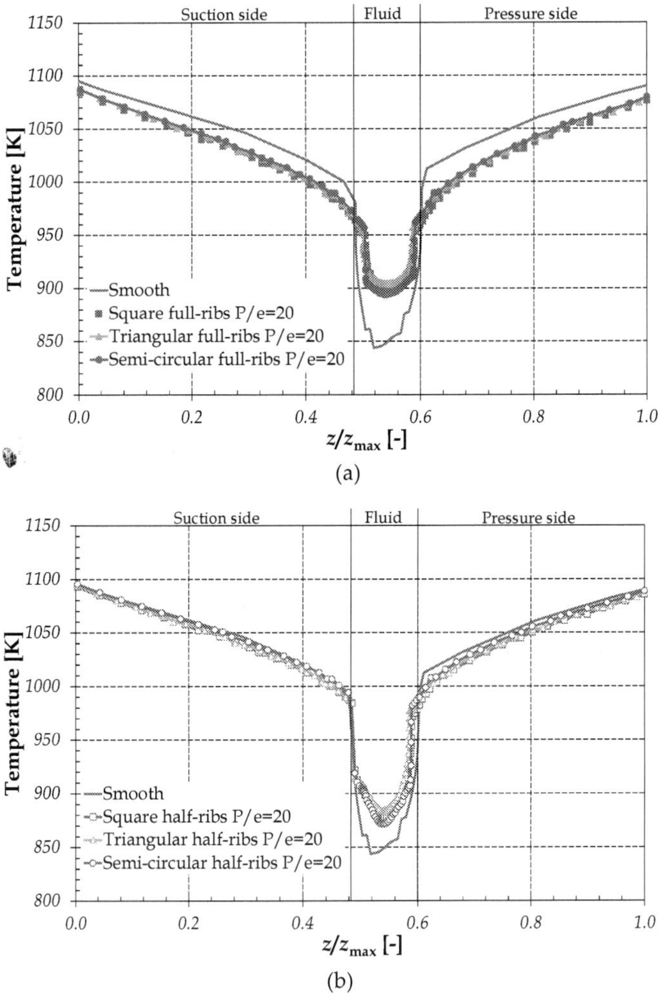

Figure 21. Temperature distributions in the blade with ribbed cooling channels with $P/e = 20$

The temperature distributions of cooling air presented in Figures 20 and 21 have a symmetrical parabolic behaviour due to mixing flow. This is generated by placing ribs, while the profile related with the smooth channel has an asymmetrical behaviour. In this case the profile presents a tendency to attach to the pressure side due to the blade rotational force.

The effect of having a symmetric profile improves the heat transfer from the internal surface of the cooling channels to the air, due to the turbulence generated in the flow which is increased because of the ribs removing a high quantity of heat.

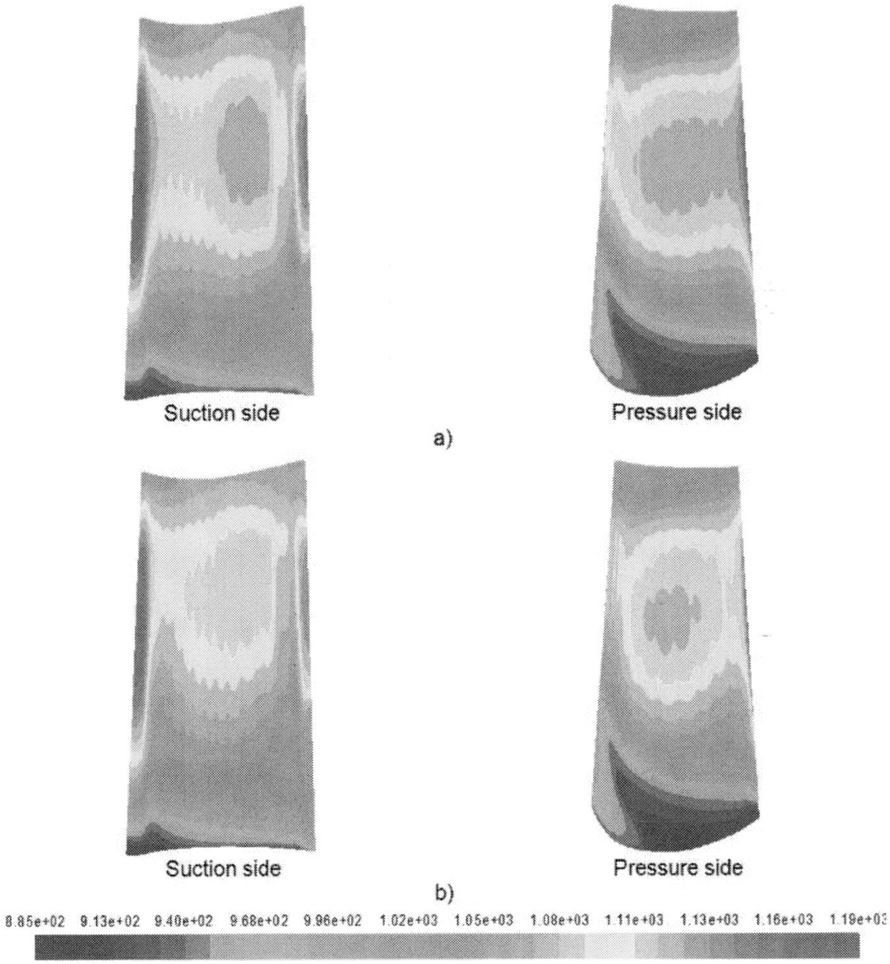

Figure 22. Surface temperature distributions [K] on the blade with a) smooth cooling channel and b) ribbed cooling channels with $P/e = 10$

3.4. Surface temperature distribution in the body blade

Figure 22 shows a comparison between the surface temperature distribution on the pressure side and the suction side of the blade for the cases of models with smooth and ribbed cooling channels. The temperature distributions of the blade with ribbed internal cooling channels correspond to the configuration with square full-ribs with a ratio P/e = 10. As can be seen in Figure 22, internal cooling generated by ribs has an effect on the blade surface temperature, since it presented a substantial decrease in surface temperature on the pressure and suction sides of the blade. Also, it is observed a reduction of the spot of maximum temperature on the leading and trailing edges of the blade generated by the parabolic temperature profile at the inlet of the hot combustion gases. Another effect which can be seen is the stain of cooling at the root of the blade on the suction side, which is generated by the flow of air entering the vane platform.

4. Conclusions

In the present work, a numerical study was performed, with the aim to assess the effect generated by the ribs in the temperature distributions inside the blade body as well as the pressure drop through the cooling channels with different types of ribs. The main conclusions are:

The validation of the numerical model by comparing the internal and external flow with experimental [22] and semi-empirical [23, 24] data was developed. The pressure drop in the internal flow obtained through the numerical solution and semi-empirical [23, 24] data, offers a close enough agreement, with an absolute difference of 3.25%.

The higher and smaller temperatures of the internal flow are presented for the configurations of full-ribs in the ribbed section, reaching temperatures from 937 K to 741 K, respectively, while the half-ribs configurations offer a smaller difference of temperature in this section. This range is between temperature of 927 K and 791 K. The configurations of full-ribs have a larger contact area than configurations of half-ribs. Due to this reason, the configurations of full-ribs remove more thermal energy from the blade body.

The configuration with full-ribs with P/e = 20, offers the best cooling with any rib type. This could be due to the fact that the flow has a high recirculation zone between the ribs, generating a hydrodynamic perturbation to provoke a separation of the boundary layer.

The acceleration and deceleration effects, which are presented in the ribbed section, play an important role in the flow behaviour of the compressible fluid, since the high velocity of the flow shows a strong influence on the variations of temperature in the flow field. The highest Mach number and velocity are obtained with the ribbed channel, whose values are 1.45 and 823 m/s, respectively, while that smooth channel presents a continuous acceleration of the flow along the channels.

The ratio between the required inlet pressure in the cooling channels and the outlet pressure increases from 4 to 4.5 times approximately for the cases with full-ribs with aspect ratios

(P/e) of 10 and 20, respectively. For the half-ribs, this ratio is between 3 to 4 times, approximately. These values are higher than the values obtained in smooth cooling channels.

The ribbed cooling channels present different pressure drops, ordered from higher to lower pressure drops, they are triangular, square and semi-circular ribs, respectively. The triangular ribs offer the highest cooling effects of the analyzed cases; however, this configuration presents the highest pressure drop when compared with any other case.

The turbulence promoters allow to obtain a maximum temperature decrease, approximately about 10 to 20 degrees, close to the internal surfaces of the blade body. This allows reducing damages by fatigue and thermal stresses.

Author details

Armando Gallegos-Muñoz, Nicolás C. Uzárraga-Rodríguez and Francisco Elizalde-Blancas
Department of Mechanical Engineering, University of Guanajuato, Salamanca, Gto., Mexico

5. References

[1] Je-Chin Han, Sandip Dutta , Srinath Ekkad (2000) Gas Turbine Heat Transfer and Cooling Technology. Chapter 1, 1-10. Taylor and Francis.

[2] A. K. Sleiti, J. S. Kapat (2006) Comparison between EVM and RSM Turbulence Models in Predicting Flow and Heat Transfer in Rib-Roughened Channels. Journal of Turbulence. 7(29): 1-21.

[3] Marcel León de Paz, B.A. Jubran (2011) Numerical Modeling of Multi Micro Jet Impingement Cooling of a Three Dimensional Turbine Vane. Heat Mass Transfer. 47: 1561-1579.

[4] Hojjat Saberinejad, Adel Hashiehbaf, Ehsan Afrasiabian (2010) A Study of Various Numerical Turbulence Modeling Methods in Boundary Layer Excitation of a Square Ribbed Channel. World Academy of Science, Engineering and Technology. 71: 338-344.

[5] N. Cristobal Uzarraga-Rodriguez, Armando Gallegos-Muñoz, J. Cuauhtemoc Rubio-Arana, Alfonso Campos-Amezcua Mazur Zdzislaw (2009) Study of the Effect of Turbulence Promoters in Circular Cooling Channels. Proceedings of 2009 ASME Summer Heat Transfer Conference. 1: 1-14.

[6] Gongnan Xie, Weihong Zhang, Bengt Sunden (2012) Computational Analysis of the Influences of Guide Ribs/Vanes on Enhanced Heat Transfer of a Turbine Blade Tip-Wall. International Journal of Thermal Sciences. 51: 184-194.

[7] Kyung Min Kim, Hyun Lee, Beom Seok Kim, Sangwoo Shin, Dong Hyung Lee, Hyung Hee Cho (2009) Optimal Design of Angled Rib Turbulators in a Cooling Channel. Heat Mass Transfer. 45: 1617-1625.

[8] Mushatet Khudheyer S. (2011) Simulation of Turbulent Flow and Heat Transfer Over a Backward-Facing Step with Ribs Turbulators. Thermal Science. 15: 245-255.

[9] Smith Eiamsa-ard, Wayo Changcharoen (2011) Analysis of Turbulent Heat Transfer and Fluid Flow in Channels with Various Ribbed Internal Surfaces. Journal of Thermal Science. 20: 260-267.

[10] Hossein Shokouhmand, Mohammad A. Esmaeili Koohyar Vahidkhah (2011) Numerical Simulation of Conjugated Heat Transfer Characteristics of Laminar Air Flows in Parallel-Plate Dimpled Channels. World Academy of Science, Engineering and Technology. 73: 218-225.

[11] Pongjet Promvonge, Withada Jedsadaratanachai, Sutapat Kwankaomeng, Chinaruk Thianpong (2012) 3D Simulation of Laminar Flow and Heat Transfer in V-Baffled Square Channel. International Communications in Heat and Mass Transfer. 39: 85-93.

[12] H. K. Versteeg, W. Malalasekera (1995) An Introduction to Computational Fluid Dynamics. Chapter 2, 10-26. Longman Scientific & Technical.

[13] Fluent Inc. Products (2006). FLUENT 6.3 Documentation User's Guide.

[14] Gambit 2.4.6 (2006). Fluent Inc. Products Documentation User's Guide.

[15] A. Campos Amezcua, A. Gallegos Muñoz, Mazur Z., Vicente Pérez, Arturo Alfaro-Ayala, J.M. Riesco-Avila, J.J. Pacheco-Ibarra (2007a) Análisis Dinámico-Térmico-Estructural de un Alabe de Turbina de Gas con Enfriamiento Interno. 4to. Congreso Internacional, 2do Congreso Nacional de Métodos Numéricos en Ingeniería y Ciencias Aplicadas UMSNH-SMMNI-CIMNE, 1: 1-11.

[16] Z. Mazur, R. A. Luna, I.J. Juarez, A. Campos (2005) Failure Analysis of a Gas Turbine made of Inconel 738LC Alloy. Engineering Failure Analysis. 12: 474-486.

[17] Nicolás C. Uzárraga-Rodríguez (2009) Análisis de Promotores de Turbulencia para mejorar el Enfriamiento en Álabes de Turbinas de Gas. Thesis.

[18] A. Campos-Amezcua (2007b) Análisis y Optimización Termomecánica, en Estado Transitorio, de la Primera Etapa de Alabes en una Turbina de Gas con Enfriamiento Interno. Thesis.

[19] Frank P. Incropera, David P. DeWitt, Theodore L. Bergman, Theodore L. Bergman (2007) Fundamentals to Heat and Mass Transfer, 6th Ed.. Appendix A, Table A.4, page 941. John Wiley.

[20] S. V. Patankar (1980) Numerical Heat Transfer and Fluid Flow. Chapter 2, 11-24. Hemisphere..

[21] B. E. Launder, D. R. Spalding (1974) The Numerical Computation of Turbulent Flow. Computer Methods in Applied Mechanics and Engineering. 3(2): 269-289. Elsevier

[22] J. S. Kwak, J. Ahh J., Han, Lee, C. Pang, R. S. Bunker, R. Boyle, R. Gaugler (2003) Heat Transfer Coefficient on the Squealer Tip and Near Squealer Tip Region of Gas Turbine Blade. Journal of Turbomachinery. 125: 778-787.

[23] J. Nikuradse (1937) Law of Flow in Rough Pipes. National Advisory Commission for Aeronautics, Washington, DC, USA.

[24] R. L. Webb, E.R.G. Eckert, R.J. Goldstein (1971) Heat Transfer and Friction in Tubes with Repeated-Rib Roughness. International Journal of Heat and Mass Transfer. 14: 601-617.

[25] Yunus A. Cengel, M. A. Boles (2006) Thermodynamics an Engineering Approach, 5th Ed.. Chapter 17, Figure 17-52, page 862. McGraw Hill.

Chapter 17

Heat Transfer to Separation Flow in Heat Exchangers

S. N. Kazi, Hussein Togun and E. Sadeghinezhad

Additional information is available at the end of the chapter

http://dx.doi.org/10.5772/51331

1. Introduction

Separation flow is appeared over and behind a body surface when it is separated from that surface. In separation flow the region is relatively small compared to the body and enclosed by the separating stream line and points of separation and reattachment. Separation flows are formed at the upstream of a forward facing step downstream of a rearward facing step, within a cutout in a body surface and also on the upper surface of an airfoil.

The step separated flow is one of wedge-type separated flows, cutout flow (cavity-type separated flows) and the separated region over an airfoil (separation bubbles). A relatively small incidence angle between the separated flow and the body at points of separation and reattachment represents the wedge-type separated flow. On the other hand in cavity type separated flow the body boundaries at separation and reattachment are in general approximately perpendicular to the flow direction. An abrupt change of geometrical configuration of the body surface causes these two types of flow separation. The separation flow is strongly dependent on the nature of flow, such as laminar, transitional or turbulent.

In practice the separated flows are caused by flaps for deflection, spoiler control, rocket nozzle of over expanded type, leeward side of an object inclined at a large angle of attack etc. In practical cases the vortices of separated flow are unsteady and it is difficult to experimentally study. In a cutout the simulation of practical situation is considered and the understanding of the mechanics of real vortices and noise caused could be achieved. The present study highlights the separation flow mechanism, heat transfer to separated flow with subsequent pressure loss and possible use in practice.

2. Control of separation flow

Separation flow is performed to ratio the efficiency or to improve the performance of equipment, vehicles and machineries involving many engineering applications. The separation of flow may be controlled in two ways, such as (i) by prevention or delay of the onset of separation regions and (ii) with the help of provoking localized separation flow by utilizing the separated flow characteristics.

3. Prevention of delay of separation of flow

Adverse pressure gradient and viscosity are the two governing factors of flow separation. By changing or maintaining the structure of viscosity flow the control of separation can be achieved. Pressure gradient and viscosity ultimately prevent or delay the separation. Further, by designing the geometrical configuration the separation flow may be controlled, such as a pump could be used to suck the boundary layer flow away for suppressing the flow.

4. Retardation of the delay of separation by geometrical design of the body surface

Geometrical configurations include the basic body surface configuration, slot, vortex generator, notch, heating edge extension, step passage etc. These arrangements are adequately installed with respect to the body configurations.

Methods of computation for potential pressure distribution, boundary layer development, separation criteria etc. should be well understood for obtaining the controlled separation by basic body design from analysis. These analytical predictions on body configurations are not regular accessable, so additional geometrical shaping are employed if the design of the basic body geometry is not adequate to control the separation.

5. Reduction of heat transfer in separation region and delay of separation by cooling

Reduction of heat transfer in separation flow of laminar flow phenomena is obtained by injecting gas into that region. This technique may be practically applied in special case. Charwat and Dewey [1] computed the recovery factor in separation flow as a function of mass injection. They presented recovery factor as a function of dimensionless mass flow injection where, parameter Pr=1, 0.72 and 0.55 represents width of two-dimensional flow, l the length of separated mixing layer, c is constant of proportionally between viscosity and temperature.

ζ Represents dimensionless mass-flow variable defined by equation (1).

$$\zeta = \Psi_{*}/\sqrt{X_{*}} \tag{1}$$

Ψ_{*} is a transformed stream function

$$\Psi_{*} = \Psi/\sqrt{\upsilon_{e} u_{e} lc} \tag{2}$$

$$\rho u = \rho_{e} \frac{\partial \Psi}{\partial y} \tag{3}$$

$$\rho v = -\rho_e \frac{\partial \Psi}{\partial x} \tag{4}$$

Where,

L is length of separated mixing layer, c is constant of proportionality between viscosity and temperature. X_*=X/L. X is coordinate parallel to direction of flow along the dividing stream line ($\Psi = 0$) within the mixing layer, m_i is the injected mass flux which is equal to density x velocity x area

The recovery factor is expressed as (5),

$$q = (h_{aw}\text{-}h_e)/(u_e^2/2) \tag{5}$$

Where,

H_{aw} is enthalpy per unit mass at adiabatic wall conditions.

At Pr=1, the recovery factor is independent of mass injection. When Pr<1, the recovery factor drops with injection or mixing in boundary layer. The dotted line represent the computed value at r as obtained by low [2].

The pressure gradient deepens and the extent of the separated region is delayed and vice versa when the wall of the body is cooled. Illingworth [3] observed that the separation distance of laminar has flow reduced by 16 percent by raising the wall temperature from that of room to the boiling point for a retarding laminar flow.

Recirculation flows with separation generates high pressure losses accompanied by enhancement of turbulence and argumentation of heat and mass transfer rate.

6. Thermal effects on separation flow

At subsonic speed, the heat transfer of separation flows is given emphasis due to the existing design of heat transfer equipment. Attention is drawn by the phenomena of separation flow which holds existence of a hot spot in reattachment region. With laminar and turbulent flow, specially in laminar flow separation have become concern.

6.1. Heat transfer in separation flow

The heat transfer of separated flows at subsonic speeds is important for the design of heat transfer equipment. In flow separation, turbulence augments heat transfer in general but in particular region due to flow reattachment heat transfers and hot spot develops. Investigators are attracted to the point of reattachment. Flow separation, mainly laminar flow and its analytical study has become a concern. The onset of separation flow and its characteristics have not been understood yet. From the analysis, some knowledge has been gathered. Gadd [4] has presented heat transfer effects in separation air flow on the relevant pressure gradient extended to the separated region.

6.1.1. Heat transfer in the separated flow through sudden expansion

a. Experimental studies (Turbulent range)

There are many researchers, who investigated experimental study of turbulent heat transfer in separation flow with different geometries such as separation flow for sudden expansion in passage, backward or forward facing step, blunt body, rib channels, and swirl generators. The summary of references has been selected from the earliest studies to recent studies and included more detail about turbulent heat transfer in separation flow.

Sudden expansion

Boelter et al. [5] presented results of investigations conducted to observe the distributions of the heat transfer rate of air flowing in a circular pipe in the separated and reattached regions downstream of an orifice. The investigation was performed in Re range of 17,000 to 26,400 for internal pipe flow. They used various entrance sections as shown in Figure 1 which makes variation of point of reattachment at entrance pipe. From the experimental data obtained, they found maximum heat transfer coefficient near the point of reattachment which is about four times away of the length of fully developed flow.

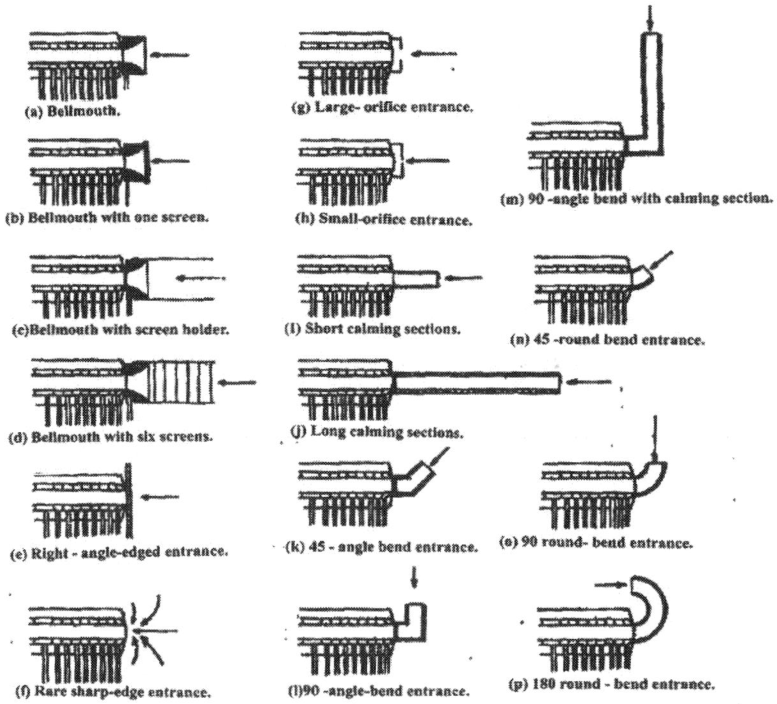

Figure 1. Circular pipes with various entrance sections.

Ede et al. [6] have investigated the effect of an abrupt convergence of straight pipe on the local heat-transfer coefficient for flowing water. The experimental covered for Reynolds numbers from 800 to 100,000 corresponding to a smaller pipe of diameter 1 inch. They have also determined effect of an abrupt divergence at Reynolds numbers from 3,700 to 45,000 and observed a considerable variation in local heat transfer coefficient. Koram and Sparrow [7] had performed experimental study of turbulent heat transfer of water flowing in circular pipe with unsymmetric blockage (segmental orifice plate) for range of Reynolds number from 10,000 to 60,000.

Krall and Sparrow [8] conducted experiments to determine the effect of flow separation on the heat transfer characteristics of a turbulent pipe flow. The water flow separation was driven by an orifice situated at the inlet of an electrically heated circular tube. The degree of flow separation was varied by employing orifices of various bore diameters. The Reynolds and Prandtl numbers were varied from 1000 to 130000 and from 3 to 6 respectively and ratios of the orifice to the tube diameter ranged from 2/3 to 1/4. Results show that the augment of the heat transfer coefficient due to flow separation accentuated with the decrease of Reynolds number decreases as shown in Figure 2. They have also found the effect of expansion ratio on the distribution temperature. The point of flow reattachment, corresponding to maximum value of the heat transfer coefficient was found to occur from 1.25 to 2.5 pipe diameter from the onset of flow separation. Suzuki et al. [9] performed experiment to study heat transfer and visulation of flow and surface temperature in the recirculating flow of an orifice in tube where they obtained results similar to Krall and Sparrow [8].

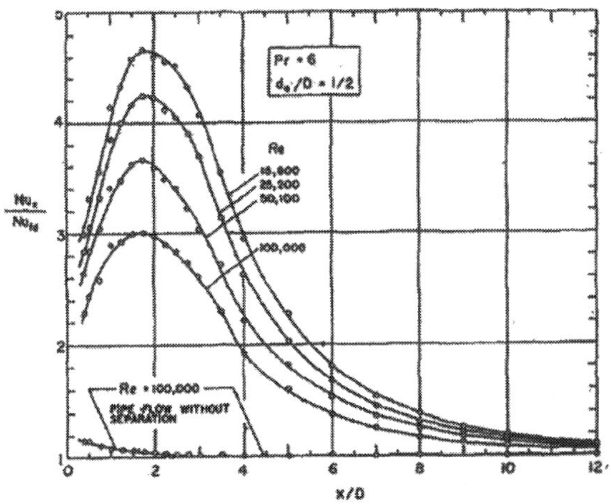

Figure 2. Local Nusselt number distribution

Filetti and Kays [10] presented experimental study on separation flow and heat transfer through flat duct with double step. They showed the highest heat transfer occurs in both long and short stall on sides at the point of reattachment, followed by decay towards the achievement of fully developed duct flow. They have also observed at different distances on the two walls of the duct the boundary-layer reattachment occurred and these distances are independent of Reynolds numbers. Figure 3 presents Nusselt number versus normalized distance downstream of step for expansion ratio (2.125). Also Seki et al. [11], Seki et al. [12] have performed experimental study on effect of stall length and turbulent fluctuations on turbulent heat transfer for separation flow behind double step at the entrance of flat duct.

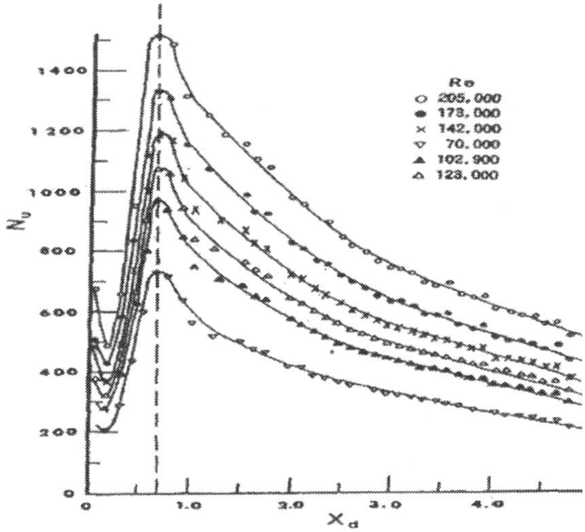

Figure 3. Local Nusselt number distribution.

Zemanick and Dougall [13] carried out experimental study of heat transfer to fluid flow in abrupt expansion of circular channel. Three expansions maintained during the test ratios of upstream –to- downstream diameter of magnitudes 0.43, 0.54 and 0.82 were considered with air as the working fluid. They have presented data for Reynolds numbers from 4000 to 90000, corresponding to the geometry of the ducting. For the range of the Reynolds numbers and expansion ratios studied, the following conclusions could be drawn from the test data:

1. The flow beyond an abrupt expansion in a circular duct shows a significant augmentation (over the fully developed value) of the average convective heat transfer coefficient in the separated flow region.
2. The degree of heat transfer coefficient enhancement increases with the increase of diameter ratio.
3. The location of highest heat transfer moves downstream as the ratio of downstream to upstream diameter increases.
4. The peak Nusselt number shown an apparent dependence on upstream.
5. In the **Nu-Re** expression, the Reynolds number exponent raised to a magnitude of approximately 2/3. The equation (6) reasonably represents the maximum Nusselt number data of all three geometries tested:

$$Nu_{max} = 0.20Re_d^{0.667} \quad (6)$$

Smyth [14] conducted experiments to study the physical effects of separation and the associated reattachment and redevelopment, on the heat transfer characteristics of turbulent flow in pipes and compared the results of these flow conditions with the fully developed one-dimensional flow condition. He also compared results with a recently developed numerical technique for the solution of recirculation flows. Separation of the flow was induced in a 4 ft length of 2 in. internal diameter tube of wall thickness 0.001 in. by means of a sudden enlargement of diameter at the entry of the tube. The tube was electrically heated by the passage of a current along its length. The first 25 in. of the tube was monitored by thermocouples which gave the wall temperatures and from these the local heat transfer rates were measured at Reynolds numbers up to 5 x 10^4 using air as the working fluid. There is a comparison between prediction and experiment as presented in Figure 4, a modest agreement at points near the peak of Nusselt number is obtained but a greatly poorer agreement in the developed region after reattachment of the flow.

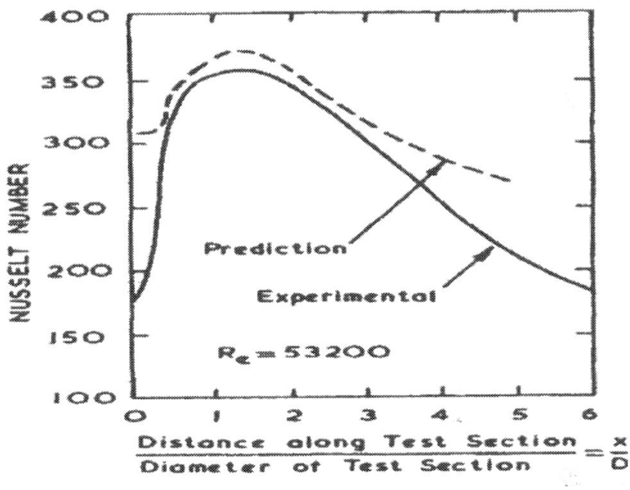

Figure 4. Experimental and predicated Nusselt numbers.

Baughn et al. [15] presented experimental study of the local heat transfer coefficient to an air flow at the downstream of an axisymmetric abrupt expansion in a circular channel with constant heat flux. They used a range of expansion ratios (d/D), small diameter to large diameter of 0.266 to 0.800 over the Reynolds number range of 5,300 to 87,000. From the experimental data they obtained for all expansion ratios and for d/D=0.266 the value of Nu/Nu_{DB} falls monotonically with the increase of Reynolds number as shown in Figure 5. This behavior is qualitatively satisfying the water tests [6] but differs with the out come of air study [13] for downstream Reynolds numbers above 30,000, where the ratio of peak to fully developed Nusselt numbers became independent of Re. Further details are shown in Figure 6. They have obtained the Reynolds number dependency on the maximum Nusselt number compared to the equation suggested by some authors [13] using Red instead of ReD in order to find the effect of expansion ratios on the Nusselt number. Also Runchal [16], Baughn et al. [15], Baughn et al. [17], Baughn et al. [18], Baughn et al. [19], Baughn et al. [20], Iguchi and Sugiyama [21] and Habib et al. [22], Habib et al. [23] have investigated effects of Schmidt numbers, Reynolds number, expansion ratio, velocity, segmented baffles, baffle spacing, baffle material and heat flux on the local heat transfer coefficient where they obtained the augmentations of heat transfer.

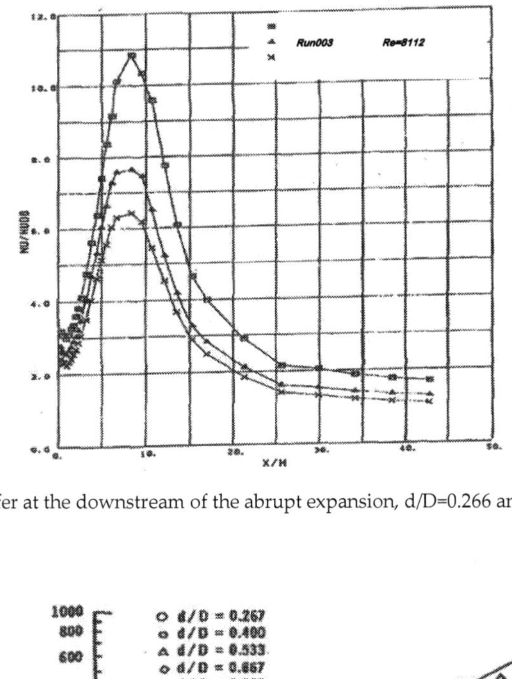

Figure 5. Heat transfer at the downstream of the abrupt expansion, d/D=0.266 and various Re.

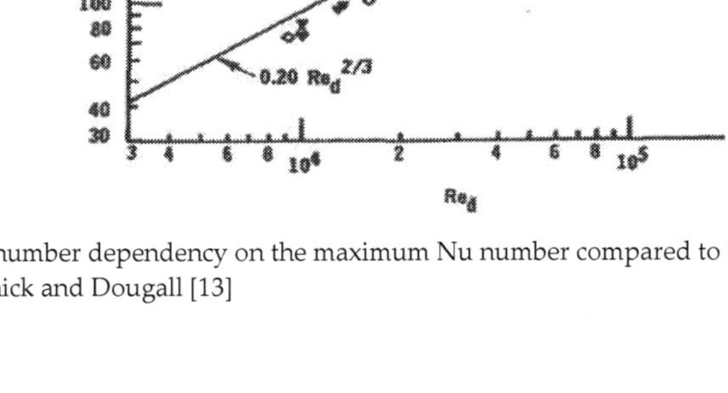

Figure 6. Reynolds number dependency on the maximum Nu number compared to an equation suggested by Zemanick and Dougall [13]

Hussein et al. (24) carried out experimental study of turbulent heat transfer and separated flow in an annular passage. They found the effect of separation flow on the average and local convective heat transfer and observed augmentation of local heat transfer coefficient occurred with the increase of heat flux and or Re while step height effect was clear at separation region and obtained increase in the local heat transfer coefficient with increase of step height.

b. Numerical studies (Turbulent range)

In the present chapter emphasis is given on the references which used various numerical methods to analyze turbulent heat transfer in separation flow for sudden expansion in passage, backward or forward facing step, blunt body, rib channels, and swirl generators.

Sudden expansion

Various models have been adopted to conduct heat transfer analysis in separated flow with sudden expansion. Chieng and Launder [24] performed numerical study of turbulent heat transfer and flow in separation region with an abrupt pipe expansion by using standard k-ε model and they obtained a good agreement with experimental data of Zemanick and Dougall [13].

Gooray et al. [25] presented numerical calculations for heat transfer in recirculation flow over two dimensional, rearward- facing steps and sudden pipe expansions by using the standard k-∊model at low Reynolds number range from 500-10,000. They have considered turbulent modeling for two dimensional back-step and pipe expansion geometries. The results obtained could be compared with the experimental data of Zemanick and Dougall [13], Aung and Goldstein [26], and Sparrow and O'Brien [27], and with experimental fluid dynamic data of Moss and Baker [28], and Eaton and Johnston [29] as shown in Figures (7, 8, 9, 10). It is observed that the improved k-∊procedure is capable of providing an insight into complex phenomena having turbulent separated and reattachment flow with heat transfer.

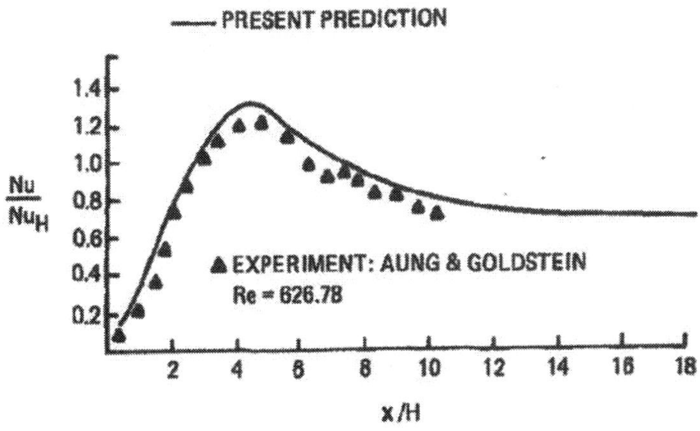

Figure 7. Streamwise variation of Nusselt number at the downstream of step for two-pass procedure

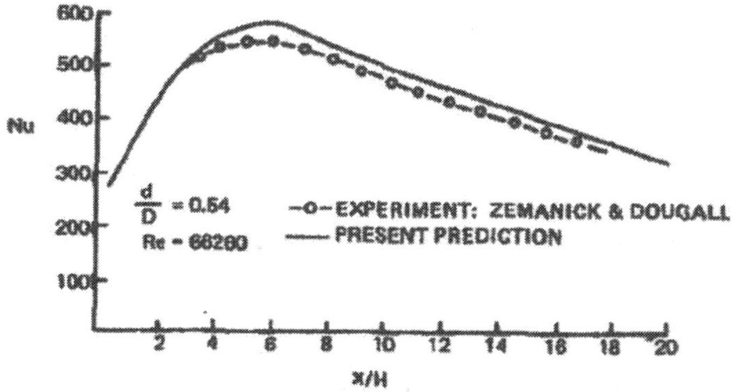

Figure 8. Streamwise variation of Nusselt number at the downstream of sudden expansion for two-pass procedure.

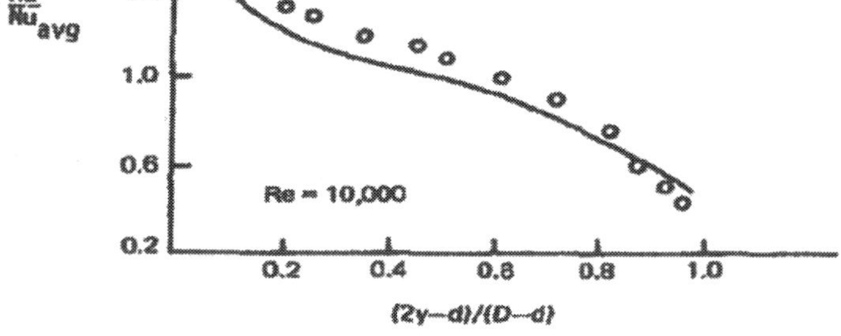

Figure 9. Cross-stream variation of Nusselt number on the downstream face of the pipe expansion.

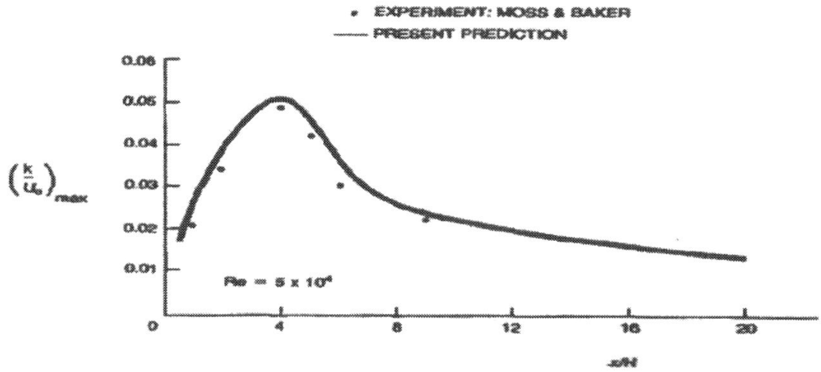

Figure 10. Streamwise variation of cross-stream maximum in turbulent kinetic energy for step.

Habib et al. [30] performed numerical study on effect of Reynolds and Prandtl numbers, element spacing and length in channels with streamwise periodic flow on heat transfer and turbulent flow to solve the time-averaged conservation equations of mass, momentum, and energy by using k-ϵ turbulence model and wall functions for high Reynolds number based on a finite-control-volume method. With the increase of Reynolds number and element height and decrease of baffle spacing they obtained augmentation of heat transfer. They have also obtained a good agreement with experimental data as reported by Habib et al. [31] and Berner et al. [32]. Kim and Lee [33] carried out prediction study of turbulent flow and heat transfer characteristics at downstream of a sudden circular pipe expansion with the full Reynolds stress model by adopting finite volume method with power-law scheme to discrete the governing differential equations. They found the Reynolds stress model is much better than k-ϵ model in the predictions of velocity and temperature fields as well as the heat transfer coefficients as shown in Figure 11.

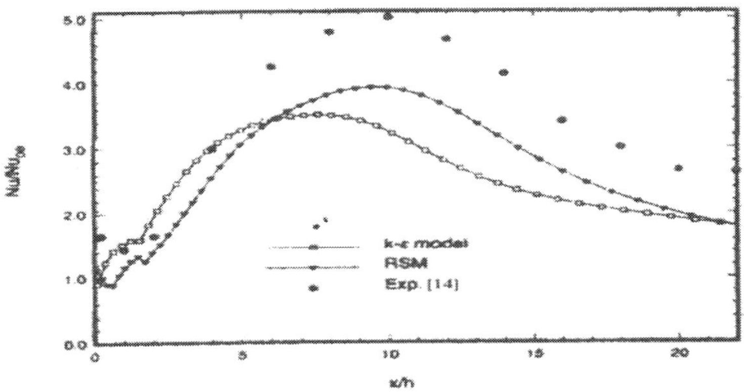

Figure 11. Nu distribution by predictions with Re stress model, k-ϵ model and experiment by Baughn et al. [17].

Chung and Jia [34] studied numerically turbulent heat transfer in abrupt expansion by using new K-∊model and they found enhanced turbulent kinetic energy and velocity and noticed a good agreement with previous experimental data. Hsieh and Chang [35] observed numerically the wall heat transfer in pipe expansion turbulent flows by applying a new modified low-Reynolds number k-∊turbulence [Chang, Hsieh and Chen (CHC)] model and compared the performance of nine other conventional low-Reynolds number k-∊models developed earlier by Launder and Sharma [36], Lam and Bremhorst [37], Chien [38], Nagano and Hishida [39], Myong and Kasagi [41], Nagano and Tagawa [42], Yang and Shih [43], Abe et al. [40], and Chang et al. [41] (Table 1) provided a summary of model functions. From comparisons of the predicted distributions of the Nusselt number obtained through the low-Reynolds number k-∊models with data of Baughn et al. [17] (Figure 12) and Zemanick and Dougall [13] (Figure 13) they found only CHC model generates correct trend.

Guo et al. [45] performed numerical study of thermal effect in the recirculation zone due to sudden expansion for gas flow and they have the result compared with some experimental data to check the heating effect on the corner recirculation zone (CRZ) (Table 2) [13, 46, 47]. Experimental and numerical data are showing Reynolds number and the expansion ratio on the CRZ for the sudden expansion turbulent flows.

Sugawara et al. [49] analyzed numerically the large eddy simulation (LES) method. They have studied the three dimensional turbulent heat transfer and separated flow in a symmetric expansion plane channel by applying LES. They used smagorinsky model and fundamental equations based on finite difference method and found a good agreement between numerical data with previously published experimental data.

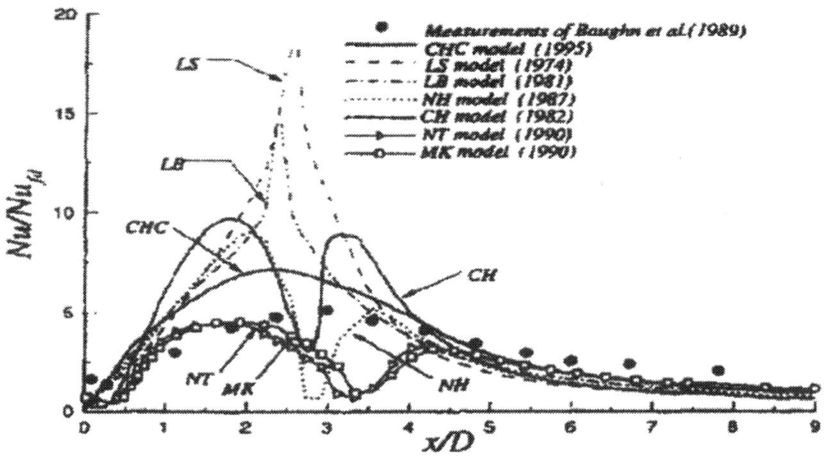

Figure 12. Comparison of the predicted Nusselt number distributions obtained through various Low-Reynolds number models with the measurements (Re=17300 and EPR= 2.5).

Model	Fμ	f1	f2
Standard	1.0	1.0	1.0
Launder-	exp [- 3.4/(1 - Rt/50) ^{2}]	1.0	*1.0-0.3exp (-R t^2)*
Sharma.[36]	[1.0- exp(- 0.0165Rk)]^2 (1 +	1.0 + (0.05/fμ)^3	*1.0-0.3exp (-R t^2)*
Lam and	20.5/Rt)	1.0	*1.0-0.22exp (-R t ^{2} /36)*
Bremhorst [42]	1.0 - exp(- 0.0115 y^+)	1.0	*1.0-0.3exp (-R t ^{2})*
Chien.[38]	[1.0- exp(- y^+/26)]^2	1.0	*{1.0-2/9exp[-(R t /6)^2]}[1-exp(-y^+/5)] ^{2}*
Naganty-Hishida	[1.0- exp(-y^+/70)](1 + 3.45/Rt$^{1/2}$)	1.0	*{1.0-0.3exp[-(R t /6.5)^2]}[1-exp(-y^+/6)]^2*
[39]	[1.0- exp(-y^+/26)]2(1 +4.1/Rt$^{3/4}$)	1.0	*1.0*
Myong-Kasagi [43]	[1.0- exp(-1.5 x 10^{-4} Rk-5.0 x 10 -		
Nagano-Tagawa	^{7} Rk^3		
[39]	- 1.0 × 10 $^{-10}$ Rk^5)$^{1/2}$	1.0	*{1.0-0.3exp[-(R t /6.5) ^{2}]}[1-exp(-*
Yang-Shih [44]			*y*/3.1)] ^{2}*
	[1.0- exp (-y*/14)]^2 {1 + 5.0/Rt$^{3/4}$	1.0	
	exp [-(Rt /200)^2]}		*[1.0-0.01exp (-R t ^{2}] [1-exp(-0.0631*
Abe et al. [40]	[1.0- exp (-0.0215Rk)]^2(1		*Rk)]*
	+31.66/Rt$^{5/4}$)		
Chang et al. [41]			

Table 1. Comparison of the developed models of low Reynolds number

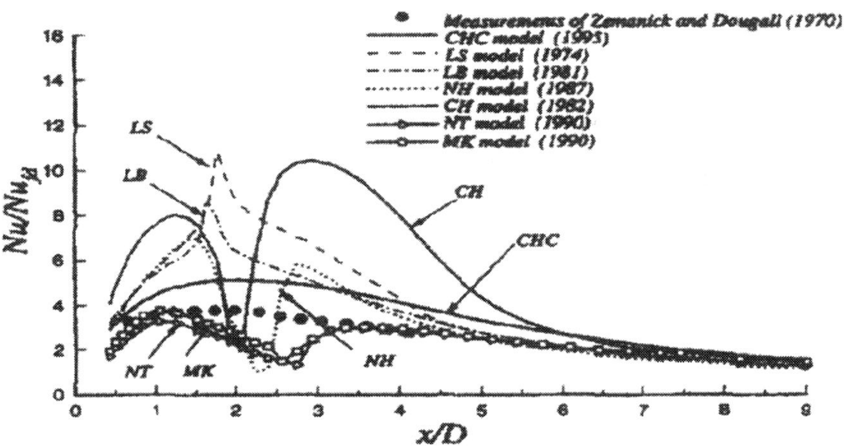

Figure 13. Comparison of the predicted Nusselt number distributions obtained through various Low-Reynolds number models with the measurements (Re=48090 and EPR=1.86).

Authors	R/r	Re	Lr/h
Zemanick and Dougall experimental [13]	1.22	$(44.36\text{-}88.78)\times 10^3$	7.2
	1.85	$(4.86\text{-}66.26)\times 10^3$	7.6
	2.32	$(4.18\text{-}47.64)\times 10^3$	9-8
Zeng et al calculated [48]	2	2.0×10^5	7.31
Prud'homme and Elghobashi calculated [46]	1.85	6.6×10^4	6.52
Moon and Rudinger experimental [47]	1.43	$1.0\times103\text{-}1.0\times10^6$	6-9

Table 2. Effect of expansion ratio and Reynolds number on separation length

Numerical analysis of the unsteady and steady separated flow [50] is the concern of several researchers [51] both of them employed Navier-Stokes and energy equation based on finite difference method. They informed that increase of Reynolds number causes the recirculation region and separated shear layer unstable and the vortices created become significant and have greater influence on the heat transfer rate. While OTA et al. [52] observed numerically two and three dimensional heat transfer and separated flow in enlarged channel by using Navier-Stokes and energy equations based on finite difference method and informed about the effect of aspect of ratio on both the results of two and three dimensional on local heat transfer rate and the transitions from symmetric to asymmetric flow with the increase of the aspect ratio.

c. Experimental studies (Laminar range)

Sudden expansion

The earliest and most comprehensive and qualitative investigations on laminar flow in sudden expansion was carried out by Macagno and Hung [53]. For Reynolds number range ≤ 200 in one to two axisymmetric expansion (based on pipe diameters). Durst et al. [54] conducted flow visualtion and laser –anemometry measurements in the flow at the downstream of a plan 3:1 symmetric expansion in a duct with an aspect ratio of 9.2:1. The flow was found to be markedly dependent on Reynolds number up to the lowest measurable velocities. Symmetric velocity profiles existed from the expansion to a fully developed parabolic profile for downstream in spite of the substantial three dimensional effects in the vicinity of the separation regions. The velocity profiles were in good agreement with those obtained by solving the two–dimensional momentum equation.

Goharzadeh and Rodgers [55] have investigated experimentally the laminar water flow through a confined annular channel with sudden expansion. They measured velocity and length of separation by using particle image velocimetry (PIV) and refractive index matching (RIM). They have also reported that separation regions increases with the increase of Reynolds number as shown in Figure (14,15,16) and also they have obtained good agreement with the numerical results reported by Nag and Datta [56].

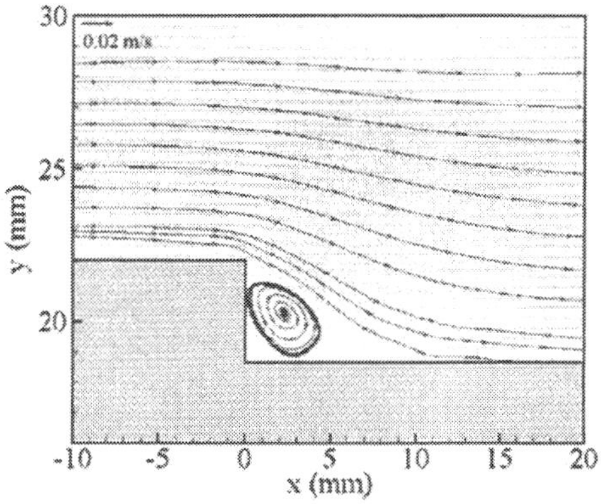

Figure 14. Measurements of velocity flow field at Re= 100

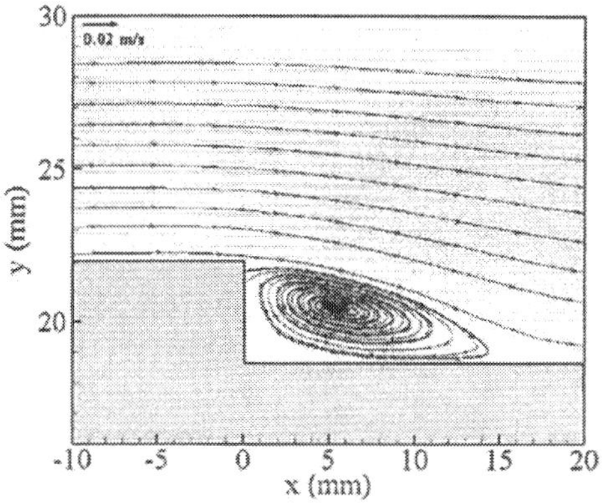

Figure 15. Measurements of velocity flow field at Re= 300

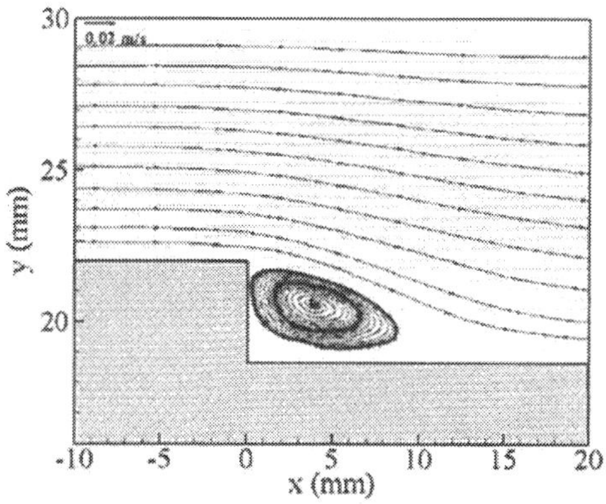

Figure 16. Measurements of velocity flow field at Re= 500

d. Numerical study (Laminar range)

Sudden expansion

Chiang et al. [57] have studied the effect of side wall on the laminar incompressible fluid flow over a plane symmetric sudden expansion. In their analysis, 14 aspect ratios were considered from 3 to 48, and Re=60 for three dimensional analysis and Re=60 and 140 for two dimensional analysis. The numerical results have given a good agreement with experimental results conducted by Fearn et al. [58] for symmetric flow at Re=26 and asymmetric flow at Re up to 36.

Boughamoura et al. [59] presented numerical date of heat transfer and laminar flow in a piston driven in a cylinder with sudden expansion by using finite element method with simpler algorithm of pressure –velocity coupling, and staggered and moved grid. The numerical results showed a good agreement with pervious experimental data where they observed the increase of expansion ratio leads to decrease the recirculating zone of the step plane. They have noticed also there are three regions created by moving the piston with sudden expansion namely primary vortex, secondary positive vortex, and secondary negative vortex. Pinho et al. [60] have numerically evaluated the pressure losses of laminar non Newtonian flow in sudden expansion having diameter ratio of 1 to 2.6. They used power law model to find irreversible pressure losses coefficient by equation 7.

$$C_1 = \frac{\Delta P_I}{\frac{1}{2}\rho U_1^2} = \frac{\Delta P - \Delta P_R - \Delta P_F}{\frac{1}{2}\rho U_1^2} \tag{7}$$

The numerical results revealed the decrease of separation length with shear thinning and they also showed the variation of separation length is linear at high Reynolds number and become asymptote to constant value for creeping flow condition.

Thiruvengadam et al. [61] investigated three-dimensional mixed convection in vertical duct with sudden expansion. The effects of aspect ratio, heat flux, and buoyancy force on the laminar flow in separated regions were presented in their studies. They obtained sharp increase in the recirculation regions with the increase of aspect ratio as shown in Figure 17. Linear increasing of streamwise distribution and independent aspect ratio after recirculation region as shown in Figure 18.

6.1.2. Heat transfer in the separated flow over backward and forward step

a. Experimental studies (Turbulent range)

Forward and backward facing step

Seban et al. [62] studied experimentally heat transfer in the separated, reattached, and redeveloping flow regions around the downward facing step. They found that the heat

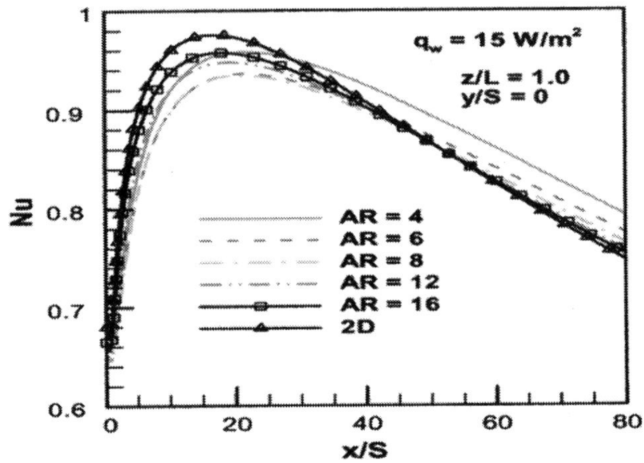

Figure 17. Distribution of local Nusselt number with aspect ratio

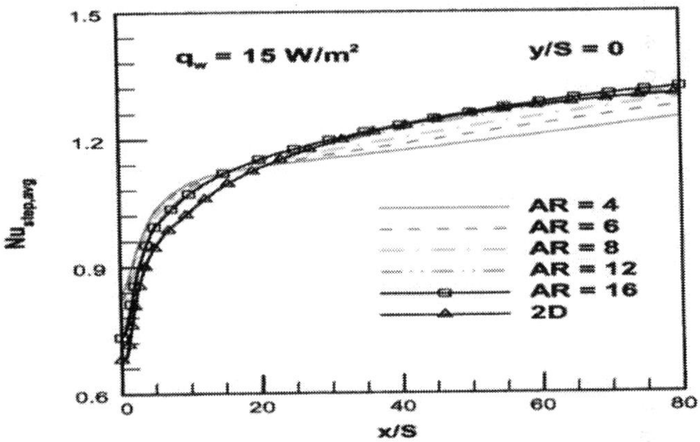

Figure 18. Distribution of streamwise of the average Nusselt number with aspect ratio

transfer coefficient reached a highest value at the reattachment point, and after that decreases as shown in Figure 19. Seban [63] investigated experimentally the relative effect of the suction and injection on the heat transfer and fluid flow in a separated region at downstream of a backward facing step for fixed rates of suction and injection through a slot at the base of the step, with air velocities in the free stream varied in the range of 15 to 45 m/s. They found that with the fixed suction the length of the separated region decreases as the free–stream velocity is reduced and the maximum value of the group (h/k) $(\upsilon/Ua)^{0.8}$ is also increased. With injection, there is no region of separated flow at the wall when the free–stream velocity is of the same order of the injection velocity and the local heat transfer is at first influenced primarily by the injection velocity [64].

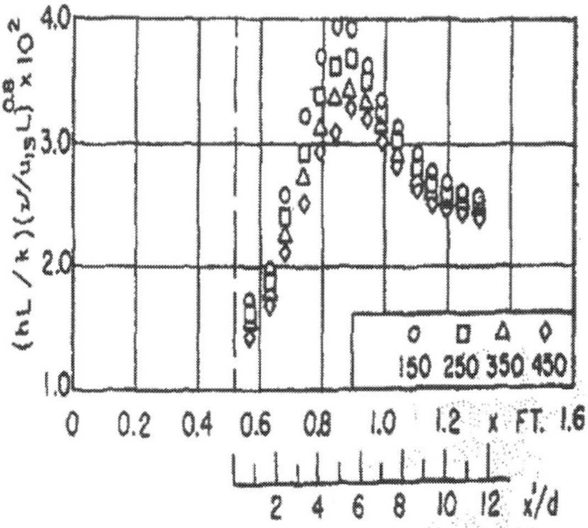

Figure 19. Local heat transfer coefficient.

Sogin [65] performed experiments on a bluff flat plate stripe in two dimensional flows and measured the local heat flux and temperature distribution under steady–state conditions. Different configurations were obtained by changing angle of incidence or modifying its cross–sectional profile. Air speed varied from 12 m/s to 47.5 m/s and the Reynolds number based on 170 mm chord length ranged from 100000 to 440000 and the nominal blockage ratio was maintained (0.21) for the angle of incidence 90 degree. The local heat transfer by forced convection from the base surface of a bluff obstacle is presented for a variety of configurations. The data were satisfactorily represented by equation of type (8).

$$Nu = C. Re^{0.667} \tag{8}$$

For measuring details of the heat transfer near the reattached point of the separated flows, Mori et al. [66] used the thermal tuft probe while Kawamura et al. [67], Kawamura et al. [68] designed new heat flux probe to determine time and spatial characteristics of heat transfer at the reattachment region of a two dimensional backward facing step deign.

Abu-Mulaweh [69] studied experimentally the turbulent fluid flow and heat transfer of mixed convection boundary-layer of air flowing over an isothermal two-dimensional, vertical forward step. They studied the effect of forward–facing step heights on local Nusselt number distribution as shown in Figure 20, and obtained the increase of local Nusselt number with the increase of step height and the greatest value was obtained at the reattachment region. The present results indicate that the increase of step height leads to increase of intensity of temperature fluctuations, the reattachment length, transverse velocity fluctuations and the turbulence intensity in the stream.

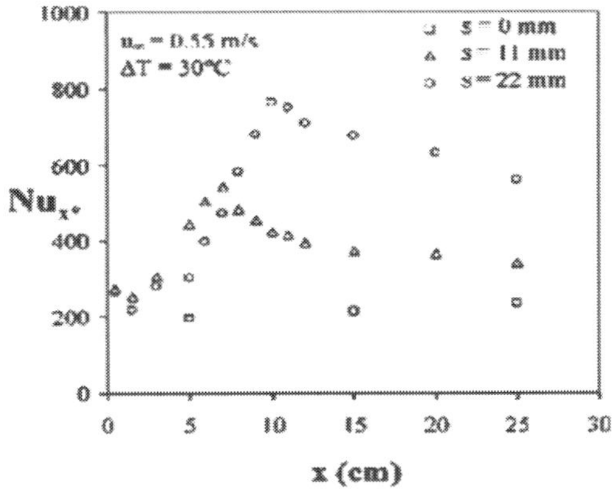

Figure 20. Local Nusselt number variation at the downstream of the step.

Sano et al. [70] presented experimental results of the turbulent channel flow over a backward-facing step by using suction through a slit at the bottom corner of the step and the direction of the suction was perpendicular and horizontal to the main flow. They measured local heat transfer coefficient and wall static pressure behind the backward-facing step and obtained the enhancement of the heat transfer coefficient in the recirculating region with the increase of suction flow and reduction of the pressure drop.

b. Numerical studies (Turbulent range)

Forward and backward facing step

Shisnov et al. [71] studied the heat transfer in the recirculating region formed by a backward-facing step. They used a vertical structure of a recirculating flow formed by a backward-facing step with the comparable predicted separation zone lengths from the experimental data obtained by Etheridge and Kemp [72], Moss et al. [73], Kim et al. [75] Abbott and Kline [74], Tani et al. [75], Eaton and Johnston [76], Smyth [77], Seban [78], and Filetti and Kays [10] (see Table 3).

Abe et al. [40] performed studies to the predict heat transfer and fluid flow in separation and reattachment over backward facing step and suggested a new two equation heat transfer model obtained from the model of Youssef et al. [79]. Thus they presented a new model of low Reynolds number by using Kolmogorov velocity scale $u\varepsilon = (\nu\varepsilon)^{1/4}$ instead of the friction velocity uT, which could predict heat transfer and fluid flow at downstream of backward facing step. While Heyerichs and Pollard [80] used mathematical model with the aid of the k-εand k-w turbulence models to study heat transfer with both separated and impinging turbulent flow. They showed that the K-w model is more simple numerical method to estimate heat transfer in complex turbulent flows. All models have predicted reasonable results for the channel flow test case and have shown a strong agreement with the log law and temperature data of Kader [81]. This result can be anticipated since they are derived on the basis of the log-law and temperature log-law [82]. All low Reynolds number models except LB have predicted C_f within 5% error. When low Reynolds number approach is considering them a strong correlation between a models ability to predict the log-law, C_f and Nu. The models of LB [42] and MK Myong and Kasagi [43] underpredict the log law, overestimate Cf and Nu is obtained. The models of CH [38], and LS [39] overestimate the log-law, under predict Cf and Nu. The WCP(W) Wolfshtein [83] and (CH) Chein et al. [38] models show the best fit to the log-law and Nu as predicted within 1% error of the output from Dittus-Boelter correlation. When the low Reynolds number approach is used a strong correlation is obtained between the models ability to predict the log-law, C_f and Nu.

Reference	Experimental xR	Predicated xR	Percentage error(%)
Shishov et al. [71]	6.55	6.6	1
Eitheridge and Kemp [72]	6.6	6.7	1
Eitheridge and Kemp [72]	5.30	5.45	3
Moss et al. [73]	5-6	5.5	9
Kim et al [84]	7	7.1	1.5
Abott and Klin [74]	7.5	6.82	9.9
Thani et al. [75]	6	5.58	7.5
Eaton and Johnston [76]	7.97-8.2	6.7	20
Smith [77]	6	6.6	10
Seban [78]	6	5.85	2.5
Filetti and Kays [10]	6	6.3	5
Eitheridge and Kemp [72]	8	8.8	10

Table 3. Comparison separation length for experimental and numerical studies.

Some researchers have used finite volume methods to analyze turbulent heat transfer and separated flow, such as Zigh [83] presented computational study of simultaneous heat and mass transfer in turbulent separated flows. It is observed that the location of the maximum Nusselt number is better predicated by the RNG based nonlinear K-∊model, which has a good agreement with results of Zemanick and Dougall [13] as presented in Figure 21.

Yin et al. [85] numerically calculated the turbulent heat transfer in high Prandtl number fluids by using the two equations turbulence model developed from the model of Nagano and Kim [86] and have compared the numerical results with existing experimental data on water, aqueous ethylene glycol, oil and also obtained a good agreement with experimental data especially with the data of water. Dutta and Acharya [87] performed comparative study on the standard k-t model, nonlinear k-∊ turbulence model, and the modified k-∊ turbulence model for analysis of heat transfer and flow in a backstep, where they found the nonlinear k-∊ and modified models agree batter with the experimental data than the standard k-t model. Rhee and Sung [88] also developed a low Reynolds number $k\theta$-$\varepsilon\theta$ model that dealt with turbulent separated flow and heat transfer [89] and they obtained satisfactory numerical results compared to the experimental data. Tsay et al. [90] performed numerical study on the effect of baffle height, baffle thickness, and distance between backward facing step and baffle on temperature, Nusselt number, and flow structure. They found a maximum augmentation of Nusselt number about 190% for heating step and 150% for heating section of bottom plate.

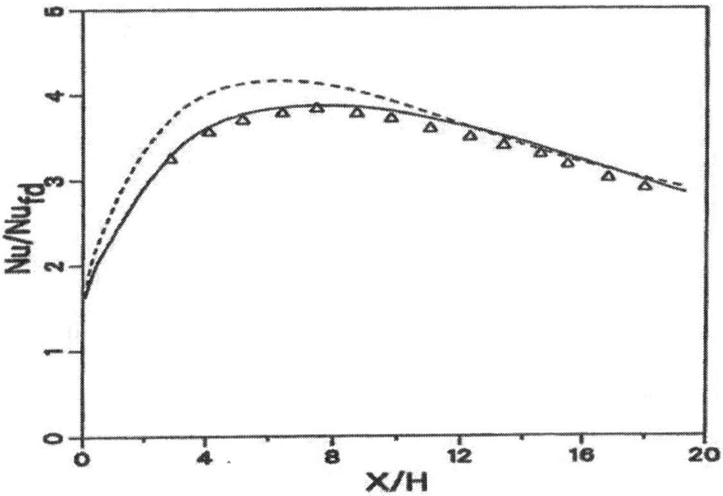

Figure 21. Normalized Nusselt number for turbulent flow along symmetric expansion with constant heat transfer at Re= 4.8 × 10 4, Δ Zemanick & Dougall experimental [13], standard K-∈model, ✤✤ RNG based on nonlinear K-∈ model.

c. Experimental studies (Laminar range)

Forward and backward facing step

Armaly et al. [91] have conducted experimental and theoretical study on two-dimensional on effects Reynolds number at reattachment length on flow at downstream of a backward facing step in a rectangular channel by using laser Doppler measurement where the range of Reynolds number varied between 70 to 8000 and aspect ratio 1:36. The experimental and predicted results are agreed on increase of reattachment length with increase of Reynolds number.

Barton [92] used model of Eulerian-Lagrangian for study of laminar flows over backward facing step in stream of hot particle. He focused in his investigation the effect of heat transfer and thermal characteristic on recirculation regions and noticed the streamlines in the separation region as 10 times smaller than the streamlines in the free flow.

Lee and Mateescu [93] studied two-dimensional air flows over backward facing step. They measured lengths of separation and reattachment on lower and upper part of duct by using hot wire probe with expansion ratio 1.17 and 2 and Re below 3000. The numerical and experimental results agree with previous experimental data for separation and reattachment lengh and locations on the lower and upper wall of the duct. Stüer et al. [94] presented experimental study of laminar separation flow on forward facing step by using particle tracking velocimetry to get more information about separation phenomena. They observed the increase of distance between the breakthroughs in span with decrease of Reynolds number. Armaly et al. [95] adopted experimental measurements of velocity in three dimensional laminar separated flows on backward facing step by using two components laser Doppler velocimeter with expansion ratio of 2.02 and the range of Reynolds number from 98.5 to 525 and observed increase of recirculation regions with increase of Reynolds number. Velazquez et al. [96] presented study of enhancements of heat transfer for laminar flow over a backward facing step by using pulsating flow. They showed at Re=100, the maximum time average Nusselt number was 55% higher in comparison to steady case.

d. Numerical studies (Laminar range)

Forward and backward facing step

Han et al. [97] carried out numerical study of incompressible laminar flow through 90 degree diversions of rectangular cross section. They showed that, with the increase of aspect ratio of duct the distance between top and bottom of symmetry plane effect the flow. Rouizi et al. [98] investigated numerically, the two-dimensional incompressible laminar flows over backward facing step. They employed model reduction and identification technique with finite element method and observed that with the increase model order the accuracy increases. They have also validated the length of recirculation bubble by reduced order model and obtained good agreement with previous studies.

Li and Armaly [99] presented result of laminar mixed convection in 3D backward facing step by solving the fill elliptic 3 coupled governing equation by using finite volume method. They found the effect of buoyancy force and temperature on reattachment length.

Saldana et al. [100] conducted numerical study on three-dimensional mixed laminar air flow over horizontal backward step by using finite volume method. The bottom wall of channel was heated at constant temperature and the other walls were adiabatic, the aspect ratio maintained as 4 and range of Richardson number (Ri) varied between 0 to 3. The numerical results showed the decrease of size of primary recirculation region and rarefaction of maximum value of average Nusselt number with the increase of Richardson number.

Lima et al. [101] investigated two-dimension laminar air flows over backward facing step by using two CFD commercial codes, the first one based on finite element method (comsol multiphysics) and the other one finite volume method (Fluent) at a range of Reynolds number. The numerical results agreed with the previous experimental data.

7. Remarks

Separation flow along with heat transfer and pressure drop phenomena are described in the present chapter. Different channel configurations for laminar and turbulent flow have studied to find the optimum heat transfer and pressure loss configurations. Numerical and experimental studies have been conducted to find out selected models for good estimation of heat transfer and friction loss.

Nomenclature

3D	three-dimensional	
A	drag surface of the obstacle	m^2
As	Surfaces area	m^2
CRZ	Corner recirculation zone	
C_f	Friction factor coefficient	
D	Diameter of heated tube in the sudden expansion region	m
ER	Expansion ratio	
h	Heat transfer coefficient	W/ m^2.K
H	Height	mm
Hr	Height ratio	
K	Thermal conductivity	W/m.K
Ka	Thermal conductivity of air	W/m.K
Nu	Nusselt number	
NuDB	Nusselt number determined by by Dittus-Boelter	
Pr	Prandtl number	
Re	Reynold number	
Tb	Air bulk temperature	°C
Ts	Surface temperature	
U_{ref}	Free stream velocity	m/sec
V_s	Voltage	Volt
δ	turbulent boundary layer thickness	M
ε	energy dissipation rate	m^2/s^3
ζ	$\zeta = x/\delta_s$	
ξ	$\xi = p/p_s$	
φ	Cylinder angle measured from stagnation point	
Ψ_*	$\Psi_* = \Psi/\sqrt{v_e u_e l_c}$	

Author details

S. N. Kazi, Hussein Togun and E. Sadeghinezhad
Department of Mechanical Engineering,
Faculty of Engineering, University of Malaya, Kuala Lumpur, Malaysia

8. References

[1] Charwat, A. and C. Dewey, *An investigation of separated flows part II: flow in the cavity and heat transfer.* Journal of Aerospace Seience, 1961. 28(1): p. 513-527.

[2] Low, G.M., *The compressible laminar boundary layer with fluid injection.* NACA TN, 1955. 3404.

[3] Illingworth, C., *The effect of heat transfer on the separation of a compressible laminar boundary layer.* The Quarterly Journal of Mechanics and Applied Mathematics, 1954. 7(1): p. 8-34.

[4] Gadd, G., *Boundary Layer Separation in the Presence of Heat Transfer*, 1960, DTIC Document.

[5] Boelter, L., G. Young, and H. Iversen, *An Investigation of Aircraft Heaters, XXVII±Distribution of Heat-transfer Rate in the Entrance Section of a Circular Tube, National Advisory Committee for Aeronautics*, 1948, Technical Note.

[6] Ede, A., C. Hislop, and R. Morris, *Effect on the local heat-transfer coefficient in a pipe of an abrupt disturbance of the fluid flow: abrupt convergence and divergence of diameter ratio 2/1.* Proceedings of the Institution of Mechanical Engineers, 1956. 170(1): p. 1113-1130.

[7] Koram, K. and E. Sparrow, *Turbulent heat transfer downstream of an unsymmetric blockage in a tube.* Journal of heat transfer, 1978. 100: p. 588.

[8] Krall, K. and E.M. Sparrow, *Turbulent heat transfer in the separated, reattached, and redevelopment regions of a circular tube.* Journal of heat transfer, 1966. 88: p. 131.

[9] Suzuki, K., et al., *Heat Transfer in the Downstream Region of an Orifice in a Tube.* Trans. JSME B, 1982. 48: p. 132-140.

[10] Filetti, E. and W.M. Kays, *Heat transfer in separated, reattached, and redevelopment regions behind a double step at entrance to a flat duct.* Journal of heat transfer, 1967. 89: p. 163.

[11] Seki, N., S. Fukusako, and T. Hirata, *Effect of stall length on heat transfer in reattached region behind a double step at entrance to an enlarged flat duct.* International Journal of Heat and Mass Transfer, 1976. 19: p. 700-702.

[12] Seki, N., S. Fukusako, and T. Hirata, *Turbulent fluctuations and heat transfer for separated flow associated with a double step at entrance to an enlarged flat duct.* Journal of heat transfer, 1976. 98: p. 588.

[13] Zemanick, P.P. and R.S. Dougall, *Local heat transfer downstream of abrupt circular channel expansion.* Journal of heat transfer, 1970. 92: p. 53.

[14] Smyth, R., *Heat transfer in turbulent separated flow.* J. Nucl. Sci. Technol.(Tokyo), v. 11, no. 12, pp. 545-553, 1974. 11(12).

[15] Baughn, J., et al., *Local heat transfer downstream of an abrupt expansion in a circular channel with constant wall heat flux.* Journal of heat transfer, 1984. 106: p. 789.

[16] Runchal, A., *Mass transfer investigation in turbulent flow downstream of sudden enlargement of a circular pipe for very high Schmidt numbers.* International Journal of Heat and Mass Transfer, 1971. 14(6): p. 781-792.

[17] Baughn, J., et al., *Heat transfer, temperature, and velocity measurements downstream of an abrupt expansion in a circular tube at a uniform wall temperature.* Journal of heat transfer, 1989. 111: p. 870.

[18] Baughn, J., et al. *Turbulent velocity and temperature profiles downstream of an abrupt expansion in a circular duct with a constant wall temperature.* 1987.

[19] Baughn, J., et al., *Heat transfer downstream of an abrupt expansion in the transition Reynolds number regime.* Journal of heat transfer, 1987. 109: p. 37.

[20] Baughn, J., et al., *Local heat transfer measurements using an electrically heated thin gold-coated plastic sheet.* Journal of heat transfer, 1985. 107: p. 953.

[21] Iguchi, M. and T. Sugiyama, *Study on the flow near the wall of the larger pipe, downstream of a sudden expansion.* Trans JSME B, 1988. 54(507): p. 3010-3015.

[22] Habib, M., et al., *An experimental investigation of heat-transfer and flow in channels with streamwise-periodic flow.* Energy, 1992. 17(11): p. 1049-1058.

[23] Habib, M., et al., *Experimental investigation of heat transfer and flow over baffles of different heights.* ASME Transactions Journal of Heat Transfer, 1994. 116: p. 363-368.

[24] Chieng, C. and B. Launder, *On the calculation of turbulent heat transport downstream from an abrupt pipe expansion.* Numerical Heat Transfer, 1980. 3(2): p. 189-207.

[25] Gooray, A., C. Watkins, and W. Aung, *Turbulent heat transfer computations for rearward-facing steps and sudden pipe expansions.* Journal of heat transfer, 1985. 107: p. 70.

[26] Aung, W. and R. Goldstein, *Heat transfer in turbulent separated flow downstream of a rearward-facing step.* Israel journal of technology, 1972. 10(1-2): p. 35-41.

[27] Sparrow, E. and J. O'Brien, *Heat Transfer Coefficients on the Downstream Face of an Abrupt Enlargement or Inlet Constriction in a Pipe.* Journal of heat transfer, 1980. 102: p. 408.

[28] Moss, W. and S. Baker, *Re-circulating flows associated with two-dimensional steps.* Aeronautical Quarterly, 1980. 31: p. 151-172.

[29] Eaton, J.K. and J. Johnston, *Turbulent flow reattachment: an experimental study of the flow and structure behind a backward-facing step*, 1980, Stanford University.

[30] Habib, M., A. Attya, and D. McEligot, *Calculation of turbulent flow and heat transfer in channels with streamwise-periodic flow.* Journal of turbomachinery, 1988. 110: p. 405.

[31] Habib, M., F. Durst, and D. McEligot. *Streamwise-periodic flow around baffles.* 1985.

[32] Berner, C., F. Durst, and D. McEligot, *Flow around baffles.* Journal of heat transfer, 1984. 106: p. 743.

[33] Kim, K.Y. and Y.J. Lee, *Prediction of turbulent heat transfer downstream of an abrupt pipe expansion.* Journal of Mechanical Science and Technology, 1994. 8(3): p. 248-254.

[34] Chung, B.T.F. and S. Jia, *A turbulent near-wall model on convective heat transfer from an abrupt expansion tube.* Heat and Mass Transfer, 1995. 31(1): p. 33-40.

[35] Hsieh, W. and K. Chang, *Calculation of wall heat transfer in pipeexpansion turbulent flows.* International Journal of Heat and Mass Transfer, 1996. 39(18): p. 3813-3822.

[36] Launder, B. and B. Sharma, *Application of the energy-dissipation model of turbulence to the calculation of flow near a spinning disc.* Letters Heat Mass Transfer, 1974. 1: p. 131-137.

[37] Bremhorst, K., *Modified form of the kw model for predicting wall turbulence.* Journal of Fluid Engineering, 1981. 103: p. 456-460.

[38] Chien, K.Y. *Predictions of channel and boundary-layer flows with a low-Reynolds-number two-equation model of turbulence.* 1980.

[39] Nagano, Y. and M. Hishida, *Improved form of the k-∊model for wall turbulent shear flows.* Journal of fluids engineering, 1987. 109: p. 156.

[40] Abe, K., T. Kondoh, and Y. Nagano, *A new turbulence model for predicting fluid flow and heat transfer in separating and reattaching flows—II. Thermal field calculations.* International Journal of Heat and Mass Transfer, 1995. 38(8): p. 1467-1481.

[41] Chang, K., W. Hsieh, and C. Chen, *A modified low-Reynolds-number turbulence model applicable to recirculating flow in pipe expansion.* Journal of fluids engineering, 1995. 117: p. 417.

[42] Lam, C.K.G. and K. Bremhorst, *A modified form of the k-epsilon model for predicting wall turbulence.* ASME, Transactions, Journal of Fluids Engineering, 1981. 103: p. 456-460.

[43] Myong, H.K. and N. Kasagi, *A new approach to the improvement of k-ε turbulence model for wall-bounded shear flows.* JSME Int. J., 1990. 33: p. 63–72.

[44] Shih, T.H., et al., *A new k-[epsilon] eddy viscosity model for high reynolds number turbulent flows.* Computers & Fluids, 1995. 24(3): p. 227-238.

[45] Guo, Z.Y., D.Y. Li, and X.G. Liang, *Thermal effect on the recirculation zone in sudden-expansion gas flows.* International Journal of Heat and Mass Transfer, 1996. 39(13): p. 2619-2624.

[46] Prud'Homme, M. and S. Elghobashi, *Turbulent heat transfer near the reattachment of flow downstream of a sudden pipe expansion.* Numerical Heat Transfer, Part A: Applications, 1986. 10(4): p. 349-368.

[47] Moon, L. and G. Rudinger, *Velocity distribution in an abruptly expanding circular duct.* Journal of Fluids Engineering, 1977. 99: p. 226.

[48] Huang, Z., et al., *Combustion behaviors of a compression-ignition engine fuelled with diesel/methanol blends under various fuel delivery advance angles.* Bioresource Technology, 2004. 95(3): p. 331-341.

[49] Sugawara, K., H. Yoshikawa, and T. Ota. *LES of Three-Dimensional Separated Flow and Heat Transfer in a Symmetric Expansion Plane Channel.* 2004. Japan Heat Transfer Society; 1999.

[50] Yoshikawa, H., K. Ishikawa, and T. Ota. *Numerical Simulation of Three-Dimensional Separated Flow and Heat Transfer in an Enlarged Rectangular Channel.* 2004. Japan Heat Transfer Society; 1999.

[51] Yoshikawa, H., M. Yoshikawa, and T. Ota. *Numerical Simulation of Heat Transfer in Unsteady Separated and Reattached Flow Around a Symmetric Sudden Expansion Channel.* 2002. Japan Heat Transfer Society; 1999.

[52] OTA, T., et al., *Numerical Analysis of Separated Flow and Heat Transfer in an Enlarged Channel.* Transactions of the Japan Society of Mechanical Engineers. B, 2000. 66(648): p. 2109-2116.

[53] Macagno, E.O. and T.K. Hung, *Computational and experimental study of a captive annular eddy.* Journal of fluid Mechanics, 1967. 28(01): p. 43-64.

[54] Durst, F., A. Melling, and J. Whitelaw, *Low Reynolds number flow over symmetric sudden expansion.* 1974.

[55] Goharzadeh, A. and P. Rodgers, *Experimental Measurement of Laminar Axisymmetric Flow Through Confined Annular Geometries With Sudden Inward Expansion.* Journal of Fluids Engineering, 2009. 131: p. 124501.

[56] Nag, D. and A. Datta, *On the eddy characteristics of laminar axisymmetric flows through confined annular geometries with inward expansion.* Proceedings of the Institution of Mechanical Engineers, Part C: Journal of Mechanical Engineering Science, 2007. 221(2): p. 213.

[57] Chiang, T., T.W.H. Sheu, and S. Wang, *Side wall effects on the structure of laminar flow over a plane-symmetric sudden expansion.* Computers & fluids, 2000. 29(5): p. 467-492.

[58] Fearn, R., T. Mullin, and K. Cliffe, *Nonlinear flow phenomena in a symmetric sudden expansion.* J. Fluid Mech, 1990. 211(595-608): p. C311.

[59] Boughamoura, A., H. Belmabrouk, and S. Ben Nasrallah, *Numerical study of a piston-driven laminar flow and heat transfer in a pipe with a sudden expansion.* International Journal of Thermal Sciences, 2003. 42(6): p. 591-604.

[60] Pinho, F., P. Oliveira, and J. Miranda, *Pressure losses in the laminar flow of shear-thinning power-law fluids across a sudden axisymmetric expansion.* International Journal of Heat and Fluid Flow, 2003. 24(5): p. 747-761.

[61] Thiruvengadam, M., B. Armaly, and J. Drallmeier, *Three dimensional mixed convection in plane symmetric-sudden expansion: Symmetric flow regime.* International Journal of Heat and Mass Transfer, 2009. 52(3-4): p. 899-907.

[62] Seban, R., A. Emery, and A. Levy, *Heat transfer to separated and reattached subsonic turbulent flows obtained downstream of a surface step.* J. Aerospace Sci, 1959. 26: p. 809-814.

[63] Seban, R., *The Effect of Suction and Injection on the Heat Transfer and Flow in a Turbulent Separated Airflow.* Journal of heat transfer, 1966. 88: p. 276.

[64] Togun, H., S. Kazi, and A. Badarudin, *A Review of Experimental Study of Turbulent Heat Transfer in Separated Flow.* Australian Journal of Basic and Applied Sciences, 2011. 5(10): p. 489-505.

[65] Sogin, H.H., *A summary of experiments on local heat transfer from the rear of bluff obstacle to a law speed Air stream.* Aournal of heat transfer-transactions of the ASME, 1964. 86: p. 200-202.

[66] Mori, Y., Y. Uchida, and K. Sakai, *A study of the time and spatial micro-structure of heat transfer performance near the reattaching point of separated flows.* Trans. JSME Ser. B, 1986. 52: p. 481.

[67] Kawamura, T., et al., *Temporal and spatial characteristics of heat transfer at the reattachment region of a backward-facing step.* Experimental Heat Transfer, 1987. 1(4): p. 299-313.

[68] Kawamura, T., et al. *Time and spatial characteristics of heat transfer at the reattachment region of a two-dimensional backward-facing step*. 1991. American Society of Mechanical Engineers.

[69] Abu-Mulaweh, H.I., *Turbulent mixed convection flow over a forward-facing step—the effect of step heights.* International Journal of Thermal Sciences, 2005. 44(2): p. 155-162.

[70] Sano, M., I. Suzuki, and K. Sakuraba, *Control of turbulent channel flow over a backward-facing step by suction.* Journal of Fluid Science and Technology, 2009. 4(1): p. 188-199.

[71] Shisnov, E., et al., *Heat transfer in the recirculating region formed by a backward-facing step.* International Journal of Heat and Mass Transfer, 1988. 31(8): p. 1557-1562.

[72] Etheridge, D. and P. Kemp, *Measurements of turbulent flow downstream of a rearward-facing step.* Journal of Fluid Mechanics, 1978. 86(03): p. 545-566.

[73] Moss, W., S. Baker, and L. Bradbury. *Measurements of mean velocity and Reynolds stresses in some regions of recirculating flow*. 1977.

[74] Abbott, D. and S. Kline, *Experimental investigation of subsonic turbulent flow over single and double backward facing steps.* Journal of basic engineering, 1962. 84: p. 317.

[75] Tani, I., M. Iuchi, and H. Komoda, *Experimental investigation of flow separation associated with a step or a groove.* Aeronautical Res. Inst. Univ. of Tokyo, Rept, 1961. 364.

[76] Eaton, J. and J. Johnston. *A review of research on subsonic turbulent-flow reattachment.* 1980.

[77] Smyth, R., *Turbulent flow over a plane symmetric sudden expansion.* Journal of fluids engineering, 1979. 101: p. 348.

[78] Seban, R., *Heat transfer to the turbulent separated flow of air downstream of a step in the surface of a plate.* Journal of heat transfer, 1964. 86: p. 259.

[79] Youssef, M., Y. Nagano, and M. Tagawa, *A two-equation heat transfer model for predicting turbulent thermal fields under arbitrary wall thermal conditions.* International Journal of Heat and Mass Transfer, 1992. 35(11): p. 3095-3104.

[80] Heyerichs, K. and A. Pollard, *Heat transfer in separated and impinging turbulent flows.* International Journal of Heat and Mass Transfer, 1996. 39(12): p. 2385-2400.

[81] Kader, B., *Temperature and concentration profiles in fully turbulent boundary layers.* International Journal of Heat and Mass Transfer, 2006. 24(9): p. 1541-1544.

[82] Jayatillaka, C., *The influence of prandtl number and surface roughness on the resistances of the laminar sub-layer to momentum and heat transfer.* Prog. Heat Mass Transfer, 1969. 1: p. 193-329.

[83] Zigh, A., *Computational study of simultaneous heat and mass transfer in turbulent separated flows*. 1993.

[84] Kim, J., S. Kline, and J. Johnston, *Investigation of a reattaching turbulent shear layer: flow over a backward-facing step.* Journal of Fluids Engineering, 1980. 102: p. 302.

[85] Yin, Y., Y. Nagano, and M. Tagawa, *Numerical prediction of turbulent heat transfer in high-Prandtl-number fluids.* JSME Transactions, 1992. 58: p. 2254-2260.

[86] Nagano, Y. and C. Kim, *A two-equation model for heat transport in wall turbulent shear flows.* Journal of heat transfer, 1988. 110: p. 583.

[87] Dutta, S. and S. Acharya, *Heat Transfer and Flow Past a Backstep with the Nonlinear k-∈ Turbulence Model and the Modified k-∈Turbulence Model.* Numerical Heat Transfer, Part A Applications, 1993. 23(3): p. 281-301.

[88] Rhee, G.H. and H.J. Sung, *A nonlinear low-Reynolds-number k-∈ model for turbulent separated and reattaching flows - II. Thermal field computations.* International Journal of Heat and Mass Transfer, 1996. 39(16): p. 3465-3474.

[89] Park, T.S. and H.J. Sung, *A nonlinear low-Reynolds-number κ-ε model for turbulent separated and reattaching flows—I. Flow field computations.* International Journal of Heat and Mass Transfer, 1995. 38(14): p. 2657-2666.

[90] Tsay, Y.L., T. Chang, and J. Cheng, *Heat transfer enhancement of backward-facing step flow in a channel by using baffle installation on the channel wall.* Acta mechanica, 2005. 174(1): p. 63-76.

[91] Armaly, B., et al., *Experimental and theoretical investigation of backward-facing step flow.* Journal of Fluid Mechanics, 1983. 127(1): p. 473-496.

[92] Barton, I., *Laminar flow over a backward-facing step with a stream of hot particles.* International Journal of Heat and Fluid Flow, 1997. 18(4): p. 400-410.

[93] Lee, T. and D. Mateescu, *Experimental and numerical investigation of 2-D backward-facing step flow.* Journal of Fluids and Structures, 1998. 12(6): p. 703-716.

[94] Stüer, H., A. Gyr, and W. Kinzelbach, *Laminar separation on a forward facing step.* European Journal of Mechanics Series B Fluids, 1999. 18: p. 675-692.

[95] Armaly, B., A. Li, and J. Nie, *Measurements in three-dimensional laminar separated flow.* International Journal of Heat and Mass Transfer, 2003. 46(19): p. 3573-3582.

[96] Velazquez, A., J. Arias, and B. Mendez, *Laminar heat transfer enhancement downstream of a backward facing step by using a pulsating flow.* International Journal of Heat and Mass Transfer, 2008. 51(7): p. 2075-2089.

[97] Han, J., L. Glicksman, and W. Rohsenow, *An investigation of heat transfer and friction for rib-roughened surfaces.* International Journal of Heat and Mass Transfer, 1978. 21(8): p. 1143-1156.

[98] Rouizi, Y., et al., *Numerical model reduction of 2D steady incompressible laminar flows: Application on the flow over a backward-facing step.* Journal of Computational Physics, 2009. 228(6): p. 2239-2255.

[99] Li, A. and B. Armaly, *Mixed Convection Adjacent to 3-D Backward-Facing Step.* ASME-PUBLICATIONS-HTD, 2000. 366: p. 51-58.

[100] Saldana, J.G.B., N. Anand, and V. Sarin, *Numerical simulation of mixed convective flow over a three-dimensional horizontal backward facing step.* Journal of heat transfer, 2005. 127: p. 1027.

[101] Lima, R., C. Andrade, and E. Zaparoli, *Numerical study of three recirculation zones in the unilateral sudden expansion flow.* International Communications in Heat and Mass Transfer, 2008. 35(9): p. 1053-1060.